Birkhäuser

Carlo Bardaro • Paul L. Butzer • Ilaria Mantellini •
Gerhard Schmeisser

Mellin Analysis, Transform Theory, and Applications

A Self-Contained Approach

Birkhäuser

Carlo Bardaro
Department of Mathematics and Informatics
University of Perugia
Perugia, Italy

Paul L. Butzer
Lehrstuhl A für Mathematik
RWTH Aachen University
Aachen, Germany

Ilaria Mantellini
Department of Mathematics and Informatics
University of Perugia
Perugia, Italy

Gerhard Schmeisser
Department Mathematik
Friedrich-Alexander-Universität
Erlangen-Nürnberg (FAU)
Erlangen, Germany

ISBN 978-3-031-96671-2 ISBN 978-3-031-96672-9 (eBook)
https://doi.org/10.1007/978-3-031-96672-9

Mathematics Subject Classification: 44Axx, 41A58, 26A33, 30D20, 40A30, 41A35, 46E30, 65D30, 30G30, 35A22, 41A55, 42A85, 44A05, 44A15, 94A20

This book is published under the imprint Birkhäuser, www.birkhauser-science.com by the registered company Springer Nature Switzerland AG
The registered company address is: Gewerbestrasse 11, 6330 Cham, Switzerland

To our parents, families, close friends for their profound influence on our lives, giving us the gift of dreams and the ability to realize them

Dedicated to Professors R.G. Mamedov, Z. Szmydt, and B. Ziemian for paving the way to an independent study of Mellin analysis

Preface

The aim of this monograph is to evolve the theory of Mellin transform and the consequential Mellin analysis in a rigorous and unified way as a self-contained discipline being, in particular, fully independent of related integral transforms such as the Fourier and the Laplace transform. A justification for such a project arises from the numerous applications of the Mellin transform. In fact, it is used in the study of certain boundary value problems, in sampling theory and signal reconstruction, in number theory, in finance, and in connection with numerical analysis in various fields of scientific research such as potential theory, optical physics, electromagnetism, stochastic processes, and several others.

A first book devoted fully to the Mellin transform appeared at the beginning of the 1990s of the last century. It was authored by R.G. Mamedov who had collected a series of previous results obtained in collaboration with G.N. Orudhzev. A few years later, P.L. Butzer and S. Jansche turned to this topic and studied it in a very general and systematic way in several papers. The present book is based on that work and its numerous continuations with special emphasis on our own contributions.

In our thorough studies some new notions helped us to reveal the beauty of Mellin analysis. For example, we realized that a modified concept of analyticity, called *polar analyticity*, provides powerful tools, comparable with those of classical complex analysis in the Fourier theory, and allows us to present some important results in a simple and elegant way. Likewise a new metric, referred to as the *Mellin distance*, leads to efficient error estimates for approximate formulae.

We do not claim that the present theory is complete. Our book can also be seen as an invitation to young researchers to supplement not only this work but also to establish a multidimensional theory.

The book has 14 chapters, a bibliography, lists of notation, Mellin transform rules, Mellin transforms of particular functions, and an index. Each chapter ends with a section entitled *Notes* in which we disclose sources, discuss related work, and sometimes give suggestions for further investigations.

In Chap. 1, we fix the basic notations and collect some fundamental notions and results from real and complex analysis, Lebesgue spaces, functional analysis, special functions, and basic Fourier transform theory.

Chapter 2 presents a modified concept of analyticity. It leads to the notion of *polar-analytic functions* which turn out to be a simple and useful tool for the development of the Mellin transform theory.

Mellin analysis starts with Chap. 3 in which we introduce the basic function spaces (weighted Lebesgue spaces X_c^p and Y_c^p), the notion of Mellin translation with application to Mellin–Lipschitz classes, the notion of Mellin differential operators, also in polar form with a further development of the theory of polar-analytic functions, and Mellin versions of Sobolev spaces, Bernstein spaces, and Hardy spaces.

Chapter 4 deals with the notion of convolution in the Mellin setting, which is of fundamental importance for the development of the theory, in particular for the study of convergence of Mellin type integral operators.

Chapter 5 contains the theory of the finite Mellin transform and the Mellin–Fourier series. These are the counterparts of the finite Fourier transform and the Fourier series in the Mellin setting.

The Mellin transform over the multiplicative group of the positive real numbers is treated in Chap. 6 for the spaces $X_c^1 \equiv X_c$. The first three sections confer the basic theory of Mellin transform while Sect. 6.4 contains results on pointwise and almost everywhere convergence of Mellin type integral operators. The next four sections are devoted to inversion theory, a uniqueness theorem, an equivalence theorem, and the fundamental theorem of differential and integral calculus in the Mellin setting. Section 6.9 presents the Mellin–Poisson summation formula, the counterpart of the classical Poisson formula of Fourier analysis.

In Chap. 7, we introduce the Mellin transform for function of the spaces X_c^p, with $p \in]1, 2]$. The presented results are essentially new. The Mellin transform is now defined via the Riesz–Thorin interpolation theorem, stated in Chap. 1. We do not discuss an extension of the Mellin transform to functions belonging to X_c^p with $p > 2$. In order to do that we would have to introduce a distributional version of the Mellin transform, analogously to the case of the Fourier transform. It can be done by modifying the notion of the Schwartz space of test functions suitably, but we will not treat this instance. The reader interested in Mellin distributions may consult the notes of Chap. 7.

Chapter 8 is devoted to a result in the spirit of the important Paley–Wiener theorem of Fourier analysis, which states the equivalence between the Bernstein space B_σ^2 comprising all the entire functions of exponential type σ whose restriction to the real axis belongs to $L^2(\mathbb{R})$ with the Paley–Wiener space $\widehat{B_\sigma^2}$ comprising all $L^2(\mathbb{R})$ functions whose Fourier transform is compactly supported. Here the Mellin theory turns out to be of a very different nature. Indeed, it can be shown that a non-trivial function $f \in X_c^2$ whose Mellin transform is compactly supported cannot be extended to an entire function of exponential type. Therefore, we have to change the notion of Bernstein spaces by introducing the so-called Mellin–Bernstein spaces.

We give three different approaches to a Mellin–Paley–Wiener theorem: one uses the Riemann surfaces of the (complex) logarithm and analytical branches; a second concerns one part of the theorem only for which a fundamental inequality, named Mellin–Bernstein inequality, is used for a short proof; a third based on the notion of polar analyticity. It is remarkable that in the second approach only one-dimensional real analysis is needed.

Then we generalize the Mellin–Paley–Wiener theorem by extending it to a characterization of functions with an exponentially decaying Mellin transform at infinity. This extension is a powerful tool in the study of quadrature formulae over the positive real line.

Chapter 9 deals with the use of the Mellin transform in fractional calculus. Fractional integration is based on Hadamard fractional integrals, here called *Mellin fractional integrals*. The content of the present chapter comprises a thorough investigation of the domain of the fractional integral operators and a study of two notions of fractional derivative, the second in a pointwise and also in a strong sense. Some of these results will be used later for a treatment of partial fractional differential equations.

In Chap. 10, we introduce a new notion of distance by means of the inversion formula of the Mellin transform. It enables us to establish efficient error estimates for some basic formulae of Mellin analysis in their approximate form such as the exponential sampling formula, the Mellin reproducing kernel formula, the Bernstein inequality, and others. These estimates are asymptotically best possible and can be employed for characterizing certain function spaces by the speed of convergence to zero of error terms. The considered function spaces include Mellin–Paley–Wiener spaces, Mellin–Hardy spaces, and Mellin–Sobolev spaces.

In Chap. 11, we apply the theory developed in some of the previous chapters to the solutions of certain partial differential equations also in the fractional instance. We discuss the wave and the heat equation in a half-plane, using the Mellin transform method. The fractional instance is treated in case of an integro-differential equation and a fractional equation with a second order derivative. Further applications of the Mellin transform to the treatment of partial differential equations can be found in several books and papers (see the notes of Chap. 11).

Chapters 12 and 13 are devoted to applications of Mellin analysis to sampling theory. We begin with the (classical) exponential sampling formula (Sect. 12.1), first used in a formal way by physicists and engineers for investigating certain phenomena in optical physics such as light scattering and diffraction, and then rigorously explored in the work of Butzer and Jansche. This formula is the counterpart in Mellin analysis of the classical sampling theorem of Whittaker–Kotel'nikov–Shannon, which has been extensively studied in the literature (see the notes of Chap. 12). In Sect. 12.2, we present an analogue of Valiron's sampling formula for exponentially spaced sample points. It implies an efficient sampling formula for polar Mellin derivatives. In Sect. 12.3, we extend the classical exponential sampling formula by involving in addition samples of the Mellin derivatives up to a given order.

When the sampled function f does not satisfy the hypotheses of the exponential sampling formula, the latter may still be useful in an approximate sense. This situation is studied in Sect. 12.4, where pointwise and uniform error estimates are established.

The exponential sampling formula may be seen as the discrete form of a reproducing kernel formula in the Mellin setting. The latter is investigated in both, its exact and its approximate form (Sects. 12.5–12.6). Section 12.7 exhibits interconnections between the exact and the approximate exponential sampling formulae, which lead to a new result on convergence in the norm. Finally, in Sect. 12.8, the exponential sampling formula is employed as an essential tool for deducing a useful Parseval formula in the Mellin setting.

The classical exponential sampling formula is based on a function, denoted by lin, which we call its *kernel*. Although it is an elegant formula, its hypotheses are quite strong and its series may converge very slowly. Therefore, it is of interest to consider generalizations, in which the role of the lin function is taken by some other suitable kernel functions φ. This is the topic of Chap. 13. In the first four sections, we introduce a class Φ of admissible kernel functions φ, study pointwise and uniform convergence as well as convergence in the norm of X_c^p for $p \geq 1$ and illustrate some of the results by examples. In Sect. 13.5, we consider a subclass of Φ comprising kernel functions of the form ψlin, that is, the kernel of the classical exponential sampling formula is modified by some function ψ as a multiplier. Such a generalized exponential sampling operator is always interpolating. Depending on the choice of the multiplier ψ, one can approximate and even reconstruct functions from wide classes including functions that are unbounded on the positive real line. With a multiplier based on a Gaussian function, one can establish an algorithm of high precision. It even has the rate of convergence of an optimal algorithm (see the notes of Chap. 13).

Finally, in Chap. 14, we consider applications to quadrature over the positive real line. The Mellin theory leads us in a natural way to a class of quadrature formulae with exponentially spaced nodes. First we study the precision of these formulae for certain function spaces. Then, by considering converse results, we arrive at a characterization of function spaces by the speed of convergence to zero of remainders. We also point out that within Mellin analysis, these quadrature formulae have features of Gaussian formulae. For a bibliography on this subject, see the notes of Chap. 14.

It is our pleasure to express thanks to Professor Rudolf Stens for several important discussions on the matter, and to Professors Laura Angeloni, Danilo Costarelli, Anna Rita Sambucini, Marco Seracini, Gianluca Vinti, and Luca Zampogni for providing the graphical files and encouragements. We also express gratitude to all peoples who contributed to the development of Mellin analysis, starting with the pioneering work of Professor R.G. Mamedov. Last but not the least we want to thank the Senior Editor of Mathematics Birkhäuser D. Mazlum and the production team for their encouragement, patience, and valuable advice.

Perugia, Italy	Carlo Bardaro
Aachen, Germany	Paul L. Butzer
Perugia, Italy	Ilaria Mantellini
Erlangen, Germany	Gerhard Schmeisser
October 2025	

A Historical Overview

We start with a short history of the origin of the theory of Mellin transform. A detailed introduction can be found in Butzer and Jansche [89].

Robert Hjalmar Mellin was born in the parish Tyrnävä, Finland, on June 19, 1854, and studied from 1875 to 1882 under Mittag-Leffler at the University of Helsingfors, receiving the doctorate. He spent some time in Berlin attending lectures of Weierstrass and at the Stockholm Högskola. In 1884 he was appointed Head Teacher at the Helsingfors Polytechnical Institute; later he held a professorship at the Helsingfors Institute of Technology and retired in 1926. Apart from mathematics, he also did some research in relativity theory. He was an editorial board member of the famous journal *Acta Mathematica* from 1908 until his death on April 5, 1933. For a biography of Mellin, see Elfving [145] and Lindelöf [198].

The origin of the transform which bears his name can be found in the theory of the Euler gamma function, defined by

$$\Gamma(s) = \int_0^\infty e^{-x} x^{s-1} dx \qquad (s \in \mathbb{C}),$$

and the inversion formula

$$e^{-x} = \frac{1}{2\pi i} \int_{c-i\infty}^{c+i\infty} \Gamma(s) x^{-s} ds \qquad (x > 0).$$

The first formula is due to Euler [149] while the second formula goes back to the Italian mathematician Pincherle [243], as Mellin reported in his paper [218]. However, Mellin was the first to realize the deep connections between the two formulae, replacing the role of the reciprocal exponential function more generally by some function f and that of the gamma function by an associated function $[f]^\wedge_M$, now called the *Mellin transform* of f. That is, he considered the pair of formulae

$$[f]^\wedge_M(s) := \int_0^\infty f(x) x^{s-1} dx \qquad (s \in \mathbb{C}) \qquad \text{(a)}$$

and

$$f(x) = \frac{1}{2\pi i} \int_{c-i\infty}^{c+i\infty} [f]^{\wedge}_{M}(s) x^{-s} ds \qquad (x > 0). \tag{b}$$

When, prior to Mellin, Riemann in his memoir [259] on the zeta function (see, e.g., Chaen [121], Titchmarsh [287])

$$\zeta(s) = \sum_{k=1}^{\infty} \frac{1}{k^s} \qquad (\Re s > 1)$$

gave two proofs of the famous functional equation

$$\pi^{-s/2}\Gamma(s/2)\zeta(s) = \pi^{-(1-s)/2}\Gamma\big((1-s)/2\big)\zeta(1-s)$$

(Mellin [222, p. 8]), he used already the pair of formulae (a) and (b) in the special cases

$$f(x) = \frac{1}{e^x - 1} \quad \text{and} \quad g(x) = \sum_{k=1}^{\infty} e^{-\pi k^2 x},$$

where

$$[f]^{\wedge}_{M}(s) = \Gamma(s)\zeta(s) \quad \text{and} \quad [g]^{\wedge}_{M}(s) = \pi^{-s}\Gamma(s)\zeta(2s),$$

respectively; also see Titchmarsh [287].

Perron [238] and Mellin [218–225] were the first to investigate formula (a) and its inverse (b) in a rigorous and systematic way, giving precise conditions for their validity. For a history of the Mellin transform, see Doetsch [137–139] and Titchmarsh [286].

In recent times, integral transform methods based on the Mellin transform gained considerable importance in the solution of initial and boundary value problems for ordinary and partial differential equations (see Annaby and Butzer [16], Sneddon [275, 276], Colombo [128], Colombo and Lavoine [129], Zemanian [310]) along with its practical applications (see Sasiela [268]); in number theory (see Apostol [17, 18], Edwards [144], Hamburger [164], Hardy and Littlewood [165, 166], Ivic [178], Rademacher [251], Weil [298]), sampling theory (see Bardaro et al. [52, 54, 57], Bertero and Pike [70], Butzer and Jansche [90, 92], Casasent [125], Gori [159], Ostrowsky et al. [233]), numerical analysis (see Bardaro et al. [60], Elschner and Graham [146], Prössdorf and Silbermann [246], Schmeisser [270]), and financial mathematics (see Panini and Srivastav [235], Frontczak [153], Garcia [154]).

The Mellin transform has several other useful features: It is a very convenient tool for deriving expansions of integrals (see, e.g., Davies [135, p. 199]), for summing infinite series in a closed form (see, e.g., Sneddon [275], Tithcmarsh [286]), and for the study of stochastic processes via the theory of moment type operators, which are a particular case of Mellin type convolution operators (see Boccuto et al. [74]).

Despite all these features and applications, the Mellin transform was very often treated as a short section of monographs on Fourier or Laplace transforms, due to its formal equivalence to the latter, performed by a change of variables and functions (see, e.g., Titchmarsh [286], Doetsch [137, 138], Widder [302, 303], Bracewell [76] and more recently Brychkov [81], Glaeske et al. [157]). Books dealing exclusively with the Mellin transform are, e.g., Oberhettinger [230], Colombo [128], Bertrand et al. [71], Mamedov [208], Brychkov [79]. But again, their Mellin results are mostly obtained via other integral transforms, thereby disguising Mellin immanent hypotheses and providing by no means a unified and systematic approach. The first attempt to establish a theory of Mellin transform that is fully independent of other integral transforms was made in Butzer and Jansche [88, 89].

The present book treats the *Mellin analysis*, the counterpart of the Fourier analysis, in a self-contained way without significant support from the Fourier theory. We shall see that the two theories have very different features despite their formal equivalence.

Contents

Chapter 1
Preliminaries

In this chapter we introduce notations, definitions and basic prerequisites which will be used throughout this book.

1.1 Basic Notations for Number Systems and Functions

We begin with notations for the fundamental number systems.

- $\mathbb{N}$ is the set of positive integers, and $\mathbb{N}_0 := \mathbb{N} \cup \{0\}$
- $\mathbb{Z}$ is the set of integers
- $\mathbb{R}$ is the set of real numbers
- $\mathbb{C}$ is the set of complex numbers

Further, we shall use the notations $\mathbb{R}^+$ for the set of positive real numbers and $\mathbb{R}_0^+$ for the set of nonnegative real numbers.

For any real number r, we define $r^+ := \max\{r, 0\}$ and $r^- := \max\{-r, 0\}$. Then $r = r^+ - r^-$ and $|r| = r^+ + r^-$. This notation is extended to functions. For a real-valued function f, we define the positive and negative parts of f by $f^+(x) := \max\{f(x), 0\}$ and $f^-(x) := \max\{-f(x), 0\}$, respectively. Thus, $f = f^+ - f^-$ and $|f| = f^+ + f^-$.

Next we recall two functions that map real numbers r to neighboring integers: The floor function $\lfloor r \rfloor$ produces the integer part of r, that is, the greatest integer less than or equal to r. Similarly, the ceiling function $\lceil r \rceil$ produces the least integer greater than or equal to r.

By the symbol i, we always indicate the complex unit $i = (0, 1)$ or $i = \sqrt{-1}$. For any complex number z, we write $z = \Re z + i\Im z$, where $\Re z$ and $\Im z$ are the real and imaginary parts of z, respectively.

An open and a closed interval of $\mathbb{R}$ will be denoted by $]a, b[$ and $[a, b]$, respectively. The meaning of the notations $]a, b]$ or $[a, b[$ should be clear. For

C. Bardaro et al., *Mellin Analysis, Transform Theory, and Applications*,
https://doi.org/10.1007/978-3-031-96672-9_1

unbounded intervals we shall use the standard notations $]a, \infty[$, $[a, \infty[$, $]-\infty, a[$ and $]-\infty, a]$. In particular, $]0, \infty[= \mathbb{R}^+$ and $[0, \infty[= \mathbb{R}_0^+$.

While functions are usually denoted by letters like f, g, h, φ, ψ, it is sometimes desirable to indicate the place of a variable by a dot. This is of particular importance when we have several variables and apply an operator which acts with respect to one of them. For example, we write $\|f(\cdot, y)\|_p$ when f has two variables and the norm is built with respect to the first one while the second variable y takes the role of a parameter. Likewise the symbol $(\cdot)^r f(\cdot)$ denotes the function $x \mapsto x^r f(x)$ and $f(\cdot\, t)$ means that the argument of f is scaled by a factor t giving the function $x \mapsto f(xt)$.

Occasionally we shall use the Landau symbols "big Oh" and "little oh" which may be introduced as follows. Let f and g be two functions defined in a neighborhood U of a point x_0 such that $g(x) \neq 0$ for $x \in U \setminus \{x_0\}$. Then

$$f(x) = \mathcal{O}(g(x)) \quad (x \to x_0) \qquad \text{if and only if} \qquad \limsup_{x \to x_0, x \neq x_0} \left| \frac{f(x)}{g(x)} \right| < \infty$$

and

$$f(x) = o(g(x)) \quad (x \to x_0) \qquad \text{if and only if} \qquad \lim_{x \to x_0, x \neq x_0} \frac{f(x)}{g(x)} = 0.$$

In particular, $f(x) = o(1)$ for $x \to x_0$ means that $f(x)$ approaches zero as $x \to x_0$.

Finally, we define and denote the *sign function* by $\operatorname{sgn} x := x/|x|$ for $x \neq 0$ and $\operatorname{sgn} 0 = 0$.

1.2 Basic Function Spaces

In this section we report on some function spaces which are basic for the theory developed in this book. Section 1.7 Notes contains some references.

1.2.1 L^p Spaces and Their Basic Properties

For a Lebesgue measurable set $A \subset \mathbb{R}$ and $p \in [1, \infty[$ we shall denote by $L^p(A)$ the space of all Lebesgue measurable functions $f : A \to \mathbb{C}$ such that $|f|^p$ is Lebesgue integrable on A. The Lebesgue space $L^p(A)$ for $p \in [1, \infty[$ endowed with the norm

$$\|f\|_{L^p(A)} = \|f\|_p := \left\{ \int_A |f(u)|^p du \right\}^{1/p}$$

is a Banach space. For $p = \infty$ the space $L^\infty(A)$ is the space of all Lebesgue measurable functions which are essentially bounded, endowed with the norm $\|f\|_\infty := \operatorname{ess\,sup}_{u\in A} |f(u)|$. The space $L^\infty(A)$ is also a Banach space.

Sometimes we have estimates involving an effective supremum for bounded functions $f : A \to \mathbb{C}$. Therefore we denote by $\|\cdot\|_\infty^*$ the norm defined by

$$\|f\|_\infty^* := \sup_{x\in A} |f(x)|.$$

For continuous and bounded functions, the two norms $\|\cdot\|_\infty$ and $\|\cdot\|_\infty^*$ coincide.

For an unbounded interval $I \subset \mathbb{R}$, we denote by $L^1_{\mathrm{loc}}(I)$ the space comprising all locally integrable functions, i.e., all functions f such that $f \in L^1(K)$ for any bounded interval $K \subset I$.

For any interval $A \subset \mathbb{R}$, we denote by $C(A)$ the space of all continuous functions $f : A \to \mathbb{C}$. Next, by $C_{\mathrm{u}}(A)$ we denote the subspace of $C(A)$ containing all uniformly continuous function on A and by $C_{\mathrm{ub}}(A)$ the subspace of $C_{\mathrm{u}}(A)$ containing all bounded functions. For $k \in \mathbb{N}$, the notation $C^k(A)$, is used for the subspace of all k-times differentiable functions with $f^{(k)} \in C(A)$. Analogously we define the subspace $C^k_{\mathrm{u}}(A)$. Finally, $C_{\mathrm{comp}}(A)$ and $C^k_{\mathrm{comp}}(A)$ will denote those subspaces of $C(A)$ and $C^k(A)$, respectively, which contain only the functions of compact support on A. We recall that the support of a function $f : A \to \mathbb{C}$, denoted by $\operatorname{supp} f$, is defined as the closure of the set $\{x \in A : f(x) \neq 0\}$. Obviously, the two spaces with subscript comp are subspaces of $C_{\mathrm{ub}}(A)$. Finally, we denote by $C^\infty(A)$ the space of all infinitely differentiable functions on A and by $C^\infty_{\mathrm{comp}}(A)$ the subspace of $C^\infty(A)$ comprising all functions with compact support. The latter is also known as the space of test functions.

Let $\mathbb{A}$ be an infinite, countable set of integers and let $p \geq 1$. Then $\ell^p(\mathbb{A})$ denotes the space of all sequences $(a_n)_{n\in\mathbb{A}}$ of complex numbers such that

$$\sum_{n\in\mathbb{A}} |a_n|^p < \infty,$$

endowed with the norm

$$\|(a_n)\|_p := \left\{\sum_{n\in\mathbb{A}} |a_n|^p\right\}^{1/p},$$

and for $p = \infty$, the space $\ell^\infty(\mathbb{A})$ comprises all sequences such that $\|(a_n)\|_\infty := \sup_{n\in\mathbb{A}} |a_n| < \infty$.

For $p = 2$ the space $\ell^2(\mathbb{A})$ is a Hilbert space under the scalar product

$$\langle a, b\rangle_{\ell^2} = \sum_{k\in\mathbb{A}} a_k \overline{b_k},$$

where $a = (a_k)_{k\in\mathbb{A}}$ and $b = (b_k)_{k\in\mathbb{A}}$.

The following result is the well-known Riemann–Lebesgue lemma, which is a basic tool in Fourier analysis.

Proposition 1.1 *Let $f \in L^1_{\text{loc}}(\mathbb{R})$. Then for every compact interval $I \subset \mathbb{R}$ we have*

$$\lim_{\rho \to \infty} \int_I f(x) e^{i\rho x} dx = 0.$$

Moreover, if $f \in L^1(\mathbb{R})$, then

$$\lim_{\rho \to \infty} \int_{-\infty}^{\infty} f(x) e^{i\rho x} dx = 0.$$

In the following, we denote by $|A|$ the Lebesgue measure of the set A. Among the basic results on L^p spaces, we mention the following ones:

Proposition 1.2 (Fatou's Lemma) *Let $(f_n)_{n\in\mathbb{N}}$ be a sequence of nonnegative functions on a (Lebesgue) measurable set $A \subset \mathbb{R}$. If $\liminf_{n\to\infty} f_n(x) = f(x)$ almost everywhere on A, then*

$$\int_A f(x)dx \leq \liminf_{n\to\infty} \int_A f_n(x)dx.$$

Proposition 1.3 (Beppo Levi's Theorem) *Let $(f_n)_{n\in\mathbb{N}}$ be a sequence of measurable functions $f_n : A \to \mathbb{R}_0^+$ such that*

$$0 \leq f_1(x) \leq f_2(x) \leq \cdots \qquad (x \in A).$$

If the sequence of integrals $\int_A f_n(x)dx$ is bounded, then $\lim_{n\to\infty} f_n(x) =: f(x)$ exists almost everywhere on A as a measurable function f and

$$\lim_{n\to\infty} \int_A f_n(x)dx = \int_A f(x)dx. \tag{1.1}$$

Proposition 1.4 (Lebesgue's Dominated Convergence Theorem) *Let $(f_n)_{n\in\mathbb{N}} \subset L^1(A)$, and suppose that $\lim_{n\to\infty} f_n(x) = f(x)$ almost everywhere. If there exists $g \in L^1(A)$ such that $|f_n(x)| \leq g(x)$ almost everywhere on A for all n, then $f \in L^1(A)$ and*

$$\lim_{n\to\infty} \int_A f_n(x)dx = \int_A f(x)dx. \tag{1.2}$$

Another basic convergence theorem for L^p spaces is given by Vitali's theorem.

Proposition 1.5 (Vitali's Convergence Theorem) *Let $p \in [1, \infty[$, let $(f_n)_{n\in\mathbb{N}} \subset L^p(A)$ and let f be a measurable function. Then $f \in L^p(A)$ and $\lim_{n\to\infty} \|f_n - f\|_p = 0$ if and only if the following conditions hold:*

(a) *f_n converges to f in measure;*
(b) *for every $\varepsilon > 0$ there exists a $\delta > 0$ such that for any (Lebesgue) measurable set $E \subset A$ with $|E| < \delta$ one has*

$$\int_E |f_n(x)|^p dx < \varepsilon,$$

uniformly with respect to n;
(c) *for every $\varepsilon > 0$ there exists a measurable set $E_\varepsilon \subset A$ with finite measure such that*

$$\int_{A \setminus E_\varepsilon} |f_n(x)|^p dx < \varepsilon$$

for every $n \in \mathbb{N}$.

The triangular inequality for the L^p norms is given by the following basic result, known as *Minkowski's inequality*

Proposition 1.6 (Minkowski's Inequality) *Let $A \subset \mathbb{R}$ be a measurable set, and let $f, g \in L^p(A)$, where $p \in [1, \infty]$. Then*

$$\|f + g\|_p \leq \|f\|_p + \|g\|_p.$$

If $p \in [1, \infty]$ the *conjugate exponent* of p, denoted by p', is defined through the relation $(1/p) + (1/p') = 1$, where we set $p' = \infty$ if $p = 1$ and $p' = 1$ if $p = \infty$. The following basic inequality holds.

Proposition 1.7 (Hölder's Inequality) *Let $A \subset \mathbb{R}$ and let $f \in L^p(A)$, $g \in L^{p'}(A)$. Then $fg \in L^1(A)$ and $\|fg\|_1 \leq \|f\|_p \|g\|_{p'}$.*

In particular, for $p = 2$, we obtain $p' = 2$ and then Hölder's inequality is also known as *Cauchy–Schwartz–Bunjakowski inequality*.

Corresponding inequalities hold for the sequence spaces $\ell^p(\mathbb{A})$.

Let $A, B \subset \mathbb{R}$ be two measurable sets. The following result is a very useful tool for the study of double integrals.

Proposition 1.8 (Fubini's Theorem) *Let $f : A \times B \to \mathbb{C}$ be measurable with respect to the two-dimensional Lebesgue measure on $A \times B$.*

(a) *Suppose that $f \in L^1(A \times B)$, that is the double integral $\iint_{A \times B} f(x, y) dxdy$ is absolutely convergent. Then, for almost all $x \in A$, we have $f(x, \cdot) \in L^1(B)$. Moreover, $\int_B f(\cdot, y) dy \in L^1(A)$ and*

$$\int_A \left\{ \int_B f(x, y) dy \right\} dx = \int_B \left\{ \int_A f(x, y) dx \right\} dy = \iint_{A \times B} f(x, y) dxdy.$$

(b) *Suppose that $x \mapsto \|f(x, \cdot)\|_1 \in L^1(A)$. Then $f \in L^1(A \times B)$ and*

$$\iint_{A\times B} f(x,y)dxdy = \int_A \left\{\int_B f(x,y)dy\right\}dx = \int_B \left\{\int_A f(x,y)dx\right\}dy.$$

The next result is a generalization of Minkowski's inequality.

Proposition 1.9 (Generalized Minkowski Inequality) *Let $A, B \subset \mathbb{R}$ be two measurable sets, let $f : A \times B \to \mathbb{C}$ be a measurable function with respect to the two-dimensional Lebesgue measure and let $p \geq 1$. If the function $y \mapsto \|f(\cdot, y)\|_{L^p(A)}$ belongs to $L^1(B)$, then*

$$\left\| \int_B f(\cdot, y)dy \right\|_{L^p(A)} \leq \int_B \|f(\cdot, y)\|_{L^p(A)}dy.$$

The generalized Minkowki inequality for infinite sums can be stated as follows.

Proposition 1.10 *Let $(a_{n,k})_{k,n\in\mathbb{Z}}$ be a double sequence of complex numbers. For $p \geq 1$, let $(A_n)_{n\in\mathbb{Z}}$ be the sequence defined by*

$$A_n := \|(a_{k,n})_{k\in\mathbb{Z}}\|_{\ell^p(\mathbb{Z})} \qquad (n \in \mathbb{Z}).$$

If $(A_n)_{n\in\mathbb{Z}} \in \ell^1(\mathbb{Z})$, then

$$\left(\sum_{n=-\infty}^{\infty} \left| \sum_{k=-\infty}^{\infty} a_{k,n} \right|^p \right)^{1/p} \leq \sum_{n=-\infty}^{\infty} \left(\sum_{k=-\infty}^{\infty} |a_{k,n}|^p \right)^{1/p}.$$

Another useful tool is an inequality due to Jensen. Here we give a version for L^1 spaces and for sums.

Proposition 1.11 (Jensen's Inequality for Integrals) *Let $f, g \in L^1(A)$ be two nonnegative functions and let $\|g\|_1 = 1$. If $\varphi : [0, \infty[\to [0, \infty[$ is a convex function, then*

$$\varphi\left(\int_A f(x)g(x)dx \right) \leq \int_A \varphi(f(x))g(x)dx.$$

Proposition 1.12 (Jensen's Inequality for Sums) *Let $\mathbb{A}$ be an infinite, countable set of integers. Let $(a_n)_{n\in\mathbb{A}}$ and $(b_n)_{n\in\mathbb{A}} \in \ell^1(\mathbb{A})$ be two sequences of nonnegative real numbers, such that $\|(b_n)\|_1 = 1$. If $\varphi : [0, \infty[\to [0, \infty[$ is a convex function, then*

$$\varphi\left(\sum_{n\in\mathbb{A}} a_n\, b_n \right) \leq \sum_{n\in\mathbb{A}} \varphi(a_n)\, b_n.$$

For any $f \in L^p(\mathbb{R})$, where $p \in [1, \infty]$, let us denote by $\mathcal{T}_h f(x) := f(x+h)$ with $h \in \mathbb{R}$ the classical translation operator. It is clear that $\mathcal{T}_h$ maps $L^p(\mathbb{R})$ onto

$L^p(\mathbb{R})$ and $\|\mathcal{T}_h f\|_p = \|f\|_p$ for any $h \in \mathbb{R}$. A fundamental result is the continuity property of the translation operator, which may be stated as follows.

Proposition 1.13 *Let* $f \in L^p(\mathbb{R})$, *where* $p \in [1, \infty[$. *Then*

$$\lim_{h \to 0} \|\mathcal{T}_h f - f\|_p = 0.$$

Note that this result does not extend to $p = \infty$ as one can see by easy examples. However, it holds for uniformly continuous and bounded functions.

We mention the following density result for L^p spaces. Here, we will use the following general notation. Let H be a subspace of a normed linear space X. Then we denote by $\mathrm{cls}_X(H)$ the closure of H with respect to the norm topology of X.

Proposition 1.14 *Let* $A \subset \mathbb{R}$ *be a measurable set. Then, for every* $p \in [1, \infty[$, *we have* $\mathrm{cls}_{L^p(A)}\big(C^{\infty}_{\mathrm{comp}}(A)\big) = L^p(A)$.

Finally, we state the well-known *Lebesgue differentiation theorem*, which will later be useful for obtaining certain results concerning the almost everywhere convergence of nets of Mellin convolution integrals (see Sect. 6.4.6).

Proposition 1.15 *Let* $f \in L^1_{\mathrm{loc}}(\mathbb{R}^+)$. *Take any* $x \in \mathbb{R}^+$ *and let* $\{J_r\}_{r>0}$ *be a family of intervals all containing* x *such that* $\lim_{r \to r_0} |J_r| = 0$. *Then*

$$\lim_{r \to r_0} \frac{1}{|J_r|} \int_{J_r} |f(x) - f(t)| dt = 0$$

for almost every $x \in \mathbb{R}^+$.

1.2.2 Functions of Bounded Variation

Let $I = [a, b] \subset \mathbb{R}$ be an interval. Let $P_n := \{x_0, x_1, \ldots, x_n\}$ be a partition of I, which means that $a = x_0 < x_1 < \cdots < x_n = b$. A function $f : I \to \mathbb{C}$ is said to be of *bounded variation* on I if

$$\mathrm{V}(f, I) := \sup_{P_n, n \in \mathbb{N}} \left\{ \sum_{j=1}^{n} |f(x_j) - f(x_{j-1})| \right\} < \infty,$$

where the supremum extends over all partitions P_n of I and all $n \in \mathbb{N}$. The number $\mathrm{V}(f, I)$ is called the *total variation* of f on I.

A function $f : \mathbb{R} \to \mathbb{C}$ is said to be of *bounded variation* on $\mathbb{R}$ if

$$V_{\mathbb{R}} f := \sup_{I} V(f, I) < \infty,$$

where the supremum extends over all intervals $I = [a, b] \subset \mathbb{R}$. The number $V_{\mathbb{R}} f$ is called the *total variation* of f on $\mathbb{R}$.

For $J = [a, b]$ or $J = \mathbb{R}$, we denote by $\mathrm{BV}(J)$ the space comprising all functions of bounded variation on J.

Functions of bounded variation have several interesting properties. For example, $\mathrm{BV}(J) \subset L^\infty(J)$ and any function $f \in \mathrm{BV}(J)$ can be expressed as a difference of two monotone increasing functions. In particular, any such function f is continuous except for a denumerable set of points of J. Moreover, any $f \in \mathrm{BV}(J)$ is differentiable almost everywhere on J.

For functions $f : \mathbb{R} \to \mathbb{C}$, the notion of total variation can be introduced in the following alternative way. Let $(x_j)_{j\in\mathbb{Z}}$ be a strictly increasing sequence of real numbers. Then $\{x_j\}_{j\in\mathbb{Z}} := \{x_j : j \in \mathbb{Z}\}$ is called a partition of $\mathbb{R}$. Denoting by $\mathscr{P}$ the class of all partitions of $\mathbb{R}$, we have

$$V_{\mathbb{R}} f = \sup_{\{x_j\}\in\mathscr{P}} \sum_{j\in\mathbb{Z}} |f(x_j) - f(x_{j-1})|.$$

For functions on $\mathbb{R}$ we shall also use the following weaker form of the notion of bounded variation. A function $g : \mathbb{R} \to \mathbb{C}$ is said to be of *bounded coarse variation* if for every $\varepsilon > 0$,

$$\mathrm{V}_\varepsilon(g) := \sup_{\{x_j\}\in\mathscr{P}_\varepsilon} \sum_{j\in\mathbb{Z}} |g(x_j) - g(x_{j-1})| < \infty,$$

where now the supremum extends only over those partitions $\{x_j\}_{j\in\mathbb{Z}} \in \mathscr{P}$ for which $x_j - x_{j-1} \geq \varepsilon$ for every $j \in \mathbb{Z}$. They form a subclass $\mathscr{P}_\varepsilon$, called the *ε-separated* elements of $\mathscr{P}$. We denote by $\mathrm{BCV}(\mathbb{R})$ the space of all functions g of bounded coarse variation on $\mathbb{R}$.

If $g \in \mathrm{BV}(\mathbb{R})$, then $g \in \mathrm{BCV}(\mathbb{R})$ since in this case $\sup_{\varepsilon>0} \mathrm{V}_\varepsilon(g) < \infty$. However, there exist functions $g \in \mathrm{BCV}(\mathbb{R})$ which are not of bounded variation.

Among the properties of the functions in $\mathrm{BCV}(\mathbb{R})$ we mention the following statement.

Proposition 1.16 *Let $\{x_j\}_{j\in\mathbb{Z}} \in \mathscr{P}$ be a partition of $\mathbb{R}$ such that*

$$0 < \inf_j (x_j - x_{j-1}) \leq \sup_j (x_j - x_{j-1}) < \infty.$$

Suppose that $f : \mathbb{R} \to \mathbb{C}$ is Riemann integrable on every compact interval I. If $f \in \mathrm{BVC}(\mathbb{R})$, then the series

$$\sum_{j\in\mathbb{Z}} f(x_j^*)(x_j - x_{j-1})$$

converges for any $x_j^* \in [x_{j-1}, x_j]$.

The notion of bounded coarse variation allows a necessary and sufficient condition for the approximability of an improper integral on $\mathbb{R}$ by (infinite) Riemann sums. We shall only use the following partial result.

Proposition 1.17 *Let $f : \mathbb{R} \to \mathbb{C}$ be improperly Riemann integrable on $\mathbb{R}$. If $f \in \mathrm{BVC}(\mathbb{R})$, then*

$$\int_{\mathbb{R}} f(x)dx = \lim_{\delta \to 0+} \sum_{j \in \mathbb{Z}} f(x_j^*)(x_j - x_{j-1}),$$

where the limit is performed by using partitions $\{x_j\}_{j \in \mathbb{Z}} \in \mathscr{P}$ such that

$$0 < \inf_{j \in \mathbb{Z}} (x_j - x_{j-1}) \le \sup_{j \in \mathbb{Z}} (x_j - x_{j-1}) \le \delta$$

and $x_j^* \in [x_{j-1}, x_j]$.

For a reference on the notion of bounded coarse variation and the previous propositions, see Sect. 1.7 Notes.

1.2.3 Sobolev Spaces

First we introduce the notion of absolute continuity and its specifications. Let $[a, b]$ be an interval and $f : [a, b] \to \mathbb{C}$. We say that f is *absolutely continuous* on $[a, b]$ if it has a representation

$$f(x) = f(a) + \int_a^x g(t)dt \qquad (x \in [a, b]),$$

where $g \in L^1([a, b])$. If f is absolutely continuous, the derivative $f'(x)$ exists almost everywhere and $f'(x) = g(x)$ almost everywhere.

For a positive integer k, we say that f is *k-times absolutely continuous* on $[a, b]$ if f is k-times differentiable and the derivatives $f^{(j)}$ are absolutely continuous on $[a, b]$ for every $0 \le j \le k$.

A function $f : \mathbb{R} \to \mathbb{C}$ is *k-times locally absolutely continuous on $\mathbb{R}$ if f is k-times absolutely continuous on every interval $I = [a, b] \subset \mathbb{R}$.*

For $r \in \mathbb{N}$, we denote by $AC_{\mathrm{loc}}^{r-1}(\mathbb{R})$ the class of all functions that are $(r-1)$-times locally absolutely continuous on $\mathbb{R}$ and define

$$W^{r,p}(\mathbb{R}) := \{f : \mathbb{R} \to \mathbb{C} : f = \varphi \text{ a.e, } \varphi \in AC_{\mathrm{loc}}^{r-1}(\mathbb{R}),\ \varphi^{(k)} \in L^p(\mathbb{R}),\ 0 \le k \le r\}.$$

For $f \in W^{r,p}(\mathbb{R})$ we may use the equivalent function φ to define $\|f^{(k)}\|_p := \|\varphi^{(k)}\|_p$, for $k = 1, \ldots, r$. Endowed with the norm

$$\|f\|_{W^{r,p}(\mathbb{R})} := \left\{ \sum_{k=0}^{r} \|f^{(k)}\|_p^p \right\}^{1/p},$$

the space $W^{r,p}(\mathbb{R})$ is a Banach space, which we call a *Sobolev space*.

1.3 Fundamentals of Functional Analysis

In this section we report on some basic theorems on normed linear spaces which we will be used later.

Let X and Y be two normed linear spaces, endowed with their norms $\|\cdot\|_X$ and $\|\cdot\|_Y$, respectively. Let $T : D \to Y$ be a linear operator defined on a (linear) subspace D of X. The operator T is called *bounded* if there exists a constant $M > 0$ such that $\|Tf\|_Y \leq M\|f\|_X$ for every $f \in D$. This implies that T is *Lipschitzian*, i.e., $\|Tf - Tg\|_Y \leq M\|f - g\|_X$ for every $f, g \in D$. We shall also say that T is a continuous linear operator. The *(uniform) norm* of the operator T is defined by

$$\|T\|_{[D,Y]} := \sup_{f \neq O} \frac{\|Tf\|_Y}{\|f\|_X} = \sup_{\|f\|_X \leq 1} \|Tf\|_Y = \sup_{\|f\|_X = 1} \|Tf\|_Y, \tag{1.3}$$

where O denotes the null element of X. Therefore T is a bounded operator if and only if $\|T\|_{[D,Y]} < \infty$ and

$$\|Tf\|_Y \leq \|T\|_{[D,Y]}\|f\|_X \qquad (f \in D).$$

We denote by $[D, Y]$ the normed linear space comprising all the bounded linear operators $T : D \to Y$, endowed with the norm $\|T\|_{[D,Y]}$. If Y is a Banach space, then $[D, Y]$ is also a Banach space.

A sequence $(T_n)_{n \in \mathbb{N}} \subset [D, Y]$ is said to *converge strongly* to a (unique) operator $T \in [D, Y]$ if

$$\lim_{n \to \infty} \|T_n f - Tf\|_Y = 0$$

for every $f \in D$. It is said to be *strongly Cauchy* if

$$\lim_{n,m \to \infty} \|T_n f - T_m f\|_Y = 0.$$

Moreover, we shall say that $(T_n)_{n \in \mathbb{N}} \subset [D, Y]$ is *uniformly convergent* to $T \in [D, Y]$, if

$$\lim_{n\to\infty} \|T_n - T\|_{[D,Y]} = 0,$$

and the notion of uniform Cauchy sequence is analogous. The operator $T \in [D, Y]$ is an isometry (or it is isometric) if $\|Tf\|_Y = \|f\|_X$ for every $f \in D$.

A subset $A \subset X$ is said to be *dense* in X if for any $f \in X$ there exists a sequence $(f_n)_{n\in\mathbb{N}} \subset A$ such that $\lim_{n\to\infty} \|f - f_n\|_X = 0$. The subset $A \subset X$ is said to be *fundamental* in X if the set of all finite linear combinations of elements of A is dense in X.

The following Proposition is a particular version of the classical Hahn–Banach theorem.

Proposition 1.18 *Let A be a dense linear subspace of a Banach space X and suppose that T_0 is a bounded linear operator of A into the Banach space Y with norm $\|T_0\|_{[A,Y]}$. Then T_0 can be uniquely extended to a bounded linear operator T of X into Y such that $Tf = T_0 f$ for every $f \in A$ and $\|T\|_{[X,Y]} = \|T_0\|_{[A,Y]}$.*

Next, we give the statement of the classical uniform boundedness principle.

Proposition 1.19 *Let $(T_n)_{n\in\mathbb{N}}$ be a sequence of bounded linear operators of a Banach space X into the normed linear space Y. If the sequence $(\|T_n f\|_Y)_{n\in\mathbb{N}}$ is bounded for each $f \in X$ separately, i.e., for every $f \in X$ there exists a constant $M_f > 0$ such that*

$$\|T_n f\|_Y \le M_f \tag{1.4}$$

for all $n \in \mathbb{N}$, then the sequence $(\|T_n\|_{[X,Y]})_{n\in\mathbb{N}}$ is bounded, i.e., there exists a constant $M > 0$ such that

$$\|T_n f\|_Y \le M\|f\|_X \tag{1.5}$$

for all $n \in \mathbb{N}$ and $f \in X$.

As an immediate consequence, one obtains results on the convergence of sequences of bounded linear operators. Here we quote the classical Banach–Steinhaus theorem.

Proposition 1.20

(1) A sequence $(T_n)_{n\in\mathbb{N}}$ of bounded linear operators of a Banach space X into a Banach space Y converges strongly to a bounded linear operator T of X into Y if and only if there exist

(i) a constant $M > 0$ such that $\|T_n\|_{[X,Y]} \le M$ for all $n \in \mathbb{N}$ and
(ii) a dense subset $A \subset X$ such that the sequence $\{T_n\}_{n\in\mathbb{N}}$ is strongly Cauchy in A.

(2) Let X be a Banach space and $(T_n)_{n\in\mathbb{N}}$ a sequence of bounded linear operators of X into itself. Then, for every $f \in X$, we have

$$\lim_{n\to\infty} \|T_n f - f\|_X = 0 \tag{1.6}$$

if and only if there exist

(i) *a constant* $M > 0$ *such that* $\|T_n\|_{[X,X]} \leq M$ *for all* $n \in \mathbb{N}$ *and*
(ii) *a dense subspace* $A \subset X$ *such that (1.6) holds for every* $g \in A$.

The next result is an important interpolation theorem for L^p spaces, known as the *Riesz–Thorin interpolation theorem.*

First we make some agreements. If (R, μ) is an arbitrary (Hausdorff topological) measure space with a regular measure μ, we denote by $M(R, \mu)$ the space of all measurable functions f defined on R. For any $1 \leq p \leq \infty$, we denote by $L^p(\mu)$ the corresponding Lebesgue space with norm

$$\|f\|_{L^p(\mu)} \equiv \|f\|_p := \left\{\int_R |f(t)|^p d\mu(t)\right\}^{1/p} \qquad (1 \leq p < \infty)$$

and $\|f\|_\infty := \inf\{a \geq 0 : \mu(\{x \in R : |f(x)| > a\}) = 0\}$, which is denoted by $\|f\|_\infty = \operatorname{ess\,sup}_{x\in R} |f(x)|$.

Definition 1.21 Let $T_0 : M(R, \mu_1) \to M(R_2, \mu_2)$ be a linear operator. We say that T_0 is of *strong type* $(1/p, 1/q)$, with $0 \leq p, q \leq 1$, if there exists a constant $M > 0$ such that

$$\|T_0 f\|_{1/q} \leq M\|f\|_{1/p} \qquad (f \in L^{1/p}(\mu_1)).$$

Here we set $1/p = \infty$ if $p = 0$.

Now we are ready to state the Riesz–Thorin interpolation theorem.

Proposition 1.22 *Let* T_0 *be simultaneously of strong type* $(1/p_1, 1/q_1)$ *and* $(1/p_2, 1/q_2)$ *with* $(p_1, q_1), (p_2, q_2) \in [0, 1] \times [0, 1]$ *and constants* M_1, M_2, *respectively. Then* T_0 *is also of strong type* $(1/p, 1/q)$ *with constant* $M_1^{1-t} M_2^t$ *for every* $p = (1-t)p_1 + tp_2$ *and* $q = (1-t)q_1 + tq_2$, *where* $t \in]0, 1[$, *i.e.,*

$$\|T_0 f\|_{1/q} \leq M_1^{1-t} M_2^t \|f\|_{1/p} \qquad \big(f \in L^{1/p}(\mu_1)\big).$$

We say that a function f defined on R is a *simple* function if it assumes a finite number of values and has compact support. The proof of the Riesz–Thorin interpolation theorem can be performed by using the density of the space of all simple functions in the spaces $L^p(\mu)$ (see, e.g., [150, Chap. 6, Sect. 5, p. 193]). Therefore, very often Proposition 1.22 is formulated for operators T_0 which are defined on the space S_1 of simple functions with compact support (see Sect. 1.7 Notes). Nevertheless, for operators acting on Lebesgue spaces whose elements are functions defined over a measurable subset $R_1 \subset \mathbb{R}$, one can also employ the space of test functions $C_{\text{comp}}^{\infty}(R_1)$ which is dense in $L^p(R_1)$ for $1 \leq p < \infty$ (Proposition 1.14).

1.4 Fundamentals of Complex Analysis

In this section we report on some basic results from complex analysis along with certain particular associated spaces and properties. When in the following we have terms of the form $x+iy$ we tacitly agree that x and y are real. Moreover, the complex number $x + iy$ will sometimes be identified with the point (x, y) in $\mathbb{R}^2$.

1.4.1 Basic Definitions and Properties

Let $A \subset \mathbb{C}$ be a subset of the complex plane and let $f : A \to \mathbb{C}$ be a function

$$f(z) = f(x+iy) = u(x, y) + iv(x, y) = \Re f(z) + i\Im f(z) \qquad (z = x + iy \in A).$$

Then we say that f is a *complex function*. Such a function can be interpreted as a vector-valued function defined on a subset of the plane $\mathbb{R}^2$ with values in $\mathbb{R}^2$.

If z_0 is an accumulation point of A, we say that f has a (unique) limit $\zeta \in \mathbb{C}$ for $z \to z_0$ and we write

$$\lim_{z \to z_0} f(z) = \zeta$$

if and only if for every $\varepsilon > 0$, there exists $\delta > 0$ such that for every $z \in A$ with $0 < |z - z_0| < \delta$, we have $|f(z) - \zeta| < \varepsilon$.

Setting $f(z) = u(x, y) + iv(x, y)$, $z = x + iy$, $z_0 = x_0 + iy_0$ and $\zeta = \zeta_1 + i\zeta_2$, then

$$\lim_{z \to z_0} f(z) = \zeta \text{ if and only if } \lim_{(x,y)\to(x_0,y_0)} u(x, y) = \zeta_1, \quad \lim_{(x,y)\to(x_0,y_0)} v(x, y) = \zeta_2.$$

If $z_0 \in A$ is an accumulation point of A, we say that f is *continuous* at z_0 if

$$\lim_{z \to z_0} f(z) = f(z_0).$$

We shall also need some topological notions for the complex plane. Let $I := [a, b] \subset \mathbb{R}$ be any bounded, closed interval and let $\gamma(t) := x(t) + iy(t)$, where $t \in I$, be a continuous function. Then we call γ an *arc* or a *curve*. In the following specifications we continue by using the word arc as a synonym for curve. We say that γ is a *simple arc* if $\gamma(t_1) \neq \gamma(t_2)$ whenever $t_1, t_2 \in I$ with $t_1 \neq t_2$ and

$\{t_1, t_2\} \neq \{a, b\}$. Geometrically, this means that the arc does not meet or cross itself except for the possibility of coinciding endpoints. If $\gamma(a) = \gamma(b)$, we say that the arc γ is *closed*. An arc γ is called *smooth* if γ is continuously differentiable on I. It is said to be *piecewise smooth* on I if there exists a partition $a = t_0 < t_1 < \cdots < t_{n-1} < t_n = b$ such that the restrictions of γ to the subintervals $[t_{i-1}, t_i]$ are smooth arcs for $i = 1, \ldots, n$. Smooth and piecewise smooth arc have a well-defined finite length.

A non-empty open subset $A \subset \mathbb{C}$ is called *connected* if for any two points $z_1, z_2 \in A$ there is a piecewise smooth arc γ with endpoints z_1, z_2 which lies entirely in A. For example, one can consider a polygonal path consisting of a finite number of line segments.

We shall say that $A \subset \mathbb{C}$ is a *domain* or a *region* if it is a non-empty, open and connected set. From now on, we assume that A is a domain.

We say that f is continuous on A if it is continuous at every point of A. This implies that f is continuous on A if and only if the functions $u(x, y)$ and $v(x, y)$ are continuous on A interpreted as a subset of $\mathbb{R}^2$.

Let $z_0 \in A$ be a point in $A \subset \mathbb{C}$. Then f is said to be *complex differentiable* at z_0 if the limit

$$\lim_{z \to z_0} \frac{f(z) - f(z_0)}{z - z_0} =: f'(z_0)$$

exists and is the same howsoever z approaches z_0 within A. We call $f'(z_0)$ the (complex) derivative of f at z_0. Derivatives of higher order are defined in the same way. We say that f is analytic on A if it is complex differentiable at each point of A. Obviously, every analytic function on a domain A is also continuous on A.

The following result characterizes complex differentiability by real differentiability.

Proposition 1.23 *A function $f = u + iv : A \to \mathbb{C}$ is complex differentiable at $z_0 = x_0 + iy_0 \in A$ if and only if the functions u and v are real differentiable at (x_0, y_0) and*

$$\frac{\partial u}{\partial x}(x_0, y_0) = \frac{\partial v}{\partial y}(x_0, y_0), \qquad \frac{\partial u}{\partial y}(x_0, y_0) = -\frac{\partial v}{\partial x}(x_0, y_0). \tag{1.7}$$

Moreover,

$$f'(z_0) = \frac{\partial f}{\partial x}(z_0) = \frac{1}{i}\frac{\partial f}{\partial y}(z_0).$$

Equations (1.7) are known as *Cauchy–Riemann equations*.

In practice it is sometimes more convenient to work with partial derivatives only. Therefore the following consequences of Proposition 1.23 are of interest:

For $f = u + iv$ to be complex differentiable at $x_0 + iy_0 \in A$ it is necessary that u and v have partial derivatives at (x_0, y_0) which satisfy Eqs. (1.7).

For $f = u + iv$ to be complex differentiable at $x_0 + iy_0 \in A$ it is sufficient that u and v have continuous partial derivatives in a neighborhood of (x_0, y_0) *which satisfy Eqs.* (1.7).

The Cauchy–Riemann equations impose severe restrictions on the functions u, v in order that $f = u+iv$ is analytic. Indeed, given two functions u, v with any degree of regularity, then the complex function $f = u + iv$ is not analytic in general. For example the function $f(z) = \overline{z}$, where $\overline{z}$ denotes the conjugate of z, is not analytic since the Cauchy–Riemann equations are not both satisfied.

When $f : \mathbb{C} \to \mathbb{C}$ is analytic on the whole $\mathbb{C}$, then f is said to be an *entire function*. In this case the Cauchy–Riemann equations hold for all points of $\mathbb{C}$.

The Cauchy–Riemann equations can be written in terms of polar coordinates. Let us denote by $\mathbb{H}$ the polar plane, namely $\mathbb{H} := \{(r, \theta) : r \in \mathbb{R}^+, \theta \in \mathbb{R}\}$. If

$$f(x + iy) = u(x, y) + iv(x, y),$$

then for $(r, \theta) \in \mathbb{H}$, we have

$$f(re^{i\theta}) = u(r\cos\theta, r\sin\theta) + iv(r\cos\theta, r\sin\theta) =: U(r, \theta) + iV(r, \theta),$$

with real-valued functions U and V. Writing the above Cauchy–Riemann equations in terms of U, V, r and θ, we obtain their polar form as

$$\frac{\partial U}{\partial \theta}(r, \theta) = -r\frac{\partial V}{\partial r}(r, \theta), \quad \frac{\partial V}{\partial \theta}(r, \theta) = r\frac{\partial U}{\partial r}(r, \theta), \tag{1.8}$$

which will be of interest later on (see Chapter 2).

As some examples of analytic functions, we mention polynomials of a complex variable, the exponential function e^z defined by its power series and the functions $\sin z$, $\cos z$ defined by the classical Euler equations.

The complex logarithm is a multi-valued function given by

$$\log z = \log|z| + i\arg z \qquad (z \neq 0),$$

where $\arg z$ denotes the argument of the complex number z, which is determined up to additive integer multiples of 2π. The principal value of the logarithm is defined by

$$\operatorname{Log} z := \log|z| + i\operatorname{Arg} z,$$

where $-\pi < \operatorname{Arg} z \leq \pi$. The function Log is analytic on the domain $A = \mathbb{C} \setminus \{z : \Re z \leq 0\}$.

1.4.2 Integration and Taylor's Series

The basic result of complex analysis is Cauchy's integral theorem for analytic functions. First we introduce the line integral of a complex function.

Let $f : A \to \mathbb{C}$ be a continuous function on a domain A, and let C be a smooth curve in A given by $z(t) = x(t) + iy(t)$ for $t \in [a, b]$. We define the line integral of f on C as

$$\int_C f(z)dz := \int_a^b f(z(t))z'(t)dt.$$

Writing $f(z(t)) = u(x(t), y(t)) + iv(x(t), y(t))$ and $z'(t) = x'(t) + iy'(t)$, we obtain

$$\int_C f(z)dz = \int_a^b [u(x(t), y(t))x'(t) - v(x(t), y(t))y'(t)]dt$$

$$+ i \int_a^b [v(x(t), y(t))x'(t) + u(x(t), y(t))y'(t)]dt.$$

Therefore, the line integral $\int_C f(z)dz$ is given by the integrals of two differential forms, namely

$$\int_C f(z)dz = \int_C (udx - vdy) + i \int_C (vdx + udy).$$

This integral is linear and satisfies the additive property for arcs. Moreover, there holds the Darboux estimate

$$\left| \int_C f(z)dz \right| \leq \int_C |f||dz| \leq M\ell(C),$$

where $M := \max_{z \in C} |f(z)|$ and $\ell(C)$ denotes the length of the curve C.

By using the additive property, the above definition can be extended in a natural way to piecewise smooth curves. If C is composed by a finite number of smooth arcs $C_1, \ldots, C_n$, we set

$$\int_C f(z)dz = \sum_{k=1}^{n} \int_{C_k} f(z)dz.$$

Finally, by a *regular domain* $D \subset \mathbb{C}$ we mean a bounded domain in the complex plane whose boundary, denoted by ∂D, consists of a finite number of simple, closed, piecewise smooth curves, each curve being oriented such that D lies on the left-hand side. Now Cauchy's integral theorem can be stated as follows.

Theorem 1.24 (Cauchy's Integral Theorem) *Let f be analytic on a domain A. If $D \subset A$ is a regular domain, then*

$$\int_{\partial D} f(z)dz = 0. \tag{1.9}$$

The following theorem is of a converse form. It provides a useful sufficient condition for analyticity.

Theorem 1.25 (Morera's Theorem) *Let $f : A \to \mathbb{C}$ be a continuous function on a domain A. Suppose that for every rectangular domain $R \subset A$ with sides parallel to the axes, there holds*

$$\int_{\partial R} f(z)dz = 0.$$

Then f is analytic on A.

The following proposition tells us that the derivative of an analytic function is again analytic. Hence analytic functions are infinitely differentiable.

Proposition 1.26 (Cauchy's Integral Formulae) *Let f be analytic on a domain A. If $D \subset A$ is a regular domain, then for every $z_0 \in D$ we have*

(a)

$$f(z_0) = \frac{1}{2\pi i} \int_{\partial D} \frac{f(z)}{z - z_0} dz.$$

(b) Moreover, f is infinitely differentiable on D and

$$f^{(n)}(z_0) = \frac{n!}{2\pi i} \int_{\partial D} \frac{f(z)}{(z - z_0)^{n+1}} dz \qquad (n \in \mathbb{N}).$$

Note that the formulae of Proposition 1.26 extend to any point $z_0 \in A$ since D can be chosen as a sufficiently small disk centered at z_0.

Another basic result on analytic functions is the following Taylor theorem which provides an expansion by a power series.

Theorem 1.27 *Let $f : A \to \mathbb{C}$ be an analytic function on a domain A. Let z_0 be any point in A and denote by $B_r(z_0) \subset A$ any disk centered at z_0 with radius r. Then there holds the representation*

$$f(z) = \sum_{n=0}^{\infty} \frac{f^{(n)}(z_0)}{n!} (z - z_0)^n \qquad (z \in B_r(z_0)). \tag{1.10}$$

As a consequence, we can deduce an *identity principle of the analytic functions.* First we present some preliminary facts. A point z_0 such that $f(z_0) = 0$ is called a *zero* of f. The smallest integer k such that $f^{(k)}(z_0) \neq 0$ is called the *order* of the zero z_0.

Lemma 1.28 *Let f be an analytic function on a domain A and let $z_0 \in A$ be a point such that there exists a disk $B_r(z_0) \subset A$ with $f(z) = 0$ for every $z \in B_r(z_0)$. Then $f(z) = 0$ for all $z \in A$.*

This lemma has an immediate consequence: Let f be a non-trivial analytic function. If $z_0 \in A$ is a zero of f, then it has finite order. Indeed, if we had $f^{(n)}(z_0) = 0$ for all $n \in \mathbb{N}_0$ then, by Theorem 1.27 and Lemma 1.28, we would obtain that f is identically zero on A, which is a contradiction. Hence the Taylor expansion implies that zeros cannot accumulate. We obtain the following result.

Theorem 1.29 (Identity Principle) *Let f be an analytic function on a domain A. If the set of all the zeros of f has a finite cluster point in A, then $f(z) = 0$ for all $z \in A$.*

From this theorem, we deduce that if f and g are two analytic functions defined on a domain A which coincide on an infinite set $B \subseteq A$ with a cluster point in A, then $f(z) = g(z)$ for all $z \in A$.

Another useful result is the following maximum principle.

Theorem 1.30 (Maximum Principle) *Let f be analytic on a domain A and let γ be a closed, piecewise smooth curve that is the boundary of a convex domain $D \subset A$. Then*

$$|f(z)| \leq \max_{w \in \gamma} |f(w)| \qquad (z \in D).$$

We shall also need the following results on analytic functions. The first one is a Montel type theorem for strips.

Proposition 1.31 (Montel) *Let $a, b, c \in \mathbb{R}$ with $b < c$, and let $A := \{z = x+iy \in \mathbb{C} : x \geq a,\ y \in [b, c]\}$. If $f : A \to \mathbb{C}$ is a bounded and analytic function such that $\lim_{x\to\infty} f(x + iy_0) = \zeta$ for some $y_0 \in]b, c[$, then $\lim_{x\to\infty} f(x + iy) = \zeta$ uniformly for $y \in [b + \delta, c - \delta]$ with any $\delta \in]0, (c - b)/2[$.*

The next result concerns entire functions. It is easily deduced from Proposition 1.26 (b) for $n = 1$.

Proposition 1.32 (Liouville) *A bounded entire function on $\mathbb{C}$ is a constant.*

Now let $A \subset \mathbb{C}$ be a domain and let $z_0 \in A$. If there exists a neighborhood $U \subset A$ of z_0 such that a function $f : A \to \mathbb{C}$ is analytic on $U \setminus \{z_0\}$, then z_0 is called an *isolated singularity* of f. If $\lim_{z\to z_0} f(z) =: \zeta$ exists, the isolated singularity is called *removeable*. Then f can be extended to an analytic function on U by defining $f(z_0) = \zeta$. If there is a positive integer n such that $\lim_{z\to z_0}(z - z_0)^n f(z)$ exists and is different from zero, then z_0 is said to be a *pole of order* n. In all other cases the isolated singularity is called *essential.*

If z_0 is a pole of order n, the *residue* of f at the point z_0 is defined as

$$\mathrm{Res}(f, z_0) := \frac{1}{(n-1)!} \lim_{z \to z_0} \frac{d^{n-1}}{dz^{n-1}} (z - z_0)^n f(z).$$

In particular, for $n = 1$ the pole z_0 is called simple and then

$$\mathrm{Res}(f, z_0) = \lim_{z \to z_0} (z - z_0) f(z).$$

For a removeable isolated singularity the residue is defined as zero. In the following fundamental result of complex analysis we disregard essential singularities since they will not be needed in the development of Mellin analysis..

Theorem 1.33 (Cauchy's Residue Theorem) *Let* $A \subset \mathbb{C}$ *be a domain and let* $f : A \to \mathbb{C}$ *be analytic on* A *except for a finite number of poles* $z_1, \ldots, z_k$*. Suppose that* $D \subset A$ *is a regular domain containing the poles. Then*

$$\int_{\partial D} f(z) dz = 2\pi i \sum_{n=1}^{k} \mathrm{Res}(f, z_n). \tag{1.11}$$

In evaluating a line integral over a general closed, piecewise smooth curve γ, the following concept is very useful. Let $a \notin \gamma$. Then the integral

$$n(\gamma, a) := \frac{1}{2\pi i} \int_{\gamma} \frac{dz}{z - a} \tag{1.12}$$

is called the *winding number* of the curve γ with respect to the point a. It is known that the winding number is always an integer. As an example, if γ represents an oriented circle $\gamma(t) = z_0 + re^{it}$ with $t \in [0, 2\pi]$, then

$$n(\gamma, a) = \begin{cases} 1 & \text{if } a \text{ is inside the circle,} \\ 0 & \text{if } a \text{ is outside the circle.} \end{cases}$$

If $\gamma(t) = z_0 + re^{it}$ with $t \in [0, 2k\pi]$ and a is inside the circle, then $n(\gamma, a) = k$.

1.4.3 Bernstein Spaces

Now we introduce a class of entire functions which is important in several branches of analysis as well as in signal processing.

Definition 1.34 For $p \in [1, \infty]$ and $\sigma \in \mathbb{R}^+$, the *Bernstein space* B_σ^p comprises all entire functions with the following properties:

(i) $|f(z)| \le A \exp(\sigma |\Im z|)$ for all $z \in \mathbb{C}$ with a constant $A > 0$ depending on f but not on z;
(ii) the restriction of f to the real line belongs to $L^p(\mathbb{R})$.

It can be shown that

$$B_\sigma^1 \subset B_\sigma^p \subset B_\sigma^r \subset B_\sigma^\infty \qquad (1 < p < r < \infty).$$

The following result describes the behaviour of a function $f \in B_\sigma^p$ for large values of $|\Re z|$.

Proposition 1.35 *Let $f \in B_\sigma^p$, where $p \in [1, \infty]$ and $\sigma \in \mathbb{R}^+$. Then*

$$\|f(\cdot + iy)\|_p \le e^{\sigma |y|} \|f\|_p \qquad (y \in \mathbb{R}).$$

Moreover, if $p \neq \infty$, then $\lim_{|x| \to \infty} f(x) = 0$.

The next proposition gives a representation of the derivative of a function $f \in B_\sigma^p$ by a series involving samples of f.

Proposition 1.36 (Boas' Differentiation Formula) *Let $f \in B_\sigma^p$, where $p \in [1, \infty]$ and $\sigma \in \mathbb{R}^+$. Then*

$$f'(x) = \frac{4\sigma}{\pi^2} \sum_{k \in \mathbb{Z}} \frac{(-1)^k}{(2k+1)^2} f\left(x + \frac{(2k+1)\pi}{2\sigma}\right) \qquad (x \in \mathbb{R}).$$

Applying the L^p norm on both sides, using the triangular inequality for norms on the right-hand side and noting that $\sum_{k \in \mathbb{Z}} 1/(2k+1)^2 = \pi^2/4$, we find the following sharp result.

Corollary 1.37 (Bernstein's Inequality) *Let $f \in B_\sigma^p$, where $p \in [1, \infty]$ and $\sigma \in \mathbb{R}^+$. Then $\|f'\|_p \le \sigma \|f\|_p$.*

For $f \in B_\sigma^p$, Bernstein's inequality shows that f' is an entire function which satisfies condition (ii) of Definition 1.34. But f' also satisfies condition (i). Indeed, by Cauchy's integral theorem for derivatives (see Proposition 1.26), we may write

$$f'(z) = \frac{1}{2\pi i} \int_C \frac{f(\zeta)}{(\zeta - z)^2} \, d\zeta$$

with C being the positively oriented circle with center at $z = x + iy$ and radius 1. Knowing that (ii) holds for f, we may estimate

$$|f'(z)| \le \frac{1}{2\pi} \int_0^{2\pi} \left| f\big(x + \cos t + i(y + \sin t)\big) \right| dt \le A e^{\sigma} e^{\sigma |y|},$$

which is (i) for f'. Hence B_σ^p is closed under differentiation and therefore Bernstein's inequality may be iterated and extended as follows.

Proposition 1.38 *Let $f \in B_\sigma^p$, where $p \in [1, \infty]$ and $\sigma \in \mathbb{R}^+$. Then*

$$\|f^{(r)}\|_p \leq \sigma^r \|f\|_p \qquad (r \in \mathbb{N}). \tag{1.13}$$

Another interesting inequality for functions belonging to a Bernstein space is the following statement.

Proposition 1.39 (Nikol'skii's Inequality) *Let $f \in B_\sigma^p$, where $p \in [1, \infty[$ and $\sigma \in \mathbb{R}^+$. Then for any positive h, we have*

$$\left\{h \sum_{k\in\mathbb{Z}} |f(hk)|^p\right\}^{1/p} \leq (1 + \sigma h)\|f\|_p.$$

1.4.4 *Hardy Spaces for Strips*

An intermediate space between the Bernstein spaces and the Sobolev spaces is the following Hardy space. For $d > 0$, we denote by $\mathcal{S}_d$ the strip

$$\mathcal{S}_d := \{z \in \mathbb{C} \, : \, |\Im z| < d\}.$$

Definition 1.40 For $p \in [1, \infty[$ we define the *Hardy space* for the strip $\mathcal{S}_d$, denoted by $H^p(\mathcal{S}_d)$, as the space of all analytic functions on $\mathcal{S}_d$ such that

$$\sup_{0<y<d} \int_{\mathbb{R}} \big(|f(t - iy)|^p + |f(t + iy)|^p\big) dt < \infty,$$

endowed with the norm

$$\|f\|_{H^p(\mathcal{S}_d)} := \left\{\sup_{0<y<d} \int_{\mathbb{R}} \frac{|f(t - iy)|^p + |f(t + iy)|^p}{2} dt\right\}^{1/p}.$$

Now we state some useful properties of the space $H^p(\mathcal{S}_d)$. The first result can be seen as a weak form of property (i) of Bernstein spaces (see Definition 1.34).

Proposition 1.41 *Let $f \in H^p(\mathcal{S}_d)$ and let $d_1 \in]0, d[$. Then*

$$|f(z)| \leq \left(\frac{4}{\pi(d - d_1)}\right)^{1/p} \|f\|_{H^p(\mathcal{S}_d)} \qquad (z \in \mathcal{S}_{d_1}). \tag{1.14}$$

The next result is related to Nikol'skii's inequality (Proposition 1.39).

Proposition 1.42 *Let $f \in H^p(\mathcal{S}_d)$ and let $(t_n)_{n\in\mathbb{Z}}$ be a sequence in $\mathbb{R}$ such that $t_{n+1} - t_n > 2\delta$ for all $n \in \mathbb{Z}$, where $\delta \in]0, d[$. Then*

$$\left\{\sum_{n\in\mathbb{Z}} |f(t_n)|^p\right\}^{1/p} \leq \left(\frac{2}{\pi\delta}\right)^{1/p} \|f\|_{H^p(\mathcal{S}_d)}. \tag{1.15}$$

The following result is a strengthened form of Proposition 1.31 for Hardy spaces.

Proposition 1.43 *Let $f \in H^p(\mathcal{S}_d)$ and let $d_1 \in]0, d[$. Then $\lim_{t\to\pm\infty} f(t+iy) = 0$ uniformly for $y \in [-d_1, d_1]$.*

1.4.5 The Riemann Surface of the Complex Logarithm

Here we want to describe a visual approach to the Riemann surface of the logarithm. As we have mentioned already, the logarithm is multi-valued. In order to define it as a single-valued analytic function, we need a domain in which the origin cannot be encircled. For example, we may choose an $\alpha \in [-\pi, \pi[$ and slit the complex plane $\mathbb{C}$ from 0 to ∞ along the ray

$$R_\alpha := \left\{z = re^{i\alpha} \,:\, r \in \mathbb{R}_0^+\right\}.$$

The slit produces a domain $D_0 := \mathbb{C} \setminus \mathbb{R}_\alpha$ which we call a sheet. By the slitting, the ray R_α builds two coinciding edges of the sheet, which we distinguish by the notation

$$R_{\alpha+} := \lim_{\varepsilon\to 0+} R_{\alpha+\varepsilon} \quad \text{and} \quad R_{\alpha-} := \lim_{\varepsilon\to 0+} R_{\alpha-\varepsilon}.$$

For every $z \in \mathbb{C} \setminus \{0\}$, there exists a uniquely determined $\theta_z \in]0, 2\pi]$ such that

$$z = |z| \exp\big(i(\alpha + \theta_z)\big).$$

On D_0 we define a complex logarithm as a single-valued analytic function g_0 by putting

$$g_0(z) := \log|z| + i(\alpha + \theta_z).$$

Since for $z = re^{i\alpha}$ we have $\theta_z = 2\pi$, the definition of g_0 extends continuously to the edge $R_{\alpha-}$ and gives $g_0(re^{i\alpha}) = \log r + i(\alpha + 2\pi)$. However, for $z \in D_0$ approaching $re^{i\alpha}$ as a point on the edge $R_{\alpha+}$, we obtain $g_0(z) \to \log r + i\alpha$. This shows that continuous continuation of g_0 across the ray R_α is not possible.

Note that for $\alpha = -\pi$, the function g_0 is the principal value Log of the logarithm (see p. 15). Cases where α is chosen as 0 can be found in the study of Mellin–Bernstein spaces (see Sects. 3.6, 8.1–8.2).

Now, for every $k \in \mathbb{Z}$ let D_k be a copy of D_0. On all these sheets we have the same system of polar coordinates for their points, but we consider these sheets as being separated in $\mathbb{R}^3$ such that D_{k+1} is parallel to D_k and is lying above the latter. The edges of D_k are again denoted by $R_{\alpha+}$ and $R_{\alpha-}$. On D_k we define

$$g_k(z) := g_0(z) + i2k\pi \qquad (k \in \mathbb{Z}).$$

It is easily verified that $\exp(g_k(z)) = z$ for all $k \in \mathbb{Z}$. Therefore the functions g_k are inverses of the exponential functions and so they are indeed logarithms. We call them the *branches* of the logarithm with respect to the ray R_α.

When $z \in D_k$ approaches $re^{i\alpha}$ understood as a point of the edge $R_{\alpha+}$, then

$$g_k(z) \longrightarrow \log|z| + i(\alpha + 2k\pi)$$

and when z approaches $re^{i\alpha}$ understood as a point of the edge $R_{\alpha-}$, then

$$g_k(z) \longrightarrow \log|z| + i(\alpha + 2(k+1)\pi).$$

Now the idea is to glue the edge $R_{\alpha-}$ of the sheet D_k with the edge $R_{\alpha+}$ of the sheet D_{k+1} for all $k \in \mathbb{Z}$. This can be done by bending the sheets appropriately without affecting the coordinates of their points. The result is a connected surface in $\mathbb{R}^3$ which spirals around a vertical axis. When we are on D_k and cross the glued edges in the counterclockwise sense, we reach D_{k+1} where g_{k+1} is seen to be a continuous continuation of g_k. The constructed surface is a model of the Riemann surface $S_{\log}$ of the logarithm. On this model the various branches are composed to a single-valued analytic function which attains all possible values of the logarithm. For a rigorous mathematical approach one introduces a notion of neighborhoods and describes the abstract Riemann surface as a connected Hausdorff space with certain local properties. In this setting, the models are topologically equivalent and, other than the branches, they do not depend on the choice of the ray R_α. For a simple explicit model of $S_{\log}$ and the definition of analytic functions on it, see Sect. 2.1.

1.5 Some Basic Special Functions

In this section we recall the definitions and the main basic properties of some special functions: the Euler gamma and beta functions, the Riemann zeta function and the Möbius function. They will be useful in the development of the theory.

1.5.1 Gamma and Beta Functions

The *Euler gamma function* is defined on the right complex half-plane $\Re z > 0$ by the formula

$$\Gamma(z) := \int_0^\infty e^{-t} t^{z-1} dt.$$

As we shall see later, this function occurs naturally in Mellin analysis, being the Mellin transform of the function e^{-t}.

By elementary calculations, we find that $\Gamma(1) = 1$, and $\Gamma(1/2) = \sqrt{\pi}$. The following statement shows that the gamma function extends and interpolates the factorials $n!$.

Proposition 1.44 *For any $z \in \mathbb{C}$ with $\Re z > 0$, we have*

$$\Gamma(z+1) = z\Gamma(z).$$

In particular, for any $n \in \mathbb{N}$,

$$\Gamma(n+1) = n!.$$

The gamma function is analytic on the set $\Re z > 0$ and, using Proposition 1.44, it has an analytic continuation to the set $\mathbb{C} \setminus \{0, -1, -2, \dots, \}$. At each point $z = -n$ with $n \in \mathbb{N}_0$, it has a pole of order 1. The residue at the point $-n$ is equal to $(-1)^n/n!$. The function $1/\Gamma(z)$ can be extended to the whole complex plane as an entire function.

An interesting identity is contained in the following statement.

Proposition 1.45 *For any $z \in \mathbb{C}$ such that $\Re z \in]0, 1[$, we have*

$$\Gamma(z)\,\Gamma(1-z) = \frac{\pi}{\sin(\pi z)}.$$

The following proposition gives a useful representation of the gamma function, known as *Euler's formula*.

Proposition 1.46 *For $z \in \mathbb{C} \setminus \{0, -1, -2, \dots\}$, there holds*

$$\Gamma(z) = \lim_{j\to\infty} \frac{j!\, j^z}{z(z+1)\cdots(z+j)}. \tag{1.16}$$

For $\alpha, \beta \in \mathbb{C}$ with $\Re\alpha > 0$ and $\Re\beta > 0$, the function

$$B(\alpha, \beta) := \int_0^1 t^{\alpha-1}(1-t)^{\beta-1} dt,$$

is called the *Euler beta function*. It is analytic with respect to α and β on the prescribed sets. The beta function may be expressed in terms of the gamma function by the following formula.

Proposition 1.47 *For any* $\alpha,\ \beta \in \mathbb{C}$ *with* $\Re\alpha > 0$ *and* $\Re\beta > 0$, *there holds*

$$B(\alpha, \beta) = \frac{\Gamma(\alpha)\,\Gamma(\beta)}{\Gamma(\alpha+\beta)}.$$

An alternative representation of the beta function may be stated as follows.

Proposition 1.48 *For any* $\alpha,\ \beta \in \mathbb{C}$ *with* $\Re\alpha > 0$ *and* $\Re\beta > 0$, *there holds*

$$B(\alpha, \beta) = 2\int_0^{\pi/2} \sin^{2\alpha-1}\theta\ \cos^{2\beta-1}\theta\, d\theta.$$

1.5.2 The Riemann Zeta Function

The *Riemann zeta function*, denoted by $\zeta(z)$, is defined by the Dirichlet series

$$\zeta(z) := \sum_{k=1}^{\infty} \frac{1}{k^z} \qquad (\Re z > 1).$$

This function is of special interest in number theory since it is important in the study of the distribution of prime numbers.

The Riemann zeta function can be extended beyond the domain $\Re z > 1$ as an analytic function on the whole complex plane except for the point $z = 1$ which is a simple pole for this extension.

An interesting representation of the function $\zeta(z)$ in terms of the gamma function is given by the following proposition.

Proposition 1.49 *For* $z \in \mathbb{C} \setminus \{1\}$, *there holds*

$$\zeta(z) = \frac{1}{\Gamma(z)} \int_0^{\infty} \frac{t^{z-1}}{e^t - 1} dt.$$

1.5.3 The Möbius Function

The *Möbius function* $\mu : \mathbb{N} \to \{-1, 0, 1\}$, defined by

$$\mu(k) = \begin{cases} 1 & \text{if } k = 1, \\ (-1)^n & \text{if } k = p_1 \cdots p_n \text{ with distinct primes}, \\ 0 & \text{if } k \text{ is divisible by a square of a prime}, \end{cases}$$

is another fundamental function in number theory. It appears in various interesting relations. For example, the reciprocal of the Riemann zeta function can be represented as

$$\frac{1}{\zeta(z)} = \sum_{k=1}^{\infty} \frac{\mu(k)}{k^z} \qquad (\Re z > 1).$$

An important tool involving the function $\mu(\cdot)$ is the classical Möbius inversion formula for certain transforms by finite sums. It has been extended in several ways. In this book, we shall only need a special case of an extension to infinite sums, which may be stated as follows.

Proposition 1.50 *Let $(a_n)_{n\in\mathbb{N}}$ and $(b_n)_{n\in\mathbb{N}}$ be two sequences of complex numbers such that*

$$\sum_{n\in\mathbb{N}} |a_n| < \infty \quad \textit{and} \quad \sum_{n\in\mathbb{N}} |b_n| < \infty.$$

If the two sequences are linked by

$$b_n = \sum_{k\in\mathbb{N}} a_{nk} \qquad (n \in \mathbb{N}), \tag{1.17}$$

then

$$a_n = \sum_{k\in\mathbb{N}} \mu(k) b_{nk} \qquad (n \in \mathbb{N}). \tag{1.18}$$

One may interpret Eq. (1.17) as a linear discrete transform that maps the sequence $(a_n)_{n\in\mathbb{N}}$ to the sequence $(b_n)_{n\in\mathbb{N}}$. Then Eq. (1.18) describe its inversion.

1.6 Fundamentals of Fourier Transforms

This section is devoted to a short review of the basic notions of the Fourier transforms and some fundamental results which are of interest for the present book.

1.6.1 Basic Definitions and Properties

For a function $f \in L^1([-\pi, \pi])$, we define the *finite Fourier transform* of f as the sequence

$$k \longrightarrow \widehat{f_k} := \frac{1}{2\pi} \int_{-\pi}^{\pi} f(x) e^{-ikx} dx \qquad (k \in \mathbb{Z}).$$

The function f is extended as a 2π-periodic function to the whole real line by modifying, if necessary, the values of f at the endpoints of the interval $[-\pi, \pi]$.

The numbers $\widehat{f_k} \in \mathbb{C}$ are called the *Fourier coefficients* of f. They are well-defined as Lebesgue integrals. Among the basic properties of the finite Fourier transform we recall the following results.

Proposition 1.51 *The finite Fourier transform is a linear operator with*

$$|\widehat{f_k}| \leq \frac{\|f\|_1}{2\pi} \qquad (k \in \mathbb{Z}).$$

Moreover, $\lim_{|k| \to \infty} \widehat{f_k} = 0$.

For a function $f \in L^1([-\pi, \pi])$, the series

$$(Sf)(x) := \sum_{k=-\infty}^{\infty} \widehat{f_k} e^{ikx} \qquad (x \in \mathbb{R})$$

is called the *Fourier series of* f. Its symmetric partial sums are denoted by

$$(S_n f)(x) := \sum_{k=-n}^{n} \widehat{f_k} e^{ikx} \qquad (n \in \mathbb{N},\ x \in \mathbb{R}).$$

It is known that $S_n f$ does not converge to the function f in general.

For a function $f \in L^2([-\pi, \pi])$, we define the finite Fourier transform of f in the same way as before since $f \in L^1([-\pi, \pi])$ by Hölder's inequality. In this instance, the sequence of Fourier coefficients $(\widehat{f_k})_{k \in \mathbb{Z}}$ belongs to $\ell^2(\mathbb{Z})$ and the *Parseval equation*

$$\|f\|_2 = \sqrt{2\pi}\, \|\widehat{f}\|_{\ell^2(\mathbb{Z})}$$

holds. Moreover, the following convergence result for the Fourier series of $f \in L^2([-\pi, \pi])$ is of importance.

Proposition 1.52 *For any* $f \in L^2([-\pi, \pi])$, *we have*

$$\lim_{n \to \infty} \|S_n f - f\|_2 = 0.$$

Using the Riesz–Thorin interpolation theorem (see Proposition 1.22), one can extend the finite Fourier transform to functions $f \in L^p([-\pi, \pi])$ with $1 < p < 2$, obtaining that $\widehat{f} \in \ell^{p'}$, where p' is the conjugate exponent of p, and the *Hausdorff–Young inequality*

$$\|f\|_p \leq (2\pi)^{1/p'} \|\widehat{f}\|_{\ell^{p'}(\mathbb{Z})}$$

holds.

Concerning the pointwise convergence of the Fourier series of a function f, the following result, known as the *Jordan test*, will be used.

Proposition 1.53 *Let f be a 2π-periodic function of bounded variation on every bounded interval. Then the Fourier series of f converges pointwise at any point. In particular, it converges to $f(x)$ at any continuity point x. Moreover its partial sums are uniformly bounded.*

Remark 1.54 Note that this proposition remains valid for functions that are T-periodic with $T > 0$, under an obvious modification of the definitions of Fourier coefficients and Fourier series; see, e.g., [312, Chap. 1].

For a function $f \in L^1(\mathbb{R})$, we define the (normalized) *Fourier transform* of f as

$$\widehat{f}(v) := \frac{1}{\sqrt{2\pi}} \int_{-\infty}^{\infty} f(x) e^{-ivx} dx \qquad (v \in \mathbb{R}).$$

The integral is well-defined as a Lebesgue integral and the following statement holds.

Proposition 1.55 *The Fourier transform is a linear operator on $L^1(\mathbb{R})$ with range in the space $C(\mathbb{R})$ and*

$$\|\widehat{f}\|_\infty \leq \frac{\|f\|_1}{\sqrt{2\pi}}.$$

Moreover, $\lim_{|v| \to \infty} \widehat{f}(v) = 0$.

For a function $f \in L^2(\mathbb{R})$, we cannot define the Fourier transform as before since in this instance the integral may be divergent. Therefore we change the definition in the following way. The Fourier transform $F^2[f]$ of $f \in L^2(\mathbb{R})$ is the unique L^2 function such that

$$\lim_{\rho \to \infty} \left\| F^2[f](\cdot) - \frac{1}{\sqrt{2\pi}} \int_{-\rho}^{\rho} f(x) e^{-(\cdot)x} dx \right\|_2 = 0. \qquad (1.19)$$

It is known that this definition is well-posed in the sense that the limit in L^2 exists. Moreover, the fundamental *consistency property* holds, saying that $F^2[f](v) =$

$\widehat{f}(v)$ almost everywhere for $v \in \mathbb{R}$ when $f \in L^1(\mathbb{R}) \cap L^2(\mathbb{R})$. As a remarkable property, we state:

Proposition 1.56 *For any $f \in L^2(\mathbb{R})$, the Fourier transform $F^2[f]$ belongs to $L^2(\mathbb{R})$ and satisfies the equality*

$$\|f\|_2 = \frac{\|F^2[f]\|_2}{\sqrt{2\pi}}. \tag{1.20}$$

Note that there is an important difference between the Fourier transforms for L^1 and L^2 functions. In the first case, the Fourier transform is a continuous function. As such it is defined everywhere, but it does not belong to $L^1(\mathbb{R})$ in general. In the second case, the Fourier transform is defined only almost everywhere as an L^2 function.

Using again the Riesz–Thorin interpolation theorem, it is possible to define the Fourier transform of a function $f \in L^p(\mathbb{R})$ when $1 < p < 2$. Denoting by p' the conjugate exponent of p, it is introduced as the unique function $F^p[f] \in L^{p'}(\mathbb{R})$ such that

$$\lim_{\rho \to \infty} \left\| F^p[f](\cdot) - \frac{1}{\sqrt{2\pi}} \int_{-\rho}^{\rho} f(x) e^{-i(\cdot)x} dx \right\|_{p'} = 0.$$

As in the L^2 case, the above limit exists and so this definition is well-posed. A counterpart of the Hausdorff–Young inequality also holds for the Fourier transform in L^p, namely the *Titchmarsh inequality*

$$\|f\|_p \le (\sqrt{2\pi})^{1/p'} \|F^p[f]\|_{L^{p'}(\mathbb{R})}.$$

Moreover, the consistency property is again valid, that is, $F^p[f](v) = \widehat{f}(v)$ almost everywhere for $v \in \mathbb{R}$ when $f \in L^p(\mathbb{R}) \cap L^1(\mathbb{R})$.

For the inversion theory of the Fourier transform, which represents in some sense the counterpart of the Fourier series of f in the periodic case, we have the following "symmetric" definition. For $g \in L^1(\mathbb{R})$, the (normalized) *inverse Fourier transform* of g is defined by

$$g^\vee(x) := \frac{1}{\sqrt{2\pi}} \int_{-\infty}^{\infty} g(v) e^{ivx} dv \qquad (x \in \mathbb{R}).$$

This integral is well-defined for any $g \in L^1(\mathbb{R})$ and we have

$$\|g^\vee(x)\|_\infty \le \frac{\|g\|_1}{\sqrt{2\pi}}.$$

We recall the following result.

Proposition 1.57 *Let $f \in L^1(\mathbb{R})$. If $\widehat{f} \in L^1(\mathbb{R})$, then*

$$f(x) = [\widehat{f}]^{\vee}(x) \qquad (\text{a.e. } x \in \mathbb{R}).$$

Moreover, if f is continuous, then this equality holds everywhere.

Analogously, we define the inverse Fourier transform of a function $g \in L^2(\mathbb{R})$ as the unique L^2 function $F^{2,-1}[g]$ such that

$$\lim_{\rho\to\infty} \left\| F^{2,-1}[g](\cdot) - \frac{1}{\sqrt{2\pi}} \int_{-\rho}^{\rho} g(v)e^{iv(\cdot)}dv \right\|_2 = 0.$$

Again, this definition is well-posed and, putting $g = F^2[f]$ for $f \in L^2(\mathbb{R})$, we obtain

$$f(x) = F^{2,-1}\big[F^2[f]\big](x) \qquad (\text{a.e. } x \in \mathbb{R}).$$

For the Fourier transform in $L^2(\mathbb{R})$, we have the following result, known as the *Plancherel theorem*, which summarize the previous results.

Proposition 1.58 *For each $f \in L^2(\mathbb{R})$, there exists a function $F^2[f] \in L^2(\mathbb{R})$, called the Fourier transform of f, such that (1.19) holds. Moreover, every function $f \in L^2(\mathbb{R})$ is the Fourier transform of a unique element of $L^2(\mathbb{R})$ and the relation (1.20) is satisfied.*

Corresponding definitions of the inverse Fourier transform in L^p spaces with $1 < p < 2$ are given in an analogous way. For any $g \in L^{p'}(\mathbb{R})$, the inverse Fourier transform $F^{p,-1}[g]$ is the unique function in $L^p(\mathbb{R})$ such that

$$\lim_{\rho\to\infty} \left\| F^{p,-1}[g](\cdot) - \frac{1}{\sqrt{2\pi}} \int_{-\rho}^{\rho} g(v)e^{iv(\cdot)}dv \right\|_p = 0.$$

By defining $F^1[f] := \widehat{f}$ and $F^{1,-1}[f] := f^{\vee}$, the next proposition becomes an extension of Proposition 1.57.

Proposition 1.59 *Let $f \in L^p(\mathbb{R})$ with $p \in [1, 2]$. If $F^p[f] \in L^1(\mathbb{R})$, then*

$$f(x) = F^{p,-1}\big[F^p[f]\big](x) \qquad (\text{a.e. } x \in \mathbb{R}).$$

Moreover, if f is continuous, then this equality holds everywhere.

As we have seen before, for the first conclusion, concerning equality almost everywhere, the condition that $F^p[f] \in L^1(\mathbb{R})$ is not needed in case of $p = 2$.

In the following, we shall use the symbol $\widehat{f}$ for the Fourier transform not only when $f \in L^1(\mathbb{R})$ but for all functions $f \in L^p(\mathbb{R})$ with $1 \le p \le 2$. It will be clear from the context how the Fourier transform has to be built.

In connection with the Fourier transform, Sobolev spaces can be characterized as follows.

Proposition 1.60 *For $p \in [1, 2]$ we have*

$$W^{r,p}(\mathbb{R}) = \left\{ f \in L^p(\mathbb{R}) : (iv)^r \widehat{f}(v) = \widehat{g}(v),\ \ g \in L^p(\mathbb{R}) \right\}.$$

1.6.2 Paley–Wiener Spaces and the Paley–Wiener Theorem

Let $\sigma > 0$ and $p \in [1, 2]$. Then the *Paley–Wiener space* $\widehat{\mathcal{B}}^p_\sigma$ comprises all functions $f \in L^p(\mathbb{R})$ such that $\widehat{f}(v) = 0$ almost everywhere for $|v| \geq \sigma$. Since for $p = 1$ the Fourier transform of f is continuous, we see that $f \in \widehat{\mathcal{B}}^1_\sigma$ if and only if $\widehat{f}(v) = 0$ for *every* $|v| \geq \sigma$. As for any $f \in \widehat{\mathcal{B}}^p_\sigma$ the Fourier transform is compactly supported, we conclude that $\widehat{f} \in L^1(\mathbb{R})$. Therefore Proposition 1.59 is applicable. This shows that every function $f \in \widehat{\mathcal{B}}^p_\sigma$ has a representation

$$f(x) = \frac{1}{\sqrt{2\pi}} \int_{-\sigma}^{\sigma} g(v) e^{ivx} dv \qquad (\text{a.e. } x \in \mathbb{R})$$

with $g \in L^1(\mathbb{R})$, which may be taken as $g = \widehat{f}$.

The Paley–Wiener space $\widehat{\mathcal{B}}^2_\sigma$ is strictly connected with the Bernstein space B^2_σ, introduced in Definition 1.34, by the famous Paley–Wiener theorem.

Theorem 1.61 (Paley–Wiener) *If $f \in B^2_\sigma$ for some $\sigma > 0$, then the restriction of f to the real line belongs to the Paley–Wiener space $\widehat{\mathcal{B}}^2_\sigma$. Conversely, every function $f \in \widehat{\mathcal{B}}^2_\sigma$ has an extension to the complex plane as an element of the Bernstein space B^2_σ.*

Theorem 1.61 can be extended from $p = 2$ to $1 \leq p \leq 2$.

1.7 Notes

For the material concerning Lebesgue spaces, one can see Folland [150, Chap. 6] or, for a general treatment, Dunford and Schwartz [143, Chap. IV]. In particular the Lebesgue differentiation theorem (Proposition 1.15) is a special case of Folland [150, Theorem 3.21]. The basic inequalities of Hölder and Minkowski, both in integral and discrete versions can be found in Hardy et al. [167].

Functions of bounded variation are treated in many books on real analysis (see, e.g., Folland [150, Chap. 3, Sect. 5], Royden [264, Chap. 5, Sect. 2]). The notion of bounded coarse variation was first introduced and studied by Haber and Shisha in [162]. Subsequent contributions are due to Osgood [231] and Osgood and Shisha [232]. Proposition 1.16 is an obvious modification of a result in [232, Theorem

2] with the role of $\mathbb{R}^+$ being taken by $\mathbb{R}$. Proposition 1.17 is a corresponding consequence of the main theorem in [162, Theorem 3].

The definition of Sobolev spaces given in the present book, is the classical one without distributions. In this form it is often used in approximation theory; see, e.g., Butzer and Nessel [95, (1.1.16), p. 33; (3.1.48), p. 128].

For the basic theory of functional analysis, see, e.g., Dunford and Schwartz [143], Yosida [307] and Rudin [266].

The Riesz–Thorin interpolation theorem, formulated with the subset of simple functions, is stated in Katznelson [190, p. 141 ff] and Zygmund [312, Vol. II, Chap. XII, p. 93 ff].

For references on complex analysis numerous excellent books are available. Here we drew inspiration from Ahlfors [4], Bak and Newman [25], Burckel [83] and Churchill et al. [127]. Morera's theorem with rectangles can be found in [25, Sect. 7.2]. For details on Riemann surfaces associated with multi-valued functions, see, e.g., [127, pp. 115–117].

For Proposition 1.31, see the book of Boas [73, Theorem 1.4.8]. The Bernstein spaces B_σ^p are subsets of the class of *entire functions of exponential type* σ. This class has been extensively studied in [73]. Proposition 1.35 follows from [73, Theorems 6.2.4, 6.7.1]. The differentiation formula of Proposition 1.36 is due to Boas [72]; see also [73, formula 11.3.1]. It is also known as *generalized Riesz interpolation formula* since prior to Boas it was proved by Riesz [261] in the special case that f is a trigonometric polynomial. A short proof by contour integration was given by Browder [78]. It was shown by Rahman and Schmeisser [254] that Corollary 1.37 holds for all positive p, even when the L^p functional

$$\|\cdot\|_p \,:\, g \longmapsto \left(\int_{-\infty}^{\infty} |g(x)|^p\,dx\right)^{1/p}$$

is not a norm as it happens for $p \in]0, 1[$. The classical Nikol'skii inequality can be found in Nikol'skii [229, p. 123] (see also Butzer et al. [113]). In subsequent chapters we shall extend this inequality to the Mellin setting. Propositions 1.41–1.43 were proved in Butzer et al. [116].

Numerous formulae for the special functions considered in Sect. 1.5 are collected in the handbook by Abramowitz and Stegun [1, Chaps. 6, 23, 24]. For some theoretical aspects, see Bak and Newman [25, Chap. 18]. An extensive treatment is given in Rainville [258]. In particular, for the Euler gamma function, see the fine little monograph of Artin [20], for the Riemann zeta function, see Patterson [237] and Titchmarsh [287]. Proposition 1.50 is a special case of a result by Loxton and Sanders [200, Theorem 2] presented in a different notation.

The Fourier transform has an extensive literature. The fundamentals presented in Sect. 1.6 are based on the book of Butzer and Nessel [95]. However, their approach does not extend to functions $f \in L^p(\mathbb{R})$ with $p > 2$ (see Zygmund [312, Chap. XVI, Sect. 3]). For these classes, the definition of a Fourier transform is given in the setting of distribution theory; see, e.g., the books of Stein and Weiss

[277, Chap. 1], Folland [151, Chap. 9] and Vretblad [297, Chap. 8] for an instructive treatment.

Concerning the Paley–Wiener spaces and the Paley–Wiener theorem, we refer to Boas [73, Chap. 6, Theorem 6.8.1] or Higgins [173, Chap. 7] where also a distributional version of the Paley–Wiener theorem is proposed. For some further references, see Yosida [307, Chap. VI, Sect. 4] and Rudin [265, Chap. 19], [266, Chap. 7].

Chapter 2
Polar-Analytic Functions

In the development of Mellin analysis, in particular in studies in the spirit of the Paley–Wiener theorem, we shall come across the problem of characterizing functions that are originally defined on $\mathbb{R}^+$ but can be extended to a larger domain in an analytic way. For this, the classical notion of analyticity will turn out to be quite awkward. A modified concept, which we shall describe in this chapter, leads to more elegant and concise results.

2.1 The Starting Point

Let us first consider the logarithmic function. It is originally defined on $\mathbb{R}^+$ where it is infinitely differentiable. Now consider any non-zero complex number z written in polar coordinates as $z = re^{i\theta}$, where $r = |z| \in \mathbb{R}^+$ and $\theta \in \mathbb{R}$. Due to the periodicity of the exponential function, this representation is not unique. Indeed, we have $z = re^{i(\theta+2\pi k)}$ for all integers k. Therefore the complex logarithm

$$\log z = \log r + i(\theta + 2\pi k) \qquad (k \in \mathbb{Z})$$

becomes a multi-valued function on $\mathbb{C} \setminus \{0\}$. A classical way of creating a single-valued function is to define the logarithm on its Riemann surface $S_{\log}$. A simple geometric model of the latter is the helicoidal surface in $\mathbb{R}^3$, given by

$$\mathbb{E} := \Big\{(x_1, x_2, x_3) \in \mathbb{R}^3 \ : \ x_1 = r\cos\theta,\ x_2 = r\sin\theta,\ x_3 = \theta,\ r \in \mathbb{R}^+,\ \theta \in \mathbb{R}\Big\}.$$

It can be seen as an extension of $\mathbb{R}^+$ since its subset $\{(r, 0, 0) \in \mathbb{R}^3 \ : \ r \in \mathbb{R}^+\}$ may be identified with $\mathbb{R}^+$. For $\alpha,\ \beta \in \mathbb{R}$ with $\alpha < \beta$, we call the surface

C. Bardaro et al., *Mellin Analysis, Transform Theory, and Applications*,
https://doi.org/10.1007/978-3-031-96672-9_2

$$\mathbb{E}_{\alpha,\beta} := \Big\{(x_1, x_2, x_3) \in \mathbb{R}^3 \ : \ x_1 = r\cos\theta,\ x_2 = r\sin\theta,$$
$$x_3 = \theta,\ r \in \mathbb{R}^+,\ \theta \in \,]\alpha, \beta[\Big\}$$

a *segment* of $\mathbb{E}$. The projection of $\mathbb{E}_{\alpha,\beta}$ into the (x_1, x_2)-plane may be interpreted as a sector in $\mathbb{C}$, denoted by

$$S_{\alpha,\beta} := \left\{x_1 + ix_2 \in \mathbb{C} \ : \ x_1 = r\cos\theta,\ x_2 = r\sin\theta,\ r \in \mathbb{R}^+,\ \theta \in \,]\alpha, \beta[\right\}.$$

If $\beta - \alpha \in \,]0, 2\pi]$, then this projection is a bijection. In fact, given any $z \in S_{\alpha,\beta}$, there exists a unique $\theta \in \,]\alpha, \beta[$, denoted by $\theta := \arg_\alpha z$, such that $z = |z|e^{i \arg_\alpha z}$. Hence $(\Re z, \Im z, \arg_\alpha z)$ is the pre-image of z on $\mathbb{E}_{\alpha,\beta}$.

After these preparations, analytic functions, including the logarithm, can be defined on $\mathbb{E}$ as follows:

Definition 2.1 A function $\mathfrak{f} : \mathbb{E} \to \mathbb{C}$ is said to be *analytic* if for every segment $\mathbb{E}_{\alpha,\beta}$ with $\beta - \alpha \in \,]0, 2\pi]$ the function $z \mapsto \mathfrak{f}(\Re z, \Im z, \arg_\alpha z)$ is analytic on $S_{\alpha,\beta}$.

According to this definition, the function $\mathfrak{log}$, given by

$$\mathfrak{log}(x_1, x_2, x_3) = \frac{1}{2}\log(x_1^2 + x_2^2) + ix_3,$$

is analytic on $\mathbb{E}$ and reduces to $\log r$ when $(x_1, x_2, x_3) = (r, 0, 0)$ with $r \in \mathbb{R}^+$.

Given $\alpha \in \mathbb{R}$, the largest admissible β in the previous definition is $\alpha + 2\pi$. For $k \in \mathbb{Z}$, each of the sectors $S_{\alpha+2k\pi,\alpha+2(k+1)\pi}$ coincides with the complex plane slit along the ray $\{z = re^{i\alpha} \ : \ r \in \mathbb{R}^+\}$. The analytic functions induced by $\mathfrak{f}$ on these sectors are called the *analytic branches* of $\mathfrak{f}$.

Although the study of analytic functions on the Riemann surface of the logarithm is well established, it may become inconvenient in practice. We have to work with three variables x_1, x_2, x_3, which are not independent since they can be expressed by the two variables r and θ. Moreover, in order to profit from classical complex analysis, we have to work with analytic branches, which may lead to cumbersome statements and proofs. Since there exists a bijection between the helicoidal surface $\mathbb{E}$ and the right half-plane $\mathbb{H} := \{(r, \theta) \in \mathbb{R}^+ \times \mathbb{R}\}$, one may think of considering instead of the function $\mathfrak{f}$ of Definition 2.1 the function f of only two variables given by

$$f \ : \begin{cases} \mathbb{H} & \longrightarrow \mathbb{C} \\ (r, \theta) & \longmapsto \mathfrak{f}(r\cos\theta, r\sin\theta, \theta). \end{cases}$$

Obviously f attains the same values as $\mathfrak{f}$. Moreover, since $f(r, 0) = \mathfrak{f}(r, 0, 0)$, the two functions coincide on $\mathbb{R}^+$. However, we have to pay a prize for this simplification. Interpreting $\mathbb{H}$ as the right complex half-plane by writing $z = r + i\theta$, we find that f is not analytic in the classical sense. It satisfies a modified form of analyticity, which we will introduce in the next section. We call it *polar analyticity* since it can be described by the Cauchy–Riemann equations with respect to polar coordinates (see formulae (1.8)). We think that it is worthwhile to pay attention to this concept as we shall do in this chapter. As a benefit, we shall obtain more elegant statements by avoiding splittings into analytic branches.

2.2 Polar Analyticity

In the following we consider functions defined on the right half-plane $\mathbb{H}$ or subsets of it. A convenient type of subsets are the *domains*, which are defined as open, connected and non-empty subsets.

Definition 2.2 Let $\mathcal{D}$ be a domain in $\mathbb{H}$. A function $f : \mathcal{D} \to \mathbb{C}$ is said to be *polar-analytic* on $\mathcal{D}$ if for any $(r_0, \theta_0) \in \mathcal{D}$ the limit

$$\lim_{(r,\theta)\to(r_0,\theta_0)} \frac{f(r,\theta) - f(r_0,\theta_0)}{re^{i\theta} - r_0e^{i\theta_0}} =: (D_{\text{pol}} f)(r_0, \theta_0)$$

exists and is the same howsocvcr (r, θ) approaches (r_0, θ_0) within $\mathcal{D}$. We call $(D_{\text{pol}} f)(r_0, \theta_0)$ the *polar derivative* of f at (r_0, θ_0).

From this definition it is easily deduced that if two functions are polar-analytic, then so are their arithmetic combinations and the familiar laws also apply for calculating polar derivatives. A precise statement may be formulated as follows.

Proposition 2.3 *Let $\mathcal{D}$ be a domain in $\mathbb{H}$. Suppose that f and g are polar-analytic on $\mathcal{D}$. Then $f \pm g$ and fg are also polar-analytic on $\mathcal{D}$ and*

$$D_{\text{pol}}(f \pm g) = D_{\text{pol}} f \pm D_{\text{pol}} g, \qquad D_{\text{pol}}(fg) = f D_{\text{pol}} g + g D_{\text{pol}} f.$$

If, in addition, $f(r, \theta) \neq 0$ for all $(r, \theta) \in \mathcal{D}$, then $1/f$ is polar-analytic on $\mathcal{D}$ and

$$D_{\text{pol}} \frac{1}{f} = -\frac{D_{\text{pol}} f}{f^2}.$$

When in the situation of Definition 2.2 we choose $\theta = \theta_0$ and perform the limit $r \to r_0$, we obtain

$$(D_{\text{pol}} f)(r_0, \theta_0) = e^{-i\theta_0} \frac{\partial f}{\partial r}(r_0, \theta_0),$$

whereas for $r = r_0$ and $\theta \to \theta_0$, we find that

$$(D_{\text{pol}} f)(r_0, \theta_0) = \frac{e^{-i\theta_0}}{ir_0} \frac{\partial f}{\partial \theta}(r_0, \theta_0).$$

These two representations of the polar derivative imply a relation between the partial derivatives of f which must necessarily hold in case of polar analyticity. We fix these observations in the following statement.

Proposition 2.4 *Let $\mathcal{D}$ be a domain in $\mathbb{H}$. Suppose that f is polar-analytic on $\mathcal{D}$. Then the partial derivatives of f exist at all points $(r, \theta) \in \mathcal{D}$ and*

$$(D_{\mathrm{pol}} f)(r, \theta) = e^{-i\theta} \frac{\partial}{\partial r} f(r, \theta) = \frac{e^{-i\theta}}{ir} \frac{\partial}{\partial \theta} f(r, \theta), \tag{2.1}$$

which entails that

$$\frac{\partial f}{\partial \theta} = ir \frac{\partial f}{\partial r} \tag{2.2}$$

on $\mathcal{D}$.

Note that on $\mathbb{R}^+$ polar differentiation reduces to ordinary differentiation. Indeed, if $\varphi(\cdot) := f(\cdot, 0)$, then $(D_{\mathrm{pol}} f)(r, 0) = \varphi'(r)$.

When we write the function of Proposition 2.4 as $f = u + iv$, where u and v are real-valued functions on $\mathcal{D}$, then Eqs. (2.1) and (2.2) take the form

$$\begin{aligned}(D_{\mathrm{pol}} f)(r, \theta) &= e^{-i\theta} \left(\frac{\partial}{\partial r} u(r, \theta) + i \frac{\partial}{\partial r} v(r, \theta) \right) \\ &= \frac{e^{-i\theta}}{r} \left(\frac{\partial}{\partial \theta} v(r, \theta) - i \frac{\partial}{\partial \theta} u(r, \theta) \right)\end{aligned} \tag{2.3}$$

and

$$\frac{\partial u}{\partial \theta} = -r \frac{\partial v}{\partial r}, \qquad \frac{\partial v}{\partial \theta} = r \frac{\partial u}{\partial r}, \tag{2.4}$$

respectively. Note that the latter equations coincide with the Cauchy–Riemann equations of an analytic function g defined by $g(z) := u(r, \theta) + iv(r, \theta)$ for $z = re^{i\theta}$; see also (1.8).

Identifying $\mathbb{C}$ with $\mathbb{R}^2$, we may interpret $f = u + iv$ as a mapping of real analysis from a subset of the half-plane $\mathbb{H}$ into $\mathbb{R}^2$. If this mapping is differentiable in the classical sense, namely local approximable by a linear mapping with an error decaying of higher than linear order (see formula (2.5) below), we shall say that f is *real differentiable*. This notion is used in the following theorem which contains two characterizations of polar analyticity.

Theorem 2.5 *Let $\mathcal{D}$ be a domain in $\mathbb{H}$ and let $f = u + iv : \mathcal{D} \to \mathbb{C}$, where u and v are real-valued functions. Then the following statements are equivalent:*

(i) f is polar-analytic on $\mathcal{D}$;
(ii) u and v have continuous partial derivatives which satisfy Eqs. (2.4) on $\mathcal{D}$;
(iii) f is real differentiable and Eqs. (2.4) hold on $\mathcal{D}$.

Proof Let (r_0, θ_0) be any point of $\mathcal{D}$.

(i) $\Rightarrow$ (ii): In view of Proposition 2.4, it only remains to show that the partial derivatives are continuous. For this, let $\mathcal{D}_0$ be a subdomain such that

$$(r_0, \theta_0) \in \mathcal{D}_0 \subset \mathcal{D} \cap \{(r, \theta) \in \mathbb{H} \, : \, |\theta - \theta_0| < \pi\}.$$

Then there exists a bijection between $\mathcal{D}_0$ and the domain

$$G := \left\{ z = re^{i\theta} \in \mathbb{C} \, : \, (r, \theta) \in \mathcal{D}_0 \right\}.$$

For $z = re^{i\theta}$ with $(r, \theta) \in \mathcal{D}_0$, define $g(z) = f(r, \theta)$ and consider any $z_1 = r_1 e^{i\theta_1}$ with $(r_1, \theta_1) \in \mathcal{D}_0$. Then

$$\frac{g(z) - g(z_1)}{z - z_1} = \frac{g(re^{i\theta}) - g(r_1 e^{i\theta_1})}{re^{i\theta} - r_1 e^{i\theta_1}} = \frac{f(r, \theta) - f(r_1, \theta_1)}{re^{i\theta} - r_1 e^{i\theta_1}}.$$

Since f is polar-analytic, these relations show us that g is complex differentiable at z_1, and so g is analytic on G. Then, by classical complex analysis, g is infinitely continuously differentiable on G. Since

$$\frac{\partial f}{\partial r}(r, \theta) = e^{i\theta} g'(re^{i\theta}), \qquad \frac{\partial f}{\partial \theta}(r, \theta) = ire^{i\theta} g'(re^{i\theta}),$$

we conclude that the partial derivatives of u and v are continuous on $\mathcal{D}_0$ and in particular at (r_0, θ_0) which was an arbitrary point of $\mathcal{D}$.

(ii) $\Rightarrow$ (iii): By a standard result of calculus, the existence and continuity of the partial derivatives of u and v implies that

$$\begin{pmatrix} u(r, \theta) - u(r_0, \theta_0) \\ v(r, \theta) - v(r_0, \theta_0) \end{pmatrix} = J_0 \begin{pmatrix} r - r_0 \\ \theta - \theta_0 \end{pmatrix} + o\left(\left\| \begin{pmatrix} r \\ \theta \end{pmatrix} - \begin{pmatrix} r_0 \\ \theta_0 \end{pmatrix} \right\|_2 \right) \tag{2.5}$$

as $(r, \theta) \to (r_0, \theta_0)$, where J_0 is the Jacobian matrix of the mapping $\binom{u}{v}$ at (r_0, θ_0) and $\| \cdot \|_2$ is the euclidean norm for column vectors of $\mathbb{R}^2$. Hence f is real differentiable.

(iii) $\Rightarrow$ (i): With statement (iii) as hypothesis, we know that (2.5) holds and may use (2.4) to express the entries of J_0 by partial derivatives of u only. Thus we obtain

$$J_0 = \begin{pmatrix} \dfrac{\partial u}{\partial r} & \dfrac{\partial u}{\partial \theta} \\ -\dfrac{1}{r_0} \dfrac{\partial u}{\partial \theta} & r_0 \dfrac{\partial u}{\partial r} \end{pmatrix}$$

with the partial derivatives evaluated at (r_0, θ_0). By elementary calculus, we find that

$$\begin{pmatrix} r - r_0 \\ \theta - \theta_0 \end{pmatrix} = B_0 \begin{pmatrix} r\cos\theta - r_0\cos\theta_0 \\ r\sin\theta - r_0\sin\theta_0 \end{pmatrix} + o\left(\left|re^{i\theta} - r_0e^{i\theta_0}\right|\right) \tag{2.6}$$

as $(r, \theta) \to (r_0, \theta_0)$, where

$$B_0 := \begin{pmatrix} \cos\theta_0 & \sin\theta_0 \\ -\dfrac{\sin\theta_0}{r_0} & \dfrac{\cos\theta_0}{r_0} \end{pmatrix}.$$

Now we substitute (2.6) in (2.5) and write the resulting column vectors as complex numbers. Noting that the usual product (row times column) of the matrices J_0 and B_0 yields

$$J_0 B_0 = \begin{pmatrix} a & b \\ -b & a \end{pmatrix},$$

where

$$a = \cos\theta_0 \frac{\partial u}{\partial r}(r_0, \theta_0) - \frac{\sin\theta_0}{r_0}\frac{\partial u}{\partial \theta}(r_0, \theta_0),$$

$$b = \sin\theta_0 \frac{\partial u}{\partial r}(r_0, \theta_0) + \frac{\cos\theta_0}{r_0}\frac{\partial u}{\partial \theta}(r_0, \theta_0),$$

we obtain

$$f(r, \theta) - f(r_0, \theta_0) = (a - ib)\left(re^{i\theta} - r_0e^{i\theta_0}\right) + o\left(\left|re^{i\theta} - r_0e^{i\theta_0}\right|\right)$$

as $(r, \theta) \to (r_0, \theta_0)$. This shows that f is polar analytic. □

Statement (ii) is very useful for verifying polar analyticity of a function given by a concrete term.

In order to become familiar with polar-analytic functions, we now consider some examples and discuss connections with known classes of functions. When g is an entire function, then $f : (r, \theta) \mapsto g(re^{i\theta})$ is easily seen to be polar-analytic on $\mathbb{H}$ and 2π-periodic with respect to θ. For the polar-derivative we find that $(D_{\text{pol}} f)(r, \theta) = g'(z)$ with $z = re^{i\theta}$. However, the 2π periodicity of a polar-analytic function f with respect to θ does not guarantee a representation of the form $f(r, \theta) = g(re^{i\theta})$ with an entire function g. For example, $f(r, \theta) := e^{-i\theta}/r$ is polar-analytic and 2π-periodic with respect to θ on $\mathbb{H}$. Furthermore, $f(r, \theta) = h(re^{i\theta})$, where $h(z) = 1/z$. But h is analytic on $\mathbb{C} \setminus \{0\}$ only and cannot be extended to an entire function. However, if f is a polar-analytic function on $\mathbb{H}$, then $g : z = x + iy \mapsto f(e^x, y)$ is always an entire function since Eqs. (2.4) imply that the Cauchy–Riemann equations hold for g on $\mathbb{C}$.

The function $L(r, \theta) := \log r + i\theta$ is a single-valued extension of the logarithm from $\mathbb{R}^+$ to the whole of $\mathbb{H}$. By Theorem 2.5, it is easily seen to be polar-analytic everywhere, but it is not 2π-periodic with respect to θ. Using (2.3), we find for $z = re^{i\theta}$ that

$$(D_{\text{pol}}L)(r, \theta) = e^{-i\theta}\frac{1}{r} = \frac{1}{re^{i\theta}} = \frac{1}{z}.$$

Hence the polar derivative is 2π-periodic with respect to θ.

Another example is the general power $g(z) = z^a$ for positive a. For $a \in \mathbb{R}^+ \setminus \mathbb{N}$ it is multi-valued on $\mathbb{C}$. Defining $f(r, \theta) := g(re^{i\theta}) = r^a e^{ia\theta}$, we have

$$f(r, \theta) = u(r, \theta) + iv(r, \theta) := r^a \cos(a\theta) + ir^a \sin(a\theta),$$

which is an extension of r^a from $\mathbb{R}^+$ to a single-valued function on the whole of $\mathbb{H}$. For the polar derivative we obtain

$$(D_{\text{pol}}f)(r, \theta) = e^{-i\theta}\left[\frac{\partial u}{\partial r}(r, \theta) + i\frac{\partial v}{\partial r}(r, \theta)\right] = a(re^{i\theta})^{a-1} = g'(z).$$

Finally, we want to point out how polar-analytic functions are connected with functions which are analytic on the Riemann surface $S_{\log}$. For $\alpha, \beta \in \mathbb{R}$ with $\alpha < \beta$, we define the set

$$\mathbb{H}_{\alpha,\beta} := \{(r, \theta) \in \mathbb{R}^+ \times \mathbb{R} \; : \; \theta \in]\alpha, \beta[\} \tag{2.7}$$

and call it a *strip* of $\mathbb{H}$. Now with any polar-analytic function $f : \mathbb{H} \to \mathbb{C}$ we associate a function g, which is analytic on $S_{\log}$. For this we proceed as follows. For each $k \in \mathbb{Z}$, we consider the restriction of f to a strip $\mathbb{H}_{\alpha+2k\pi,\alpha+2(k+1)\pi}$, where $k \in \mathbb{Z}$ and denote it by f_k. Then the definition of polar analyticity shows immediately that $g_k(re^{i\theta}) := f_k(r, \theta)$ defines a function g_k that is analytic in the slit complex plane $\mathbb{C} \setminus \{re^{i\alpha} : r \in \mathbb{R}^+\}$. These are the analytic branches of g. Of course, when f is 2π-periodic with respect to θ all these branches are equal and we do not need the Riemann surface.

2.3 Cauchy and Morera Type Theorems

We begin with an analogue of Cauchy's fundamental theorem of complex function theory for polar-analytic functions. For the needs in this book, it suffices to consider functions defined on a convex domain $\mathcal{D} \subset \mathbb{H}$. For the following, we refer to the definition of smooth and piecewise smooth curves given in Sect. 1.4.1.

Theorem 2.6 *Let f be a polar-analytic function on a convex domain $\mathcal{D} \subset \mathbb{H}$ and let (r_1, θ_1) and (r_2, θ_2) be any two points in $\mathcal{D}$. Then the line integral*

$$\int_{\gamma} f(r, \theta)e^{i\theta}(dr + ird\theta) \tag{2.8}$$

has the same value for each piecewise smooth curve γ in $\mathcal{D}$ that starts at (r_1, θ_1) and ends at (r_2, θ_2). In particular, the integral vanishes for closed, piecewise smooth curves.

Proof As a simple consequence of Eq. (2.2), we have

$$\frac{\partial}{\partial\theta}\Big[f(r, \theta)e^{i\theta}\Big] = \frac{\partial}{\partial r}\Big[f(r, \theta)ire^{i\theta}\Big],$$

and by Theorem 2.5 these partial derivatives are continuous. Since $\mathcal{D}$ is convex, a standard result of two-dimensional calculus tells us that the integrand in (2.8) is an exact differential on $\mathcal{D}$. In other words, there exists a function $F : \mathcal{D} \to \mathbb{C}$ such that

$$\frac{\partial F}{\partial r}(r, \theta) = f(r, \theta)e^{i\theta} \quad \text{and} \quad \frac{\partial F}{\partial \theta}(r, \theta) = f(r, \theta)ire^{i\theta}.$$

This implies that the integral in (2.8) is equal to $F(r_2, \theta_2) - F(r_1, \theta_1)$, and so the assertions follow immediately. □

Next we present a converse of Theorem 2.6 in the spirit of Morera's theorem (Theorem 1.25). We restrict ourselves to functions defined on the whole of $\mathbb{H}$ which will show already the basic idea for this type of results.

Theorem 2.7 *Let $f : \mathbb{H} \to \mathbb{C}$ be a continuous function. If*

$$\int_{\partial\mathcal{R}} f(r, \theta)e^{i\theta}(dr + ird\theta) = 0$$

for the positively oriented boundary $\partial\mathcal{R}$ of every rectangle $\mathcal{R} \subset \mathbb{H}$ with sides parallel to the axes, then f is polar-analytic on $\mathbb{H}$.

Proof Obviously, $g(z) = g(x + iy) := e^{x+iy}f(e^x, y)$ is a continuous function on $\mathbb{C}$. In this proof, the notion of *rectangle* shall always include that the sides are parallel to the axes of the coordinate system. Under this agreement, the substitution $r = e^x$ and $\theta = y$ transforms each rectangle $\mathcal{R}$ in the (r, θ)-half-plane $\mathbb{H}$ into a rectangle R in the (x, y)-plane $\mathbb{R}^2$ which can be identified with the complex plane $\mathbb{C}$. Conversely, each rectangle $R \subset \mathbb{R}^2$ has a rectangle $\mathcal{R} \subset \mathbb{H}$ as pre-image of this transformation. Since

$$0 = \int_{\partial\mathcal{R}} f(r, \theta)e^{i\theta}(dr + ird\theta) = \int_{\partial R} g(x + iy)d(x + iy),$$

we conclude with the help of Morera's theorem in the form of Theorem 1.25 that g is an entire function. Hence the function $g_1 : (x+iy) \mapsto f(e^x, y)$ is also entire. As such, the real and the imaginary part of g_1 have continuous partial derivatives which satisfy the Cauchy–Riemann equations on $\mathbb{C}$. With this information, we can employ statement (ii) of Theorem 2.5 for verifying that f is polar-analytic on $\mathbb{H}$. □

2.4 An Analogue of Cauchy's Integral Formula

The aim of this section is an analogue of the classical Cauchy integral formula. We begin with a lemma in which the hypothesis of Theorem 2.6 is weakened by allowing an exceptional point (r_0, θ_0) where the function is only continuous.

Lemma 2.8 *Let $\mathcal{D}$ be a convex domain in $\mathbb{H}$. Let $g : \mathcal{D} \to \mathbb{C}$ be continuous on the whole of $\mathcal{D}$ and polar-analytic except at a point $(r_0, \theta_0) \in \mathcal{D}$. Suppose that γ is a positively oriented, closed, piecewise smooth curve in $\mathcal{D}$ that is the boundary of a convex domain* int(γ). *Then*

$$\int_\gamma g(r, \theta)e^{i\theta}(dr + ird\theta) = 0.$$

Proof In view of Theorem 2.6, it suffices to consider the case of $(r_0, \theta_0) \in \text{int}(\gamma)$ only. Then we decompose int(γ) into four parts as shown in Fig. 2.1. One part is a disk centered at (r_0, θ_0) of sufficiently small radius $\varepsilon > 0$. Its negatively oriented

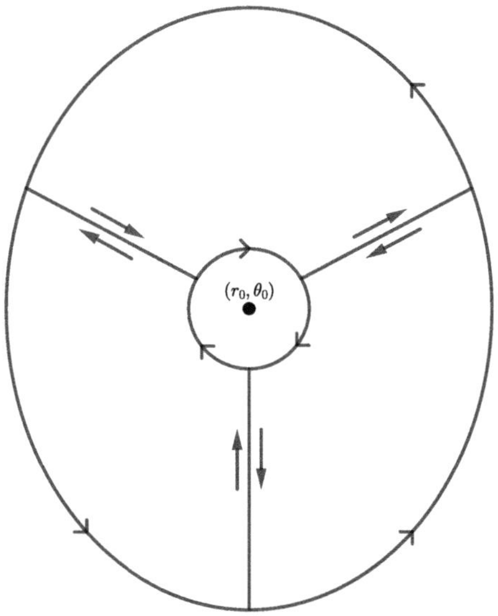

Fig. 2.1 A picture of the decomposition of int(γ) in Lemma 2.8

boundary will be denoted by λ_ε. The other three parts are obtained by inserting line segments that lie on rays starting at (r_0, θ_0) and intersecting under an angle of $2\pi/3$. The positively oriented boundaries of these three parts are denoted by β_1, β_2 and β_3. They are closed, piecewise smooth curves. Each of $\beta_1, \beta_2, \beta_3$ can be separated from (r_0, θ_0) by a line. Hence each lies in a half-plane that does not contain (r_0, θ_0). By intersecting each of these half-planes with $\mathcal{D}$, we produce convex subdomains $\mathcal{D}_j$ containing β_j for $j = 1, 2, 3$ but not (r_0, θ_0). By Theorem 2.6, it follows that

$$\int_{\beta_j} g(r,\theta)e^{i\theta}(dr + ird\theta) = 0 \qquad (j = 1, 2, 3).$$

Adding these integrals and noting that the contributions coming from the three line segments will cancel, we obtain

$$\int_\gamma g(r,\theta)e^{i\theta}(dr + ird\theta) = -\int_{\lambda_\varepsilon} g(r,\theta)e^{i\theta}(dr + ird\theta).$$

The left-hand side does not depend on ε. Since g is continuous on $\mathcal{D}$, the right-hand side converges to zero as $\varepsilon \to 0$. This completes the proof. □

An analogue of Cauchy's integral formula for polar-analytic functions can be stated as follows.

Theorem 2.9 *Let $\mathcal{D}$ be a convex domain in $\mathbb{H}$, and let $f : \mathcal{D} \to \mathbb{C}$ be polar-analytic on $\mathcal{D}$. Let γ be a positively oriented, closed, piecewise smooth curve in $\mathcal{D}$ that is the boundary of a convex domain* $\operatorname{int}(\gamma) \subset \mathcal{D}$*. Given $(r_0, \theta_0) \in \operatorname{int}(\gamma)$, define $\theta_j := \theta_0 + 2j\pi$ for $j \in \mathbb{Z}$. Suppose that none of the points (r_0, θ_j) lies on γ. Then*

$$\frac{1}{2\pi i}\int_\gamma \frac{f(r,\theta)e^{i\theta}}{re^{i\theta} - r_0e^{i\theta_0}}(dr + ird\theta) = \sum_{(r_0,\theta_j)\in\operatorname{int}(\gamma)} f(r_0, \theta_j). \tag{2.9}$$

In particular, if γ lies in a strip $\mathbb{H}_{a,b}$ with $0 < b - a < 2\pi$, then

$$\frac{1}{2\pi i}\int_\gamma \frac{f(r,\theta)e^{i\theta}}{re^{i\theta} - r_0e^{i\theta_0}}(dr + ird\theta) = f(r_0, \theta_0). \tag{2.10}$$

Proof The strips

$$\mathcal{S}_j := \{(r,\theta) \in \mathbb{H} : |\theta - \theta_j| \le \pi\} \qquad (j \in \mathbb{Z})$$

decompose $\operatorname{int}(\gamma)$ into a finite number of non-empty convex sets $\operatorname{int}(\gamma) \cap \mathcal{S}_j$ (see Fig. 2.2). Their positively oriented boundaries shall be denoted by α_j and their interiors by $\operatorname{int}(\alpha_j)$. Clearly, each α_j is a closed, piecewise smooth curve.

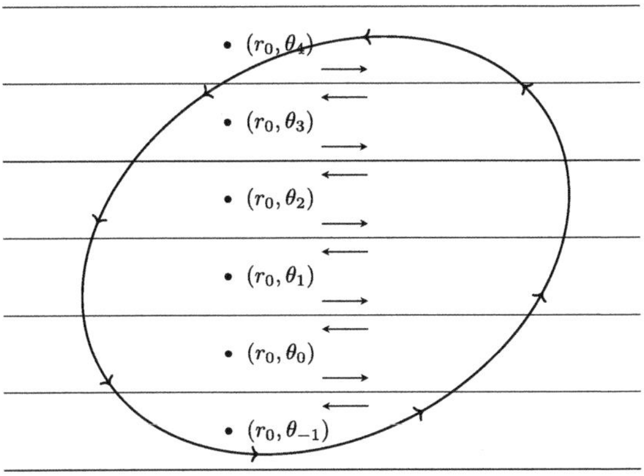

Fig. 2.2 An example of the decomposition of $\operatorname{int}(\gamma)$ in Theorem 2.9

If $\operatorname{int}(\gamma) \cap \mathcal{S}_j \neq \emptyset$ but $(r_0, \theta_j) \notin \operatorname{int}(\gamma)$, then, by Proposition 2.3,

$$g : (r, \theta) \longmapsto \frac{f(r, \theta)}{re^{i\theta} - r_0 e^{i\theta_0}}$$

is polar-analytic on a convex domain that contains $\operatorname{int}(\alpha_j)$. Hence Theorem 2.6 applies and yields

$$\frac{1}{2\pi i} \int_{\alpha_j} \frac{f(r, \theta)e^{i\theta}}{re^{i\theta} - r_0 e^{i\theta_0}} (dr + ird\theta) = 0.$$

Therefore we obtain

$$\frac{1}{2\pi i} \int_{\gamma} \frac{f(r, \theta)e^{i\theta}}{re^{i\theta} - r_0 e^{i\theta_0}} (dr + ird\theta)$$
$$= \sum_{(r_0, \theta_j) \in \operatorname{int}(\gamma)} \frac{1}{2\pi i} \int_{\alpha_j} \frac{f(r, \theta)e^{i\theta}}{re^{i\theta} - r_0 e^{i\theta_0}} (dr + ird\theta). \tag{2.11}$$

The integrals on the right-hand side may be written as

$$\frac{1}{2\pi i} \int_{\alpha_j} \frac{f(r, \theta)e^{i\theta}}{re^{i\theta} - r_0 e^{i\theta_0}} (dr + ird\theta) = \frac{1}{2\pi i} \int_{\alpha_j} g_j(r, \theta)e^{i\theta} (dr + ird\theta)$$
$$+ \frac{f(r_0, \theta_j)}{2\pi i} \int_{\alpha_j} \frac{dr + ird\theta}{r - r_0 e^{i(\theta_0 - \theta)}}, \tag{2.12}$$

where

$$g_j(r,\theta) := \begin{cases} \dfrac{f(r,\theta) - f(r_0,\theta_j)}{re^{i\theta} - r_0 e^{i\theta_0}} & \text{if } (r,\theta) \neq (r_0,\theta_j), \\ (D_{\text{pol}} f)(r_0,\theta_j) & \text{if } (r,\theta) = (r_0,\theta_j). \end{cases}$$

Since f is polar-analytic at (r_0, θ_j) and $e^{i\theta_0} = e^{i\theta_j}$, we see that g_j is continuous on a convex domain that contains $\text{int}(\alpha_j)$ and is there polar-analytic except, possibly, at the point (r_0, θ_j). By Lemma 2.8 it follows that

$$\frac{1}{2\pi i} \int_{\alpha_j} g_j(r,\theta) e^{i\theta} (dr + ir d\theta) = 0. \tag{2.13}$$

It remains to determine the second integral on the right-hand side of (2.12). Set $z = re^{i\theta}$. If (r,θ) traverses α_j once, then z moves along a piecewise continuously differentiable closed curve $\widetilde{\alpha}_j$, say, in $\mathbb{C}$ and surrounds $z_0 := r_0 e^{i\theta_0} = r_0 e^{i\theta_j}$ exactly once in the mathematically positive sense. Hence by formula (1.12) for the winding number of $\widetilde{\alpha}_j$ with respect to z_0, we find that

$$\frac{1}{2\pi i} \int_{\alpha_j} \frac{dr + ir d\theta}{r - r_0 e^{i(\theta_0 - \theta)}} = \frac{1}{2\pi i} \int_{\widetilde{\alpha}_j} \frac{dz}{z - z_0} = 1.$$

Now (2.9) follows from Eqs. (2.11)–(2.13). Formula (2.10) is an obvious consequence. □

Remark 2.10 Under the hypotheses of Theorem 2.9, it can be deduced from (2.9) by differentiating inside the integral (see the proof of the next theorem) that

$$\frac{1}{2\pi i} \int_{\gamma} \frac{f(r,\theta) e^{i\theta}}{(re^{i\theta} - r_0 e^{i\theta_0})^2} (dr + ir d\theta) = \sum_{(r_0,\theta_j) \in \text{int}(\gamma)} (D_{\text{pol}} f)(r_0, \theta_j).$$

Another interesting consequence of Theorem 2.9 is an extension of (2.10) for representing polar derivatives of higher order.

Theorem 2.11 *Under the hypotheses of Theorem 2.9, suppose that γ lies in a strip $\mathbb{H}_{a,b}$ with $0 < b - a < 2\pi$. Then f has polar derivatives of arbitrary order on $\text{int}(\gamma)$ and*

$$(D^k_{\text{pol}} f)(r,\theta) = \frac{k!}{2\pi i} \int_{\gamma} \frac{f(\rho,\phi) e^{i\phi}}{(\rho e^{i\phi} - re^{i\theta})^{k+1}} (d\rho + i\rho d\phi) \quad \big((r,\theta) \in \text{int}(\gamma),\ k \in \mathbb{N}\big). \tag{2.14}$$

Proof Since formula (2.10) holds for any $(r_0, \theta_0) \in \text{int}(\gamma)$, we may write

$$f(r,\theta) = \frac{1}{2\pi i} \int_{\gamma} \frac{f(\rho,\phi) e^{i\phi}}{\rho e^{i\phi} - re^{i\theta}} (d\rho + i\rho d\phi) \qquad \big((r,\theta) \in \text{int}(\gamma)\big). \tag{2.15}$$

It is easily seen that the integrand has polar derivatives with respect to (r, θ) of arbitrary order and, using Proposition 2.3, we find that

$$D_{\text{pol}}^{k} \frac{1}{\rho e^{i\phi} - re^{i\theta}} = \frac{k!}{(\rho e^{i\phi} - re^{i\theta})^{k+1}} \qquad (k = 1, 2, \ldots).$$

Now, as in a classical result on differentiation inside an integral, we can conclude that the operator D_{pol}^{k} applies to the right-hand side of (2.15) and may be interchanged with the integration. This gives (2.14) and completes the proof. □

2.5 A Series Expansion for Polar-Analytic Functions

In this section, we establish a series expansion for a polar-analytic function, which corresponds to the classical Taylor series with the role of the derivatives $f^{(k)}(z_0)$ being taken by the polar derivatives $D_{\text{pol}}^{k} f(r_0, \theta_0)$. We do this by empolying Theorem 2.11. Therefore the envisaged expansion is only possible within a strip of width less than 2π. Our procedure can also be interpreted as taking an analytic branch on the Riemann surface of the logarithm and expanding it as in classical complex analysis. While for the classical Taylor series the largest region of convergence is a disk, the polar-analytic analogue converges in a domain of $\mathbb{H}$ which corresponds to a disk in $\mathbb{C}$ and will now be introduced.

Let γ be a curve as in Theorem 2.11 and let $(r_0, \theta_0) \in \text{int}(\gamma)$. Then we define

$$\mathcal{D}(r_0, \theta_0; \gamma) := \left\{ (r, \theta) \in \text{int}(\gamma) \,:\, |re^{i\theta} - r_0 e^{i\theta_0}| < \min_{(\rho,\phi)\in\gamma} |\rho e^{i\phi} - r_0 e^{i\theta_0}| \right\}.$$

The significance of this set can be seen as follows. By the mapping $(r, \theta) \mapsto z := r^{i\theta}$, we can map $\text{int}(\gamma)$ to a subset G, say, of $\mathbb{C}$. Obviously $r_0 e^{i\theta_0} \in G$, and we may look for the largest disk with center at $r_0 e^{i\theta_0}$ which is contained in G. Its pre-image is the set $\mathcal{D}(r_0, \theta_0; \gamma)$.

Theorem 2.12 *Under the assumptions of Theorem 2.11, let* $(r_0, \theta_0) \in \text{int}(\gamma)$. *Then*

$$f(r, \theta) = \sum_{k=0}^{\infty} \frac{D_{\text{pol}}^{k} f(r_0, \theta_0)}{k!} (re^{i\theta} - r_0 e^{i\theta_0})^k$$

for $(r, \theta) \in \mathcal{D}(r_0, \theta_0; \gamma)$. *The series converges absolutely on* $\mathcal{D}(r_0, \theta_0; \gamma)$ *and uniformly on its compact subsets.*

Proof For $(r, \theta) \in \mathcal{D}(r_0, \theta_0; \gamma)$ and $(\rho, \phi) \in \gamma$, we find by using the geometric series that

$$\frac{1}{\rho e^{i\phi} - re^{i\theta}} = \frac{1}{\rho e^{i\phi} - r_0 e^{i\theta_0}} \frac{1}{1 - \dfrac{re^{i\theta} - r_0 e^{i\theta_0}}{\rho e^{i\phi} - r_0 e^{i\theta_0}}}$$

$$= \frac{1}{\rho e^{i\phi} - r_0 e^{i\theta_0}} \sum_{k=0}^{\infty} \left(\frac{re^{i\theta} - r_0 e^{i\theta_0}}{\rho e^{i\phi} - r_0 e^{i\theta_0}} \right)^k .$$

Note that this is an absolutely convergent expansion. Substituting it on the right-hand side of (2.15), interchanging summation and integration, and employing (2.14), we obtain

$$f(r,\theta) = \frac{1}{2\pi i} \sum_{k=0}^{\infty} \int_{\gamma} \frac{f(\rho,\phi)e^{i\phi}}{(\rho e^{i\phi} - r_0 e^{i\theta_0})^{k+1}} (d\rho + i\rho d\phi)\,(re^{i\theta} - r_0 e^{i\theta_0})^k$$

$$= \sum_{k=0}^{\infty} \frac{(D_{\text{pol}}^k f)(r_0,\theta_0)}{k!} (re^{i\theta} - r_0 e^{i\theta_0})^k ,$$

which is the desired expansion of f. The absolute convergence is verified by estimating the polar derivatives via the integral in (2.14) and keeping the definition of $\mathcal{D}(r_0, \theta_0; \gamma)$ in mind. The statement on uniform convergence is seen as in the case of power series. □

Theorem 2.12 shows again that every polar-analytic function f can be locally expressed as $f(r,\theta) = g(re^{i\theta})$, where g is analytic.

As an example, we consider the function $f(r,\theta) = \log r + i\theta$. By a straight forward calculation we find that

$$D_{\text{pol}}^k(\log r + i\theta) = \frac{(-1)^{k-1}(k-1)!}{(re^{i\theta})^k} \qquad (k \in \mathbb{N}).$$

Hence the expansion of Theorem 2.12 yields

$$\log r + i\theta = \log r_0 + i\theta_0 + \sum_{k=1}^{\infty} \frac{(-1)^{k-1}}{k} \left(\frac{re^{i\theta}}{r_0 e^{i\theta_0}} - 1 \right)^k .$$

Concerning convergence, we refer to the subsequent remark. For $r_0 = 1$ and $\theta = \theta_0 = 0$, we recover a familiar expansion of the real logarithm.

Remark 2.13 The region of convergence $\mathcal{D}(r_0, \theta_0; \gamma)$ in Theorem 2.12 depends on the choice of γ and involves a minimal distance. Conversely, we can try to prescribe the distance and construct γ suitably. For given $M > 0$ define γ (or, more precisely, the points of γ) by

$$\gamma := \left\{ (r,\theta) \in \mathbb{H} : \left| re^{i\theta} - r_0 e^{i\theta_0} \right| = M \right\} .$$

Thus the points $(r, \theta) \in \gamma$ satisfy the equation

$$r^2 + r_0^2 - 2rr_0 \cos(\theta - \theta_0) = M^2.$$

However, in order to be suitable for Theorem 2.12, we need in addition that γ lies in a strip $\mathbb{H}_{a,b}$ of width less than 2π and $\mathrm{int}(\gamma)$ is convex. By an elementary calculation one can show that these properties hold for $M \in]0, r_0[$.

2.6 Notes

The material of this chapter originates mainly from Bardaro et al. [62] with a few new details. For example, Theorem 2.5 is an extended form of [62, Proposition 3.4]. The latter consists of the implication (i) $\Rightarrow$ (iii) only. The notion of polar-analytic functions was first introduced in Bardaro et al. [58]. The main applications of the concept of polar analyticity have been established in Mellin analysis and in the realm of quadrature formulae. In Mellin analysis, polar analyticity turns out to be very helpful for an efficient approach, independent of Fourier analysis. It leads to a precise and simple analysis for functions defined on the Riemann surface of the complex logarithm via the helicoidal surface. In particular, one can obtain an analogue of the classical Paley–Wiener theorem of Fourier analysis in the Mellin setting (see Bardaro et al. [58, 59]). As we shall see in Sect. 3.8, the notion of polar-analytic function gives a simple definition of the so-called Mellin–Bernstein spaces, without resorting to Riemann surfaces and analytical branches. These spaces characterize, in a precise sense, the functions $f : \mathbb{R}^+ \to \mathbb{C}$ with compactly supported Mellin transform. Moreover, polar analyticity enables us to define in a simple way Hardy spaces in the Mellin setting (see Bardaro et al. [56]).

We mention that another definition of generalized analyticity was introduced by Ziemian in [311], using integral analogues of power series and involving Laplace transformations. It is completely different from the concept of this chapter. Further studies of analytic functions over the Riemann surface of the logarithm, identified with $\mathbb{H}$, were presented in papers dealing with Dirichlet's problem in certain domains, the Riemann mapping theorem, o-minimality and the extension of analytic functions on a domain with a boundary having prescribed properties (see Kaiser [182, 183] and Kaiser et al. [184]). However, the notion of derivative introduced by these authors is different from the one of this chapter (see [184, Definition 2.1]) and is rather related to the Mellin derivative of a function (see Mamedov [208, Chap. 2, p. 61], Butzer and Jansche [89]).

For sake of simplicity and since it is sufficient for the needs in this book, we stated the theorems of Cauchy and Morera type for convex domains only. However, as in classical complex analysis, this assumption is not necessary. Indeed, the proof of Cauchy's integral theorem (Theorem 2.6) can be extended to *simply connected* domains (see, e.g., Bak and Newman [25, Chap. 8]). Generally, the presented theory of polar-analytic functions is far from being complete. For example, one could establish a polar version of Laurent series.

A remarkable geometric property of an analytic function g is that at any point z_0 with $g'(z_0) \neq 0$ it preserves angles and orientation. We cannot expect that this extends to polar-analytic functions. But one may ask if the change of angles is the same for all polar-analytic functions. In [62, Sect. 7], the authors studied this question. We give a short account of their results.

Let γ_1 and γ_2 be two smooth arcs in $\mathbb{H}$ which intersect at (r_0, θ_0) with non-vanishing derivatives under an angle α. Let

$$\begin{pmatrix} r \\ \theta \end{pmatrix} = \begin{pmatrix} r_0 \\ \theta_0 \end{pmatrix} + t \begin{pmatrix} c_j \\ s_j \end{pmatrix} \qquad (t \in \mathbb{R};\ j = 1, 2),$$

where

$$c_j = \cos\phi_j, \quad s_j = \sin\phi_j \qquad \left(\phi_j \in \left]-\frac{\pi}{2}, \frac{\pi}{2}\right];\ j = 1, 2\right),$$

be the two tangents at (r_0, θ_0). Then $\alpha = |\phi_2 - \phi_1|$, and it follows that

$$\cos\alpha = c_1c_2 + s_1s_2 = \frac{c_1c_2 + s_1s_2}{\sqrt{c_1^2 + s_1^2} \cdot \sqrt{c_2^2 + s_2^2}}. \tag{2.16}$$

Now let f be any function that is polar-analytic in a neighborhood of (r_0, θ_0) and suppose that $(D_{\text{pol}} f)(r_0, \theta_0) \neq 0$. Then it is shown that the images $f \circ \gamma_1$ and $f \circ \gamma_2$ of the arcs intersect at $f(r_0, \theta_0)$ under an angle β with

$$\cos\beta = \frac{c_1c_2 + r_0^2 s_1s_2}{\sqrt{c_1^2 + r_0^2 s_1^2} \cdot \sqrt{c_2^2 + r_0^2 s_2^2}}. \tag{2.17}$$

Thus the change of the angle does not depend on f and θ_0. It only depends on the local geometric situation described by ϕ_1, ϕ_2 and r_0. Note that the right-hand side of (2.17) is obtained from that of (2.16) by multiplying the numbers s_1 and s_2 by r_0. At all the points $(1, \theta)$ with $\theta \in \mathbb{R}$ polar-analytic functions with non-vanishing polar derivative preserve angles.

Another result may be of interest for numerical methods using grids; see [62, Proposition 7.2]: *Let $\mathcal{D}$ be a domain in $\mathbb{H}$, and let $f : \mathcal{D} \to \mathbb{C}$ be polar-analytic with $D_{\text{pol}} f \neq 0$. Denote by $\mathcal{N}$ an orthogonal net of lines parallel to the axes of $\mathbb{H}$. Then f maps $\mathcal{N} \cap \mathcal{D}$ onto an orthogonal net of smooth curves in $f(\mathcal{D})$.*

Chapter 3
The Foundation of Mellin Analysis

In this chapter we introduce the notion of the Mellin differential operator. It is based on the notion of Mellin translation which represents the Mellin counterpart of the classical translation operator $\mathcal{T}_h$ defined by $\mathcal{T}_h f(x) = f(x+h)$. We begin a section about the basic theory of the spaces on which Mellin analysis is built.

3.1 Weighted Lebesgue Spaces X_c^p and Y_c^p, c-Recurrent Functions

For a given $c \in \mathbb{R}$ and $p \in [1, \infty[$, we will denote by X_c^p the space of all Lebesgue measurable functions $f : \mathbb{R}^+ \to \mathbb{C}$ such that the function $(\cdot)^{c-1/p} f(\cdot) \in L^p(\mathbb{R}^+)$, endowed with the norm

$$\|f\|_{X_c^p} := \|(\cdot)^{c-1/p} f(\cdot)\|_p = \left\{\int_0^\infty u^{cp-1} |f(u)|^p du\right\}^{1/p}.$$

For $p = 1$, we simply write X_c for X_c^1 and the norm in X_c is given by

$$\|f\|_{X_c} = \int_0^\infty u^{c-1} |f(u)| du.$$

For $p = \infty$, we define X_c^∞ as the space of all Lebesgue measurable functions $f : \mathbb{R}^+ \to \mathbb{C}$ such that $\|f\|_{X_c^\infty} := \text{ess sup}_{u \in \mathbb{R}^+} u^c |f(u)| < \infty$. For $c = 0$ we have $X_0^\infty = L^\infty(\mathbb{R}^+)$ and we can denote its norm by $\|f\|_{X_0^\infty} = \|f\|_{L^\infty(\mathbb{R}^+)} = \|f\|_\infty$ (see Sect. 1.2.1). As in Sect. 1.2.1, for functions $f : \mathbb{R}^+ \to \mathbb{C}$ such that $u^c f(u)$ is bounded, we denote by $\|\cdot\|^*_{X_c^\infty}$ the norm defined by

C. Bardaro et al., *Mellin Analysis, Transform Theory, and Applications*,
https://doi.org/10.1007/978-3-031-96672-9_3

$$\|f\|^*_{X_c^\infty} := \sup_{u>0} u^c |f(u)|.$$

If f is also continuous, then the two norms $\|\cdot\|_{X_c^\infty}$ and $\|\cdot\|^*_{X_c^\infty}$ coincide.

As in the case of the Lebesgue spaces $L^p(\mathbb{R}^+)$, the spaces X_c^p are Banach spaces with respect to their norms.

Occasionally, we will consider the following related spaces: For any interval $I \subset \mathbb{R}^+$, we define

$$X_I := \bigcap_{c\in I} X_c, \quad X_I^p := \bigcap_{c\in I} X_c^p \qquad (p \in]1, \infty[).$$

A norm for the above spaces is given by $\|\cdot\|_{X_c^p}$ for any $c \in I$. In particular, for a closed interval $I = [a, b]$ it is easy to see that for every $p \geq 1$, one has

$$\|f\|_{X_c^p} \leq \|f\|_{X_a^p} + \|f\|_{X_b^p} \qquad (c \in [a, b]). \tag{3.1}$$

Note that the spaces X_I^p are Banach spaces if I is compact, but X_I^p is not a Banach space for general intervals I.

For $p = 2$, the space X_c^2 is a Hilbert space endowed with the scalar product

$$\langle f, g\rangle_{X_c^2} := \int_0^\infty f(u)\overline{g(u)}u^{2c}\frac{du}{u} \qquad (f, g \in X_c^2).$$

Finally we define a local version of the space X_c. By $X_{c,\mathrm{loc}}$, we denote the space of all measurable functions $f : \mathbb{R}^+ \to \mathbb{C}$ such that

$$\int_0^a |f(x)|x^{c-1}dx < +\infty$$

for every $a > 0$.

Remark 3.1 The spaces X_c^p may be characterized in the following way. Denoting by $\mu : \mathcal{B}(\mathbb{R}^+) \to \mathbb{R}_0^+$ the measure defined on the Borel σ-algebra $\mathcal{B}(\mathbb{R}^+)$ of $\mathbb{R}^+$ as

$$\mu(A) := \int_A \frac{1}{u}\,du \qquad (A \in \mathcal{B}(\mathbb{R}^+)),$$

we easily see that X_c^p is the space of all functions $f : \mathbb{R}^+ \to \mathbb{C}$ such that $(\cdot)^c f(\cdot) \in L^p(\mathbb{R}^+, \mu)$, where the last symbol denotes the L^p space with respect to the measure μ. Note that this measure is invariant with respect to the multiplicative structure of the group $\mathbb{R}^+$. This is a basic fact in Mellin analysis.

For the spaces X_c^p, the following version of Hölder's inequality holds.

Proposition 3.2 *Let $f \in X_c^p$ and $g \in X_c^{p'}$, where $(1/p) + (1/p') = 1$. Then $fg \in X_{2c}$ and*

$$\|fg\|_{X_{2c}} \leq \|f\|_{X_c^p} \|g\|_{X_c^{p'}}. \tag{3.2}$$

Proof The proof is an easy consequence of the classical Hölder inequality for L^p spaces, applied to the functions $f(\cdot)(\cdot)^{c-1/p}$ and $g(\cdot)(\cdot)^{c-1/p'}$. □

For $c = 0$ we obtain the inequality

$$\|fg\|_{X_0} \leq \|f\|_{X_0^p} \|g\|_{X_0^{p'}}.$$

Another basic result is the following X_c^p version of the generalized Minkowski inequality of L^p spaces.

Proposition 3.3 *Let $f(x, y)$ be a measurable function on $\mathbb{R}^2$ such that $\|f(\cdot, y)\|_{X_c^p} \in X_c$, with $c \in \mathbb{R}$ and $p \geq 1$. Then we have*

$$\left\| \int_0^\infty f(\cdot, y) dy \right\|_{X_c^p} \leq \int_0^\infty \|f(\cdot, y)\|_{X_c^p} dy. \tag{3.3}$$

Proof The proof is a simple consequence of the definition of the spaces X_c^p, using the generalized Minkowski inequality for L^p spaces. □

The following proposition gives a useful density result for the spaces X_c^p, which represents the counterpart of Proposition 1.14. As before, for a subspace $H \subset X_c^p$, the notation $\mathrm{cls}_{X_c^p}(H)$ denotes the closure of H with respect to the norm topology of X_c^p.

Proposition 3.4 *For every $c \in \mathbb{R}$ and $p \geq 1$, we have*

$$\mathrm{cls}_{X_c^p}\big(C_{\mathrm{comp}}^\infty(\mathbb{R}^+)\big) = X_c^p.$$

Proof Let $f \in X_c^p$. Then, by definition, setting $g(x) := x^{c-1/p} f(x)$, $x \in \mathbb{R}^+$, one has $g \in L^p(\mathbb{R}^+)$. Hence for a fixed $\varepsilon > 0$, there exists a compact interval $[\alpha, \beta] \subset \mathbb{R}^+$ such that

$$\int_{\mathbb{R}^+ \setminus [\alpha,\beta]} |g(x)|^p dx = \int_{\mathbb{R}^+ \setminus [\alpha,\beta]} x^{cp-1} |f(x)|^p dx < \varepsilon.$$

By Proposition 1.14, there exists a sequence $(g_n)_{n\in\mathbb{N}} \subset C_{\mathrm{comp}}^\infty(\mathbb{R}^+)$ of functions with supports in $[\alpha, \beta]$ such that

$$\lim_{n\to\infty} \|g_n - g\|_{L^p([\alpha,\beta])} = 0. \tag{3.4}$$

Therefore, setting $h_n(x) := x^{-c+1/p} g_n(x)$, we have $h_n \in C^{\infty}_{\text{comp}}(\mathbb{R}^+)$ and

$$\|g_n - g\|^p_{L^p[\alpha,\beta]} = \int_\alpha^\beta x^{cp-1}|h_n(x) - f(x)|^p dx.$$

Hence

$$\|h_n - f\|^p_{X^p_c} = \int_\alpha^\beta x^{cp-1}|h_n(x) - f(x)|^p dx + \int_{\mathbb{R}^+\setminus[\alpha,\beta]} x^{cp-1}|f(x)|^p dx$$
$$\leq \|g_n - g\|^p_{L^p([\alpha,\beta])} + \varepsilon.$$

Passing to the limit, we obtain by (3.4) that

$$\limsup_{n\to\infty} \|h_n - f\|^p_{X^p_c} \leq \varepsilon.$$

The assertion follows from the arbitrarity of ε. □

Our next aim are the spaces Y^p_c. For this, we first introduce the following notions.

Definition 3.5 A function $f : \mathbb{R}^+ \to \mathbb{C}$ is said to be *recurrent*, if $f(x) = f(e^{2\pi}x)$ for all $x \in \mathbb{R}^+$. For $c \in \mathbb{R}$, the function f is said to be *c-recurrent* if the function $g(x) := x^c f(x)$ is recurrent, that is, if $f(x) = e^{2\pi c} f(e^{2\pi}x)$ for all $x \in \mathbb{R}^+$. In particular, recurrent and 0-recurrent is the same.

Note that if f is c-recurrent, then $f(x) = e^{2\pi jc} f(e^{2\pi j}x)$ for all $x \in \mathbb{R}^+$ and $j \in \mathbb{Z}$. Moreover, the product of a c_1-recurrent function and a c_2-recurrent function gives a $(c_1 + c_2)$-recurrent function. The proofs are simple consequences of the definition.

The notion of recurrence is the counterpart of the periodicity in the setting of the multiplicative group $\mathbb{R}^+$. Note that a generalized recurrence with the role of the factor $e^{2\pi}$ being taken by any other positive real number could also be introduced.

Definition 3.6 We define the space

$$Y_c := \{f \in L^1_{\text{loc}}(\mathbb{R}^+) \ : \ f \text{ is c-recurrent and } \|f\|_{Y_c} < \infty\},$$

where the norm $\|\cdot\|_{Y_c}$ is given by

$$\|f\|_{Y_c} := \int_{e^{-\pi}}^{e^\pi} |f(u)|u^{c-1} du.$$

For $c = 0$, we simply write Y instead of Y_0.

The space Y_c endowed with the norm $\|\cdot\|_{Y_c}$ is a Banach space. The following invariance holds.

Proposition 3.7 *Let $f \in Y_c$ with $c \in \mathbb{R}$. Then, for any $k \in \mathbb{Z}$ and $\lambda \in \mathbb{R}^+$, we have*

$$\int_{e^{-\pi}}^{e^{\pi}} f(u)u^{c+ik-1}du = \int_{\lambda e^{-\pi}}^{\lambda e^{\pi}} f(u)u^{c+ik-1}du.$$

Proof By definition the integrals are meaningful. For any $\lambda > 0$, we can choose $\gamma \in [e^{-\pi}, e^{\pi}[$ and $\ell \in \mathbb{Z}$ such that $\lambda = e^{2\pi\ell}\gamma$. Then, taking into account that $e^{i2\pi k} = 1$ for $k \in \mathbb{Z}$, using the c-recurrency of f and substituting $x = e^{-2\pi\ell}u$, we obtain

$$\int_{\lambda e^{-\pi}}^{\lambda e^{\pi}} f(u)u^{c+ik}\frac{du}{u} = \int_{\lambda e^{-\pi}}^{\lambda e^{\pi}} e^{-2\pi\ell c} f(e^{-2\ell\pi}u)u^{c+ik}\frac{du}{u} = \int_{\gamma e^{-\pi}}^{\gamma e^{\pi}} f(x)x^{c+ik}\frac{dx}{x}.$$

Again with the help of the c-recurrency of f and the substitution $e^{2\pi}x = u$, the latter integral can be rewritten as

$$\begin{aligned}
&\left\{\int_{\gamma e^{-\pi}}^{e^{-\pi}} + \int_{e^{-\pi}}^{\gamma e^{\pi}}\right\} f(x)x^{c+ik}\frac{dx}{x} \\
&= \int_{\gamma e^{-\pi}}^{e^{-\pi}} f(e^{2\pi}x)(e^{2\pi}x)^{c+ik}\frac{dx}{x} + \int_{e^{-\pi}}^{\gamma e^{\pi}} f(x)x^{c+ik}\frac{dx}{x} \\
&= \left\{\int_{\gamma e^{\pi}}^{e^{\pi}} + \int_{e^{-\pi}}^{\gamma e^{\pi}}\right\} f(u)u^{c+ik}\frac{du}{u} \\
&= \int_{e^{-\pi}}^{e^{\pi}} f(u)u^{c+ik}\frac{du}{u},
\end{aligned}$$

which gives the assertion. □

The fundamental interval for c-recurrent functions can be taken as $[e^{-\pi}, e^{\pi}]$ or more generally as $[\lambda e^{-\pi}, \lambda e^{\pi}]$ with $\lambda > 0$. These are the counterparts of the intervals $[-\pi, \pi]$ or $[-\pi + \alpha, \pi + \alpha]$ with $\alpha \in \mathbb{R}$ in the periodic case.

Remark 3.8 The connection between the spaces Y_c for different values of c can be characterized as follows: Let $c, \widetilde{c} \in \mathbb{R}$. Then $f \in Y_c$ if and only if $x^{c-\widetilde{c}}f(x) \in Y_{\widetilde{c}}$ and

$$\|f\|_{Y_c} = \|(\cdot)^{c-\widetilde{c}}f(\cdot)\|_{Y_{\widetilde{c}}}. \tag{3.5}$$

This relation can be used in order to transform results obtained for Y_0, to Y_c.

Example 3.9 With i being the imaginary unit, we observe that every polynomial

$$p(x^i) := \sum_{j=-n}^{n} a_j x^{ij} \qquad (a_j \in \mathbb{C},\ |j| \le n)$$

belongs to Y. Indeed, $p(x^i)$ is 0-recurrent since each x^{ij} is so and

$$\|p((\cdot)^i)\|_Y \leq \sum_{j=-n}^{n} |a_j| \int_{e^{-\pi}}^{e^{\pi}} |u^{ij}| u^{-1} du = 2\pi \sum_{j=0}^{n} |a_j| < \infty.$$

Also note that the function

$$x^{-c} p(x^i) = \sum_{j=-n}^{n} a_j x^{-c+ij} \qquad (a_j \in \mathbb{C},\ |j| \leq n)$$

is c-recurrent and belongs to Y_c.

Further examples can be obtained by composition of a periodic function with the logarithm. More precisely, let $F \in L^1_{\rm loc}(\mathbb{R})$ be 2π-periodic and let $c \in \mathbb{R}$. Then the function f, defined by

$$f(x) := x^{-c} F(\log x) \qquad (x \in \mathbb{R}^+),$$

belongs to Y_c. In fact,

$$e^{2\pi c} f(e^{2\pi} x) = e^{2\pi c} (e^{2\pi} x)^{-c} F(\log(e^{2\pi} x)) = x^{-c} F(\log x) = f(x)$$

and so f is c-recurrent. Furthermore, $\|f\|_{Y_c} = \|F\|_{L^1([-\pi,\pi])} < \infty$, as is seen by substituting $\log x = u$ in the integral defining the norm on the left-hand side.

Analogously to Definition 3.6, we introduce spaces Y_c^p for $p > 1$.

Definition 3.10 For $c \in \mathbb{R}$ and $1 < p < \infty$, the Banach space Y_c^p is defined by

$$Y_c^p := \{f : \mathbb{R}^+ \to \mathbb{C} \ :\ f \in L^1_{\rm loc}(\mathbb{R}^+),\ f \text{ is c-recurrent},\ \|f\|_{Y_c^p} < \infty\}$$

with the norm

$$\|f\|_{Y_c^p} := \left\{ \int_{e^{-\pi}}^{e^{\pi}} |f(u) u^c|^p \frac{du}{u} \right\}^{1/p}.$$

The space Y_c^2 is a Hilbert space under the inner product

$$\langle f, g\rangle_{Y_c^2} := \int_{e^{-\pi}}^{e^{\pi}} f(u)\overline{g(u)} u^{2c} \frac{du}{u}. \tag{3.6}$$

Finally, for $p = \infty$, we denote by Y_c^∞ the space of all the c-recurrent functions f such that

$$\|f\|_{Y_c^\infty} := \operatorname*{ess\,sup}_{u \in [e^{-\pi}, e^{\pi}]} u^c |f(u)|.$$

3.2 The Mellin Translation Operator

A suitable concept of a translation is of fundamental importance in any integral transform method.

Definition 3.11 Let $f : \mathbb{R}^+ \to \mathbb{C}$ be a function and let $h \in \mathbb{R}^+$. For $c \in \mathbb{R}$, the *Mellin translation operator* τ_h^c applied to f is defined by

$$(\tau_h^c f)(x) := h^c f(hx) \qquad (x \in \mathbb{R}^+).$$

For $c = 0$ we simply set $\tau_h^0 = \tau_h$ and so $(\tau_h^c f)(x) = h^c(\tau_h f)(x)$.

The following proposition is an easy consequence of the definition.

Proposition 3.12 *The Mellin translation operator* $\tau_h^{\widetilde{c}} : X_c^p \to X_c^p$ *for* $p \in [1, \infty[$, $c, \widetilde{c} \in \mathbb{R}$ *and* $h \in \mathbb{R}^+$ *is an isomorphism with* $(\tau_h^{\widetilde{c}})^{-1} = \tau_{1/h}^{\widetilde{c}}$ *and*

$$\|\tau_h^{\widetilde{c}} f\|_{X_c^p} = h^{\widetilde{c}-c}\|f\|_{X_c^p} \qquad (f \in X_c^p). \tag{3.7}$$

Likewise, $\tau_h^{\widetilde{c}} : X_I^p \to X_I^p$ *for any interval* $I \subset \mathbb{R}$ *and* $h \in \mathbb{R}^+$ *is an isomorphism.*

Proof We only prove formula (3.7). Using the substitution $hu = v$, we see that

$$\|\tau_h^{\widetilde{c}} f\|_{X_c^p}^p = \int_0^\infty h^{\widetilde{c}p}|f(hu)|^p u^{cp}\,\frac{du}{u} = h^{p(\widetilde{c}-c)}\int_0^\infty |f(v)|^p v^{cp}\,\frac{dv}{v},$$

which is (3.7). □

Note that the factor h^c in the definition of the translation τ_h^c is chosen to preserve the norm in (3.7) in case of $c = \widetilde{c}$.

The Mellin translation operator has the following properties.

Proposition 3.13 *Let* $f \in X_c^p$, *where* $c \in \mathbb{R}$ *and* $p \in [1, \infty[$. *Then, for* $\widetilde{c} \in \mathbb{R}$, *we have*

$$\lim_{h\to 1} \|\tau_h^{\widetilde{c}} f - f\|_{X_c^p} = 0. \tag{3.8}$$

Moreover, if $f \in X_{]a,b[}^p$, *then*

$$\lim_{h\to 1} \|\tau_h^c f - f\|_{X_c^p} = 0, \tag{3.9}$$

uniformly with respect c *on every compact subinterval of* $]a, b[$.

Proof Let us first assume that $f \in C_{\text{comp}}^\infty(\mathbb{R}^+)$. Then for any h sufficiently close to 1, the supports of the functions f and $\tau_h^{\widetilde{c}} f$ are contained in an interval $[c, d] \subset \mathbb{R}^+$, say. Therefore

$$\begin{aligned}\|\tau_h^{\widetilde{c}} f - f\|_{X_c^p}^p &= \int_0^\infty |h^{\widetilde{c}} f(hu) - f(u)|^p u^{cp} \frac{du}{u} \\ &\leq |h^{\widetilde{c}} - 1|^p \int_0^\infty |f(hu)| u^{cp} \frac{du}{u} \\ &\quad + \int_c^d |f(hu) - f(u)|^p u^{cp} \frac{du}{u} =: I_1 + I_2.\end{aligned}$$

For I_1, we easily see that $I_1 \leq |h^{\widetilde{c}} - 1|^p h^{-cp} \|f\|_{X_c^p}$ and therefore $I_1 \to 0$ as $h \to 1$. As to the term I_2, let any $\varepsilon > 0$ be given. Since f is uniformly continuous, there exists $\delta > 1$ such that $|f(hu) - f(u)| < \varepsilon$ for $h \in [1/\delta, \delta]$ and every $u \in [c, d]$. Setting

$$D := \int_c^d u^{cp} \frac{du}{u} < \infty,$$

we obtain $I_2 \leq D\varepsilon$ for $h \in [1/\delta, \delta]$. Therefore (3.8) follows for $f \in C_{\text{comp}}^\infty(\mathbb{R}^+)$.

Next we turn to the general case. For $f \in X_c^p$ and $\varepsilon > 0$, there exists by Proposition 3.4 a function $g \in C_{\text{comp}}^\infty(\mathbb{R}^+)$ such that $\|f - g\|_{X_c^p} < \varepsilon$. Then, by (3.7),

$$\begin{aligned}\|\tau_h^{\widetilde{c}} f - f\|_{X_c^p} &\leq \|\tau_h^{\widetilde{c}} f - \tau_h^{\widetilde{c}} g\|_{X_c^p} + \|\tau_h^{\widetilde{c}} g - g\|_{X_c^p} + \|g - f\|_{X_c^p} \\ &\leq (h^{\widetilde{c} - c} + 1)\varepsilon + \|\tau_h^{\widetilde{c}} g - g\|_{X_c^p}.\end{aligned}$$

Now the assertion for the spaces X_c^p follows from the first part of the proof.

Concerning the spaces $X_{]a,b[}^p$, let $[\alpha, \beta]$ be a closed interval contained in $]a, b[$. For $c \in [\alpha, \beta]$, we find by employing (3.1) that

$$\begin{aligned}\|\tau_h^c f - f\|_{X_c^p} &\leq |h^c - 1| \|f\|_{X_c^p} + h^c \|\tau_h f - f\|_{X_c^p} \\ &\leq |h^c - 1| \{\|f\|_{X_\alpha^p} + \|f\|_{X_\beta^p}\} + h^c \{\|\tau_h f - f\|_{X_\alpha^p} + \|\tau_h f - f\|_{X_\beta^p}\}.\end{aligned}$$

Noting that h^c converges uniformly with respect to $c \in [\alpha, \beta]$ as $h \to 1$, we obtain the assertion. □

Remark 3.14 As for Lebesgue spaces, the assertion of Proposition 3.13 does not hold in general for functions in X_c^∞. As a simple counter-example for the space X_0^∞, one can take a step function. However, the assertion trivially holds for the subspace $C_{\text{ub}}(\mathbb{R}^+)$. In this case Proposition 3.13 expresses uniform continuity in the Mellin setting, taking the usual L^∞ norm, i.e.,

$$\lim_{h \to 1} \|\tau_h f - f\|_{X_0^\infty} = 0.$$

Based on the previous remark, we introduce the following definition.

Definition 3.15 A function $f : \mathbb{R}^+ \to \mathbb{C}$ is *log-uniformly continuous* on $\mathbb{R}^+$ if for every $\varepsilon > 0$, there exists $\delta_\varepsilon > 0$ such that $|f(u) - f(v)| < \varepsilon$ whenever $|\log u - \log v| < \delta_\varepsilon$.

Note that the notion of log-uniform continuity is equivalent to the continuity of the operator τ_h, that is, $|f(hx) - f(x)| < \varepsilon$ for $|h| < \delta_\varepsilon$ uniformly with respect to $x \in \mathbb{R}^+$.

It should be noted that this notion is different from the usual uniform continuity. For example, the function $f(u) = \sin u$ is uniformly continuous but not log-uniformly continuous on $\mathbb{R}^+$, while the function $g(u) = \sin(\log u)$ is log-uniformly continuous but not uniformly continuous on $\mathbb{R}^+$. However, the two notions are equivalent on every bounded interval $[a, b]$ with $a > 0$. In the following, we denote by $\mathcal{C}(\mathbb{R}^+)$ the space comprising all the log-uniformly continuous functions $f : \mathbb{R}^+ \to \mathbb{C}$.

For the spaces of recurrent functions, we have the same results for the translation operator.

Proposition 3.16 *Let $f \in Y_c^p$, where $c \in \mathbb{R}$ and $p \in [1, \infty[$. Then, for $\widetilde{c} \in \mathbb{R}$, we have*

$$\lim_{h \to 1} \|\tau_h^{\widetilde{c}} f - f\|_{Y_c^p} = 0. \tag{3.10}$$

3.3 Mellin–Lipschitz Classes

In this section the notion of Mellin translation is used for defining Mellin–Lipschitz classes. They are important for an asymptotic study of approximations.

We begin with the definitions of differences of integer order and an appropriate modulus of smoothness. For a function $f \in X_c^p$, $r \in \mathbb{N}$ and $h > 0$, we define the *difference of order* r of f by

$$(\Delta_h^{r,c} f)(u) := \sum_{j=0}^{r} (-1)^{r-j} \binom{r}{j} (\tau_{h^j}^c f)(u) = \sum_{j=0}^{r} (-1)^{r-j} \binom{r}{j} f(h^j u) h^{jc}.$$

Note that the recursion formula

$$(\Delta_h^{r,c} f)(u) = (\Delta_h^{1,c} \Delta_h^{r-1,c} f)(u) \qquad (r = 1, 2, \ldots) \tag{3.11}$$

holds. For $\delta > 0$, the *modulus of smoothness* of $f \in X_c^p$ is defined by

$$\omega_r(f, \delta, X_c^p) := \sup_{|\log h| \le \delta} \|\Delta_h^{r,c} f\|_{X_c^p}.$$

In particular for $p = 1$,

$$\omega_r(f, \delta, X_c) := \sup_{|\log h| \leq \delta} \|\Delta_h^{r,c} f\|_{X_c}.$$

The following proposition lists some basic properties of ω_r.

Proposition 3.17 *The modulus of smoothness $\omega_r(f, \delta, X_c^p)$ has the following properties:*

(i) $\omega_r(f, \cdot, X_c^p)$ *is a non-decreasing function on* $\mathbb{R}^+$;
(ii) $\omega_r(f, \delta, X_c^p) \leq 2^r \|f\|_{X_c^p} \quad (\delta > 0)$;
(iii) for any positive λ *and* δ, *one has*

$$\omega_r(f, \lambda\delta, X_c^p) \leq (1 + \lambda)^r \omega_r(f, \delta, X_c^p).$$

Proof Assertion (i) is obvious. Assertion (ii) follows with the help of the relation

$$\sum_{j=0}^{r} \binom{r}{j} = 2^r$$

and a change of variables.

Concerning (iii), we first prove that for any integer $n \in \mathbb{N}$, we have

$$\omega_r(f, n\delta, X_c^p) \leq n^r \omega_r(f, \delta, X_c^p). \tag{3.12}$$

We start with $r = 1$, considering

$$(\Delta_h^{1,c} f)(u) = (\tau_h^c f)(x) - f(x).$$

For the case of $n = 2$, we assume that $h > 0$ with $|\log h| \leq 2\delta$. Setting $t := \log h$, we can write $t = t_1 + t_2$ with $|t_j| \leq \delta$ for $j = 1, 2$. Thus there exist (unique) $h_j > 0$ with $t_j = \log h_j$ for $j = 1, 2$ and $h = h_1 h_2$. Therefore we can write

$$\begin{aligned}(\tau_h^c f)(x) - f(x) &= f(h_1 h_2 x) h_1^c h_2^c - f(x) \\ &= \big(f(h_1 h_2 x) h_1^c h_2^c - f(h_1 x) h_1^c\big) + \big(f(h_1 x) h_1^c - f(x)\big).\end{aligned}$$

Taking the norm in X_c^p, we obtain by the Minkowski inequality that

$$\begin{aligned}\|\Delta_h^{1,c} f\|_{X_c^p} &\leq \left\{\int_0^\infty |f(h_1 h_2 x) h_1^c h_2^c - f(h_1 x) h_1^c|^p x^{cp} \frac{dx}{x}\right\}^{1/p} \\ &\quad + \left\{\int_0^\infty |f(h_1 x) h_1^c - f(x)|^p x^{cp} \frac{dx}{x}\right\}^{1/p} \\ &=: I_1 + I_2.\end{aligned}$$

As to I_1, we find by the change of variable $u = h_1 x$ that

$$I_1 = \|\Delta_{h_2}^{1,c} f\|_{X_c^p} \le \omega_1(f, \delta, X_c^p).$$

Analogously we obtain

$$I_2 = \|\Delta_{h_1}^{1,c} f\|_{X_c^p} \le \omega_1(f, \delta, X_c^p).$$

Thus the assertion follows for $r = 1$ and $n = 2$. The previous approach can be easily extended to the case of general n. Finally, using (3.11) and an induction argument, we deduce (3.12) for general r.

Now let $\lambda > 0$ be fixed and let n be an integer such that $n \le \lambda < n + 1$. Then, by (i) and (3.12), we have

$$\omega_r(f, \lambda\delta, X_c^p) \le \omega_r(f, (n+1)\delta, X_c^p) \le (n+1)^r \omega_r(f, \delta, X_c^p) \le (\lambda+1)^r \omega_r(f, \delta, X_c^p)$$

and the proof is complete. □

Now we are ready to define the Mellin–Lipschitz classes.

Definition 3.18 Let $r \in \mathbb{N}$. Then, for $\alpha \in \,]0, r]$, the *Mellin–Lipschitz class* is defined by

$$\mathrm{Lip}_r(\alpha, X_c^p) := \left\{ f \in X_c^p \ : \ \omega_r(f; \delta; X_c^p) = \mathcal{O}(\delta^\alpha), \ \delta \to 0+ \right\}.$$

3.4 The Mellin Differential Operator

The Mellin differential operator is introduced as the limit of the difference quotient involving the Mellin translation operator (see Definition 3.11). Thus, if f' exists, we have

$$\begin{aligned}\lim_{h\to 1} \frac{\tau_h^c f(x) - f(x)}{h-1} &= \lim_{h\to 1} \left\{ h^c x \frac{f(hx) - f(x)}{hx - x} + \frac{h^c - 1}{h-1} f(x) \right\} \\ &= x f'(x) + c f(x).\end{aligned}$$

Definition 3.19 The *Mellin differential operator*, denoted by Θ_c and the *Mellin derivative* $\Theta_c f$ of a function $f : \mathbb{R}^+ \to \mathbb{C}$ with $c \in \mathbb{R}$, are defined by

$$\Theta_c f(x) := x f'(x) + c f(x) \qquad (x \in \mathbb{R}^+) \tag{3.13}$$

provided that f' exists almost everywhere on $\mathbb{R}^+$.

The Mellin differential operator of order $r \in \mathbb{N}$ is recursively defined by

$$\Theta_c^1 := \Theta_c, \qquad \Theta_c^r := \Theta_c(\Theta_c^{r-1}).$$

For convenience, we set $\Theta^r := \Theta_0^r$ for $c = 0$ and $\Theta_c^0 := I_d$, where I_d is the identity operator $I_d f = f$.

For example, the first three Mellin derivatives are given by

$$\begin{aligned}
\Theta_c f(x) &= xf'(x) + cf(x)\\
\Theta_c^2 f(x) &= x^2 f^{(2)}(x) + (2c+1)xf'(x) + c^2 f(x)\\
\Theta_c^3 f(x) &= x^3 f^{(3)}(x) + (3c+3)x^2 f^{(2)}(x) + (3c^2+3c+1)xf'(x) + c^3 f(x).
\end{aligned}$$

Remark 3.20 Calculations with Mellin derivatives turn out to be more complex than those with ordinary derivatives. For example, the familiar Leibniz rule is not valid. In general, $\Theta_c(fg) \neq f\Theta_c g + g\Theta_c f$ but we have

$$\Theta_c(fg) = f\Theta_c g + g\Theta_0 f = f\Theta_0 g + g\Theta_c f.$$

Now we introduce the notion of Stirling numbers which help us to establish the main connection between the Mellin derivatives and the ordinary ones. For a fixed $r \in \mathbb{N}$, we recursively define the *generalized Stirling numbers of the second kind* as

$$S_c(r, 0) := c^r, \quad S_c(r, r) := 1, \tag{3.14}$$

$$S_c(r+1, k) := S_c(r, k-1) + (c+k)S_c(r, k) \quad (k \in \mathbb{N}). \tag{3.15}$$

They lead us to the following relations.

Proposition 3.21 *For $c \in \mathbb{R}$, the Mellin derivative of order $r \in \mathbb{N}$ has the representation*

$$\Theta_c^r f(x) = \sum_{k=0}^{r} S_c(r, k)x^k f^{(k)}(x). \tag{3.16}$$

In particular, for $c = 0$,

$$\Theta^r f(x) = \sum_{k=0}^{r} S(r, k)x^k f^{(k)}(x),$$

where $S(r, k) := S_0(r, k)$ are the classical Stirling numbers of the second kind.

Proof First let $c = 0$. The case $r = 1$ is obvious. Assuming that the formula holds for $r \in \mathbb{N}$, we have

$$\begin{aligned}\Theta^{r+1}f(x) = x\frac{d}{dx}\Theta^r f(x) &= x\sum_{k=0}^{r} S(r,k)\left(kx^{k-1}f^{(k)}(x) + x^k f^{(k+1)}(x)\right)\\ &= \sum_{k=0}^{r} S(r,k)\left(kx^k f^{(k)}(x) + x^{k+1}f^{(k+1)}(x)\right).\end{aligned}$$

By rearranging the terms on the right-hand side, we obtain

$$\begin{aligned}\Theta^{r+1}f(x) &= \sum_{k=1}^{r}\left(kS(r,k) + S(r,k-1)\right)x^k f^{(k)}(x) + x^{r+1}f^{(r+1)}(x)\\ &= \sum_{k=1}^{r} S(r+1,k)x^k f^{(k)}(x) + S(r+1,r+1)x^{r+1}f^{(r+1)}(x)\\ &= \sum_{k=0}^{r+1} S(r+1,k)x^k f^{(k)}(x).\end{aligned}$$

Thus we have proved the case of $c = 0$ by induction on r.

For general $c \in \mathbb{R}$ and $r = 1$, we have

$$\sum_{k=0}^{1} S_c(r,k)x^k f^{(k)}(x) = S_c(1,0)f(x) + S_c(1,1)xf'(x) = xf'(x) + cf(x) = \Theta f(x).$$

Now assume that (3.16) holds for $r \in \mathbb{N}$. Since

$$x\frac{d}{dx}\Theta_c^r f(x) = \sum_{k=0}^{r} S_c(r,k)(kx^k f^{(k)}(x) + x^{k+1}f^{(k+1)}(x)),$$

we obtain

$$\begin{aligned}\Theta_c^{r+1}f(x) &= x\frac{d}{dx}\Theta_c^r f(x) + c\Theta_c^r f(x)\\ &= \sum_{k=0}^{r} S_c(r,k)\left[(k+c)x^k f^{(k)}(x) + x^{k+1}f^{(k+1)}(x)\right]\\ &= \sum_{k=0}^{r+1} S_c(r+1,k)x^k f^{(k)}(x).\end{aligned}$$

Hence also the more general formula (3.16) has been verified by induction on r. □

Conversely, one can obtain a representation of ordinary derivatives in terms of Mellin derivatives by using the *generalized Stirling numbers of the first kind*. They are recursively defined as follows: For $r \in \mathbb{N}$ and $0 \le k \le r$, set

$$s_c(r,0) := (-1)^r r! \binom{r+c-1}{r}, \qquad s_c(r,r) := 1,$$

$$s_c(r+1,k) := s_c(r,k-1) - (c+r)s_c(r,k).$$

Now we are ready for a converse formula.

Proposition 3.22 *Let $r \in \mathbb{N}$. The r-th derivative $f^{(r)}$ of a function f can be represented in terms of the Mellin derivatives $\Theta_c^k f$ for $k = 0, \ldots, r$ as*

$$f^{(r)}(x) = x^{-r} \sum_{k=0}^{r} s_c(r,k) \Theta_c^k f(x).$$

Proof The proof is based on the orthogonality relation (see [89, 244])

$$\sum_{j=k}^{r} s_c(r,j) S_c(j,k) = \delta_{r,k}$$

with $\delta_{r,k}$ being Kronecker's delta. Thus an application of (3.16) yields

$$x^r f^{(r)}(x) = \sum_{k=0}^{r} \delta_{r,k} x^k f^{(k)}(x) = \sum_{k=0}^{r} \sum_{j=k}^{r} s_c(r,j) S_c(j,k) x^k f^{(k)}(x)$$
$$= \sum_{j=0}^{r} s_c(r,j) \Theta_c^j f(x)$$

from which the assertion follows. □

3.5 Mellin–Sobolev Spaces

In order to define the underlying spaces on which Θf is meaningful, we shall use the notion of a *locally absolutely continuous function* on $\mathbb{R}^+$. We recall that the space of these functions, denoted by $AC_{\text{loc}}(\mathbb{R}^+)$, is the set of all $f : \mathbb{R}^+ \to \mathbb{C}$ which have a representation

$$f(x) = \alpha + \int_1^x g(u)du \qquad (x \in \mathbb{R}^+),$$

where $g \in L^1_{\mathrm{loc}}(\mathbb{R}^+)$ and α is a complex constant. Thus for $f \in AC_{\mathrm{loc}}(\mathbb{R}^+)$, the derivative of f exists almost everywhere and $f' = g$ almost everywhere.

Definition 3.23 For $c \in \mathbb{R}$, $p \geq 1$ and $r \in \mathbb{N}$, the *Mellin–Sobolev spaces* $W_c^{r,p}(\mathbb{R}^+)$ is defined as the set of all functions $f \in X_c^p$ which are equivalent to a function $g \in C^{r-1}(\mathbb{R}^+)$ with $g^{(r-1)} \in AC_{\mathrm{loc}}(\mathbb{R}^+)$ such that $\Theta_c^r g \in X_c^p$. Correspondingly, the Mellin–Sobolev spaces $W_I^{r,p}(\mathbb{R}^+)$ are defined for an interval $I \subset \mathbb{R}$ on setting

$$W_I^{r,p}(\mathbb{R}^+) := \bigcap_{c \in I} W_c^{r,p}(\mathbb{R}^+).$$

The following proposition provides an integral representation of differences in terms of Mellin derivatives.

Proposition 3.24 *Let $r \in \mathbb{N}$, $c \in \mathbb{R}$ and $f \in W_c^{r,1}(\mathbb{R}^+)$. Then, for the differences $\Delta_h^{r,c} f$, the following representation holds*

$$(\Delta_h^{r,c} f)(x) = \sum_{j=0}^{r} (-1)^{r-j} \binom{r}{j} f(h^j x) h^{jc}$$
$$= \int_{1/h}^{1} \cdots \int_{1/h}^{1} \Theta_c^r f\left(\frac{x}{u_1 \cdots u_r}\right) (u_1 \cdots u_r)^{-c} \frac{du_r}{u_r} \cdots \frac{du_1}{u_1} \qquad (3.17)$$

for every $x, h \in \mathbb{R}^+$.

Proof We verify (3.17) by induction on r. Taking into account that $(d/du)(f(u)u^c) = u^{c-1}\Theta_c f(u)$ and using the substitution $u = x/v$, we see that

$$(\tau_h^c f)(x) - f(x) = x^{-c} \int_x^{hx} u^c \Theta_c f(u) \frac{du}{u} = \int_{1/h}^{1} \Theta_c f\left(\frac{x}{v}\right) v^{-c} \frac{dv}{v},$$

which shows that (3.17) holds for $r = 1$.

Now assume (3.17) holds for some k in place of r, where $1 \leq k < r$. Noting that

$$\tau_{h^{j+1}}^c = \tau_h^c \tau_{h^j}^c, \qquad \binom{k+1}{j} = \binom{k}{j} + \binom{k}{j-1} \quad (1 \leq j \leq k)$$

and applying (3.17) for $r = k$ to $\Theta_c f((\cdot)/u_1)$ in the last of the following five steps, we conclude that

$$\begin{aligned}
&\sum_{j=0}^{k+1}(-1)^{k+1-j}\binom{k+1}{j}(\tau_{h^j}^c f)(x)\\
&\quad=\sum_{j=0}^{k}(-1)^{k-j}\binom{k}{j}\big(\tau_h^c(h^{cj}f(h^jx))-h^{cj}f(h^jx)\big)\\
&\quad=\sum_{j=0}^{k}(-1)^{k-j}\binom{k}{j}x^{-c}\int_{h^jx}^{h^{j+1}x}u^c\Theta_c f(u)\frac{du}{u}\\
&\quad=\int_{1/h}^{1}\sum_{j=0}^{k}(-1)^{k-j}\binom{k}{j}h^{cj}u_1^{-c}\Theta_c f\Big(h^j\frac{x}{u_1}\Big)\frac{du_1}{u_1}\\
&\quad=\int_{1/h}^{1}\sum_{j=0}^{k}(-1)^{k-j}\binom{k}{j}(\tau_{h^j}^c\Theta_c f)\Big(\frac{x}{u_1}\Big)u_1^{-c}\frac{du_1}{u_1}\\
&\quad=\int_{1/h}^{1}\cdots\int_{1/h}^{1}\Theta_c^{k+1}f\Big(\frac{x}{u_1\cdots u_{k+1}}\Big)(u_1\cdots u_{k+1})^{-c}\frac{du_{k+1}}{u_{k+1}}\cdots\frac{du_1}{u_1},
\end{aligned}$$

and so the assertion follows. □

3.6 Mellin–Bernstein Spaces for Functions Analytic on a Riemann Surface

We first recall some facts and introduce some notation. Consider a function g which is analytic on the Riemann surface $S_{\log}$ of the logarithm (see Sect. 1.4.5). Then g can be split into branches g_k $(k \in \mathbb{Z})$ that are analytic on the slit complex plane $\Omega := \{z \in \mathbb{C} : z \notin \mathbb{R}_0^+\}$. They are connected on the slit by the following properties:

(i) For $x > 0$ the limits

$$g_k^+(x) := \lim_{\varepsilon\to 0+} g_k(x+i\varepsilon) \quad\text{and}\quad g_k^-(x) := \lim_{\varepsilon\to 0+} g_k(x-i\varepsilon)$$

exist and

$$g_k^-(x) = g_{k+1}^+(x), \qquad g_0^+(x) = g(x) \quad (x > 0). \tag{3.18}$$

(ii) For $x > 0$, let U_x be an open disk in the right half-plane with center at x. Then the function $\psi_k : U_x \to \mathbb{C}$, defined by

$$\psi_k(z) := \begin{cases} g_k(z) & \text{for } z \in U_x, \ \Im z < 0, \\ g_k^-(z) & \text{for } z \in U_x \cap \mathbb{R}, \\ g_{k+1}(z) & \text{for } z \in U_x, \ \Im z > 0, \end{cases}$$

is analytic.

After these preparations, the Mellin–Bernstein spaces can be introduced as follows.

Definition 3.25 For $c \in \mathbb{R}$ and $T \in \mathbb{R}^+$, the *Mellin–Bernstein space* $\widetilde{B}^2_{c,T}$ comprises all functions $f \in X^2_c$ for which $g(x) := x^c f(x)$ has an analytic extension on the Riemann surface $S_{\log}$ with branches g_k on $\Omega = \{z \in \mathbb{C} : z \notin \mathbb{R}_0^+\}$ satisfying the following (additional) conditions:

There exists a constant $C > 0$ such that for all $k \in \mathbb{Z}$ and $\theta \in [0, 2\pi]$

$$|g_k(re^{i\theta})| \le Ce^{T|2\pi k+\theta|} \qquad (r > 0). \tag{3.19}$$

For $\theta \in [0, 2\pi]$, we have

$$\lim_{r\to 0} g_k(re^{i\theta}) = \lim_{r\to\infty} g_k(re^{i\theta}) = 0 \tag{3.20}$$

uniformly with respect to θ.

In (3.19) and (3.20) the value $g_k(re^{i\theta})$ has to be defined as $g_k^+(r)$ when $\theta = 0$ and as $g_k^-(r)$ when $\theta = 2\pi$.

Our aim is an analogue of the classical Bernstein inequality for Mellin derivatives. Later we shall present another approach to the Bernstein inequality with an alternative version of the Mellin–Bernstein spaces (see Sect. 3.8).

We begin with two lemmas. The first lemma shows a connection between the Mellin–Bernstein space $\widetilde{B}^2_{c,T}$ and the Bernstein space B^2_T (see Definition 1.34).

Lemma 3.26 *Let $f \in \widetilde{B}^2_{c,T}$, where $c \in \mathbb{R}$ and $T \in \mathbb{R}^+$. Define $h(x) := e^{cx} f(e^x)$ for $x \in \mathbb{R}$. Then h has an extension to an entire function $\widetilde{h} \in B^2_T$ and $\|\widetilde{h}\|_2 = \|f\|_{X^2_c}$.*

Proof For $k \in \mathbb{Z}$, consider the strips

$$S_k := \{z \in \mathbb{C} : 2k\pi < \Im z < 2(k+1)\pi\},$$

and note that the exponential function maps each S_k onto the complex plane slit along $\mathbb{R}_0^+$. Using the notation of Definition 3.25 and the preceding considerations, we define a function $\widetilde{h} : \mathbb{C} \to \mathbb{C}$ by setting

$$\widetilde{h}(z) := \begin{cases} g_k(e^z) \ \text{if } z \in S_k \\ g_k^+(e^z) \ \text{if } \Im z = 2k\pi \end{cases} \qquad \text{for } k \in \mathbb{Z}.$$

As a composition of analytic functions, the restriction of $\widetilde{h}$ to S_k is analytic for each $k \in \mathbb{Z}$. Now let z_0 be a point on the line $\{z \in \mathbb{C} : \Im z = 2k\pi\}$ for some $k \in \mathbb{Z}$. Since $g_{k-1}^{-}(x) = g_k^{+}(x)$ for $x \in \mathbb{R}^+$, we see by property (ii) of the branches of an analytic function on the Riemann surface $S_{\log}$ that there exists an open neighborhood V_{z_0} of z_0 on which $\widetilde{h}(z) = \psi_{k-1}(e^z)$ with ψ_{k-1} being analytic on $\exp(V_{z_0})$. This shows that the restriction of $\widetilde{h}$ to V_{z_0} is analytic as well. Hence $\widetilde{h}$ is analytic in all points of $\mathbb{C}$, and so it is an entire function.

Next we show that $\widetilde{h}$ satisfies the exponential growth condition (i) of Definition 1.34. Take any $x, y \in \mathbb{R}$. Then y has a unique representation $y = 2k\pi + y_0$ with $k \in \mathbb{Z}$ and $0 \leq y_0 < 2\pi$. Employing (3.19) and recalling the subsequent convention, we find that

$$\left|\widetilde{h}(x+iy)\right| = \left|g_k(e^x e^{iy_0})\right| \leq C e^{T|2k\pi + y_0|} = C e^{T|y|}.$$

Finally, we note that

$$\widetilde{h}(x) = g_0^{+}(e^x) = g(e^x) = e^{cx} f(e^x) = h(x) \qquad (x \in \mathbb{R}).$$

Hence $\widetilde{h}$ is a continuation of h to $\mathbb{C}$ and

$$\|\widetilde{h}\|_2^2 = \int_{-\infty}^{\infty} e^{2cx} \left|f(e^x)\right|^2 dx = \int_0^{\infty} u^{2c} \left|f(u)\right|^2 \frac{du}{u} = \|f\|_{X_c^2}^2 < \infty.$$

This completes the proof. □

Lemma 3.27 *Let $f \in \widetilde{B}_{c,T}^2$, where $c \in \mathbb{R}$ and $T \in \mathbb{R}^+$. Then f is infinitely differentiable on $\mathbb{R}^+$ and*

$$\Theta_c^r f(u) = u^{-c} h^{(r)}(\log u) \qquad (r \in \mathbb{N},\ u \in \mathbb{R}^+), \tag{3.21}$$

where $h(x) := e^{cx} f(e^x)$ for $x \in \mathbb{R}$.

Proof Substituting $x = \log u$, we can express f on $\mathbb{R}^+$ in terms of h as

$$f(u) = u^{-c} h(\log u) \qquad (u \in \mathbb{R}^+).$$

Since by Lemma 3.26 the function h has an extension to an entire function, the right-hand side is infinitely differentiable on $\mathbb{R}^+$ and formula (3.21) is easily verified by induction on r. □

Now we are ready to prove the announced Mellin–Bernstein inequality.

Proposition 3.28 (Mellin–Bernstein Inequality) *Let $f \in \widetilde{B}_{c,T}^2$, where $c \in \mathbb{R}$ and $T \in \mathbb{R}^+$. Then for each $r \in \mathbb{N}$, we have*

$$\|\Theta_c^r f\|_{X_c^2} \leq T^r \|f\|_{X_c^2}. \tag{3.22}$$

Proof Again, let $h(x) := e^{cx} f(e^x)$ for $x \in \mathbb{R}$. Then Lemma 3.26 in conjunction with Proposition 1.38 show us that

$$\|h^{(r)}\|_2 \leq T^r \|h\|_2 = T^r \|f\|_{X_c^2}.$$

By Lemma 3.27 and obvious substitutions, the square of the left-hand side can be expressed as

$$\begin{aligned}\|h^{(r)}\|_2^2 &= \int_{-\infty}^{\infty} \left|h^{(r)}(x)\right|^2 dx = \int_0^{\infty} \left|h^{(r)}(\log u)\right|^2 \frac{du}{u} \\ &= \int_0^{\infty} \left|u^c \Theta_c^r f(u)\right|^2 \frac{du}{u} = \|\Theta_c^r f\|_{X_c^2}^2.\end{aligned}$$

This completes the proof. □

As regards Mellin–Sobolev spaces, Proposition 3.28 shows that the inclusion $\widetilde{B}_{c,T}^2 \subset W_c^{r,2}(\mathbb{R}^+)$ holds for all $r \in \mathbb{N}$.

3.7 Polar Mellin Derivatives

Here we introduce a Mellin differential operator in a polar form, using the notion of polar-analytic function on the half-plane $\mathbb{H}$ (see Chap. 2).

Definition 3.29 For a fixed real number $c \in \mathbb{R}$, we define the *Mellin polar derivative* of a polar-analytic function f by the formula

$$(\widetilde{\Theta}_c f)(r, \theta) := re^{i\theta}(D_{\text{pol}} f)(r, \theta) + cf(r, \theta), \tag{3.23}$$

provided that the polar derivative $D_{\text{pol}} f$ exists at the point $(r, \theta) \in \mathbb{H}$.

For $\varphi(r) := f(r, 0)$, we have $(\widetilde{\Theta}_c f)(r, 0) = (\Theta_c \varphi)(r)$, where Θ_c is the usual Mellin differential operator (see (3.13)).

Higher order polar Mellin derivatives can be represented with the help of the generalized Stirling numbers of the second kind as an extension of formula (3.16) of Proposition 3.21.It can be shown by induction on k that

$$(\widetilde{\Theta}_c^k f)(r, \theta) := \sum_{j=0}^{k} S_c(k, j) r^j e^{ij\theta} D_{\text{pol}}^j f(r, \theta) \qquad (k \in \mathbb{N},\ (r, \theta) \in \mathbb{H}). \tag{3.24}$$

Remark 3.30 Note that using (3.24), we can derive the useful formula

$$(\widetilde{\Theta}_c^k f)(r, 0) = \big(\Theta_c^k f(\cdot, 0)\big)(r) \qquad (r \in \mathbb{R}^+).$$

3.7.1 Taylor-Like Expansion with Polar Mellin Derivatives

Now we establish a Taylor-like expansion of polar-analytic functions involving polar Mellin derivatives. Other than the expansion of Theorem 2.12, it will be of a global nature. For this, the transformation $\tau : \mathbb{C} \to \mathbb{H}$, which maps $z = x + iy \in \mathbb{C}$ to the point $(r, \theta) := (e^x, y) \in \mathbb{H}$ will play a role. We first discuss some elementary geometrical properties.

For $z_0 = x_0 + iy_0 \in \mathbb{C}$ and $\rho > 0$, consider the disk $D(z_0, \rho) := \{z \in \mathbb{C} : |z - z_0| < \rho\}$ in $\mathbb{C}$. It is transformed by τ into the region with one axis of symmetry in $\mathbb{H}$ given by

$$E((r_0, \theta_0), \rho) := \left\{ (r, \theta) \in \mathbb{H} : \left(\log \frac{r}{r_0} \right)^2 + (\theta - \theta_0)^2 < \rho^2 \right\},$$

where $r_0 = e^{x_0}$ and $\theta_0 = y_0$. We call $E((r_0, \theta_0), \rho)$ a *polar-disk* with *center* (r_0, θ_0) and *radius* ρ. Its boundary is given by the graph of the functions

$$r = r_0 \exp\left(\pm \sqrt{\rho^2 - (\theta - \theta_0)^2} \right), \qquad \theta \in [-\rho + \theta_0, \rho + \theta_0]$$

or, alternatively, by

$$\theta = \theta_0 \pm \sqrt{\rho^2 - \left(\log \frac{r}{r_0} \right)^2}, \qquad r \in \left[r_0 e^{-\rho}, r_0 e^{\rho} \right].$$

For examples of polar-disks, see Figs. 3.1 and 3.2. It can be shown that polar-disks are convex sets if and only if $0 < \rho \le 3^{3/2}/2 = 2.598\ldots$.

Given a domain $\mathcal{D} \subset \mathbb{H}$, then

$$A := \{z = x + iy \in \mathbb{C} : (e^x, y) \in \mathcal{D}\} \tag{3.25}$$

is the pre-image of $\mathcal{D}$ under the transformation τ. Now for $z_0 = x_0 + iy_0 \in A$, consider a disk $D(z_0, \rho) \subset A$. Then the associated polar-disk $E((r_0, \theta_0), \rho)$ is fully contained in $\mathcal{D}$. Note that the largest ρ such that $D(z_0, \rho) \subset A$ is given by

$$\rho = \min_{(r,\theta) \in \partial\mathcal{D}} \sqrt{\left(\log \frac{r}{r_0} \right)^2 + (\theta - \theta_0)^2},$$

where $\partial\mathcal{D}$ denotes the boundary of $\mathcal{D}$.

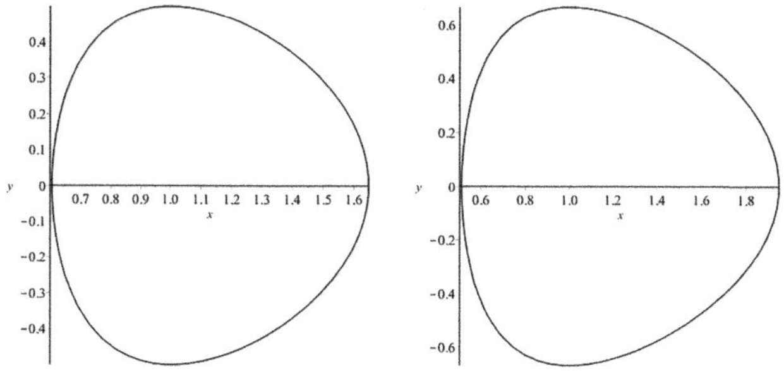

Fig. 3.1 The polar-disks around the point $(r_0, \theta_0) = (1, 0)$ with $\rho = 1/2$ and $\rho = 2/3$

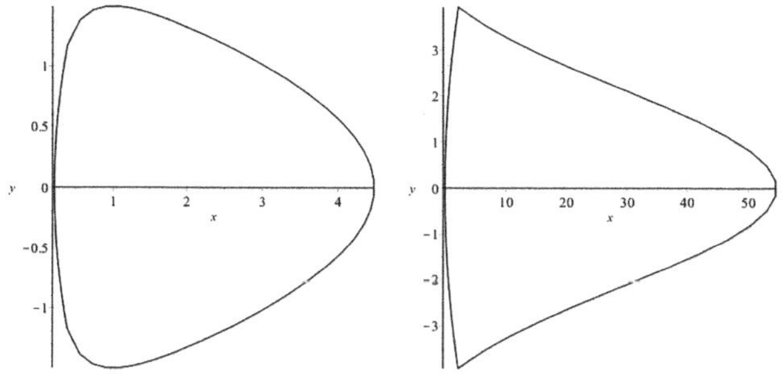

Fig. 3.2 The polar-disks around the point $(r_0, \theta_0) = (1, 0)$ with $\rho = 3/2$ and $\rho = 4$

Theorem 3.31 *Let $\mathcal{D} \subset \mathbb{H}$ be a domain and let $f : \mathcal{D} \to \mathbb{C}$ be polar-analytic on $\mathcal{D}$. If $(r_0, \theta_0) \in \mathcal{D}$, then there holds the expansion*

$$f(r, \theta) = \sum_{k=0}^{\infty} \frac{(\widetilde{\Theta}_0^k f)(r_0, \theta_0)}{k!} \left(\log \frac{r}{r_0} + i(\theta - \theta_0) \right)^k,$$

converging absolutely on every polar-disk $E((r_0, \theta_0), \rho) \subset \mathcal{D}$ and uniformly on its compact subsets.

Proof The function

$$g(z) = g(x + iy) := f(e^x, y) = u(e^x, y) + iv(e^x, y) \qquad (x, y \in \mathbb{R}),$$

where u and v are real-valued, is defined on the domain A specified in (3.25). By Theorem 2.5 we know that u and v have continuous partial derivatives on $\mathcal{D}$ which satisfy Eqs. (2.4). This implies that the real part and the imaginary part of g have continuous partial derivatives on A which satisfy the Cauchy–Riemann equations. Hence g is analytic on A. Thus, for any $z_0 = x_0 + iy_0 \in A$, the Taylor expansion

$$g(z) = \sum_{k=0}^{\infty} \frac{g^{(k)}(z_0)}{k!} (z - z_0)^k$$

exists on every disk centered at z_0 and contained in A. The series converges absolutely on these disks and uniformly on their compact subsets. Writing $z = x + iy$, $z_0 = x_0 + iy_0$ and substituting $x = \log r$, $x_0 = \log r_0$, $y = \theta$, $y_0 = \theta_0$, we obtain the expansion

$$f(r,\theta) = \sum_{k=0}^{\infty} \frac{g^{(k)}(z_0)}{k!} \left(\log \frac{r}{r_0} + i(\theta - \theta_0) \right)^k ,$$

valid on every polar-disk $E((r_0, \theta_0), \rho)$ contained in $\mathcal{D}$, the kind of convergence being inherited from that of the Taylor series of g.

Finally, using the chain rule and formulae (3.23), (3.24), we can show by induction on k that

$$g^{(k)}(z_0) = (\widetilde{\Theta}_0^k f)(r_0, \theta_0) \qquad (k \in \mathbb{N}_0).$$

This completes the proof. □

If $f : \mathbb{H} \to \mathbb{C}$ is polar-analytic, then the function g in the proof of Theorem 3.31 is an entire function. In this instance, we may choose $r_0 = 1$, $\theta_0 = 0$ and obtain

$$f(r,\theta) = \sum_{k=0}^{\infty} \frac{(\widetilde{\Theta}_0^k f)(r_0, \theta_0)}{k!} (\log r + i\theta)^k ,$$

converging everywhere on $\mathbb{H}$.

For example, the function $f(r,\theta) := \sin(re^{i\theta})$, which is 2π-periodic with respect to θ, has the expansion of Theorem 3.31 with $g(z) = \sin(\exp(z))$ but also the simpler expansion

$$f(r,\theta) = \sum_{k=0}^{\infty} \frac{(-1)^k}{(2k+1)!} r^{2k+1} e^{i(2k+1)\theta} \qquad ((r,\theta) \in \mathbb{H}).$$

As a further example, consider the function $f(r,\theta) := \log r + i\theta$. Then the expansion of Theorem 3.31 takes the trivial form

$$\log r + i\theta = \log r_0 + i\theta_0 + \left(\log \frac{r}{r_0} + i(\theta - \theta_0) \right).$$

Theorem 3.31 can be extended in the following way.

Theorem 3.32 *Let $\mathcal{D} \subset \mathbb{H}$ be a domain. Let $f : \mathcal{D} \to \mathbb{C}$ be polar-analytic on $\mathcal{D}$ and let $c \in \mathbb{R}$. If $(r_0, \theta_0) \in \mathcal{D}$, then there holds the expansion*

$$\left(re^{i\theta}\right)^c f(r,\theta) = \left(r_0 e^{i\theta_0}\right)^c \sum_{k=0}^{\infty} \frac{(\widetilde{\Theta}_c^k f)(r_0, \theta_0)}{k!} \left(\log \frac{r}{r_0} + i(\theta - \theta_0) \right)^k , \qquad (3.26)$$

converging absolutely on every polar-disk $E\big((r_0, \theta_0), \rho\big) \subset \mathcal{D}$ *and uniformly on its compact subsets.*

Proof We only have to extend the proof of Theorem 3.31 slightly. Let

$$A := \big\{z = x + iy \in \mathbb{C} \,:\, (e^x, y) \in \mathcal{D}\big\}.$$

Then the function

$$g \,:\, z = x + iy \longmapsto f(e^x, y)$$

is analytic on A and for $z_0 := \log r_0 + i\theta_0$, we have

$$g^{(k)}(z_0) \;=\; \big(\widetilde{\Theta}_0^k f\big)(r_0, \theta_0).$$

Now define $h(z) := e^{cz} g(z)$. Then h is also analytic on A and

$$h^{(k)}(z_0) = e^{cz_0} \sum_{j=0}^{k} \binom{k}{j} g^{(j)}(z_0) c^{k-j}$$

$$= \big(r_0 e^{i\theta_0}\big)^c \sum_{j=0}^{k} \binom{k}{j} \big(\widetilde{\Theta}_0^j f\big)(r_0, \theta_0) c^{k-j}.$$

It can be shown by induction on k that the sum on the right-hand side is equal to $\big(\widetilde{\Theta}_c^k f\big)(r_0, \theta_0)$. Since by the Taylor expansion

$$e^{cz} g(z) \;=\; \sum_{k=0}^{\infty} \frac{h^{(k)}(z_0)}{k!} \,(z - z_0)^k,$$

we obtain (3.26) by substituting $z = x + iy$ with $x = \log r$ and $y = \theta$. The statement on convergence is seen as in Theorem 3.31. □

As a consequence of the series expansion of Theorem (3.31), we deduce the following identity theorem for polar-analytic functions.

Theorem 3.33 *Let* $\mathcal{D}$ *be a domain in* $\mathbb{H}$ *and let* $f \,:\, \mathcal{D} \to \mathbb{C}$ *be polar-analytic. Suppose that* $(r_0, \theta_0) \in \mathcal{D}$ *is an accumulation point of distinct zeros of* f*. Then* f *is identically zero.*

For the proof we make decisive use of an observation which we state as a lemma for easy reference.

Lemma 3.34 *Under the hypotheses of Theorem 3.33, it follows that*

$$\big(\widetilde{\Theta}_0^k f\big)(r_0, \theta_0) = 0 \quad \textit{for all } k \in \mathbb{N}_0.$$

Proof For a proof by contradiction, we assume that there exists a smallest nonnegative integer k_0 such that $\big(\widetilde{\Theta}_0^{k_0} f\big)(r_0, \theta_0) \neq 0$. Then f has an expansion

$$f(r,\theta) = \left(\log\frac{r}{r_0} + i(\theta-\theta_0)\right)^{k_0}$$

$$\times \left(\frac{\big(\widetilde{\Theta}_0^{k_0} f\big)(r_0,\theta_0)}{k_0!} + \sum_{k=k_0+1}^{\infty} \frac{\big(\widetilde{\Theta}_0^{k} f\big)(r_0,\theta_0)}{k!} \left(\log\frac{r}{r_0} + i(\theta-\theta_0)\right)^{k-k_0} \right)$$

converging in a polar-disk $E((r_0, \theta_0), \rho)$ with $\rho > 0$ by Theorem 3.31. On the right-hand side, the first factor vanishes for $(r, \theta) = (r_0, \theta_0)$ only if $k_0 \geq 1$; otherwise it is identically 1. The second factor is different from zero at $(r, \theta) = (r_0, \theta_0)$. By continuity, it is also different from zero in a sufficiently small neighborhood of (r_0, θ_0). Hence (r_0, θ_0) cannot be an accumulation point of distinct zeros of f. A contradiction! □

Proof of Theorem 3.33 Let (r^*, θ^*) be an arbitrary point in $\mathcal{D}$. It suffices to show that $f(r^*, \theta^*) = 0$. Let $\gamma\ :\ [0, 1] \to \mathcal{D}$ be a Jordan arc in $\mathcal{D}$ such that $\gamma(0) = (r_0, \theta_0)$ and $\gamma(1) = (r^*, \theta^*)$. Since the trace of γ has a positive distance from the boundary of $\mathcal{D}$, there exists a number $\rho_{\text{inf}} > 0$ such that for each point (r, θ) on γ there is a polar-disk $E((r, \theta), \rho) \subset \mathcal{D}$ with $\rho \geq \rho_{\text{inf}}$.

First consider a polar-disk $E_0 := E((r_0, \theta_0), \rho_0) \subset \mathcal{D}$ with $\rho_0 \geq \rho_{\text{inf}}$. By Theorem 3.31 we have a representation

$$f(r,\theta) = \sum_{k=0}^{\infty} \frac{\big(\widetilde{\Theta}_0^{k} f\big)(r_0,\theta_0)}{k!} \left(\log\frac{r}{r_0} + i(\theta-\theta_0)\right)^{k}$$

holding for all $(r, \theta) \in E_0$. Now, by Lemma 3.34, it follows that the restriction of f to E_0 is identically zero. Thus, if $(r^*, \theta^*) \in E_0$, we have reached our aim.

Otherwise, let $(r_1, \theta_1) := \gamma(t_1)$ with $0 < t_1 \leq 1$ be a point where γ intersects the boundary of E_0. By our previous observations, there exists a polar-disk $E_1 := E((r_1, \theta_1), \rho_1) \subset \mathcal{D}$ with $\rho_1 \geq \rho_{\text{inf}}$ such that

$$f(r,\theta) = \sum_{k=0}^{\infty} \frac{\big(\widetilde{\Theta}_0^{k} f\big)(r_1,\theta_1)}{k!} \left(\log\frac{r}{r_1} + i(\theta-\theta_1)\right)^{k}$$

for all $(r, \theta) \in E_1$. Since (r_1, θ_1), being a boundary point of E_0, is an accumulation point of distinct zeros of f, we conclude with the help of Lemma 3.34 that the restriction of f to E_1 vanishes identically. If $(r^*, \theta^*) \notin E_1$, we continue this procedure with a point $(r_2, \theta_2) := \gamma(t_2)$, where $t_1 < t_2 \leq 1$, lying on the boundary of E_1. Since $\rho_{\text{inf}} > 0$, we arrive after a finite number of steps at a polar-disk E_n, say, such that the restriction of f to E_n vanishes identically and $(r^*, \theta^*) \in E_n$. This completes the proof. □

3.7.2 The Cauchy Integral Formula for Polar Mellin Derivatives

The Cauchy integral formula in Theorem 2.9 is useful in horizontal strips of width less than 2π only, since otherwise it may produce additional residues as undesired artifacts. Therefore, we establish a further analogue of the Cauchy integral formula which is free of artifacts.

Theorem 3.35 *Let $\mathcal{D}$ be a convex domain in $\mathbb{H}$, and let $f : \mathcal{D} \to \mathbb{C}$ be polar-analytic on $\mathcal{D}$. Let γ be a positively oriented, closed, piecewise smooth curve that is the boundary of a convex domain* $\operatorname{int}(\gamma) \subset \mathcal{D}$. *Then, for $(r_0, \theta_0) \in \operatorname{int}(\gamma)$, $c \in \mathbb{R}$ and $k \in \mathbb{N}_0$, we have*

$$\frac{1}{2\pi i}\int_\gamma \frac{(re^{i\theta})^{c-1} f(r,\theta) e^{i\theta}}{\big(\log(r/r_0) + i(\theta-\theta_0)\big)^{k+1}}(dr + ird\theta) = \big(r_0 e^{i\theta_0}\big)^c \frac{(\widetilde{\Theta}_c^k f)(r_0,\theta_0)}{k!}. \tag{3.27}$$

Proof We decompose $\operatorname{int}(\gamma)$ into four parts as in the proof of Lemma 2.8 (see Fig. 2.1). We recall that the disk is of radius $\varepsilon > 0$ and has a circle λ_ε as boundary, which shall be positively oriented this time. Furthermore β_1, β_2 and β_3 denote the positively oriented boundaries of the other three parts. Note that each β_j lies in a convex subset of $\mathcal{D}$ where the function

$$F : (r,\theta) \longmapsto \frac{(re^{i\theta})^{c-1} f(r,\theta)}{\big(\log(r/r_0) + i(\theta-\theta_0)\big)^{k+1}} \tag{3.28}$$

is polar-analytic. By Theorem 2.6 we have

$$\int_{\beta_j} F(r,\theta)e^{i\theta}(dr + ird\theta) = 0 \quad (j = 1, 2, 3).$$

This enables us to conclude that

$$\frac{1}{2\pi i}\int_\gamma F(r,\theta)e^{i\theta}(dr + ird\theta) = \frac{1}{2\pi i}\int_{\lambda_\varepsilon} F(r,\theta)e^{i\theta}(dr + ird\theta). \tag{3.29}$$

From Theorem 3.32 we know that

$$F(r,\theta) = \big(r_0 e^{i\theta_0}\big)^{c-1} \sum_{\ell=0}^{\infty} \frac{\big(\widetilde{\Theta}_{c-1}^\ell f\big)(r_0,\theta_0)}{\ell!}\left(\log\frac{r}{r_0} + i(\theta-\theta_0)\right)^{\ell-k-1}.$$

Note that on the right-hand side all terms with $\ell \geq k+1$ are polar-analytic as functions of (r, θ). Thus

$$\frac{1}{2\pi i}\int_{\lambda_\varepsilon} F(r,\theta)e^{i\theta}(dr + ird\theta) = \left(r_0e^{i\theta_0}\right)^{c-1}\sum_{\ell=0}^{k}\frac{(\widetilde{\Theta}_{c-1}^{\ell}f)(r_0,\theta_0)}{\ell!}I_{k+1-\ell}, \tag{3.30}$$

where

$$I_j := \frac{1}{2\pi i}\int_{\lambda_\varepsilon}\frac{e^{i\theta}(dr + ird\theta)}{\left(\log(r/r_0) + i(\theta-\theta_0)\right)^j} \qquad (j = 1, \ldots, k+1).$$

Now consider the mapping $(r, \theta) \mapsto z = re^{i\theta}$. If (r, θ) traverses λ_ε once, then, for sufficiently small $\varepsilon > 0$, the point z runs along a piecewise smooth, closed curve γ_ε, say, in $\mathbb{C}$, surrounding $z_0 = r_0e^{i\theta_0}$ exactly once in the mathematically positive sense. Therefore, by Proposition 1.26 we have

$$I_j = \frac{1}{2\pi i}\int_{\gamma_\varepsilon}\frac{dz}{(\log z - \log z_0)^j} = \frac{1}{2\pi i}\int_{\log\circ\gamma_\varepsilon}\frac{e^w dw}{(w-\log z_0)^j} = \frac{r_0e^{i\theta_0}}{(j-1)!}.$$

With this result (3.30) becomes

$$\begin{aligned}\frac{1}{2\pi i}\int_{\lambda_\varepsilon} F(r,\theta)e^{i\theta}(dr + ird\theta) &= \left(r_0e^{i\theta_0}\right)^{c}\sum_{\ell=0}^{k}\frac{(\widetilde{\Theta}_{c-1}^{\ell}f)(r_0,\theta_0)}{\ell!(k-\ell)!}\\ &= \frac{\left(r_0e^{i\theta_0}\right)^{c}}{k!}\sum_{\ell=0}^{k}\binom{k}{\ell}(\widetilde{\Theta}_{c-1}^{\ell}f)(r_0,\theta_0)\\ &= \frac{\left(r_0e^{i\theta_0}\right)^{c}}{k!}(\widetilde{\Theta}_{c}^{k}f)(r_0,\theta_0).\end{aligned}$$

The last conclusion is seen by noting that $\widetilde{\Theta}_c f = \widetilde{\Theta}_{c-1}f + f$ and using this relation repeatedly. This completes the proof. □

3.7.3 The Maximum Principle for Polar-Analytic Functions

Here we discuss polar versions of the classical maximum modulus principle and of Liouville's theorem. Note that our proofs are independent of the classical complex analysis using only tools from the theory of polar-analytic functions that were developed previously.

Theorem 3.36 *Let $\mathcal{D}$ be a domain in $\mathbb{H}$, and let $f : \mathcal{D} \to \mathbb{C}$ be polar-analytic on $\mathcal{D}$. Let γ be a positively oriented, closed, piecewise smooth curve that is the boundary of a convex domain* $\mathrm{int}(\gamma) \subset \mathcal{D}$*. Then*

$$|f(r,\theta)| \leq \max_{(\rho,\vartheta)\in\gamma} |f(\rho,\vartheta)| \quad for\ (r,\theta) \in \operatorname{int}(\gamma). \tag{3.31}$$

Proof Writing $M := \max_{(\rho,\vartheta)\in\gamma} |f(\rho,\vartheta)|$, we assume towards a contradiction that there exists a point $(r_0,\theta_0) \in \operatorname{int}(\gamma)$ such that $|f(r_0,\theta_0)| > M$. Since for any $n \in \mathbb{N}$ the function f^n also satisfies the assumptions of Theorem 3.35, we have by (3.27) with $k = c = 0$ that

$$\big(f(r_0,\theta_0)\big)^n = \frac{1}{2\pi i}\int_\gamma \frac{\big(f(r,\theta)\big)^n}{r\big(\log(r/r_0)+i(\theta-\theta_0)\big)}(dr + ird\theta).$$

Let $\gamma(t) := \big(\rho(t),\vartheta(t)\big)$, $t \in [0,1]$, be a representation of γ, where $\rho(\cdot)$ and $\vartheta(\cdot)$ are piecewise continuously differentiable functions. Then by a standard estimation,

$$|f(r_0,\theta_0)|^n \leq \frac{M^n}{2\pi}\int_0^1 \frac{\big|\rho'(t)+i\rho(t)\vartheta'(t)\big|}{\rho(t)\big|\log(\rho(t)/r_0)+i(\vartheta(t)-\theta_0)\big|}dt.$$

The integral on the right-hand side does not depend on n. Hence taking the n-th root on both sides and letting $n \to \infty$, we obtain that $|f(r_0,\theta_0)| \leq M$; a contradiction. □

Now, using Theorem 3.36, we can establish a Liouville theorem for polar-analytic functions. For this, we say that f is *polar-entire* if it is polar-analytic on the whole $\mathbb{H}$.

Theorem 3.37 *Let $f : \mathbb{H} \to \mathbb{C}$ be a bounded and polar-entire function. Then f is a constant.*

Proof Let $M > 0$ be such that $|f(r,\theta)| \leq M$ for every $(r,\theta) \in \mathbb{H}$. By Theorem 3.32 with $c = 0$, we have the expansion

$$f(r,\theta) = \sum_{k=0}^{\infty} \frac{(\widetilde{\Theta}^k f)(1,0)}{k!}\big(\log r + i\theta\big)^k \qquad \big((r,\theta) \in E_\rho\big), \tag{3.32}$$

converging on every polar-disk $E_\rho \equiv E\big((1,0),\rho\big) = \{(r,\theta) \in \mathbb{H} : \log^2 r + \theta^2 < \rho^2\}$ for all $\rho > 0$. Now consider the function

$$g(r,\theta) := \begin{cases} \dfrac{f(r,\theta) - f(1,0)}{\log r + i\theta} & \text{for } (r,\theta) \in \mathbb{H}\setminus\{(1,0)\}, \\ (\widetilde{\Theta} f)(1,0) & \text{for } (r,\theta) = (1,0). \end{cases}$$

In view of (3.32), we obviously have

$$g(r,\theta) = \sum_{k=1}^{\infty} \frac{(\widetilde{\Theta}^k f)(1,0)}{k!}(\log r + i\theta)^{k-1}.$$

The above series has the same convergence properties as the latter. Therefore g is a polar-entire function. From the definition of g we have

$$\max_{(r,\theta)\in\partial E_\rho} |g(r,\theta)| \leq \frac{2M}{\rho} \qquad (\rho > 0),$$

and so by Theorem 3.36 we have

$$|g(r,\theta)| \leq \frac{2M}{\rho} \quad \text{for } (r,\theta) \in E_\rho.$$

Letting $\rho \to \infty$, we find that $g(r,\theta) \equiv 0$ on $\mathbb{H}$. This implies that $f(r,\theta) = f(1,0)$ on $\mathbb{H}$, and so f is a constant. □

3.7.4 A Residue Theorem for Polar-Analytic Functions

First we introduce some notions.

Definition 3.38 Let $(r_0, \theta_0) \in \mathbb{H}$ and let $\mathcal{U} \subset \mathbb{H}$ be an open neighborhood of (r_0, θ_0).

(i) If $f : \mathcal{U} \setminus \{(r_0,\theta_0)\} \to \mathbb{C}$ is polar-analytic, then (r_0, θ_0) will be called an *isolated singularity* of f.
(ii) An isolated singularity (r_0, θ_0) is said to be a *pole of order k in logarithmic representation*, or simply a *pole of order k*, if $k \in \mathbb{N}$ and there exists a polar-analytic function $g : \mathcal{U} \to \mathbb{C}$ with $g(r_0, \theta_0) \neq 0$ such that

$$f(r,\theta) = \frac{g(r,\theta)}{\big(\log(r/r_0) + i(\theta-\theta_0)\big)^k} \quad \text{for } (r,\theta) \in \mathcal{U} \setminus \{(r_0,\theta_0)\}.$$

In this case

$$\big(\mathrm{res}_c f\big)(r_0,\theta_0) := \big(r_0 e^{i\theta_0}\big)^c \, \frac{\big(\widetilde{\Theta}_c^{k-1} g\big)(r_0,\theta_0)}{(k-1)!}$$

will be called the *c-residue* of f at (r_0, θ_0).

We are now in a position to formulate a residue theorem for poles in logarithmic representation which will be suitable for many applications in Mellin analysis.

Theorem 3.39 *Let $\mathcal{D}$ be a convex domain in $\mathbb{H}$ and let f be polar-analytic on $\mathcal{D}$ except for isolated singularities which are all poles. Let γ be a positively oriented, closed, piecewise smooth curve that is the boundary of a convex domain* $\mathrm{int}(\gamma) \subset \mathcal{D}$. *Suppose that no isolated singularity lies lies on γ while (r_j, θ_j) for $j = 1, \ldots, m$ are the isolated singularities lying in* $\mathrm{int}(\gamma)$. *Then, for $c \in \mathbb{R}$, there holds*

$$\int_{\gamma} \left(re^{i\theta}\right)^{c-1} f(r,\theta)e^{i\theta}\,(dr + ird\theta) \;=\; 2\pi i \sum_{j=1}^{m} \left(\mathrm{res}_c f\right)(r_j,\theta_j).$$

Proof For $(x_0, y_0) \in \mathbb{R}^2$ and $\varepsilon > 0$, we consider the vertical grid

$$V(x_0,\varepsilon) \;:=\; \left\{(r,\theta) \in \mathbb{R}^2 \;:\; \theta \in \mathbb{R},\; r = x_0 + n\varepsilon,\; n \in \mathbb{Z}\right\}$$

and the horizontal grid

$$H(y_0,\varepsilon) \;:=\; \left\{(r,\theta) \in \mathbb{R}^2 \;:\; r \in \mathbb{R},\; \theta = y_0 + n\varepsilon,\; n \in \mathbb{Z}\right\}.$$

Their union constitutes a net

$$N\big((x_0, y_0),\varepsilon\big) \;:=\; V(x_0,\varepsilon) \cup H(y_0,\varepsilon).$$

The intersection $\mathrm{int}(\gamma) \cap N\big((x_0, y_0),\varepsilon\big)$ creates a tessellation of $\mathrm{int}(\gamma)$ into squares with edges of length ε and further convex sets whose boundary contains a piece of γ. The latter may be called boundary sets. The collection of all subsets of the tessellation shall be denoted by $\mathscr{T}$. If $\varepsilon > 0$ is sufficiently small, then the boundary sets will be free of isolated singularities and different isolated singularites in $\mathrm{int}(\gamma)$ will lie in different squares of the tessellation. Furthermore, if (r_j, θ_j) lies in the interior of a square, then it will remains there under all sufficiently small variations of (x_0, y_0). But if (r_j, θ_j) lies on the boundary of a square, then there exist arbitrarily small variations such that (r_j, θ_j) goes inside. We may therefore assume that each (r_j, θ_j) lies inside a square $Q_j \in \mathscr{T}$ for $j = 1, \ldots, m$. For each set $P \in \mathscr{T}$, we denote by ∂P its positively oriented boundary. Note that each line segment in $\mathrm{int}(\gamma)$ which comes from a mesh of the net belongs to the boundaries of exactly two subsets of the tessellation where it occurs with opposite orientations. Therefore

$$\begin{aligned}
\int_{\gamma} \left(re^{i\theta}\right)^{c-1} f(r,\theta)e^{i\theta}(dr + ird\theta) &= \sum_{P\in\mathcal{T}} \int_{\partial P} \left(re^{i\theta}\right)^{c-1} f(r,\theta)e^{i\theta}(dr + ird\theta) \\
&= \sum_{j=1}^{m} \int_{\partial Q_j} \left(re^{i\theta}\right)^{c-1} f(r,\theta)e^{i\theta}(dr + ird\theta)
\end{aligned}$$

since by Theorem 2.6 the integral along ∂P vanishes if P does not contain an isolated singularity. As all our isolated singularities are poles, we conclude with the help of Theorem 3.35 and by using the notion of residue that

$$\int_{\partial Q_j} \left(re^{i\theta}\right)^{c-1} f(r,\theta)e^{i\theta}(dr + ird\theta) \;=\; 2\pi i\left(\mathrm{res}_c f\right)(r_j,\theta_j) \quad (j = 1, \ldots, m).$$

This completes the proof. □

Remark 3.40 The proof of Theorem 3.39 reveals that, if

$$f(r,\theta) = \frac{g(r,\theta)}{(\log(r/r_0) + i(\theta-\theta_0))^k} = \frac{h(r,\theta)}{(\log(r/r_0) + i(\theta-\theta_0))^{k-j}}$$

with $j \in \mathbb{N}$, $h(r_0, \theta_0) \neq 0$ and (consequently) $g(r_0, \theta_0) = 0$, then

$$\big(\mathrm{res}_c f\big)(r_0,\theta_0) := \big(r_0 e^{i\theta_0}\big)^c \, \frac{(\widetilde{\Theta}_c^{k-j-1} h)(r_0,\theta_0)}{(k-j-1)!} = \big(r_0 e^{i\theta_0}\big)^c \, \frac{(\widetilde{\Theta}_c^{k-1} g)(r_0,\theta_0)}{(k-1)!}.$$

In other words, if f has a pole of order at most k, we may naively use the formula for a pole of order k for calculating the c-residue.

3.8 Mellin–Bernstein Spaces for Polar-Analytic Functions

Now we introduce the counterpart of the classical Bernstein spaces of Fourier analysis in Mellin frame without employing the Riemann surface $S_{\log}$ of the logarithm and analytic branches, but using polar analyticity.

Definition 3.41 For $c \in \mathbb{R}$, $T \in \mathbb{R}^+$ and $p \in [1, \infty]$, the *Mellin–Bernstein space* $\mathscr{B}^p_{c,T}$ comprises all functions $f : \mathbb{H} \to \mathbb{C}$ with the following properties:

(i) f is polar-analytic on $\mathbb{H}$;
(ii) $f(\cdot, 0) \in X^p_c$;
(iii) there exists a constant $C_f > 0$ such that $r^c\,|f(r,\theta)| \le C_f e^{T|\theta|}$ for all $(r,\theta) \in \mathbb{H}$.

3.8.1 *Basic Properties*

From the definition, we have two useful consequences, which may be presented as follows.

Theorem 3.42 *Let $f \in \mathscr{B}^p_{c,T}$ with $p \in [1, \infty]$. Then the following statements hold:*

(a) $f(\cdot,\theta) \in X^p_c$ for all $\theta \in \mathbb{R}$ and $\|f(\cdot,\theta)\|_{X^p_c} \le e^{T|\theta|}\|f(\cdot,0)\|_{X^p_c}$;
(b) if $p \neq \infty$, then $\lim_{r\to 0+} r^c f(r,\theta) = \lim_{r\to\infty} r^c f(r,\theta) = 0$ uniformly with respect to θ on all compact subinterval of $\mathbb{R}$.

Proof The function $g : x + iy \mapsto f(e^x, y)$ is defined on $\mathbb{C}$. Writing $f = u + iv$ with real-valued functions u and v, we know from the polar analyticity of f that u and v have continuous partial derivatives on $\mathbb{H}$ which satisfy Eqs. (2.4). This implies that g satisfies the Cauchy–Riemann equations on $\mathbb{C}$, and so g is an entire function.

Hence

$$F : x + iy \longmapsto e^{c(x+iy)} f(e^x, y)$$

is also an entire function. Property (iii) of Definition 3.41 may now be rewritten in terms of F as

$$|F(x + iy)| \le C_f e^{T|y|}. \tag{3.33}$$

Moreover, property (ii) of Definition 3.41 yields

$$\|F\|_p = \|f(\cdot, 0)\|_{X_c^p} < \infty,$$

and so F belongs to the Bernstein space B_T^p (see Definition 1.34). Now Proposition 1.35 tells us that

$$\|F(\cdot + iy)\|_p \le e^{T|y|} \|F\|_p \tag{3.34}$$

for all $y \in \mathbb{R}$. Substituting $y = \theta$ and expressing F in terms of f, we readily obtain statement (a).

For $p \ne \infty$, we may write (3.34) as

$$\int_{-\infty}^{\infty} |F(x + iy)|^p dx \le e^{pT|y|} \int_{-\infty}^{\infty} |F(x)|^p dx$$

for all $y \in \mathbb{R}$ and know in addition that $F(x) \to 0$ as $x \to \pm\infty$. This inequality in turn implies that $F(x + iy) \to 0$ as $x \to \pm\infty$ pointwise for each $y \in \mathbb{R}$. In connection with (3.33), Proposition 1.31 guarantees that the convergence is even uniform with respect to y on compact subintervals of $\mathbb{R}$. Writing this conclusion in terms of f, we obtain statement (b). □

We now state three useful propositions on transformations in Mellin–Bernstein spaces. They are verified by straightforward calculations. To show polar analyticity, we may simply verify statement (ii) of Theorem 2.5. The first proposition affirms that $\mathscr{B}^p_{c,T}$ is invariant under Mellin translations.

Proposition 3.43 *Let $f \in \mathscr{B}^p_{c,T}$, where $p \in [1, \infty]$, $c \in \mathbb{R}$ and $T \in \mathbb{R}^+$. For $t > 0$, define*

$$g : (r, \theta) \longmapsto t^c f(tr, \theta).$$

Then $g \in \mathscr{B}^p_{c,T}$; in particular, $\|g(\cdot, 0)\|_{X_c^p} = \|f(\cdot, 0)\|_{X_c^p}$ and $r^c |g(r, \theta)| \le C_f e^{T|\theta|}$ for all $(r, \theta) \in \mathbb{H}$. Furthermore,

$$(\widetilde{\Theta}_c g)(r, \theta) = t^c (\widetilde{\Theta}_c f)(tr, \theta).$$

The second proposition concerns a transformation of T.

Proposition 3.44 *Let $f \in \mathscr{B}^p_{c,T}$, where $p \in [1, \infty]$, $c \in \mathbb{R}$ and $T \in \mathbb{R}^+$. Define*

$$h : (r, \theta) \longmapsto f(r^{1/T}, \theta/T).$$

Then $h \in \mathscr{B}^p_{c/T,1}$; in particular, $\|h(\cdot, 0)\|_{X^p_{c/T}} = \|f(\cdot, 0)\|_{X^p_c}$ and $r^{c/T}\,|h(r, \theta)| \leq C_f e^{|\theta|}$ for all $(r, \theta) \in \mathbb{H}$. Furthermore,

$$\big(\widetilde{\Theta}_{c/T} h\big)(r, \theta) = \frac{1}{T}\big(\widetilde{\Theta}_c f\big)\left(r^{1/T}, \frac{\theta}{T}\right).$$

The third proposition concerns a shift of the second argument of f. For verifying property (ii) in the definition of Mellin–Bernstein spaces, we may use Theorem 3.42.

Proposition 3.45 *Let $f \in \mathscr{B}^p_{c,T}$, where $p \in [1, \infty]$, $c \in \mathbb{R}$ and $T \in \mathbb{R}^+$. For $\alpha \in \mathbb{R}$, define*

$$\phi : (r, \theta) \longmapsto f(r, \theta + \alpha).$$

Then $\phi \in \mathscr{B}^p_{c,T}$; in particular, $\|\phi(\cdot, 0)\|_{X^p_c} \leq e^{T|\alpha|}\,\|f(\cdot, 0)\|_{X^p_c}$ and $r^c\,|\phi(r, \theta)| \leq C_f e^{T(|\alpha|+|\theta|)}$ for all $(r, \theta) \in \mathbb{H}$. Furthermore,

$$\big(\widetilde{\Theta}_c \phi\big)(r, \theta) = \big(\widetilde{\Theta}_c f\big)(r, \theta + \alpha).$$

3.8.2 Boas' Differentiation Formula and Bernstein Inequality

Here we present a formula for the polar Mellin derivatives. It is analogous to a differentiation formula of Boas for bandlimited functions.

Theorem 3.46 *Let $f \in \mathscr{B}^p_{c,T}$, where $p \in [1, \infty]$, $c \in \mathbb{R}$ and $T \in \mathbb{R}^+$. Then*

$$\big(\widetilde{\Theta}_c f\big)(r, \theta) = \frac{4T}{\pi^2}\sum_{k\in\mathbb{Z}} \frac{(-1)^k}{(2k+1)^2}\, e^{(k+1/2)\pi c/T} f\big(re^{(k+1/2)\pi/T}, \theta\big) \tag{3.35}$$

for $(r, \theta) \in \mathbb{H}$. Multiplied by r^c, the series converges absolutely and uniformly on strips of bounded width parallel to the r-axis in $\mathbb{H}$.

Proof For simplicity, we first suppose that $T = 1$. Consider the function

$$F : (r, \theta) \longmapsto \frac{f(r, \theta)}{(\log r + i\theta)^2 \cos(\log r + i\theta)}.$$

It is polar-analytic on $\mathbb{H}$ except at the points where the denominator vanishes. Writing $r_k := e^{(k+1/2)\pi}$ for short, we obtain the exceptional points as $(1, 0)$ and $(r_k, 0)$ for $k \in \mathbb{Z}$. The point $(1, 0)$ is a pole of order at most two while all the other exceptional points are poles of order at most one.

Next, for $n \in \mathbb{N}$, let $\mathcal{R}_n$ be the rectangle with vertices at $(e^{\pm n\pi}, \pm n\pi)$ and denote by $\partial\mathcal{R}_n$ its positively oriented boundary. Employing Theorem 3.39, we find that

$$\frac{1}{2\pi i}\int_{\partial\mathcal{R}_n}\left(re^{i\theta}\right)^{c-1}F(r,\theta)e^{i\theta}(dr + ird\theta) = \left(\operatorname{res}_c F\right)(1,0) + \sum_{k=-n}^{n-1}\left(\operatorname{res}_c F\right)(r_k,0). \tag{3.36}$$

Introducing

$$\phi(r,\theta) := \frac{f(r,\theta)}{\cos(\log r + i\theta)},$$

we have

$$\left(\operatorname{res}_c F\right)(1,0) = \left(\widetilde{\Theta}_c\phi\right)(1,0) = \left(\widetilde{\Theta}_c f\right)(1,0).$$

For calculating the other c-residues, we factor the cosine as

$$\cos(\log r + i\theta) = \left(\log r + i\theta - (k + \tfrac{1}{2})\pi\right)\psi_k(r,\theta),$$

where ψ_k is polar-analytic in a neighborhood of $(r_k, 0)$ and

$$\psi_k(r_k,0) = \lim_{(r,\theta)\to(r_k,0)}\frac{\cos(\log r + i\theta)}{\log r + i\theta - (k+\frac{1}{2})\pi} = (-1)^{k+1}.$$

Thus,

$$\left(\operatorname{res}_c F\right)(r_k,0) = \frac{r_k^c f(r_k,0)}{\left((k+\frac{1}{2})\pi\right)^2\psi(r_k,0)} = \frac{4}{\pi^2}\frac{(-1)^{k+1}}{(2k+1)^2}r_k^c f(r_k,0).$$

With these values of the c-residues, we may rewrite (3.36) as

$$\begin{aligned}\left(\widetilde{\Theta}_c f\right)(1,0) = {}& \frac{4}{\pi^2}\sum_{k=-n}^{n-1}\frac{(-1)^k}{(2k+1)^2}r_k^c f(r_k,0)\\ &+ \frac{1}{2\pi i}\int_{\partial\mathcal{R}_n}\left(re^{i\theta}\right)^{c-1}F(r,\theta)e^{i\theta}(dr + ird\theta).\end{aligned}$$

Next we want to estimate the integral on the right-hand side. Noting that

$$|\cos(\log r + i\theta)| > \frac{e^{|\theta|}}{3}$$

on $\partial\mathcal{R}_n$ and employing property (iii) of functions belonging to $\mathscr{B}^p_{c,1}$, we see that

$$r^{c-1}\,|F(r,\theta)| \le \frac{3C_f}{r(\log^2 r + \theta^2)}$$

for $(r,\theta) \in \partial\mathcal{R}_n$. Considering the integrals along the vertical and the horizontal line segments of $\partial\mathcal{R}_n$ separately, we find that

$$\begin{aligned}
&\left|\frac{1}{2\pi i}\int_{\partial\mathcal{R}_n}\left(re^{i\theta}\right)^{c-1}F(r,\theta)e^{i\theta}(dr + ir\,d\theta)\right| \\
&\qquad \le \frac{3C_f}{\pi}\left(\int_{-n\pi}^{n\pi}\frac{d\theta}{(n\pi)^2+\theta^2} + \int_{e^{-n\pi}}^{e^{n\pi}}\frac{1}{\log^2 r + (n\pi)^2}\,\frac{dr}{r}\right) \\
&\qquad = \frac{6C_f}{n\pi^2}\int_{-1}^{1}\frac{dx}{1+x^2} \longrightarrow 0 \qquad (n \to \infty).
\end{aligned}$$

Thus

$$\left(\widetilde{\Theta}_c f\right)(1,0) = \frac{4}{\pi^2}\sum_{k=-\infty}^{\infty}\frac{(-1)^k}{(2k+1)^2}\,r_k^c f(r_k,0)$$

for $f \in \mathscr{B}^p_{c,1}$.

With the Mellin translation of Proposition 3.43, we deduce

$$\left(\widetilde{\Theta}_c f\right)(t,0) = \frac{4}{\pi^2}\sum_{k=-\infty}^{\infty}\frac{(-1)^k}{(2k+1)^2}\,r_k^c f(tr_k,0) \qquad (t>0),$$

valid for $f \in \mathscr{B}^p_{c,1}$.

Next we want to extend this formula to $f \in \mathscr{B}^p_{c,T}$ with an arbitrary $T > 0$. For this, we note that the formula is valid for the function h defined in Proposition 3.44 provided that we replace c with c/T. Expressing h in terms of f, we obtain

$$\left(\widetilde{\Theta}_c f\right)(t^{1/T},0) = \frac{4T}{\pi^2}\sum_{k=-\infty}^{\infty}\frac{(-1)^k}{(2k+1)^2}\,r_k^{c/T} f(t^{1/T}r_k^{1/T},0) \qquad (t>0),$$

now valid for $f \in \mathscr{B}^p_{c,T}$. In this formula, we may replace the variable t with t^T on both sides. Furthermore, we may employ Proposition 3.45 for extending the argument of the polar Mellin derivative from $(t, 0)$ to an arbitrary point $(t, \alpha) \in \mathbb{H}$. This leads us to

$$(\widetilde{\Theta}_c f)(t, \alpha) = \frac{4T}{\pi^2} \sum_{k=-\infty}^{\infty} \frac{(-1)^k}{(2k+1)^2} r_k^{c/T} f(tr_k^{1/T}, \alpha)$$

for $f \in \mathscr{B}^p_{c,T}$ and all $(t, \alpha) \in \mathbb{H}$, which is identical with (3.35).

It remains to justify the statement on convergence. In order to do that, we first note that by Proposition 1.36 applied to the function $f(z) = \sin z$ with $\sigma = 1$, we obtain

$$\frac{4}{\pi^2} \sum_{k=-\infty}^{\infty} \frac{1}{(2k+1)^2} = 1. \tag{3.37}$$

Furthermore, by property (iii) in the definition of $\mathscr{B}^p_{c,T}$, there holds

$$t^c r_k^{c/T} \left| f\left(tr_k^{1/T}, \alpha\right) \right| \le C_f \, e^{T|\alpha|} \qquad (k \in \mathbb{Z}).$$

Now the assertion on absolute and uniform convergence is readily verified. □

As a consequence of Theorem 3.46, we can establish a Bernstein inequality for polar Mellin derivatives.

Corollary 3.47 *Let $f \in \mathscr{B}^p_{c,T}$, where $p \in [1, \infty]$, $c \in \mathbb{R}$ and $T > 0$. Then*

$$\left\| (\widetilde{\Theta}_c f)(\cdot, \theta) \right\|_{X_c^p} \le T \, \| f(\cdot, \theta) \|_{X_c^p} \tag{3.38}$$

for any $\theta \in \mathbb{R}$.

Proof Let us write $r_k = e^{(k+1/2)\pi}$, where $k \in \mathbb{Z}$, for short. Multiplying (3.35) by r^c, taking moduli on both sides, applying the triangular inequality on the right-hand side and using (3.37), we obtain

$$\begin{aligned} r^c \left| (\widetilde{\Theta}_c f)(r, \theta) \right| &\le \frac{4T}{\pi^2} \sum_{k \in \mathbb{Z}} \frac{1}{(2k+1)^2} r^c r_k^{c/T} \left| f(rr_k^{1/T}, \theta) \right| \\ &\le \frac{4T}{\pi^2} \sum_{k \in \mathbb{Z}} \frac{1}{(2k+1)^2} \| f(\cdot, \theta) \|_{X_c^\infty} \\ &= T \, \| f(\cdot, \theta) \|_{X_c^\infty} \end{aligned} \tag{3.39}$$

for any $(r, \theta) \in \mathbb{H}$. This implies (3.38) for $p = \infty$.

Now let $1 \le p < \infty$ and let $R > 0$. Observing that

$$a^c \| f(a \cdot, \theta) \|_{X_c^p} = \| f(\cdot, \theta) \|_{X_c^p}$$

for $a > 0$, we conclude from (3.39) using the triangular inequality for norms and Beppo Levi's theorem that

$$\begin{aligned}
&\left(\int_{1/R}^{R} r^{cp} \left|(\widetilde{\Theta}_c f)(r,\theta)\right|^p \frac{dr}{r}\right)^{1/p} \\
&\leq \left(\int_{0}^{\infty} r^{cp} \left|\frac{4T}{\pi^2}\sum_{k\in\mathbb{Z}} \frac{r_k^{c/T}}{(2k+1)^2} f(r r_k^{1/T},\theta)\right|^p \frac{dr}{r}\right)^{1/p} \\
&= \frac{4T}{\pi^2}\left\|\sum_{k\in\mathbb{Z}} \frac{r_k^{c/T}}{(2k+1)^2} f(r_k^{1/T}\,\cdot,\theta)\right\|_{X_c^p} \\
&\leq \frac{4T}{\pi^2}\sum_{k\in\mathbb{Z}} \frac{r_k^{c/T}}{(2k+1)^2} \left\|f(r_k^{1/T}\,\cdot,\theta)\right\|_{X_c^p} \\
&= \frac{4T}{\pi^2}\sum_{k\in\mathbb{Z}} \frac{1}{(2k+1)^2} \left\|f(\cdot,\theta)\right\|_{X_c^p} \\
&= T\left\|f(\cdot,\theta)\right\|_{X_c^p}.
\end{aligned}$$

This guarantees that the X_c^p norm of $(\widetilde{\Theta}_c f)(\cdot,\theta)$ exists. Letting $R \to \infty$, we obtain (3.38) for $p \in [1,\infty[$. □

Corollary 3.48 *The Mellin–Bernstein space is closed under polar Mellin differentiation, that is, if* $f \in \mathscr{B}_{c,T}^p$, *where* $p \in [1,\infty]$, $c \in \mathbb{R}$ *and* $T \in \mathbb{R}^+$, *then* $\widetilde{\Theta}_c f \in \mathscr{B}_{c,T}^p$.

Proof We have to show that $\widetilde{\Theta}_c f$ satisfies properties (i)–(iii) in the definition of Mellin–Bernstein spaces. Since by Theorem 2.11 a polar-analytic function f has polar derivatives of arbitrary order, it follows immediately that $\widetilde{\Theta}_c f$ is polar-analytic on $\mathbb{H}$, and so (i) is satisfied. Property (ii) is guaranteed by Corollary 3.47. Finally, knowing that (iii) holds for f, we deduce from (3.35) that

$$r^c\left|(\widetilde{\Theta}_c f)(r,\theta)\right| \leq \frac{4T}{\pi^2}\sum_{k\in\mathbb{Z}} \frac{C_f\, e^{T|\theta|}}{(2k+1)^2} = T C_f e^{T|\theta|}.$$

Hence (iii) holds for $\widetilde{\Theta}_c f$ with $C_{\widetilde{\Theta}_c f} = T C_f$. □

Combining Corollaries 3.47 and 3.48 and applying (3.38) repeatedly, we find that

$$\left\|(\widetilde{\Theta}_c^k f)(\cdot,\theta)\right\|_{X_c^p} \leq T^k \left\|f(\cdot,\theta)\right\|_{X_c^p} \qquad (\theta \in \mathbb{R}, k \in \mathbb{N}).$$

3.9 Mellin–Hardy Spaces

In this section we introduce Hardy spaces for polar-analytic functions as the Mellin analogue of the classical Hardy spaces considered in Sect. 1.4.4. With reference to (2.7), the basic domains will be strips $\mathbb{H}_{-a,a}$ with $a > 0$ which we simply denote by $\mathbb{H}_a$.

Definition 3.49 Let $a, c, p \in \mathbb{R}$ with $a > 0$ and $p \geq 1$. The *Mellin–Hardy space* $H_c^p(\mathbb{H}_a)$ comprises all functions $f : \mathbb{H}_a \to \mathbb{C}$ that satisfy the following conditions:

(i) f is polar-analytic on $\mathbb{H}_a$;
(ii) $f(\cdot, \theta) \in X_c^p$ for each $\theta \in]-a, a[$;
(iii) there holds

$$\|f\|_{H_c^p(\mathbb{H}_a)} := \sup_{0<\theta<a} \left(\frac{\|f(\cdot,\theta)\|_{X_c^p}^p + \|f(\cdot,-\theta)\|_{X_c^p}^p}{2} \right)^{1/p} < \infty.$$

When $a \in]0, \pi]$ we can associate with each function $f \in \mathbb{H}_a$ a function g analytic on the sector $S_a := \{z \in \mathbb{C} : |\arg z| < a\}$ by defining $g(re^{i\theta}) := f(r, \theta)$. The collection of all such functions constitutes a Hardy type space $H_c^p(S_a)$, which may be identified with $H_c^p(\mathbb{H}_a)$.

In Sect. 1.4.4, we have considered the classical Hardy space $H^p(\mathcal{S}_a)$ of analytic functions on the strip $\mathcal{S}_a := \{z \in \mathbb{C} : |\Im z| < a\}$ (see Definition 1.40). For a comparison, we mention that the Mellin–Hardy space $H_c^p(\mathbb{H}_a)$ has been designed in such a way that if $g \in H^p(\mathcal{S}_a)$ and

$$f(r, \theta) := r^{-c} e^{-ic\theta} g(\log r + i\theta),$$

then $f \in H_c^p(\mathbb{H}_a)$, and conversely, if $f \in H_c^p(\mathbb{H}_a)$ and

$$g(x + iy) := e^{c(x+iy)} f(e^x, y),$$

then $g \in H^p(\mathcal{S}_a)$.

Using this correspondence, one can deduce the following propositions from the analogous ones for the classical Hardy space (see Propositions 1.41–1.43).

Proposition 3.50 *Let $f \in H_c^p(\mathbb{H}_a)$ and let $a_1 \in]0, a[$. Then*

$$|f(r, \theta)| \leq r^{-c} \left(\frac{4}{\pi(a - a_1)} \right)^{1/p} \|f\|_{H_c^p(\mathbb{H}_a)}$$

for $(r, \theta) \in \mathbb{H}_{a_1}$.

The next proposition is a Nikol'ski type inequality for the Hardy space $H_c^p(\mathbb{H}_a)$.

Proposition 3.51 *Let $f \in H_c^p(\mathbb{H}_a)$ and let $(t_n)_{n\in\mathbb{Z}}$ be a sequence on $\mathbb{R}^+$ such that $t_{n+1}/t_n > e^{2\delta}$ for all $n \in \mathbb{Z}$ with $\delta \in]0, a[$. Then*

$$\left(\sum_{n\in\mathbb{Z}} t_n^{cp}|f(t_n, 0)|^p\right)^{1/p} \leq \left(\frac{2}{\pi\delta}\right)^{1/p} \|f\|_{H_c^p(\mathbb{H}_a)}.$$

Proposition 3.52 *Let $f \in H_c^p(\mathbb{H}_a)$ and let $a_1 \in]0, a[$. Then*

$$\lim_{r\to 0+} r^c f(r, \theta) = \lim_{r\to\infty} r^c f(r, \theta) = 0$$

uniformly for $\theta \in [-a_1, a_1]$.

The proofs are left to the reader.

3.10 Notes

The weighted Lebesgue spaces on the positive real axis, as introduced in Section 3.1, are the natural function spaces for which the Mellin transform can be defined. They were extensively studied by Butzer and Jansche for $p = 1$ in [89] and for $p = 2$ in [92]. Proposition 3.4 is borrowed from Bardaro et al. [63].

The notion of recurrency was initiated by Butzer and Jansche in [91] in order to establish a suitable counterpart of periodicity in the case when the underlying spaces have the structure of a multiplicative group, the framework of Mellin analysis. The spaces Y_c^p were introduced by these authors in [91] and [92] for $p = 1$ and $p = 2$, respectively.

The Mellin translation operator, considered in Sect. 3.2, was established by Butzer and Jansche in [89]. Proposition 3.13 was proved in [89] for $p = 1$ and in [92] for $p = 2$ by a different approach. Here we have extended the statement to all $p \geq 1$ by using the density result of Proposition 3.4. The notion of log-uniformly continuous functions originates from earlier work of Mamedov [208]. It is modeled over the multiplicative structure of the positive real line, and it is strictly related to the Mellin translation operator.

The Mellin–Lipschitz classes, considered in Sect. 3.3, were introduced by Bardaro et al. in [56, Sect. 4] in connection with the Mellin distance. They are defined as a Mellin analogue of the Lipschitz spaces widely studied in many books and papers on Fourier analysis and approximation theory (see, e.g., Butzer and Nessel [95], Butzer et al. [116]).

The Mellin differential operator was first considered by Mamedov [208] for $c = 0$. The generalized form of Sect. 3.4 originates from Butzer and Jansche [89]. For the definitions and the properties of the Stirling numbers or (more generally) Stirling functions of the first and second kind, see, e.g., Platonov [244], Hauss [168], Butzer and Hauss [87], Butzer et al. [109, 111]. For $c = 0$ see also the books by Jordan [181] and Comtet [130, Chap. V, Sects. 5.1, 5.2]. The representation formulae involving the Stirling numbers of the first and second kind were proved by Butzer and Jansche in [89].

The Mellin–Sobolev spaces of Sect. 3.5 were introduced for $p = 1$ by Butzer and Jansche in [89] as the natural analogue of the classical Sobolev spaces in the Mellin setting. Proposition 3.24 is borrowed from that paper.

The definition of Mellin–Bernstein spaces involving the Riemann surface of the (complex) logarithm and Bernstein's inequality for Mellin derivatives, stated in Sect. 3.6, were performed by Bardaro et al. [55] in conjunction with a technique of Browder [78].

The complex analysis for polar Mellin derivatives, presented in Sect. 3.7, was developed by Bardaro et al. in [64, 66]. These papers generalize and complete some of the authors earlier results in [62]. More precisely, Theorem 3.31 is already contained in [62, Theorem 6.1]. The more general Taylor-like expansion of Theorem 3.32, the Cauchy integral formula and the residue theorem as well as their consequences were first obtained in [64]. The maximum principle for polar-analytic functions and the corresponding Liouville-like theorem (Theorems 3.36 and 3.37) originate from [66]. This paper also contains a polar-analytic analogue of Rouché's theorem, even in the strengthened form of Estermann [148, p. 156]. We have developed the complex analysis of polar-analytic functions only as far as it is needed in this book. For theoretical reasons further considerations may be of interest. For example, one could extend the analogues of Cauchy's integral formula and the residue theorem (Theorems 3.35 and 3.39) from convex domains to simply connected domains and admit essential isolated singularities in the latter statement.

The Mellin–Bernstein spaces for polar-analytic functions introduced in Sect. 3.8 originate from Bardaro et al. [58, Sect. 4]. Theorem 3.42 is a revised version of [58, Theorem 4.2]. Propositions 3.43–3.45, the Mellin analogue of Boas' differentiation formula, the consequent Bernstein inequality for polar Mellin derivatives and the closure of the Mellin–Bernstein spaces under polar Mellin differentiation (Theorem 3.46 and Corollaries 3.47–3.48) were again obtained in [64, Sect. 5]. In its classical form, Boas' differentiation formula originates from Boas [72]. For extensions to higher derivatives and computational applications, see Butzer et al. [116] and Schmeisser [271].

The material of Sect. 3.9 on Mellin–Hardy spaces originates from Bardaro et al. [58, Sect. 6]. For the theory of classical Hardy spaces in Fourier analysis, see Folland [152], Bakan and Kaijser [26], Rosenblum and Rovnyak [263]. Propositions 3.50–3.52 are Mellin analogues of results on classical Hardy spaces achieved by Butzer et al. [116].

The notions of Mellin derivative and of polar Mellin derivative can be viewed as particular cases of the infinitesimal generator of one-parameter groups of strongly continuous operators. For an extensive treatment of the theory of semigroups of operators and its applications to approximation theory, we refer to the book of Butzer and Berens [86], and in connection with sampling theory to the work of Pesenson in [239–241].

As to the Mellin derivative, we consider the family of operators in X_c^p defined by the Mellin translation, that is, $T(t)f = \tau_t^c f$ or $T(t)f(x) = t^c f(tx)$ for $f \in X_c^p$, $x \in \mathbb{R}^+$ and $t \in \mathbb{R}^+$. It is easy to see that $T(t\tau)f = T(t)T(\tau)f$ for every $t, \tau \in \mathbb{R}^+$. In other words, $T(t)$ is a strongly continuous group of operators (isometries). The Mellin differential operator $\Theta_c f$ is exactly the infinitesimal generator of the group $T(t)$. This follows immediately from the equality

$$\frac{d}{dt}T(t)f(x)\Big|_{t=1} = xf'(x) + cf(x) \qquad (x \in \mathbb{R}^+)$$

whenever f has a derivative in the ordinary sense.

As to the polar Mellin derivative for a function $f : \mathbb{H} \to \mathbb{C}$, we set $f_\theta(r) = f(r, \theta)$ for $r \in \mathbb{R}^+$ and $\theta \in \mathbb{R}$. If $f_\theta \in X_c^p$ for any θ, we define the group operators $U(t)f_\theta(r) = e^{ct} f_\theta(r)$ with $t \in \mathbb{R}$. The family $U(t)$ is again a strongly continuous family of operators satisfying $U(t + \tau) = U(t)U(\tau)$. Its infinitesimal generator is given by

$$\frac{d}{dt}U(t)f_\theta(r)\Big|_{t=0} = r\frac{\partial f}{\partial r}(r, \theta) + cf(r, \theta) = re^{i\theta}(D_{\mathrm{pol}}f)(r, \theta) + cf(r, \theta) = \widetilde{\Theta}_c f(r, \theta).$$

With these interpretations, Mellin analogues of Boas' differentiation formula and Bernstein's inequality can be deduced from their classical versions via the setting of a one-parameter group of isometries, as was shown by Pesenson in [239] and [240].

Based on operator theory, Pesenson also presented an approach to multidimensional Mellin analysis. From his numerous results, we mention in view of Sect. 3.8 that in (multidimensional) Mellin analysis membership of a function in a Mellin–Bernstein space can be characterized by the validity of a Bernstein inequality and also by the validity of a Boas differentiation formula (see Pesenson [241, Theorems 8.2, 8.3]).

Chapter 4
The Mellin Convolution

For the further development of the Mellin analysis, the convolution structure in the spaces X_c^p with $1 \leq p \leq \infty$ and $c \in \mathbb{R}$ and in the spaces Y_c^p of recurrent functions is very important.

4.1 The Mellin Convolution in the Spaces X_c^p

We start with the following notion.

Definition 4.1 Given two Lebesgue measurable functions $f, g : \mathbb{R}^+ \to \mathbb{C}$, the *Mellin convolution* $f * g$ is defined by

$$(f * g)(x) := \int_0^\infty f\left(\frac{x}{u}\right) g(u) \frac{du}{u} \qquad (x \in \mathbb{R}^+)$$

provided that the integral exists.

Note that in the setting of X_c^p spaces with $c \in \mathbb{R}$, we can write the convolution $f * g$ in the form

$$(f * g)(x) = \int_0^\infty (\tau_{1/u}^c f)(x) g(u) u^c \, \frac{du}{u} \qquad (x \in \mathbb{R}^+).$$

This will be useful in the proof of one of the subsequent theorems.

Remark 4.2 As we shall see later, the definition of the convolution $f * g$ in this last form gives exactly the Mellin transform of the function $u \mapsto (\tau_{1/u}^c f)(x)\, g(u)$ evaluated at the point c (see Definition 6.1).

A first basic result may be stated as follows.

C. Bardaro et al., *Mellin Analysis, Transform Theory, and Applications*,
https://doi.org/10.1007/978-3-031-96672-9_4

Theorem 4.3 *Let $f \in X_c^p$ and $g \in X_c^{p'}$, where $p \in [1, \infty]$ and p' is its conjugate defined by $(1/p)+(1/p') = 1$ with the usual interpretation when $p \in \{1, \infty\}$. Then $(f * g)(x)$ exists for all $x \in \mathbb{R}^+$, $f * g \in X_c^\infty \cap C(\mathbb{R}^+)$ and*

$$\|f * g\|_{X_c^\infty} \leq \|f\|_{X_c^p} \|g\|_{X_c^{p'}}. \tag{4.1}$$

Proof Arguing as in [171, p. 396] with $x - u$ being replaced by x/u, one can see that the function $F : (x, u) \mapsto f(x/u)u^{-c}$ is Lebesgue measurable on $(\mathbb{R}^+)^2$. Next we show that $F(x, \cdot) \in L^p(\mathbb{R}^+, \mu)$, where μ is the measure defined in Remark 3.1.

Let $1 < p < \infty$. Using the change of variable $x/u = t$, we have

$$\int_0^\infty \left|f\left(\frac{x}{u}\right)\right|^p u^{-cp} \frac{du}{u} = x^{-cp} \int_0^\infty |f(t)|^p t^{cp} \frac{dt}{t} = x^{-cp} \|f\|_{X_c^p}^p \qquad (x \in \mathbb{R}^+).$$

Therefore by the Hölder inequality for L^p spaces, we obtain

$$\begin{aligned}\int_0^\infty \left|f\left(\frac{x}{u}\right)\right| |g(u)| \frac{du}{u} &= \int_0^\infty \left|f\left(\frac{x}{u}\right)u^{-c}\right| |g(u)|u^c \frac{du}{u}\\ &\leq x^{-c} \|f\|_{X_c^p} \|g\|_{X_c^{p'}},\end{aligned}$$

Hence

$$x^c |(f * g)(x)| \leq \|f\|_{X_c^p} \|g\|_{X_c^{p'}}$$

for every $x \in \mathbb{R}^+$, which is (4.1). It remains to prove the continuity of $f * g$. Using the same reasonings as before and the Hölder inequality for L^p spaces, we have, for h near to 1

$$\begin{aligned}|(\tau_h^c(f * g))(x) - (f * g)(x)| &\leq \int_0^\infty |g(u)|u^c \left|f\left(\frac{x}{u}\right) - h^c f\left(\frac{hx}{u}\right)\right| u^{-c} \frac{du}{u}\\ &\leq x^{-c} \|g\|_{X_c^{p'}} \|\tau_h^c f - f\|_{X_c^p}.\end{aligned}$$

Thus the assertion follows from Proposition 3.13 for $1 < p < \infty$.

Writing

$$\begin{aligned}\int_0^\infty \left|f\left(\frac{x}{u}\right)\right| |g(u)| \frac{du}{u} &= x^{-c} \int_0^\infty \left|f\left(\frac{x}{u}\right)\right| \left(\frac{x}{u}\right)^c |g(u)|u^c \frac{du}{u}\\ &\leq x^{-c} \|f\|_{X_c^\infty} \|g\|_{X_c},\end{aligned}$$

we obtain (4.1) for $p = \infty$ and analogously for $p = 1$ by substituting $x/u = t$. Concerning the continuity, we may proceed as before. □

When $f \in X_c^p$, one would like to know a condition on g such that $f * g \in X_c^p$. The following theorem gives an answer.

Theorem 4.4 *Let $f \in X_c^p$, where $1 \le p < \infty$, and let $g \in X_c$. Then $f * g \in X_c^p$ and*

$$\|f * g\|_{X_c^p} \le \|g\|_{X_c} \|f\|_{X_c^p}. \tag{4.2}$$

Proof First let $1 < p < \infty$. Using (3.2) for $c = 0$ with the roles of $f(u)$ and $g(u)$ being taken by

$$|g(u)|^{1/p} \left| f\left(\frac{x}{u}\right) \right| u^{(c/p)-c} \qquad \text{and} \qquad |g(u)|^{1/p'} u^{c/p'},$$

respectively, and with p' being the conjugate exponent of p, we may write for almost all $x \in \mathbb{R}^+$,

$$\begin{aligned} |(f * g)(x)| &\le \int_0^\infty \left| f\left(\frac{x}{u}\right) \right| |g(u)| \frac{du}{u} \\ &\le \left\{ \int_0^\infty |g(u)| u^c \frac{du}{u} \right\}^{1/p'} \left\{ \int_0^\infty |g(u)| u^{c-cp} \left| f\left(\frac{x}{u}\right) \right|^p \frac{du}{u} \right\}^{1/p}. \end{aligned}$$

Then, using Fubini's theorem (Proposition 1.8) and substituting $x/u = t$, we obtain

$$\begin{aligned} \|f * g\|_{X_c^p}^p &= \int_0^\infty |(f * g)(x)|^p x^{cp} \frac{dx}{x} \\ &\le \|g\|_{X_c}^{p/p'} \int_0^\infty |g(u)| u^{c-cp} \left\{ \int_0^\infty \left| f\left(\frac{x}{u}\right) \right|^p x^{pc} \frac{dx}{x} \right\} \frac{du}{u} \\ &= \|g\|_{X_c}^{p/p'} \int_0^\infty |g(u)| u^c \frac{du}{u} \int_0^\infty |f(t)|^p t^{cp} \frac{dt}{t} \\ &= \|g\|_{X_c}^{(p+p')/p'} \|f\|_{X_c^p}^p, \end{aligned}$$

which readily shows that (4.2) holds for $p \in]1, \infty[$.

For $p = 1$ the proof is even simpler. Indeed, using again Fubini's theorem, we may write

$$\begin{aligned} \|f * g\|_{X_c} &\le \int_0^\infty |g(u)| u^c \left\{ \int_0^\infty \left| f\left(\frac{x}{u}\right) \right| \left(\frac{x}{u}\right)^c \frac{dx}{x} \right\} \frac{du}{u} \\ &= \|g\|_{X_c} \|f\|_{X_c}, \end{aligned}$$

and the theorem is completely proved. □

Under appropriate hypotheses it can be verified by substitutions and by employing Fubini's theorem that the Mellin convolution is commutative and associative. In particular,

$$f * g = g * f \quad \text{and} \quad f * (g * h) = (f * g) * h$$

for $f, g, h \in X_c$. Since, by Theorem 4.4, in this case also

$$\|f * g\|_{X_c} \leq \|g\|_{X_c} \|f\|_{X_c},$$

we see that the spaces X_c with the ordinary addition and the operation $*$ as multiplication form a Banach algebra. However, as it happens for L^1 spaces, X_c too has no unit, that is, there does not exist a function $e \in X_c$ such that $e * f = f$ for any $f \in X_c$.

An alternative version of a convolution product is given by the formula

$$(f \,\widetilde{*}\, g)(x) := \int_0^\infty (\tau_{1/u}^c \widetilde{f})(x) g(u) u^c \frac{du}{u} = \int_0^\infty f(u) g(ux) \frac{du}{u}, \tag{4.3}$$

where $\widetilde{f}(x) := f(1/x)$ for $x \in \mathbb{R}^+$.

This kind of convolution is often used in the study of *Mellin integral operators*, which are defined by

$$(T_K f)(x) := \int_0^\infty K(u) f(ux) \frac{du}{u},$$

where K, called the *kernel function*, is fixed and f belongs to a function space. For $c = 0$ there exists the following analogue of Theorem 4.4.

Proposition 4.5 *Let $K \in X_0$ be fixed, and let $f \in X_0^p$ with $p \in [1, \infty]$. Then $T_K f$ exists and belongs to X_0^p. Moreover,*

$$\|T_K f\|_{X_0^p} \leq \|K\|_{X_0} \|f\|_{X_0^p}. \tag{4.4}$$

Proof For $p = \infty$ the assertion is obvious. For $1 \leq p < \infty$, we have

$$\int_0^\infty |K(u)| \, |f(ux)|^p \frac{dx}{x} = |K(u)| \|f\|_{X_0^p}^p,$$

as is seen by substituting $ux = t$. Integrating with respect to u, we obtain

$$\int_0^\infty \frac{du}{u} \left\{ \int_0^\infty |K(u)| \, |f(ux)|^p \frac{dx}{x} \right\} = \|K\|_{X_0} \|f\|_{X_0^p}^p.$$

With the help of Fubini's theorem, we deduce that

$$\int_0^\infty \frac{dx}{x}\left\{\int_0^\infty |K(u)|\,|f(ux)|^p\,\frac{du}{u}\right\} = \|K\|_{X_0}\,\|f\|_{X_0^p}^p, \tag{4.5}$$

and so the integral

$$\int_0^\infty |K(u)|\,|f(ux)|^p \frac{du}{u}$$

exists for almost all $x \in \mathbb{R}^+$. In particular, (4.5) implies that the assertion holds for $p = 1$.

Now let $1 < p < \infty$. Denoting again by p' the conjugate exponent of p and employing Hölder's inequality, we find that

$$\begin{aligned}
|(T_K f)(x)| &\leq \int_0^\infty |K(u)|^{(1/p)+(1/p')}|f(ux)|\frac{du}{u}\\
&\leq \left\{\int_0^\infty |K(u)|\,|f(ux)|^p\,\frac{du}{u}\right\}^{1/p}\left\{\int_0^\infty |K(u)|\frac{du}{u}\right\}^{1/p'}\\
&= \|K\|_{X_0}^{1/p'}\left\{\int_0^\infty |K(u)|\,|f(ux)|^p\,\frac{du}{u}\right\}^{1/p}.
\end{aligned}$$

Taking the norm of X_0^p on both sides and using again Fubini's theorem, we obtain

$$\begin{aligned}
\|T_k f\|_{X_0^p}^p &\leq \|K\|_{X_0}^{p/p'}\int_0^\infty \frac{dx}{x}\left\{\int_0^\infty |K(u)|\,|f(xu)|^p\,\frac{du}{u}\right\}\\
&= \|K\|_{X_0}^{1+(p/p')}\,\|f\|_{X_0^p}^p.
\end{aligned}$$

This completes the proof. □

The previous result does not extend to the spaces X_c^p with $c \neq 0$. We illustrate this by two examples.

Example 4.6 Let $p = 1$, $c = 1$, $K(x) := \chi_{]0,1]}(x)$ and $f(x) := x\chi_{]0,1]}(x)$, where $\chi_{]0,1]}$ denotes the characteristic function of the interval $]0, 1]$. Then both K and f belong to X_1 and

$$(T_K f)(x) = \begin{cases} x & \text{for } x \in]0,1[,\\ 1 & \text{for } x \geq 1.\end{cases}$$

Thus, $(T_K)f(x)$ exists for every $x \in \mathbb{R}^+$ but $T_K f \notin X_1$.

The second example shows that if $K \in X_1$ and $f \in X_1$, then $(T_K f)(x)$ may even be infinite for every x.

Example 4.7 Let $p = 1, c = 1$ and $\alpha, \beta \in]0, 1[$. Define

$$K(x) := x^{-\alpha}\chi_{]0,1]}(x) \quad \text{and} \quad f(x) := u^{-\beta}\chi_{]0,1]}(x) \quad (x > 0).$$

Then $K, f \in X_1$ but

$$(T_K f)(x) = x^{-\beta} \int_0^{\min\{1,1/x\}} u^{-\alpha-\beta-1} du = \infty.$$

Similar examples can also be obtained for general c.

However, Proposition 4.5 can be generalized as follows.

Proposition 4.8 *Let $K \in X_c$ be fixed and let $f \in X^p_{-c}$, where $c \in \mathbb{R}$ and $p \in [1, \infty]$. Then $T_K f$ exists and belongs to X^p_{-c}. Moreover,*

$$\|T_K f\|_{X^p_{-c}} \le \|K\|_{X_c} \|f\|_{X^p_{-c}}. \tag{4.6}$$

Proof For $p = \infty$, we easily see that

$$\begin{aligned}
\|T_K\|_{X^\infty_{-c}} &\le \sup_{x>0} x^{-c} \int_0^\infty |K(u)|\,|f(xu)| \frac{du}{u} \\
&= \sup_{x>0} \int_0^\infty u^c|K(u)|(ux)^{-c}|f(ux)| \frac{du}{u} \\
&\le \|K\|_{X_c}\, \|f\|_{X^\infty_{-c}}.
\end{aligned}$$

For $p = 1$, we first employ Fubini's theorem and substitute $ux = t$ subsequently to conclude that

$$\begin{aligned}
\|T_K\|_{X_{-c}} &\le \int_0^\infty \left\{ \int_0^\infty |K(u)|\,|f(ux)| \frac{du}{u} \right\} x^{-c} \frac{dx}{x} \\
&= \int_0^\infty |K(u)| \left\{ \int_0^\infty |f(ux)|\, x^{-c} \frac{dx}{x} \right\} \frac{du}{u} \\
&= \int_0^\infty |K(u)| \left\{ \int_0^\infty |f(t)|\, t^{-c} \frac{dt}{t} \right\} u^c \frac{du}{u} \\
&= \|K\|_{X_c} \|f\|_{X_{-c}}.
\end{aligned}$$

Now let $1 < p < \infty$. Denoting again by p' the conjugate exponent of p and applying Hölder's inequality, we find that

$$\int_0^\infty |K(u)|\,|f(ux)|\,\frac{du}{u}$$

$$= \int_0^\infty |K(u)|^{1/p'}\,u^{c/p'}\,|K(u)|^{1/p}|f(ux)|\big(u^{-cp/p'}\big)^{1/p}\,\frac{du}{u}$$

$$\leq \left\{\int_0^\infty |K(u)|u^c\,\frac{du}{u}\right\}^{1/p'}\left\{\int_0^\infty |K(u)|\,|f(ux)|^p u^{-cp/p'}\,\frac{du}{u}\right\}^{1/p}.$$

Taking the norm of X_{-c}^p on both sides and using Fubini's theorem, we obtain

$$\|T_K f\|_{X_{-c}^p}^p \leq \|K\|_{X_c}^{p/p'}\int_0^\infty |K(u)|u^{-cp/p'}\left\{\int_0^\infty |f(ux)|^p x^{-cp}\,\frac{dx}{x}\right\}\frac{du}{u}.$$

Finally, substituting $ux = t$ in the integration with respect to x, we arrive at

$$\|T_K f\|_{X_{-c}^p}^p \leq \|K\|_{X_c}^{p/p'}\int_0^\infty |K(u)|u^{cp-(cp/p')}\,\frac{du}{u}\int_0^\infty |f(t)|^p t^{-cp}\,\frac{dt}{t}$$

$$= \|K\|_{X_c}^{p/p'}\int_0^\infty |K(u)|u^c\,\frac{du}{u}\int_0^\infty |f(t)|^p t^{-cp}\,\frac{dt}{t}$$

$$= \|K\|_{X_c}^{(p+p')/p'}\|f\|_{X_{-c}^p}^p.$$

This completes the proof. □

As a corollary we state:

Corollary 4.9 *Let $K \in X_{]-\infty,\infty[}$. Then for every $p \in [1,\infty]$, $c \in \mathbb{R}$ and $f \in X_c^p$, there holds $T_K f \in X_c^p$ and*

$$\|T_K f\|_{X_c^p} \leq \|K\|_{X_{-c}}\|f\|_{X_c^p}.$$

Remark 4.10 There are interesting relations between the two notions of Mellin convolutions. First, note that in general the convolution (4.3) is not commutative. However, if the function f is Mellin-even, that is, $\widetilde{f}(u) := f(1/u) = f(u)$ for every $u \in \mathbb{R}^+$, then

$$(f \,\widetilde{*}\, g)(x) = (f * g)(x)$$

almost everywhere on $\mathbb{R}^+$. In the case of Mellin-even g, we have $(f \,\widetilde{*}\, g)(x) = (\widetilde{f} * \widetilde{g})(x)$ almost everywhere on $\mathbb{R}^+$.

4.2 The Mellin Convolution in Spaces Y_c^p

Now we introduce a convolution structure for the spaces Y_c^p, where $c \in \mathbb{R}$ and $p \geq 1$.

Definition 4.11 For $f, g \in Y_c^p$, the finite Mellin convolution is defined by

$$(f * g)(x) := \int_{e^{-\pi}}^{e^{\pi}} f\left(\frac{x}{u}\right) g(u) \frac{du}{u} \qquad (x \in \mathbb{R}^+)$$

provided that the integral exists.

Remark 4.12 As in the case of the X_c^p spaces, we can also introduce the finite Mellin convolution in Y_c^p with the help of the Mellin translation operator by writing

$$(f * g)(x) = \int_{e^{-\pi}}^{e^{\pi}} (\tau_{1/u}^c f)(x)\, g(u)\, u^c \frac{du}{u}.$$

Analogously to Remark 4.2, this last form gives exactly the finite Mellin transform of the function $u \mapsto (\tau_{1/u} f)(x)\, g(u)$ evaluated at the point c when c is an integer (see Definition 5.1).

The former Theorems 4.3 and 4.4 have the following analogues which can be proved in a similar way.

Theorem 4.13 *Let $f \in Y_c^p$ and $g \in Y_c^{p'}$, where $p \in [1, \infty]$ and p' is its conjugate defined by $(1/p) + (1/p') = 1$ with the usual interpretation when $p \in \{1, \infty\}$. Then $(f * g)(x)$ exists for all $x \in \mathbb{R}^+$, $f * g \in Y_c^\infty \cap C(\mathbb{R}^+)$ and*

$$\|f * g\|_{Y_c^\infty} \leq \|f\|_{Y_c^p} \|g\|_{Y_c^{p'}}.$$

Theorem 4.14 *Let $f \in Y_c^p$, $1 \leq p < \infty$, and let $g \in Y_c$. Then $f * g \in Y_c^p$ and*

$$\|f * g\|_{Y_c^p} \leq \|g\|_{Y_c} \|f\|_{Y_c^p}. \tag{4.7}$$

As before, under appropriate hypotheses, the finite Mellin convolution is commutative and associative. In particular,

$$f * g = g * f \quad \text{and} \quad (f * g) * h = f * (g * h)$$

for $f, g, h \in Y_c$. Thus, in view of Theorem 4.14 for $p = 1$, the spaces Y_c with the ordinary addition and the operation $*$ as multiplication form a Banach algebra which has no unit.

Remark 4.15 Also in the setting of Y_c^p spaces one can introduce an alternative version of a convolution by defining

$$(f \widetilde{*} g)(x) := \int_{e^{-\pi}}^{e^{\pi}} f(u)g(xu)\frac{du}{u},$$

provided that the integral exists.

4.3 Notes

In their full generality, the results of this chapter seem to have not been mentioned in the literature before. For Theorems 4.3 and 4.4 in the special case of $p = 2$, see Butzer and Jansche [92, Lemma 2.2]. For the corresponding special cases of Theorems 4.13 and 4.14, see [91, Lemma 3.1]. The inspiration for an extension to arbitrary $p \in [1, \infty]$ came from analogous results in Fourier analysis.

The Mellin convolution for functions defined on the positive real axis is the counterpart of the classical convolution in Fourier analysis, being based on the translations on $\mathbb{R}$, which constitute the (additive) line group; see Butzer and Nessel [95, Sect. 0.2]. In the Mellin setting, the convolution is performed by using the dilation operators on the positive real axis, which constitute a multiplicative group. This convolution is of fundamental importance in the theory of Mellin transform, which in turn is a basic tool for the solution of integral equations of the type

$$g(x) = \int_0^{\infty} K\left(\frac{x}{y}\right) f(y)\frac{dy}{y},$$

see, e.g., Titchmarsh [286, Chap. XI, Sect. 11.4], Sneddon [276], Zemanian [310, Chap. 4, Sect. 4.7].

Mellin convolutions were used in the theory of boundary value problems on wedge-shaped domains in the plane, such as $W = \{(r, \theta) : 0 < r < \infty, 0 < \theta < \theta_0\}$ with $\theta_0 \in]0, 2\pi[$. For example, for a Dirichlet problem with boundary conditions of type $\lim_{\theta\to\theta_0+} w(r, \theta) = f(r)$ and $\lim_{\theta\to\theta_0-} w(r, \theta) = g(r)$, the solution is given by a sum of two convolution integrals,

$$w(r, \theta) = \int_0^{\infty} f\left(\frac{r}{u}\right) K_1(u, \theta)\frac{du}{u} + \int_0^{\infty} g\left(\frac{r}{u}\right) K_2(u, \theta)\frac{du}{u},$$

where K_1 and K_2 are suitable kernel functions. For details, see, e.g., Kolbe and Nessel [194], Davies [135, Part III, Sect. 12.4], and Zemanian [310].

The alternative definition of Mellin convolution considered in formula (4.3) provides a useful tool for studying convergence of sequences of Mellin convolution integral operators of the type

$$(T_n f)(x) = \int_0^\infty K_n(xt) f(t) \frac{dt}{t},$$

where $\{K_n\}$ is a family of kernel functions. Convergence results for these operators are described in Sect. 6.4.

Luchko [201] uses a convolution of the Mellin type for deducing various old and new integral transforms of f by applying it to f and an appropriately chosen kernel function. For an application of the Mellin convolution in probability theory, see Mainardi et al. [207].

Chapter 5
The Finite Mellin Transform and the Mellin–Fourier Series

In this chapter, we first introduce the finite Mellin transform of functions in Y_c and present its basic properties. Then we consider an inversion of this transform which leads us to Mellin–Fourier series. After establishing approximations by partial sums represented by finite Mellin convolutions with appropriate kernels, we study convergence of Mellin–Fourier series in the spaces Y_c and Y_c^2.

5.1 The Finite Mellin Transform in Y_c

We begin with the following notions.

Definition 5.1 The *finite Mellin transform* $\mathscr{M}^c$ of $f \in Y_c$, where $c \in \mathbb{R}$, is defined by

$$\mathscr{M}^c[f](c+ik) \equiv [f]^\wedge_{\mathscr{M}^c}(k) := \int_{e^{-\pi}}^{e^{\pi}} f(u)u^{c+ik-1}du \qquad (k \in \mathbb{Z}).$$

The numbers $\mathscr{M}^c[f](c+ik)$ are called the *Mellin–Fourier coefficients of* $f \in Y_c$.

The basic operational properties of the finite Mellin transform are presented in the following proposition.

Proposition 5.2 *Let $c \in \mathbb{R}$ and $\alpha \in \mathbb{R}^+$. Then the following statements hold:*

(a) The finite Mellin transform of $f \in Y_c$ is a well-defined linear and bounded operator $\mathscr{M}^c : Y_c \to \ell^\infty(\mathbb{Z})$, $f \mapsto \left([f]^\wedge_{\mathscr{M}^c}(k)\right)_{k\in\mathbb{Z}}$ with

$$|[f]^\wedge_{\mathscr{M}^c}(k)| \le \|f\|_{Y_c} \qquad (k \in \mathbb{Z})$$

and its operator norm is 1.

C. Bardaro et al., *Mellin Analysis, Transform Theory, and Applications*,
https://doi.org/10.1007/978-3-031-96672-9_5

(b) For $f \in Y_c$, one has

$$[f]^{\wedge}_{\mathscr{M}^c}(k) = [(\cdot)^c f(\cdot)]^{\wedge}_{\mathscr{M}^0}(k) \qquad (k \in \mathbb{Z}).$$

(c) If $f \in Y_c$, then

$$[f(\alpha(\cdot))]^{\wedge}_{\mathscr{M}^c}(k) = \alpha^{-(c+ik)}[f]^{\wedge}_{\mathscr{M}^c}(k) \qquad (k \in \mathbb{Z}).$$

(d) If $f \in Y_c$, then $(\cdot)^{ij} f(\cdot) \in Y_c$ for $j \in \mathbb{Z}$ and

$$[(\cdot)^{ij} f(\cdot)]^{\wedge}_{\mathscr{M}^c}(k) = [f(\cdot)^{\wedge}_{\mathscr{M}^c}(k+j) \qquad (k \in \mathbb{Z}).$$

Proof (a) The linearity is an obvious consequence of the definition. Since

$$\left|[f]^{\wedge}_{\mathscr{M}^c}(k)\right| \leq \int_{e^{-\pi}}^{e^{\pi}} |f(u)| u^{c-1} du = \|f\|_{Y_c},$$

we immediately see that the operator $\mathscr{M}^c$ is bounded. Equality is attained for $g = |f|$ and $k = 0$, which shows that the operator norm is 1.

As to (b), we simply note that

$$[f(\cdot)]^{\wedge}_{\mathscr{M}^c}(k) = \int_{e^{-\pi}}^{e^{\pi}} f(u) u^c u^{ik-1} du = [(\cdot)^c f(\cdot)]^{\wedge}_{\mathscr{M}^0}(k).$$

Concerning (c), we first observe that f being c-recurrent implies that $f(\alpha(\cdot))$ is c-recurrent for all $\alpha > 0$. Under the change of variable $\alpha u = t$, we obtain by Proposition 3.7,

$$\int_{e^{-\pi}}^{e^{\pi}} |f(\alpha u)| u^{c-1} du = \alpha^{-c} \|f\|_{Y_c} < \infty,$$

and so, by (a), both sides of the equation in (c) exist for all $\alpha > 0$. Now, again by Proposition 3.7,

$$[f(\alpha(\cdot)]^{\wedge}_{\mathscr{M}^c}(k) = \alpha^{-(c+ik)} \int_{\alpha e^{-\pi}}^{\alpha e^{\pi}} f(u) u^{c+ik-1} du = \alpha^{-(c+ik)}[f(\cdot)]^{\wedge}_{\mathscr{M}^c}(k),$$

which is the desired equation.

As to (d), since the function $h(x) := x^{ij}$ is 0-recurrent for every $j \in \mathbb{Z}$, the product hf is c-recurrent. The asserted equation is verified by a straightforward calculation. □

The following proposition gives a formula for the Mellin–Fourier coefficients of a convolution of c-recurrent functions.

Proposition 5.3 *Let $f, g \in Y_c$, where $c \in \mathbb{R}$. Then*

$$[f * g]^{\wedge}_{\mathscr{M}^c}(k) = [f]^{\wedge}_{\mathscr{M}^c}(k)\,[g]^{\wedge}_{\mathscr{M}^c}(k) \qquad (k \in \mathbb{Z}).$$

Proof Using Fubini's theorem and statement (c) of Proposition 5.2, we see that

$$\begin{aligned}[f * g]^{\wedge}_{\mathscr{M}^c}(k) &= \int_{e^{-\pi}}^{e^{\pi}} \left(\int_{e^{-\pi}}^{e^{\pi}} (\tau_{1/u} f)(x) g(u) \frac{du}{u} \right) x^{c+ik-1} dx \\ &= \int_{e^{-\pi}}^{e^{\pi}} [\tau_{1/u} f]^{\wedge}_{\mathscr{M}^c}(k)\, g(u) \frac{du}{u} = [f]^{\wedge}_{\mathscr{M}^c}(k)\,[g]^{\wedge}_{\mathscr{M}^c}(k),\end{aligned}$$

as was to be shown. □

The following proposition is the counterpart of the classical Riemann–Lebesgue lemma for the Mellin–Fourier coefficients.

Proposition 5.4 *Let $f \in Y_c$, where $c \in \mathbb{R}$. Then* $\lim_{|k|\to\infty} [f]^{\wedge}_{\mathscr{M}^c}(k) = 0$.

Proof Set $g(u) := u^c f(u)$. Then, in view of statement (b) of Proposition 5.2, it suffices to show that $\lim_{|k|\to\infty} [g]^{\wedge}_{\mathscr{M}^0}(k) = 0$. With $h_k = e^{\pi/k}$ for $k \in \mathbb{Z} \setminus \{0\}$, we have by statement (c) of Proposition 5.2 that

$$[g]^{\wedge}_{\mathscr{M}^0}(k) = -e^{-i\pi}[g]^{\wedge}_{\mathscr{M}^0}(k) = -[\tau_{h_k} g]^{\wedge}_{\mathscr{M}^0}(k).$$

Hence, by statement (a) of Proposition 5.2,

$$2\left|[g]^{\wedge}_{\mathscr{M}^0}(k)\right| = \left|[\tau_{h_k} g - g]^{\wedge}_{\mathscr{M}^0}(k)\right| \le \|\tau_{h_k} g - g\|_{Y_0}.$$

Now the proof is completed by using Proposition 3.16. □

For $c \in \mathbb{R}$ and $k \in \mathbb{Z}$, we define $\varphi_{c,k}(x) := x^{-(c+ik)}$. It is easily verified that $\varphi_{c,k} \in Y_c$ for every $k \in \mathbb{Z}$. The following proposition states an orthogonality property for the family $(\varphi_{0,k})_{k\in\mathbb{Z}}$.

Proposition 5.5 *For $c \in \mathbb{R}$, $k \in \mathbb{Z}$ and $x \in \mathbb{R}^+$, let $\varphi_{c,k}(x) := x^{-(c+ik)}$. Then $\{\varphi_{c,k} : k \in \mathbb{Z}\} \subset Y_c$ and*

$$[\varphi_{c,j}]^{\wedge}_{\mathscr{M}^c}(k) = 2\pi\,\delta_{j,k} \qquad (j, k \in \mathbb{Z}) \tag{5.1}$$

with Kronecker's delta. Furthermore, $(\varphi_{0,k})_{k\in\mathbb{Z}}$ is an orthogonal system such that

$$\int_{e^{-\pi}}^{e^{\pi}} \varphi_{0,k}(u)\overline{\varphi_{0,j}(u)}\,\frac{du}{u} = 2\pi\,\delta_{j,k} \qquad (j, k \in \mathbb{Z}).$$

Proof The left-hand sides of these equations lead to the same type of integral, namely

$$[\varphi_{c,j}]^{\wedge}_{\mathscr{M}^c}(k) = \int_{e^{-\pi}}^{e^{\pi}} u^{i(k-j)-1} du$$

and

$$\int_{e^{-\pi}}^{e^{\pi}} \varphi_{0,k}(u)\overline{\varphi_{0,j}(u)}\,\frac{du}{u} = \int_{e^{-\pi}}^{e^{\pi}} u^{i(j-k)-1} du.$$

The integrals on the right-hand sides are immediately seen to be equal to $2\pi\delta_{j,k}$. □

In Mellin analysis the system $(\varphi_{0,k})_{k\in\mathbb{Z}}$ plays the same role as the orthogonal system $(e^{ikx})_{k\in\mathbb{Z}}$ in Fourier analysis.

5.2 Mellin–Fourier Series in Y_c

We begin with the following definition.

Definition 5.6 Let $f \in Y_c$, where $c \in \mathbb{R}$. We associate with $f(x)$ the series

$$\frac{x^{-c}}{2\pi} \sum_{k=-\infty}^{\infty} [f]^{\wedge}_{\mathscr{M}^c}(k) x^{-ik} \qquad (x \in \mathbb{R}^+) \tag{5.2}$$

and call it the *Mellin–Fourier series of f* with the *Mellin–Fourier coefficients* $[f]^{\wedge}_{\mathscr{M}^c}(k)$ for $k \in \mathbb{Z}$. In the following, we will also briefly speak of *Mellin series* and *Mellin coefficients*.

The n-th partial sum of the Mellin series is given by

$$(S_n^c f)(x) := \frac{x^{-c}}{2\pi} \sum_{k=-n}^{n} [f]^{\wedge}_{\mathscr{M}^c}(k) x^{-ik} \qquad (x \in \mathbb{R}^+,\ n \in \mathbb{N}_0). \tag{5.3}$$

We wish to express the partial sums of the Mellin series by a finite convolution integral involving a suitable kernel function.

Definition 5.7 For $c \in \mathbb{R}$, the *(finite) Mellin–Dirichlet kernel* $\{D_n^c\}_{n\in\mathbb{N}_0}$ is defined by

$$D_n^c(x) := \frac{x^{-c}}{2\pi} \sum_{k=-n}^{n} x^{-ik} \qquad (x \in \mathbb{R}^+). \tag{5.4}$$

The following lemma provides some useful formulae for the Mellin–Dirichlet kernel.

Lemma 5.8 *Let $c \in \mathbb{R}$.*

(a) The Mellin–Dirichlet kernel belongs to Y_c and has the representations

$$D_n^c(x) = \begin{cases} \dfrac{x^{-c}}{2\pi} \dfrac{x^{i(2n+1)/2} - x^{-i(2n+1)/2}}{x^{i/2} - x^{-i/2}}, & x \neq e^{2\pi j} \\[2ex] \dfrac{2n+1}{2\pi} e^{-2\pi j c}, & x = e^{2\pi j} \end{cases} \quad (j \in \mathbb{Z}), \tag{5.5}$$

$$D_n^c(x) = \begin{cases} \dfrac{x^{-c}}{4\pi} \left\{ (x^{ni} - x^{-ni}) \dfrac{x^{i/2} + x^{-i/2}}{x^{i/2} - x^{-i/2}} + x^{ni} + x^{-ni} \right\}, & x \neq e^{2\pi j} \\[2ex] \dfrac{2n+1}{2\pi} e^{-2\pi j c}, & x = e^{2\pi j} \end{cases} \quad (j \in \mathbb{Z}) \tag{5.6}$$

and

$$D_n^c(x) = \begin{cases} \dfrac{x^{-c}}{2\pi} \dfrac{\sin\left(\frac{2n+1}{2} \log x\right)}{\sin\left(\frac{1}{2} \log x\right)}, & x \neq e^{2\pi j} \\[2ex] \dfrac{2n+1}{2\pi} e^{-2\pi j c}, & x = e^{2\pi j} \end{cases} \quad (j \in \mathbb{Z}). \tag{5.7}$$

Further, $D_n^0(x) = D_n^0(1/x)$ for all $n \in \mathbb{N}_0$ and $x \in \mathbb{R}^+$.

(b) If $f \in Y_c$, then $S_n^c f \in Y_c$ for all $n \in \mathbb{N}_0$ and

$$(S_n^c f)(x) = (f * D_n^c)(x) \qquad (x \in \mathbb{R}^+). \tag{5.8}$$

Proof

(a) Clearly, the function D_n^c belongs to Y_c for $c \in \mathbb{R}$ and $n \in \mathbb{N}_0$ since it is given by a finite sum of functions in Y_c. If $x \in \{e^{2\pi j} : j \in \mathbb{Z}\}$, then $x^{-ik} = 1$ for all $k \in \mathbb{Z}$ and (5.4) provides the value of $D_n^c(x)$ as stated in the assertions. If $x \notin \{e^{2\pi j} : j \in \mathbb{Z}\}$, then, summing the finite geometrical series in (5.4), we find that

$$\begin{aligned} \sum_{k=-n}^{n} x^{-ik} &= -1 + \sum_{k=0}^{n} (x^{ik} + x^{-ik}) \\ &= -1 + \frac{x^{-i/2} - x^{i(2n+1)/2}}{x^{-i/2} - x^{i/2}} + \frac{x^{i/2} - x^{-i(2n+1)/2}}{x^{i/2} - x^{-i/2}}. \end{aligned}$$

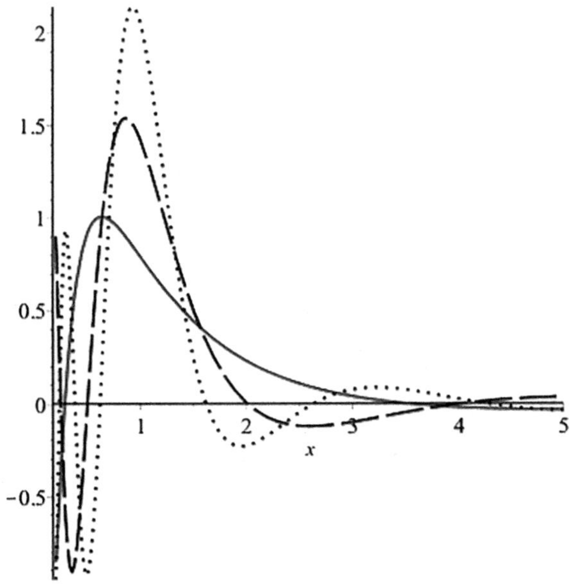

Fig. 5.1 The Mellin–Dirichlet kernel, with $c = 1$ and $n = 2$ (solid line), $n = 4$ (dashed line) and $n = 6$ (dotted line)

Now the proof of (5.5) is easily completed. The second representation (5.6) can be deduced from (5.5) by a simple calculation. The third representation (5.7) follows from (5.5) with the help of the Euler equation for the sine function. The rest of statement (a) is obvious.

(b) Clearly, the finite sum $S_n f$ belongs to Y_c. Writing its Mellin coefficients as integrals of the finite Mellin transform and interchanging summation and integration, we see that for $x \in \mathbb{R}^+$ and $n \in \mathbb{N}_0$, we have

$$\begin{aligned}(S_n^c f)(x) &= \int_{e^{-\pi}}^{e^{\pi}} f(u) \left\{ \frac{x^{-c}}{2\pi} \sum_{k=-n}^{n} x^{-ik} u^{c+ik-1} \right\} du \\ &= \int_{e^{-\pi}}^{e^{\pi}} f(u) \left\{ \frac{1}{2\pi} \sum_{k=-n}^{n} \left(\frac{x}{u}\right)^{-c-ik} \right\} \frac{du}{u} \\ &= (f * D_n^c)(x),\end{aligned}$$

which proves (5.8) (Fig. 5.1).

□

Next, we illustrate some convergence theorems for Mellin series which represent the Mellin counterpart of results for Fourier series. We begin with the following proposition that shows the reason why convergence may fail in the space Y_c.

Proposition 5.9 *For the norm of the Mellin–Dirichlet kernel there holds*

$$\|D_n^c\|_{Y_c} = \frac{4}{\pi^2}\log n + \mathcal{O}(1) \qquad (n \to \infty). \tag{5.9}$$

Proof Employing statement (a) of Lemma 5.8, we may write

$$\begin{aligned}\|D_n^c\|_{Y_c} &= 2\int_1^{e^\pi} |D_n^0(x)|\frac{dx}{x} \\ &= \frac{1}{2\pi}\int_1^{e^\pi} \left|(x^{ni} - x^{-ni})\frac{x^{i/2} + x^{-i/2}}{x^{i/2} - x^{-i/2}} + x^{ni} + x^{-ni}\right| \frac{dx}{x}.\end{aligned} \tag{5.10}$$

In order to evaluate this integral asymptotically, we first note that

$$\int_1^{e^\pi} |x^{ni} + x^{-ni}|\,\frac{dx}{x} = \mathcal{O}(1) \qquad (n \to \infty). \tag{5.11}$$

The quotient

$$\frac{x^{i/2} + x^{-i/2}}{x^{i/2} - x^{-i/2}} = -i\,\frac{\cos(\frac{1}{2}\log x)}{\sin(\frac{1}{2}\log x)}$$

is not bounded on $[1, e^\pi]$ since its denominator vanishes at $x = 1$, but

$$\lim_{x\to 1^+}\left(\frac{x^{i/2} + x^{-i/2}}{x^{i/2} - x^{-i/2}} + \frac{2i}{\log x}\right) = \lim_{t\to 0^+} i\left(\frac{1}{t} - \frac{\cos t}{\sin t}\right) = 0,$$

as can be seen by substituting $t = \frac{1}{2}\log x$ and using l'Hospital's rule or Taylor expansion. This allows us to conclude that

$$\int_1^{e^\pi} |x^{ni} - x^{-ni}|\left(\frac{x^{i/2} + x^{-i/2}}{x^{i/2} - x^{-i/2}} + \frac{2i}{\log x}\right)\frac{dx}{x} = \mathcal{O}(1).$$

Thus, by (5.10), we obtain

$$\|D_n^c\|_{Y_c} = \frac{1}{\pi}\int_1^{e^\pi} \left|\frac{x^{ni} - x^{-ni}}{\log x}\right|\frac{dx}{x} + \mathcal{O}(1). \tag{5.12}$$

The numerator of the integrand in (5.12) has sign changes exactly at $x = e^{k\pi/n}$ for $k \in \mathbb{Z}$. Therefore, by splitting up the interval of integration and substituting $x = e^{k\pi/n}u$, we find that

$$\begin{aligned}
\|D_n^c\|_{Y_c} &= \frac{1}{\pi}\sum_{k=0}^{n-1}\int_{e^{k\pi/n}}^{e^{(k+1)\pi/n}}\left|\frac{x^{ni}-x^{-ni}}{\log x}\right|\frac{dx}{x} + \mathcal{O}(1)\\
&= \frac{1}{\pi}\sum_{k=1}^{n-1}\int_{1}^{e^{\pi/n}}\left|\frac{e^{k\pi i}u^{ni}-e^{-k\pi i}u^{-ni}}{(k\pi)/n+\log u}\right|\frac{du}{u} + \mathcal{O}(1)\\
&= \frac{1}{\pi}\int_{1}^{e^{\pi/n}}\left|u^{ni}-u^{-ni}\right|\sum_{k=1}^{n-1}\frac{1}{(k\pi)/n+\log u}\,\frac{du}{u} + \mathcal{O}(1).
\end{aligned}$$

Now for $u \in [1, e^{\pi/n}]$, we have

$$\begin{aligned}
\sum_{k=1}^{n-1}\frac{1}{(k\pi)/n+\log u} &= \frac{n}{\pi}\sum_{k=1}^{n-1}\frac{1}{k+(n\log u)/\pi}\\
&= \frac{n}{\pi}\left\{1+\frac{1}{2}+\cdots+\frac{1}{n-1}+\mathcal{O}(1)\right\} = \frac{n}{\pi}\left\{\log n + \mathcal{O}(1)\right\}.
\end{aligned}$$

Furthermore, by a change of variable,

$$\int_{1}^{e^{\pi/n}}\left|u^{ni}-u^{-ni}\right|\frac{du}{u} = 2\int_{0}^{\pi/n}\sin(nt)dt = \frac{4}{n}.$$

Thus we arrive at

$$\|D_n^c\|_{Y_c} = \frac{4}{\pi^2}\log n + \mathcal{O}(1) \qquad (n\to\infty),$$

as was to be shown. □

Remark 5.10 Statement (b) of Lemma 5.8 in conjunction with Theorem 4.14 tells us that

$$\sup_{f\in Y_c\setminus\{0\}}\frac{\|S_n^c f\|_{Y_c}}{\|f\|_{Y_c}} \le \|D_n^c\|_{Y_c} = \frac{4}{\pi^2}\log n + \mathcal{O}(1) \qquad (n\to\infty). \tag{5.13}$$

It was mentioned by Butzer and Jansche [91, Remark 5.3] with reference to the book of Butzer and Nessel (see [95, Propositions 1.3.1, 1.3.2]) that equality occurs in (5.13), that is, the norm of S_n^c as an operator mapping Y_c into Y_c is equal to $\|D_n^c\|_{Y_c}$, which is not bounded as $n \to \infty$. Therefore, by the Banach–Steinhaus theorem (Proposition 1.20), there exists a function $f \in Y_c$ for which the partial sums of its associated Mellin series do not converge to f.

In order to obtain convergence, we now consider the arithmetical means of the partial sums, following a famous idea of Fejér in the case of Fourier series.

Definition 5.11 For $c \in \mathbb{R}$, the *(finite) Mellin–Fejér kernel* $\{F_n^c\}_{n\in\mathbb{N}_0}$ is defined by

$$F_n^c(x) := \frac{1}{n+1}\sum_{k=0}^{n} D_k^c(x) \qquad (x \in \mathbb{R}^+,\ n \in \mathbb{N}_0), \tag{5.14}$$

and the (first) *Mellin-arithmetic*, or $(C, 1)$ *means of the Mellin series of* $f \in Y_c$, also to be called the *Mellin–Cesaro* or *Mellin–Fejér means of* f, are given by

$$(\sigma_n^c f)(x) := \frac{1}{n+1}\sum_{k=0}^{n}(S_k^c f)(x) \qquad (x \in \mathbb{R}^+,\ n \in \mathbb{N}_0). \tag{5.15}$$

The following lemma provides some useful formulae for the (finite) Mellin–Fejér kernel analogous to those of the Mellin–Dirichlet kernel.

Lemma 5.12 *Let* $c \in \mathbb{R}$.

(a) The Mellin–Fejér kernel $\{F_n^c\}_{n\in\mathbb{N}_0}$ *has the representations*

$$F_n^c(x) = \frac{x^{-c}}{2\pi}\sum_{k=-n}^{n}\left(1 - \frac{|k|}{n+1}\right)x^{-ik} \qquad (x \in \mathbb{R}^+), \tag{5.16}$$

$$F_n^c(x) = \begin{cases} \dfrac{x^{-c}}{2\pi}\dfrac{1}{n+1}\left(\dfrac{x^{i(n+1)/2} - x^{-i(n+1)/2}}{x^{i/2} - x^{-i/2}}\right)^2, & x \neq e^{2\pi j} \\[2ex] \dfrac{n+1}{2\pi}e^{-2\pi jc}, & x = e^{2\pi j} \end{cases} \qquad (j \in \mathbb{Z}) \tag{5.17}$$

and

$$F_n^c(x) = \begin{cases} \dfrac{x^{-c}}{2\pi}\dfrac{1}{n+1}\left[\dfrac{\sin\left(\frac{n+1}{2}\log x\right)}{\sin\left(\frac{1}{2}\log x\right)}\right]^2, & x \neq e^{2\pi j} \\[2ex] \dfrac{n+1}{2\pi}e^{-2\pi jc}, & x = e^{2\pi j} \end{cases} \qquad (j \in \mathbb{Z}). \tag{5.18}$$

Further, $F_n^0(x) = F_n^0(1/x)$ *for all* $n \in \mathbb{N}_0$ *and* $x \in \mathbb{R}^+$.

(b) For $f \in Y_c$*, the Mellin–Cesaro means* $\sigma_n^c f$ *belong to* Y_c *for all* $n \in \mathbb{N}_0$ *and the following representations hold:*

$$(\sigma_n^c f)(x) = (f * F_n^c)(x) \qquad (x \in \mathbb{R}^+) \tag{5.19}$$

and

$$(\sigma_n^c f)(x) = \frac{x^{-c}}{2\pi} \sum_{k=-n}^{n} \left(1 - \frac{|k|}{n+1}\right) [f]^{\wedge}_{\mathscr{M}^c}(k) x^{-ik} \qquad (x \in \mathbb{R}^+). \quad (5.20)$$

Proof Using the definition of the Mellin–Dirichlet kernel and interchanging the finite sums, we obtain

$$\begin{aligned} F_n^c(x) &= \frac{1}{n+1} \sum_{j=0}^{n} \left\{ \frac{x^{-c}}{2\pi} \sum_{k=-j}^{j} x^{-ik} \right\} \\ &= \frac{x^{-c}}{2\pi} \sum_{j=0}^{n} \frac{1}{n+1} \left\{ -1 + \sum_{k=0}^{j} (x^{-ik} + x^{ik}) \right\} \\ &= \frac{x^{-c}}{2\pi} \frac{1}{n+1} \left\{ -(n+1) + \sum_{k=0}^{n} (n+1-k)(x^{-ik} + x^{ik}) \right\} \\ &= \frac{x^{-c}}{2\pi} \sum_{k=-n}^{n} \left(1 - \frac{|k|}{n+1}\right) x^{-ik}, \end{aligned}$$

which is (5.16). In the next two representations the values of $F_n^c(x)$ for $x = e^{2\pi j}$ with $j \in \mathbb{Z}$ are obvious. Now let $x \notin \{e^{2\pi j} : j \in \mathbb{Z}\}$. Concerning (5.17), we first note that

$$(x^{i/2} - x^{-i/2})(x^{i(2k+1)/2} - x^{-i(2k+1)/2}) = x^{i(k+1)} + x^{-i(k+1)} - (x^{ik} + x^{-ik}).$$

Thus, multiplying the Mellin–Dirichlet kernel (5.5) by $x^{i/2} - x^{-i/2}$ yields a telescoping sum and so

$$\begin{aligned} (x^{i/2} - x^{-i/2}) F_n^c(x) &= \frac{1}{2\pi} \frac{x^{-c}}{n+1} \sum_{k=0}^{n} \frac{x^{i(k+1)} + x^{-i(k+1)} - (x^{ik} + x^{-ik})}{x^{i/2} - x^{-i/2}} \\ &= \frac{1}{2\pi} \frac{x^{-c}}{n+1} \frac{x^{i(n+1)} + x^{-i(n+1)} - 2}{x^{i/2} - x^{-i/2}} \\ &= \frac{1}{2\pi} \frac{x^{-c}}{n+1} \frac{(x^{i(n+1)/2} - x^{-i(n+1)/2})^2}{x^{i/2} - x^{-i/2}}, \end{aligned}$$

which leads us to (5.17). The representation (5.18) follows by an application of the Euler equation for the sine function. The rest of statement (a) is obvious.

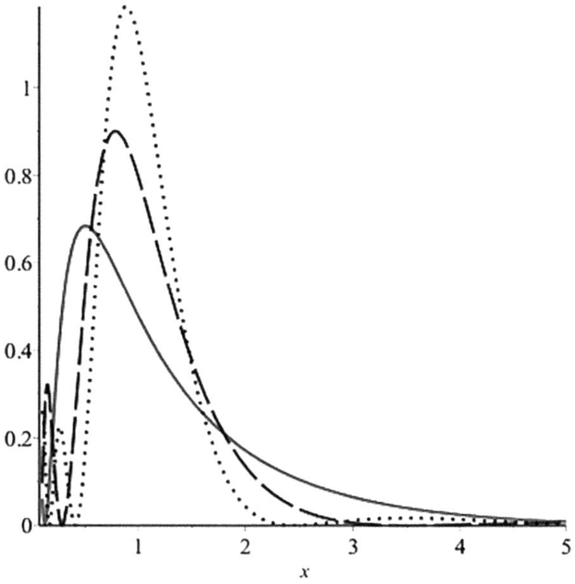

Fig. 5.2 The Mellin–Fejér kernel, with $c = 1$ and $n = 2$ (solid line), $n = 4$ (dashed line) and $n = 6$ (dotted line)

The representation (5.19) is an immediate consequence of (5.8). Finally, (5.19) in conjunction with (5.16) yields

$$
\begin{aligned}
(\sigma_n^c f)(x) &= \frac{1}{2\pi}\int_{e^{-\pi}}^{e^{\pi}} f(u)\sum_{k=-n}^{n}\left(1-\frac{|k|}{n+1}\right)\left(\frac{x}{u}\right)^{-c-ik}\frac{du}{u} \\
&= \frac{1}{2\pi}\sum_{k=-n}^{n}\left(1-\frac{|k|}{n+1}\right)x^{-c-ik}[f]^{\wedge}_{\mathscr{M}^c}(k),
\end{aligned}
$$

which completes the proof (Fig. 5.2). □

The next proposition presents some properties of the (finite) Mellin–Fejér kernel.

Proposition 5.13 *The Mellin–Fejér kernel* $\{F_n^c\}_{n\in\mathbb{N}_0}$ *consists of real-valued, non-negative, continuous functions with*

$$[F_n^c]^{\wedge}_{\mathscr{M}^c}(0) = \|F_n^c\|_{Y_c} = 1 \qquad (n \in \mathbb{N}_0). \tag{5.21}$$

Moreover, if $E_\delta := \{x \in [e^{-\pi}, e^{\pi}] : |1-x| \geq \delta\}$ *for arbitrary* $0 < \delta < 1 - e^{-\pi}$, *then*

$$\lim_{n\to\infty}\int_{E_\delta} F_n^c(u)u^{c-1}du = 0. \tag{5.22}$$

Proof First, by (5.18), it is immediately seen that F_n^c is a real-valued, nonnegative function. Also the continuity can be easily verified. Thus, in view of the nonnegativity and Proposition 5.5, we obtain

$$\|F_n^c\|_{Y_c} = [F_n^c]^\wedge_{\mathscr{M}^c}(0) = \frac{1}{2\pi}\sum_{k=-n}^{n}\left(1 - \frac{|k|}{n+1}\right)[(\cdot)^{-c-ik}]^\wedge_{\mathscr{M}^c}(0) = 1,$$

which is (5.21).

Next we note that

$$x^{i/2} - x^{-i/2} = 2i \sin\left(\frac{\log x}{2}\right)$$

is continuous and different from zero on the compact set E_δ. Hence there exists a positive constant $K = K(\delta) > 0$ such that $|x^{i/2} - x^{-i/2}| \geq K$ for $x \in E_\delta$. Therefore, by (5.17),

$$\int_{E_\delta} F_n^c(u)u^{c-1}du \leq \frac{1}{2\pi}\frac{1}{n+1}\int_{e^{-\pi}}^{e^\pi}\frac{4}{K^2}\frac{du}{u} = \frac{4}{(n+1)K^2},$$

which implies (5.22). □

The following theorem shows an important approximation property of the Mellin–Cesaro means.

Theorem 5.14 *Let $f \in Y_c$ with $c \in \mathbb{R}$. Then*

$$\lim_{n\to\infty} \|\sigma_n^c f - f\|_{Y_c} = 0. \tag{5.23}$$

Proof For E_δ as introduced in Proposition 5.13, define $\widetilde{E}_\delta := [e^{-\pi}, e^\pi] \setminus E_\delta$. Then, by Proposition 3.16, for each $\varepsilon > 0$ there exists a $\delta \in]0, 1 - e^{-\pi}[$ such that

$$\|\tau^c_{1/u} f - f\|_{Y_c} < \varepsilon \qquad (u \in \widetilde{E}_\delta). \tag{5.24}$$

By (5.19), (5.21) and Fubini's theorem, we obtain

$$\begin{aligned}\|\sigma_n^c f - f\|_{Y_c} &= \int_{e^{-\pi}}^{e^\pi}\left|\int_{e^{-\pi}}^{e^\pi}\big((\tau^c_{1/u} f)(x) - f(x)\big)F_n^c(u)u^{c-1}du\right|x^{c-1}dx \\ &\leq \left\{\int_{E_\delta} + \int_{\widetilde{E}_\delta}\right\}\|\tau^c_{1/u} f - f\|_{Y_c} F_n^c(u)u^{c-1}du.\end{aligned}$$

Next we note that

$$\|\tau^c_{1/u} f - f\|_{Y_c} \leq \|\tau^c_{1/u} f\|_{Y_c} + \|f\|_{Y_c} = 2\|f\|_{Y_c}.$$

Now (5.24) and (5.22) allow us to conclude that

$$\limsup_{n\to\infty} \|\sigma_n^c f - f\|_{Y_c} \leq \limsup_{n\to\infty} \left\{ 2\|f\|_{Y_c} \int_{E_\delta} F_n^c(u)u^{c-1}du + \varepsilon \|F_n^c\|_{Y_c} \right\} = \varepsilon.$$

This completes the proof. □

An important consequence of Theorem 5.14 is the following identity theorem for the finite Mellin transform.

Theorem 5.15 *Let* $f_1, f_2 \in Y_c$ *for* $c \in \mathbb{R}$. *If* $[f_1]^\wedge_{\mathscr{M}^c}(k) = [f_2]^\wedge_{\mathscr{M}^c}(k)$ *for all* $k \in \mathbb{Z}$, *then* $f_1 = f_2$ *almost everywhere.*

Proof Applying the representation (5.20) for the Mellin–Cesaro means to $g := f_1 - f_2$, we have

$$(\sigma_n^c g)(x) = \frac{x^{-c}}{2\pi} \sum_{k=-n}^{n} \left(1 - \frac{|k|}{n+1}\right) [g]^\wedge_{\mathscr{M}^c}(k) x^{-ik},$$

and the assumption on the Mellin coefficients implies that $(\sigma_n^c g)(x) = 0$ for all $n \in \mathbb{N}_0$. Thus the desired result follows from Theorem 5.14. □

5.3 Convergence of the Mellin–Fourier Series

We begin with a very simple example of a function f for which the Mellin series (5.2) converges to f. Let us consider the polynomial of Example 3.9:

$$p(x^i) = \sum_{j=-n}^{n} a_j x^{ij} \qquad (x \in \mathbb{R}^+).$$

As shown there, the function $(\cdot)^{-c} p((\cdot)^i)$ belongs to Y_c for $c \in \mathbb{R}$. By Proposition 5.5, its Mellin coefficients are given by

$$[(\cdot)^{-c} p((\cdot)^i)]^\wedge_{\mathscr{M}^c}(k) = 2\pi \sum_{j=-n}^{n} a_j \delta_{-j,k} \qquad (k \in \mathbb{Z}).$$

Therefore we obtain the convergence of the Mellin series

$$\frac{x^{-c}}{2\pi} \sum_{k=-\infty}^{\infty} [(\cdot)^{-c} p((\cdot)^i)]^\wedge_{\mathscr{M}^c}(k) x^{-ik} = x^{-c} \sum_{k=-\infty}^{\infty} \sum_{j=-n}^{n} a_j \delta_{-j,k} x^{-ik} = x^{-c} p(x^i).$$

In particular, for the function g defined by

$$g(x) := x^{-c}\sin(\log x) \qquad (x \in \mathbb{R}^+),$$

the Mellin series converges to g since $g(x) = ix^{-c}(x^{-i} - x^{i})/2$.

In this section we establish some general results for the convergence of the Mellin series in the pointwise sense. First we introduce a strengthened notion of convergence.

Definition 5.16 A series $\sum_{k=-\infty}^{\infty} f_k(x)$ of complex-valued functions f_k defined on a subset $A \subset \mathbb{R}^+$ is said to be *dominatedly convergent* if it converges and there exists a function $g \in Y_c$ such that for every $n \in \mathbb{N}$,

$$\left|\sum_{k=-n}^{n} f_k(x)\right| \le g(x) \qquad (\text{a.e. for } x \in A).$$

We begin with the following proposition.

Proposition 5.17 *Let $f \in Y_c$, where $c \in \mathbb{R}$.*

(a) If f has a representation

$$f(x) = \frac{x^{-c}}{2\pi}\sum_{k=-\infty}^{\infty} a_k x^{-ik} \qquad (\textit{a.e. for } x \in \mathbb{R}^+) \tag{5.25}$$

and the series converges dominatedly, then $a_k = [f]^{\wedge}_{\mathscr{M}^c}(k)$ for all $k \in \mathbb{Z}$.

(b) If the Mellin series of f converges dominatedly, then it represents f almost everywhere on $\mathbb{R}^+$.

(c) If f is continuous on $\mathbb{R}^+$ and its Mellin series converges uniformly, then it represents f everywhere on $\mathbb{R}^+$.

Proof Under the assumptions of statement (a), we may employ Lebesgue's dominated convergence theorem (Proposition 1.4) to conclude that

$$[f]^{\wedge}_{\mathscr{M}^c}(k) = \int_{e^{-\pi}}^{e^{\pi}} f(u)u^{c+ik-1}du = \frac{1}{2\pi}\sum_{j=-\infty}^{\infty} a_j \int_{e^{-\pi}}^{e^{\pi}} u^{-c-ij}u^{c+ik}\frac{du}{u} = a_k,$$

where the orthogonality relation of Proposition 5.5 has been used in the last step.

As to (b), since the series

$$\sum_{j=-\infty}^{\infty} [f]^{\wedge}_{\mathscr{M}^c}(j)x^{-c-ij}$$

converges dominatedly, Lebesgue's theorem and the orthogonality relation give for its Mellin coefficients

$$\left[\frac{1}{2\pi}\sum_{j=-\infty}^{\infty}[f]^{\wedge}_{\mathscr{M}^c}(j)(\cdot)^{-(c+ij)}\right]^{\wedge}_{\mathscr{M}^c}(k)$$

$$=\frac{1}{2\pi}\sum_{j=-\infty}^{\infty}[f]^{\wedge}_{\mathscr{M}^c}(j)\int_{e^{-\pi}}^{e^{\pi}}u^{-i(k-j)}\,\frac{du}{u}=[f]^{\wedge}_{\mathscr{M}^c}(k).$$

Since this holds for all $k\in\mathbb{Z}$, the identity theorem for the finite Mellin transform (Theorem 5.15) shows that f and its Mellin series coincide almost everywhere on $\mathbb{R}^+$.

Concerning (c), it is easily verified that uniform convergence implies dominated convergence, and so the conclusion of statement (b) holds. In addition, uniform convergence guarantees that the Mellin series of f is a continuous function. Hence it must coincide with f everywhere on $\mathbb{R}^+$. □

Corollary 5.18 *Let $f,g\in Y_c$, where $c\in\mathbb{R}$. Suppose that at least one of the associated Mellin series converges absolutely. Then the convolution $f*g$ has the representation*

$$(f*g)(x)=\frac{x^{-c}}{2\pi}\sum_{k=-\infty}^{\infty}[f]^{\wedge}_{\mathscr{M}^c}(k)[g]^{\wedge}_{\mathscr{M}^c}(k)x^{-ik}\qquad(\text{a.e. for } x\in\mathbb{R}^+)\qquad(5.26)$$

with the series converging absolutely.

Proof By Proposition 5.3, the series in (5.26) is the Mellin series of the convolution $f*g$. In view of Proposition 5.4, the absolute convergence of one of the Mellin series entails the absolute convergence of the series in (5.26). Since an absolutely convergent Mellin series is easily seen to be dominatedly convergent, the representation (5.26) follows from statement (b) of Proposition 5.17. □

The next result is the counterpart of the classical Dini theorem for pointwise convergence.

Theorem 5.19 *Let $f\in Y_c$, where $c\in\mathbb{R}$, and let $x_0\in[e^{-\pi},e^{\pi}]$ be arbitrary. If*

$$H_{x_0}(x):=\frac{(\tau^c_x f)(x_0)-f(x_0)}{x^i-1}$$

belongs to Y_0 as a function of $x\in\mathbb{R}^+$, then

$$f(x_0)=\frac{1}{2\pi}\sum_{k=-m}^{n}[f]^{\wedge}_{\mathscr{M}^c}(k)x_0^{-c-ik}-[H_{x_0}]^{\wedge}_{\mathscr{M}^0}(n+1)+[H_{x_0}]^{\wedge}_{\mathscr{M}^0}(-m),\quad(5.27)$$

for all $m, n \in \mathbb{N}$. *In particular, the Mellin series of* f *converges at the point* x_0, *i.e.,*

$$f(x_0) = \frac{1}{2\pi} \sum_{k=-\infty}^{\infty} [f]^{\wedge}_{\mathscr{M}^c}(k) x_0^{-c-ik}. \tag{5.28}$$

Proof Interpreting $f(x_0)$ as the constant function that maps $\mathbb{R}^+$ to $f(x_0)$, we see from Proposition 5.5 that $[f(x_0)]^{\wedge}_{\mathscr{M}^0}(k) = 2\pi f(x_0)\delta_{0,k}$ and so

$$f(x_0) = \frac{1}{2\pi} \sum_{k=-m}^{n} [f(x_0)]^{\wedge}_{\mathscr{M}^0}(k). \tag{5.29}$$

Now, using first statements (b) and (c) of Proposition 5.2, then combining the result with (5.29), substituting H_{x_0} and finally employing statement (d) of Proposition 5.2, we find that

$$\begin{aligned}
\frac{1}{2\pi} \sum_{k=-m}^{n} [f]^{\wedge}_{\mathscr{M}^c}(k) x_0^{-c-ik} &= \frac{1}{2\pi} \sum_{k=-m}^{n} [(\cdot)^c f(\cdot\, x_0)]^{\wedge}_{\mathscr{M}^0}(k) \\
&= f(x_0) + \frac{1}{2\pi} \sum_{k=-m}^{n} [(\cdot)^c f(\cdot\, x_0) - f(x_0)]^{\wedge}_{\mathscr{M}^0}(k) \\
&= f(x_0) + \frac{1}{2\pi} \sum_{k=-m}^{n} \left[\left((\cdot)^i - 1\right) H_{x_0}\right]^{\wedge}_{\mathscr{M}^0}(k) \\
&= f(x_0) + \frac{1}{2\pi} \sum_{k=-m}^{n} \left\{[H_{x_0}]^{\wedge}_{\mathscr{M}^0}(k+1) - [H_{x_0}]^{\wedge}_{\mathscr{M}^0}(k)\right\}.
\end{aligned}$$

The last expression is a telescoping sum, which leads us to (5.27) immediately. Since by assumption $H_{x_0} \in Y_0$, we may use Proposition 5.4 to deduce (5.28) from (5.27) by letting $m, n \to \infty$. □

Corollary 5.20 *Let* $f \in Y_c$, *where* $c \in \mathbb{R}$. *Suppose that* f *is differentiable at* $x_0 \in [e^{-\pi}, e^{\pi}]$. *Then*

$$f(x_0) = \frac{1}{2\pi} \sum_{k=-\infty}^{\infty} [f]^{\wedge}_{\mathscr{M}^c}(k) x_0^{-c-ik}.$$

Proof It suffices to show that the function H_{x_0} introduced in Theorem 5.19 belongs to Y_0. First we verify by a simple calculation that the c-recurrency of f entails the 0-recurrency of H_{x_0}. It remains to show that $\left|H_{x_0}(x)/x\right|$ is integrable on $[e^{-\pi}, e^{\pi}]$. Outside an arbitrarily small neighborhood of 1, the integrability is inherited from that of f. Using that f is differentiable at x_0, we find

$$\lim_{x\to 1} \frac{H_{x_0}(x)}{x} = \lim_{x\to 1} \frac{(\tau_x^c f)(x_0) - f(x_0)}{x^{i+1} - x}$$
$$= \lim_{x\to 1} \left\{ x^c \frac{f(xx_0) - f(x_0)}{xx_0 - x_0} \frac{xx_0 - x_0}{x^{i+1} - x} + f(x_0) \frac{x^c - 1}{x^{i+1} - x} \right\}$$
$$= -ix_0 f'(x_0) - icf(x_0).$$

This guarantees integrability on the whole of $[e^{-\pi}, e^{\pi}]$. □

Remark 5.21 Note that for a differentiable function in Y_c, the classical derivative f' is not necessarily in Y_c because it may fail to be c-recurrent. On the other hand, the Mellin derivative Θ_c of Definition 3.19 retains the c-recurrency of f. Indeed, we have

$$\Theta_c f(x) = xf'(x) + cf(x) = \lim_{h\to 1} \frac{(\tau_h^c f)(x) - f(x)}{h - 1}, \tag{5.30}$$

and the last quotient is seen to be c-recurrent.

Corollary 5.22 *Let $f \in Y_c$ for $c \in \mathbb{R}$ be a locally absolutely continuous on $\mathbb{R}^+$ such that the Mellin derivative $\Theta_c f$ is essentially bounded on $[e^{-\pi}, e^{\pi}]$. Then the partial sums of the Mellin series are uniformly bounded on $[e^{-\pi}, e^{\pi}]$, more precisely,*

$$\left| \frac{1}{2\pi} \sum_{k=-m}^{n} [f]^{\wedge}_{\mathscr{M}^c}(k) x^{-c-ik} \right| \le x^{-c} \left(|f(1)| + M \operatorname*{ess\,sup}_{u\in[e^{-\pi}, e^{\pi}]} |u^c \Theta_c f(u)| \right) \tag{5.31}$$

for all $m, n \in \mathbb{N}$ with a universal constant $M > 0$, and

$$f(x) = \frac{1}{2\pi} \sum_{k=-\infty}^{\infty} [f]^{\wedge}_{\mathscr{M}^c}(k) x^{-c-ik} \tag{5.32}$$

for all $x \in \mathbb{R}^+$.

Proof We want to deduce (5.32) from Eq. (5.27) of Theorem 5.19 with the roles of x_0 and $H_{x_0}(x)$ being taken by an arbitrary $x \in \mathbb{R}^+$ and $H_x(y)$, respectively. For this, we have first to show that $H_x(y)$ belongs to Y_0 as a function of $y \in \mathbb{R}^+$. Again, the 0-recurrency of H_x is an easy consequence of the c-recurrency of f. Since

$$\frac{d}{du}\big(f(u)u^c\big) = \Theta_c f(u) u^{c-1} \qquad (u \in \mathbb{R}^+),$$

we obtain by substituting $u = x/v$,

$$(\tau_y^c f)(x) - f(x) = x^{-c} \int_x^{xy} \Theta_c f(u) u^c \frac{du}{u} = x^{-c} \int_{1/y}^{1} \left(\frac{x}{v}\right)^c \Theta_c f\left(\frac{x}{v}\right) \frac{dv}{v}.$$

Therefore the numerator of $H_x(y)$ can be estimated as

$$|(\tau_y^c f)(x) - f(x)| \leq x^{-c} \operatorname*{ess\,sup}_{u \in [e^{-\pi}, e^{\pi}]} |u^c \Theta_c f(u)| \, |\log y|. \tag{5.33}$$

Since $\lim_{y \to 1} \log y/(y^i - 1) = -i$, we see that

$$\kappa := \int_{e^{-\pi}}^{e^{\pi}} \left| \frac{\log y}{y^{i+1} - y} \right| dy < \infty.$$

Hence

$$\left| \frac{H_x(y)}{y} \right| = \left| \frac{(\tau_y^c f)(x) - f(x)}{y^{i+1} - y} \right|$$

is integrable on $[e^{-\pi}, e^{\pi}]$ with respect to y. Thus $H_x \in Y_0$ and

$$\|H_x\|_{Y_0} \leq x^{-c} \kappa \operatorname*{ess\,sup}_{u \in [e^{-\pi}, e^{\pi}]} |u^c \Theta_c f(u)|. \tag{5.34}$$

Therefore (5.27) holds with the role of x_0 being taken by any $x \in \mathbb{R}^+$. Recalling Proposition 5.4, we obtain (5.32) by letting $m, n \to \infty$.

In the extended equation (5.27), we may estimate the Mellin coefficients of H_x by $\|H_x\|_{Y_0}$, as allowed by Proposition 5.2 (a), and use (5.34) to conclude that

$$\left| f(x) - \frac{1}{2\pi} \sum_{k=-m}^{n} [f]^\wedge_{\mathscr{M}^c}(k) x^{-c-ik} \right|$$
$$\leq x^{-c} 2\kappa \operatorname*{ess\,sup}_{u \in [e^{-\pi}, e^{\pi}]} |u^c \Theta_c f(u)| \qquad \big(x \in [e^{-\pi}, e^{\pi}]\big).$$

Inequality (5.33) with $x = 1$ and the variable y now called x implies that

$$|f(x)| \leq x^{-c} \Big(|f(1)| + \pi \operatorname*{ess\,sup}_{u \in [e^{-\pi}, e^{\pi}]} |u^c \Theta_c f(u)| \Big) \qquad \big(x \in [e^{-\pi}, e^{\pi}]\big).$$

Combining these two inequalities, we obtain

$$\left| \frac{1}{2\pi} \sum_{k=-m}^{n} [f]^\wedge_{\mathscr{M}^c}(k) x^{-c-ik} \right| \leq x^{-c} \Big(|f(1)| + (2\kappa + \pi) \operatorname*{ess\,sup}_{u \in [e^{-\pi}, e^{\pi}]} |u^c \Theta_c f(u)| \Big).$$

This shows that (5.31) holds on $[e^{-\pi}, e^{\pi}]$ with $M = 2\kappa + \pi$. □

5.4 Y_c^2-theory

In this section we will consider the case of functions belonging to Y_c^p. Since by the Hölder inequality $Y_c^p \subset Y_c$ for every $c \in \mathbb{R}$ and $p \geq 1$, the theory developed in the previous sections of this chapter can be carried on to the spaces Y_c^p. Here we briefly turn to the case $p = 2$.

We recall that Y_c^2 is a Hilbert space under the scalar product

$$\langle f, g \rangle_{Y_c^2} := \int_{e^{-\pi}}^{e^{\pi}} f(u)\overline{g(u)}u^{2c}\,\frac{du}{u}.$$

By Proposition 5.5 the family of functions $\Phi := (\varphi_{c,k})_{k\in\mathbb{Z}}$, where $\varphi_{c,k}(x) = x^{-(c+ik)}$, is an orthogonal set in Y_c^2, and for $f \in Y_c^2$, there holds

$$[f]^{\wedge}_{\mathscr{M}^c}(k) = \langle f, \varphi_{c,k} \rangle_{Y_c^2} \qquad (k \in \mathbb{Z}).$$

By Theorem 5.15 the system Φ is complete, which implies a Mellin version of the Fischer–Riesz theorem (see, e.g., [95, p. 176]). From the classical theory of Hilbert spaces, we state the following theorem, which summarizes the main facts concerning the (finite) Mellin transform in Y_c^2.

Theorem 5.23 *For Y_c^2, where $c \in \mathbb{R}$, the finite Mellin transform*

$$\mathscr{M}^c : Y_c^2 \longrightarrow \ell^2(\mathbb{Z}), \qquad f \longmapsto \big([f]^{\wedge}_{\mathscr{M}^c}(k)\big)_{k\in\mathbb{Z}} =: [f]^{\wedge}_{\mathscr{M}^c}$$

forms an isomorphism from Y_c^2 onto $\ell^2(\mathbb{Z})$ preserving the inner product, i.e., for $f, g \in Y_c^2$,

$$\langle f, g \rangle_{Y_c^2} := \int_{e^{-\pi}}^{e^{\pi}} f(u)\overline{g(u)}u^{2c}\,\frac{du}{u} \tag{5.35}$$

$$= \frac{1}{2\pi}\langle [f]^{\wedge}_{\mathscr{M}^c}, [g]^{\wedge}_{\mathscr{M}^c} \rangle_{\ell^2} = \frac{1}{2\pi}\sum_{k\in\mathbb{Z}} [f]^{\wedge}_{\mathscr{M}^c}(k)\overline{[g]^{\wedge}_{\mathscr{M}^c}(k)}. \tag{5.36}$$

In particular, the mapping $f \mapsto \big([f]^{\wedge}_{\mathscr{M}^c}(k)\big)_{k\in\mathbb{Z}}$ is an isometry between Y_c^2 and $\ell^2(\mathbb{Z})$ with

$$\|f\|^2_{Y_c^2} = \frac{1}{2\pi}\sum_{k\in\mathbb{Z}} |[f]^{\wedge}_{\mathscr{M}^c}(k)|^2 = \frac{1}{2\pi}\|[f]^{\wedge}_{\mathscr{M}^c}\|^2_{\ell^2}. \tag{5.37}$$

Furthermore, each $f \in Y_c^2$ has a unique representation by its Mellin series

$$f(u) = \frac{1}{2\pi} \sum_{k \in \mathbb{Z}} [f]^{\wedge}_{\mathscr{M}^c}(k) u^{-c-ik}, \tag{5.38}$$

the convergence being understood in the Y_c^2 *norm. If* $(a_k)_{k \in \mathbb{Z}} \in \ell^2(\mathbb{Z})$, *then there exists a unique function* $f \in Y_c^2$ *such that* $[f]^{\wedge}_{\mathscr{M}^c}(k) = a_k$ *for all* $k \in \mathbb{Z}$.

Details are left to the reader.

5.5 Notes

To the best of our knowledge, Butzer and Jansche [91] initiated a systematic study of the counterparts of Fourier series and finite Fourier transform in the Mellin setting. Although the two types of transforms are related, these authors presented a direct approach to the Mellin instance, completely independent of the Fourier theory and under minimal, natural, unified assumptions on the involved functions. Also the notion of c-recurrency, being the counterpart of periodicity, is due to them. In [92, Sect. 3] they continued with an L^2 theory. Furthermore, in [92, Sect. 6], they considered the case when the c-recurrency is defined for an arbitrary interval.

The material in Sects. 5.1–5.3 is borrowed from [91] (see also Butzer and Jansche [93]) under occasional modifications of statements and proofs. In particular, this is the case for our Corollaries 5.18 and 5.22. The contents of Sect. 5.4 originates from [92, Sect. 3].

Theoretically it would be possible to gain part of the contributions of this chapter as applications of corresponding general results on harmonic analysis on locally compact abelian groups; see, e.g., Hewitt and Ross [170, Chap. 9, Sect. 34]. But it seems that such an approach has not yet been performed in the literature.

We briefly discuss some computational aspects which will lead us to a *discrete finite Mellin transform*. Let $f \in Y_c$ for $c \in \mathbb{R}$. Then $f_c := (\cdot)^c f \in Y_0$ and

$$[f]^{\wedge}_{\mathscr{M}^c}(k) = [f_c]^{\wedge}_{\mathscr{M}^0}(k) = \int_{e^{-\pi}}^{e^{\pi}} f_c(u) u^{ik} \frac{du}{u} = \int_{-\pi}^{\pi} f_c(e^t) e^{ikt} \, dt \qquad (k \in \mathbb{Z}). \tag{5.39}$$

In general, the integrals cannot be calculated in closed form. We have to use a quadrature formula and decide for the trapezoidal method. Subsequently it will turn out to be a very reasonable choice.

For $N \in \mathbb{N}$, we write

$$[f_c]^{\wedge}_{\mathscr{M}^0_N}(k) := \frac{\pi}{N} \sum_{j=-N+1}^{N} f_c(e^{\pi j/N}) e^{i\pi kj/N} \qquad (k = -N+1, \ldots, N). \tag{5.40}$$

If f_c is continuous, then the right-hand side converges to the integral in (5.39) as $N \to \infty$. Therefore we may consider $[f_c]^{\wedge}_{\mathscr{M}^0_N}(k)$ as an approximation of the Mellin coefficient $[f_c]^{\wedge}_{\mathscr{M}^0}(k)$. Equation (5.40) describes a discrete finite Mellin transform. The vector

$$\boldsymbol{f}_{c,N} := \left(f_c(e^{h(-N+1)}), \ldots, f_c(e^{hN})\right) \in \mathbb{C}^{2N} \qquad (h := \pi/N)$$

is a discretization of the function f_c and

$$[\boldsymbol{f}_{c,N}]^{\wedge}_{\mathscr{M}^0_N} := \left([f_c]^{\wedge}_{\mathscr{M}^0_N}(-N+1), \ldots, [f_c]^{\wedge}_{\mathscr{M}^0_N}(N)\right) \in \mathbb{C}^{2N}$$

is a truncated sequence of approximate Mellin coefficients. Introducing the $2N \times 2N$ matrix $\boldsymbol{A}_N$ with entries

$$a_{\ell,m} := \frac{(-1)^{N-\ell-m}}{2N} e^{i\pi\ell m/N} \qquad (\ell, m \in \{1, \ldots, 2N\}),$$

we can rewrite the discrete finite Mellin transform in matrix form as

$$\frac{1}{2\pi}\,[\boldsymbol{f}_{c,N}]^{\wedge\mathrm{T}}_{\mathscr{M}^0_N} = \boldsymbol{A}_N\,\boldsymbol{f}^{\mathrm{T}}_{c,N}, \tag{5.41}$$

where the superscript T, referring to "transpose", means that the $2N$-tuples in bold appear as columns. As a benefit of our choice of the quadrature formula, the inverse of the matrix $\boldsymbol{A}_N$ can be easily determined. In fact, $\boldsymbol{A}_N^{-1} = 2N\overline{\boldsymbol{A}}_N$, where the bar means conjugation. Therefore the inversion of the discrete finite Mellin transform is simply given by

$$\frac{\pi}{N}\boldsymbol{f}^{\mathrm{T}}_{c,N} = \overline{\boldsymbol{A}}_N\,[\boldsymbol{f}_{c,N}]^{\wedge\mathrm{T}}_{\mathscr{M}^0_N}. \tag{5.42}$$

The sum

$$\left(\Sigma_N f_c\right)(x) := \frac{1}{2\pi} \sum_{k=-N+1}^{N} [f_c]^{\wedge}_{\mathscr{M}^0_N}(k) x^{-ik}$$

is an approximation of the Mellin series of f_c that interpolates with respect to the points $e^{\pi\ell/N}$ for $\ell = -N+1, \ldots, N$. Indeed, as a consequence of (5.42), we have

$$\left(\Sigma_N f_c\right)(e^{\pi\ell/N}) = f_c(e^{\pi\ell/N}) \qquad (\ell = -N+1, \ldots, N).$$

Therefore the approximation of f_c by $\Sigma_N f_c$ can be seen as the Mellin analogue of the interpolation of a 2π-periodic function by a trigonometric polynomial of degree N.

From (5.41) or alternatively from (5.42), it follows that

$$\frac{\pi}{N} \sum_{j=-N+1}^{N} \left| f_c(e^{\pi j/N}) \right|^2 = \frac{1}{2\pi} \sum_{k=-N+1}^{N} \left| [f_c]^{\wedge}_{\mathscr{M}_N^0}(k) \right|^2 .$$

This is a discrete analogue of the isometry between Y_c^2 and $\ell^2(\mathbb{Z})$ expressed in (5.37). The left-hand side is an approximation of $\|f\|_{Y_c^2}^2 = \|f_c\|_{Y_0^2}^2$ obtained by employing the trapezoidal method with $2N$ nodes.

The described approach has a structure that allows us to gain results for the discretization in $\mathbb{C}^{2N}$ from discretizations in $\mathbb{C}^N$ in an efficient way. Consider c and N to be fixed, and let us simplify the notation by writing

$$x_j := \frac{\pi}{N} f_c(e^{\pi j/N}), \quad y_k := (-1)^k [f_c]^{\wedge}_{\mathscr{M}_N^0}(k) \qquad \big(j, k \in \{1, \dots, 2N\}\big).$$

Then (5.40) becomes

$$y_k = (-1)^k \sum_{j=-N+1}^{N} x_j e^{i\pi k j/N} = \sum_{j=1}^{2N} x_{j-N} e^{i\pi k j/N}.$$

Now we split the sum on the right-hand side into two parts by summing over the even and the odd indices j separately, which leads us to

$$\begin{aligned} y_k &= \sum_{j=1}^{N} x_{2j-N} e^{i\pi k 2j/N} + \sum_{j=1}^{N} x_{2j-1-N} e^{i\pi k(2j-1)/N} \\ &= \sum_{j=1}^{N} x_{2j-N} e^{i2\pi kj/N} + e^{-i\pi k/N} \sum_{j=1}^{N} x_{2j-1-N} e^{i2\pi kj/N} \\ &=: E_{N,k} + e^{-i\pi k/N} O_{N,k}. \end{aligned}$$

This may not look like a gain but the crux is that

$$y_{k-N} = E_{N,k} - e^{-i\pi k/N} O_{N,k}.$$

Thus we obtain the desired $2N$ coefficients $y_{-N+1}, \dots, y_N$ by computing each of the two sums $E_{N,k}$ and $e^{-i\pi k/N} O_{N,k}$ of lengths N for $k = 1, \dots, N$ only and combining for each k the two results by an addition and a subtraction. Note that a naive computation would have required the evaluation of a sum of length $2N$ for $2N$ values of k.

The sums $E_{N,k}$ and $O_{N,k}$ can be interpreted as discrete finite Mellin transforms in $\mathbb{C}^N$. If N is even, we can split them analogously. Then we have to compute four sums of lengths $N/2$ for $k = 1, \ldots, N/2$ only and obtain all the coefficients by $\pm$ combinations of the results after multiplying with an appropriate root of unity. In particular, when $N = 2^n$, we can continue with splittings recursively till we arrive at one-dimensional discretizations. This is the principle of an algorithm known as *fast Fourier transform* or just FFT; for details see, e.g., Bracewell [76, Sect. 6.10]. The previous considerations show that this algorithm can also be utilized for the discrete finite Mellin transform.

Chapter 6
The Mellin Transform in X_c

In this chapter we introduce the Mellin transform for functions in the space X_c, where $c \in \mathbb{R}$ and study its basic properties.

6.1 Definitions and Preliminary Properties

Definition 6.1 For a function $f \in X_c$, where $c \in \mathbb{R}$, the *Mellin transform* is defined by

$$\begin{aligned} M_c[f](s) \equiv [f]^\wedge_{M_c}(s) &:= \int_0^\infty f(u)u^{s-1}du \\ &= \int_0^\infty f(u)u^s \frac{du}{u} \qquad (s = c + it,\ t \in \mathbb{R}). \end{aligned} \tag{6.1}$$

In both notations M stands for Mellin. Here we shall mostly use the notation $[f]^\wedge_{M_c}(s)$. The other one has some advantages when one is not so much interested in functions but rather wants to study the Mellin transform as a linear operator M_c on function spaces. Such situations occur in the inversion theory.

Remark 6.2 Since $|u^{it}| = 1$, it is obvious that X_c comprises all functions f for which the Mellin transform $[f]^\wedge_{M_c}(c + it)$ exists for all $t \in \mathbb{R}$ as a Lebesgue integral and the norm $\|\cdot\|_{X_c}$ satisfies the inequality

$$|[f]^\wedge_{M_c}(c + it)| \leq \|f\|_{X_c} \qquad (t \in \mathbb{R}). \tag{6.2}$$

Similarly X_I, where $I =]a, b[$ or $I = [a, b]$, is exactly the space for which the Mellin transform exists as a Lebesgue integral for all $t \in \mathbb{R}$ and all $c \in I$.

C. Bardaro et al., *Mellin Analysis, Transform Theory, and Applications*,
https://doi.org/10.1007/978-3-031-96672-9_6

Let us denote by $\mathrm{St}]a,b[$ the open vertical strip $]a,b[+i\mathbb{R} \subset \mathbb{C}$ and, analogously, for a closed subinterval $[\alpha,\beta] \subset]a,b[$ by $\mathrm{St}[\alpha,\beta]$ the closed vertical strip $[\alpha,\beta] + i\mathbb{R} \subset \mathbb{C}$. We define the Mellin transform of a function $f \in X_I$, with $I =]a,b[\subset \mathbb{R}$, on setting

$$M[f](s) \equiv [f]^{\wedge}_{M}(s) := \int_0^{\infty} f(u)u^{s-1}du \quad (s \in \mathrm{St}]a,b[).$$

As to the notation, we want to make the following agreement. When we work with the Mellin transform for functions $f \in X_c$, we prefer to indicate the domain X_c by writing $M_c[f]$ or $[f]^{\wedge}_{M_c}$, while for functions $f \in X_I$, we avoid any subscripts, writing simply $M[f]$ or $[f]^{\wedge}_{M}$. The differences between the two transforms will be pointed out in Remark 6.6 below.

Next we want to state two auxiliary results. The following lemma is readily verified by combining the inequalities (3.1) and (6.2).

Lemma 6.3 *Let $f \in X_{]a,b[}$ and let $[\alpha,\beta] \subset]a,b[$ be a closed subinterval. Then the Mellin transform is uniformly bounded on the strip* $\mathrm{St}[\alpha,\beta]$. *We have*

$$\begin{aligned}|[f]^{\wedge}_{M}(s)| \le \int_0^{\infty} |f(u)u^{s-1}|du \le \|f\|_{X_\alpha} + \|f\|_{X_\beta} \\ =: K_f(\alpha,\beta) \quad (s \in \mathrm{St}[\alpha,\beta]).\end{aligned} \tag{6.3}$$

In the next lemma, we denote by $C(c+i\mathbb{R})$, where $c \in \mathbb{R}$, the space of all continuous functions on the line $c+it$ with $t \in \mathbb{R}$.

Lemma 6.4 *The Mellin transforms for $c \in \mathbb{R}$ and $]a,b[\subset \mathbb{R}$ are mappings of the type*

$$\begin{aligned} M_c &: X_c \longrightarrow C(c+i\mathbb{R}), & f &\longmapsto M_c[f](s), \\ M &: X_{]a,b[} \longrightarrow C(\mathrm{St}]a,b[), & f &\longmapsto M[f](s), \end{aligned}$$

respectively. In particular, the Mellin transform is a continuous function on the line $c+i\mathbb{R}$ and on the strip $\mathrm{St}]a,b[$, *respectively.*

Proof Let $f \in X_c$ and $t \in \mathbb{R}$. Then $|f(\cdot)(\cdot)^{c+i(t+h)-1}| = |f(\cdot)|(\cdot)^{c-1} \in L^1(\mathbb{R}^+)$ for all $h \in \mathbb{R}$. Thus, using Lebesgue's dominated convergence theorem (Proposition 1.4), we have

$$\lim_{h\to 0}[f]^{\wedge}_{M_c}(c+i(t+h)) = \lim_{h\to 0}\int_0^{\infty} f(u)u^{c+i(t+h)-1}du = [f]^{\wedge}_{M_c}(c+it),$$

proving the continuity of $[f]^{\wedge}_{M_c}$ on the line $c+i\mathbb{R}$.

Next, let $f \in X_{]a,b[}$ and let $s = c+it$ be an arbitrary point in $\mathrm{St}]a,b[$. Then there exist $\alpha,\beta \in \mathbb{R}$ with $a < \alpha < c < \beta < b$, and so $\delta := \min\{c-\alpha, \beta-c\} > 0$.

Hence for any $h \in \mathbb{C}$ with $|h| < \delta$, we have $s, s+h \in \mathrm{St}[\alpha, \beta]$ and

$$\begin{aligned} \left|[f]^{\wedge}_{M}(s+h) - |[f]^{\wedge}_{M}(s)\right| &\leq \int_0^\infty \left|f(u)(u^{s+h-1} - u^{s-1})\right| du \\ &\leq \int_0^\infty |f(u)|\, u^{c-1}\, |u^h - 1| du. \end{aligned} \tag{6.4}$$

Now, $\lim_{h\to 0} |u^h - 1| = 0$ for $u \in \mathbb{R}^+$ and

$$|f(u)| u^{c-1} |u^h - 1| \leq |f(u)|\, u^{c+\Re h-1} + |f(u)|\, u^{c-1}.$$

The right-hand side has an integrable majorant on $\mathbb{R}^+$ for all $|h| < \delta$. Therefore we may use Lebesgue's dominated convergence theorem to conclude from (6.4) that

$$\lim_{h\to 0} \left|[f]^{\wedge}_{M}(s+h) - |[f]^{\wedge}_{M}(s)\right| = 0,$$

which shows that $[f]^{\wedge}_{M}$ is continuous in the arbitrarily chosen point $s \in \mathrm{St}]a, b[$.

□

The second basic result is the differentiation of the Mellin transform as a function of the complex variable s. In what follows, we denote by $H(\mathrm{St}]a, b[)$ the space of functions that are analytic in the strip $\mathrm{St}]a, b[$.

Theorem 6.5 *The Mellin transform M is an operator mapping the space $X_{]a,b[}$ into $H(\mathrm{St}]a, b[)$, that is, for $f \in X_{]a,b[}$, the Mellin transform $M[f]$ is an analytic function on $\mathrm{St}]a, b[$. Furthermore, for any $r \in \mathbb{N}$, we have $f(\cdot)(\log(\cdot))^r \in X_{]a,b[}$ and*

$$\left(\frac{d}{ds}\right)^r [f]^{\wedge}_{M}(s) = [f(\cdot)(\log(\cdot))^r]^{\wedge}_{M}(s) \qquad (s \in \mathrm{St}]a, b[). \tag{6.5}$$

Proof In order to prove the analyticity of the function $[f]^{\wedge}_{M}$, we will use Morera's theorem for rectangles (Theorem 1.25). Let $R \subset \mathrm{St}]a, b[$ be any rectangle with sides parallel to the axes. Its boundary ∂R can be parametrized by a piecewise continuously differentiable function $\varphi : [0, 1] \to \partial R$. Then the function $(u, y) \mapsto f(u)u^{\varphi(y)-1}\varphi'(y)$ is measurable with respect to the two-dimensional Lebesgue measure on $\mathbb{R}^+ \times [0, 1]$. Clearly, there exists a closed subinterval $[\alpha, \beta] \subset\,]a, b[$ with $R \subset \mathrm{St}[\alpha, \beta]$. Then, by Lemma 6.3,

$$\int_0^1 \left\{ \int_0^\infty |f(u)u^{\varphi(y)-1}\varphi'(y)| du \right\} dy \leq K_f(\alpha, \beta) \int_0^1 |\varphi'(y)| dy < \infty.$$

This shows that the hypotheses of Fubini's theorem are satisfied for performing the following calculations:

$$\begin{aligned}\int_{\partial R} [f]^{\wedge}_{M}(s)ds &= \int_0^1 [f]^{\wedge}_{M}(\varphi(y))\varphi'(y)dy \\ &= \int_0^\infty f(u)\left\{\int_0^1 u^{\varphi(y)-1}\varphi'(y)dy\right\}du \\ &= \int_0^\infty f(u)\left\{\int_{\partial R} u^{s-1}ds\right\}du = 0,\end{aligned}$$

where in the last line the inner integral vanishes as a consequence of Cauchy's integral theorem applied to u^{s-1} as an analytic function of s. Since $R \subset \mathrm{St}]a,b[$ was arbitrary and $[f]^{\wedge}_{M}$ is continuous (see Lemma 6.4), Morera's theorem now yields the analyticity of $[f]^{\wedge}_{M}$ on $\mathrm{St}]a,b[$.

To establish the representation (6.5), let $s \in \mathrm{St}]a,b[$ be arbitrary and let $B_\varepsilon(s)$ be a circle of radius $\varepsilon > 0$, centered at s and contained in $\mathrm{St}[\alpha,\beta] \subset \mathrm{St}]a,b[$. Setting $\psi(y) := s + \varepsilon e^{iy}$, we have for $u > 0$, by Cauchy's integral formula for derivatives, that

$$\begin{aligned}(\log u)^r u^{s-1} = \left(\frac{d}{ds}\right)^r u^{s-1} &= \frac{r!}{2\pi i}\int_{B_\varepsilon(s)} \frac{u^{w-1}}{(w-s)^{r+1}}dw \\ &= \frac{r!}{2\pi i}\int_0^{2\pi} \frac{u^{\psi(y)-1}}{(\psi(y)-s)^{r+1}}\psi'(y)dy. \qquad (6.6)\end{aligned}$$

Now we note that the function $(u, y) \mapsto f(u)\frac{u^{\psi(y)-1}}{(\psi(y)-s)^{r+1}}\psi'(y)$ is measurable with respect to the two-dimensional Lebesgue measure on $\mathbb{R}^+ \times [0, 2\pi]$ and, again by Lemma 6.3,

$$\begin{aligned}&\int_0^{2\pi}\left\{\int_0^\infty \left| f(u)\frac{u^{\psi(y)-1}}{(\psi(y)-s)^{r+1}}\psi'(y)\right| du\right\}dy \\ &\le K_f(\alpha,\beta)\int_0^{2\pi}\frac{|\psi'(y)|}{|(\psi(y)-s)^{r+1}|}dy \\ &\le 2\pi\varepsilon^{-r}K_f(\alpha,\beta) < \infty.\end{aligned}$$

Hence the hypotheses of Fubini's theorem are satisfied for concluding in conjunction with (6.6) that $f(\cdot)(\log(\cdot))^r \in X_{]a,b[}$ and

$$\int_0^\infty f(u)(\log u)^r u^{s-1}du = \frac{r!}{2\pi i}\int_0^{2\pi}\frac{\psi'(y)}{(\psi(y)-s)^{r+1}}\left\{\int_0^\infty f(u)u^{\psi(y)-1}du\right\}dy.$$

The inner integral is just the Mellin transform $[f]^{\wedge}_{M}(\cdot)$ evaluated at $\psi(y)$. Since the Mellin transform is already known to be analytic on $\mathrm{St}]a,b[$, another application of Cauchy's integral formula gives (6.5) for any $r \in \mathbb{N}$. □

Remark 6.6 We emphasize that there is an essential difference between the spaces X_c, namely the functions whose transform operates on the fixed vertical line $c+i\mathbb{R}$, and the spaces $X_{]a,b[}$, namely the functions f whose transform $[f]^{\wedge}_{M}$ operates on the strip St$]a, b[$. The Mellin transform of elements of $X_{]a,b[}$ always generates analytic functions. In particular, by the identity theorem of analytic functions, there is no function $f \neq 0$ such that $[f]^{\wedge}_{M}(s) = 0$ for all s such that $|\Im s| \geq t_0$. This is not true for X_c as we will see later. If we wish to point out the analytic properties of the Mellin transform of $f \in X_{]a,b[}$, we will use the term *global Mellin transform.* Otherwise, if we are only interested to consider $[f]^{\wedge}_{M_c}(c + it)$ for a fixed $c \in \mathbb{R}$, we use the term *local Mellin transform.* In fact, the Mellin transform for the spaces X_c is comparable to the Fourier transform operating on the real axis, and that of $X_{]a,b[}$ to the bilateral Laplace transform acting on some strip.

Remark 6.7 In general, the convergence of $[f]^{\wedge}_{M}(s)$ for $s \in$ St$]a, b[$ does not imply the convergence on the closed strip St$[a, b]$. This is the main reason that the space $X_{]a,b[}$ is not complete. As a classical example, let us consider the function $f(x) := e^{-x}$ for $x \in \mathbb{R}^+$. It is easily seen that $f \in X_{]0,\infty[}$, that is, $[f]^{\wedge}_{M}(s)$ is absolutely convergent for all $s = c + it \in$ St$]0, \infty[$. However, it is not convergent if $c = 0$, since

$$[f]^{\wedge}_{M_0}(it) = \int_0^{\infty} e^{-u} u^{it-1} du$$

is divergent for all $t \in \mathbb{R}$.

In the following statement, we present some operational properties of the Mellin transform. They are easily deduced from the definition of the Mellin transform by substitutions or estimates.

Proposition 6.8 *The global Mellin transform M satisfies the following properties for $f, g \in X_{]a,b[}$.*

(a) The Mellin transform is linear and bounded. For any $\alpha, \beta \in \mathbb{C}$, we have

$$[\alpha f + \beta g]^{\wedge}_{M}(s) = \alpha[f]^{\wedge}_{M}(s) + \beta[g]^{\wedge}_{M}(s) \qquad \big(s \in \text{St}]a, b[\big)$$

and

$$|[f]^{\wedge}_{M}(s)| \leq \|f\|_{X_c} \qquad \big(s = c + it \in \text{St}]a, b[\big).$$

(b) If $\alpha \in \mathbb{R}^+$, then $f(\alpha(\cdot)) \in X_{]a,b[}$ and

$$[f(\alpha(\cdot))]^{\wedge}_{M}(s) = \alpha^{-s}[f]^{\wedge}_{M}(s) \qquad \big(s \in \text{St}]a, b[\big). \tag{6.7}$$

(c) If $\alpha \in \mathbb{C}$, then $(\cdot)^{\alpha} f(\cdot) \in X_{]a-\Re(\alpha),b-\Re(\alpha)[}$ and

$$[(\cdot)^{\alpha} f(\cdot)]^{\wedge}_{M}(s) = [f]^{\wedge}_{M}(s + \alpha) \qquad \big(s \in \text{St}]a - \Re(\alpha), b - \Re(\alpha)[\big). \tag{6.8}$$

(d) If $\alpha \in \mathbb{R}^+$*, then* $f((\cdot)^\alpha) \in X_{]\alpha a, \alpha b[}$ *and*

$$[f((\cdot)^\alpha)]^\wedge_M(s) = \alpha^{-1}[f]^\wedge_M\left(\frac{s}{\alpha}\right) \qquad \big(s =\in \mathrm{St}]\alpha a, \alpha b[\big). \tag{6.9}$$

(e) If $\alpha \in \mathbb{R}^+$*, then* $f((\cdot)^{-\alpha}) \in X_{]-\alpha b, -\alpha a[}$ *and*

$$[f((\cdot)^{-\alpha})]^\wedge_M(s) = \alpha^{-1}[f]^\wedge_M\left(-\frac{s}{\alpha}\right) \qquad \big(s \in \mathrm{St}]-\alpha b, -\alpha a[\big). \tag{6.10}$$

Corresponding assertions hold for the local Mellin transform of $f \in X_c$ *with* $c \in \mathbb{R}$.

From the properties (6.7) to (6.10) we deduce the following compound rules.

Corollary 6.9 *Let* $f \in X_{]a,b[}$*. Then there holds:*

(a) If $\alpha \in \mathbb{R}^+$ *and* $\beta \in \mathbb{C}$*, then* $(\cdot)^\beta f(\alpha(\cdot)) \in X_{]a-\Re(\beta), b-\Re(\beta)[}$ *and*

$$[(\cdot)^\beta f(\alpha(\cdot))]^\wedge_M(s) = \alpha^{-(s+\beta)}[f]^\wedge_M(s+\beta) \qquad \big(s \in \mathrm{St}]a-\Re(\beta), b-\Re(\beta)[\big).$$

(b) If $\alpha \in \mathbb{R}^+$ *and* $\beta \in \mathbb{C}$*, then* $(\cdot)^\beta f((\cdot)^\alpha) \in X_{]\alpha a-\Re(\beta), \alpha b-\Re(\beta)[}$ *and*

$$[(\cdot)^\beta f((\cdot)^\alpha)]^\wedge_M(s) = \alpha^{-1}[f]^\wedge_M\left(\frac{s+\beta}{\alpha}\right) \qquad \big(s \in \mathrm{St}]\alpha a-\Re(\beta), \alpha b-\Re(\beta)[\big).$$

(c) If $\alpha \in \mathbb{R}^+$ *and* $\beta \in \mathbb{C}$*, then* $(\cdot)^\beta f((\cdot)^{-\alpha}) \in X_{]-\alpha b-\Re(\beta), -\alpha a-\Re(\beta)[}$ *and*

$$[(\cdot)^\beta f((\cdot)^{-\alpha})]^\wedge_M(s) = \alpha^{-1}[f]^\wedge_M\left(-\frac{s+\beta}{\alpha}\right)$$
$$\big(s \in \mathrm{St}]-\alpha b-\Re(\beta), -\alpha a-\Re(\beta)[\big).$$

Corresponding assertions hold for the local Mellin transform of $f \in X_c$ *with* $c \in \mathbb{R}$.

6.2 The Mellin Transform of Translations, Convolutions and Derivatives

In this section we study the action of the Mellin transform on translations, convolutions and Mellin derivatives. The following result is an immediate consequence of (6.7). We state it as a theorem since it is of a basic nature for translations.

Theorem 6.10 *Let* $f \in X_c$*, where* $c \in \mathbb{R}$*. Then, for* $h \in \mathbb{R}^+$ *and* $\widetilde{c} \in \mathbb{R}$*, there holds*

$$[\tau_h^{\widetilde{c}} f]^\wedge_{M_c}(s) = h^{\widetilde{c}-s}[f]^\wedge_{M_c}(s) \qquad (s = c+it,\ t \in \mathbb{R}). \tag{6.11}$$

In particular,

$$[\tau_h^c f]^\wedge_{M_c}(s) = h^{-it}[f]^\wedge_{M_c}(s) \qquad (s = c + it,\ t \in \mathbb{R}).$$

The next result is the counterpart for Mellin transforms of the classical Riemann–Lebesgue Lemma.

Theorem 6.11 *If $f \in X_c$, where $c \in \mathbb{R}$, then*

$$\lim_{|t|\to\infty} [f]^\wedge_{M_c}(c + it) = 0. \tag{6.12}$$

Moreover, if $f \in X_{]a,b[}$, then (6.12) holds uniformly with respect to $c \in [\alpha, \beta] \subset]a, b[$.

Proof For $s = c + it$, $t \in \mathbb{R} \setminus \{0\}$ and $h := \exp(-\pi/t)$, we have by Theorem 6.10,

$$-[f]^\wedge_{M_c}(s) = e^{i\pi}[f]^\wedge_{M_c}(s) = h^{-it}[f]^\wedge_{M_c}(s) = [\tau_h^c f]^\wedge_{M_c}(s).$$

Therefore, by the linearity of the Mellin transform and (6.2),

$$2\left|[f]^\wedge_{M_c}(s)\right| = \left|[f]^\wedge_{M_c}(s) - [\tau_h^c f]^\wedge_{M_c}(s)\right| = \left|[f - (\tau_h^c f)]^\wedge_{M_c}(s)\right| \le \|f - \tau_h^c f\|_{X_c}.$$

Since $h \to 1$ as $|t| \to \infty$, the assertions follow from Proposition 3.13. □

The following convolution theorem for Mellin transforms holds.

Theorem 6.12 *Let $f, g \in X_c$, where $c \in \mathbb{R}$. Then*

$$[f * g]^\wedge_{M_c}(s) = [f]^\wedge_{M_c}(s)[g]^\wedge_{M_c}(s) \qquad (s = c + it,\ t \in \mathbb{R}).$$

Proof By Theorem 4.4, the convolution $f * g$ belongs to X_c. By Theorem 6.10 and Fubini's theorem, we have for $s = c + it$ with $t \in \mathbb{R}$,

$$\begin{aligned}[f * g]^\wedge_{M_c}(s) &= \int_0^\infty \left\{\int_0^\infty (\tau_{1/u} f)(x)x^{s-1}dx\right\} g(u)\frac{du}{u}\\ &= [f]^\wedge_{M_c}(s)\int_0^\infty g(u)u^{s-1}du = [f]^\wedge_{M_c}(s)[g]^\wedge_{M_c}(s),\end{aligned}$$

which is the assertion. □

Remark 6.13 For $h \in \mathbb{R}^+$, $\widetilde{c} \in \mathbb{R}$ and $x \in \mathbb{R}^+$, define

$$m_h^{\widetilde{c}}(x) := \begin{cases} x^{-\widetilde{c}}\chi_{[1/h,1]}(x), & h \ge 1,\\ -x^{-\widetilde{c}}\chi_{[1,1/h]}(x), & 0 < h < 1,\end{cases} \tag{6.13}$$

where χ_I denotes the characteristic function of the interval I. Clearly, $m_h^{\widetilde{c}} \in X_{]-\infty,\infty[}$, and a short calculation yields

$$[m_h^{\widetilde{c}}]_M^\wedge(s) = \begin{cases} \dfrac{h^{\widetilde{c}-s}-1}{\widetilde{c}-s}, & s \in \mathbb{C} \setminus \{\widetilde{c}\}, \\ \log h, & s = \widetilde{c}. \end{cases} \tag{6.14}$$

Denoting the r-fold convolution of $m_h^{\widetilde{c}}$ with itself by $(m_h^{\widetilde{c}}*)^r$, we obtain from Theorem 6.12 that

$$[(m_h^{\widetilde{c}}*)^r]_M^\wedge(s) = (\widetilde{c}-s)^{-r}(h^{\widetilde{c}-s}-1)^r \qquad (s \in \mathbb{C} \setminus \{\widetilde{c}\}). \tag{6.15}$$

Finally we study the action of the Mellin transform on the differentiation operation. We state two results for Mellin–Sobolev spaces.

Theorem 6.14 *Let $f \in W_c^{r,1}(\mathbb{R}^+)$, where $c \in \mathbb{R}$ and $r \in \mathbb{N}$. Then*

$$[\Theta_c^r f]_{M_c}^\wedge(s) = (-it)^r [f]_{M_c}^\wedge(s) \qquad (s = c + it,\ t \in \mathbb{R}). \tag{6.16}$$

Proof We want to employ Proposition 3.24. First we consider the right-hand side of (3.17). It is easily seen that, after multiplying by x^c, it can be interpreted as

$$\Big(\Big(\cdots\Big(\big(\Theta_c^r f(\cdot)\,(\cdot)^c * m_h^0\big) * m_h^0\Big)\cdots\Big) * m_h^0\Big)(x),$$

where convolutions with the function m_h^0 defined in (6.13) occur exactly r times. Thus, using the associativity and the commutativity of the convolution, we find that the second equation in (3.17) may be rewritten as

$$\begin{aligned} &\sum_{j=0}^{r}(-1)^{r-j}\binom{r}{j}(\tau_{h^j}^c f)(x) \\ &\qquad = x^{-c}\Big((m_h^0*)^r * \big(\Theta_c^r f(\cdot)(\cdot)^c\big)\Big)(x) \quad (x \in \mathbb{R}^+). \end{aligned} \tag{6.17}$$

Now we want to take the Mellin transform on both sides of (6.17). Employing Theorem 6.10 for the calculation on the left-hand side and formula (6.8) together with Theorem 6.12 for that on the right-hand side, we obtain

$$\begin{aligned} \sum_{j=0}^{r}(-1)^{r-j}\binom{r}{j}h^{-itj}[f]_{M_c}^\wedge(s) &= [(m_h^0*)^r]_{M_c}^\wedge(it)[\Theta_c^r f(\cdot)(\cdot)^c]_{M_c}^\wedge(it) \\ &= (-it)^{-r}(h^{-it}-1)^r[\Theta_c^r f]_{M_c}^\wedge(s) \end{aligned}$$

for all $s = c + it$ with $t \neq 0$. Finally, applying the binomial theorem on the left-hand side and multiplying by $(-it)^r(h^{-it} - 1)^{-r}$, we arrive at (6.16), which extends to $t = 0$ due to the continuity of the Mellin transform. □

The corresponding result for the spaces $W^{r,1}_{[a,b]}$ reads as follows.

Theorem 6.15 *Let $f \in W^{r,1}_{[a,b]}$, where $a, b \in \mathbb{R}$ with $a < b$ and $r \in \mathbb{N}$. Then, for any $c, \nu \in]a, b[$, there holds*

$$[\Theta^r_c f]^\wedge_M(\nu + it) = (c - \nu - it)^r [f]^\wedge_M(\nu + it).$$

Proof We can use the previous approach under slight modifications. Consider

$$G(x) := \big((m^0_h *)^r * (\Theta^r_c f(\cdot)(\cdot)^c)\big)(x) \qquad (x \in \mathbb{R}^+).$$

Since by the hypotheses $\Theta^r_c f \in X_\nu$, we easily see that $\Theta^r_c f(\cdot)(\cdot)^c \in X_{\nu-c}$. Hence $G \in X_{\nu-c}$ and so $(\cdot)^{-c}G(\cdot) \in X_\nu$. Now employing formulae (6.17), (6.7) and the binomial theorem, we find that

$$\begin{aligned}
[(\cdot)^{-c}G(\cdot)]^\wedge_M(\nu + it) &= \left[\sum_{j=0}^{r}(-1)^{r-j}\binom{r}{j}\tau^c_{h^j} f\right]^\wedge_M (\nu + it) \\
&= \sum_{j=0}^{r}(-1)^{r-j}\binom{r}{j} h^{jc}\left[f(h^j(\cdot))\right]^\wedge_M (\nu + it) \\
&= (h^{c-\nu-it} - 1)^r [f]^\wedge_M(\nu + it).
\end{aligned}$$

On the other hand, using formula (6.8), Theorem 6.12 and formula (6.15), we obtain

$$\begin{aligned}
[(\cdot)^{-c}G(\cdot)]^\wedge_M(\nu + it) &= [G]^\wedge_M(\nu - c + it) \\
&= [(m^0_h *)^r]^\wedge_M(\nu - c + it)[\Theta^r_c f(\cdot)(\cdot)^c]^\wedge_M(\nu - c + it) \\
&= (h^{c-\nu-it} - 1)^r (c - \nu - it)^{-r}[\Theta^r_c f]^\wedge_M(\nu + it).
\end{aligned}$$

Combining the two representations of $[(\cdot)^{-c}G(\cdot)]^\wedge_M(\nu + it)$, we arrive at the desired result. □

6.3 Kernels, Approximate Identities and Mellin Convolution Operators

In this section we will show how functions in X_c can be approximated by certain convolution integrals involving approximate identities. We will distinguish the case of the local Mellin transform on X_c and the global Mellin transform on $X]a, b[$. The results given in this section will be basic for the Mellin inversion theory.

We begin by introducing three notions.

Definition 6.16 A family $\{\kappa_\rho\}_{\rho>0} \subset X_c$, depending on a parameter $\rho > 0$, is said to be a *(local) kernel on* X_c if for every $\rho > 0$,

$$[\kappa_\rho]^\wedge_{M_c}(c) = 1 \tag{6.18}$$

$$\|\kappa_\rho\|_{X_c} \leq K \tag{6.19}$$

for a constant $K > 0$.

Definition 6.17 A (local) kernel $\{\kappa_\rho\}_{\rho>0} \subset X_c$ is said to be a *(local) approximate identity on* X_c if for each $\delta \in]0, 1[$ defining $E_\delta := \{x \in \mathbb{R}^+ : |1 - x| \geq \delta\}$, there holds

$$\lim_{\rho\to\infty} \int_{E_\delta} |\kappa_\rho(u)| u^{c-1} du = 0. \tag{6.20}$$

Definition 6.18 Let $\{\kappa_\rho\}_{\rho>0} \subset X_c$ be a (local) approximate identity on X_c. Then the family of operators $\{U_\rho\}_{\rho>0}$, defined for $x \in \mathbb{R}^+$ by

$$U_\rho : X_c \longrightarrow X_c, \qquad (U_\rho f)(x) := (f * \kappa_\rho)(x),$$

is called *a family of (local) Mellin convolution operators on* X_c generated by the family $\{\kappa_\rho\}_{\rho>0}$.

The definition of kernels together with Theorem 4.4 guarantee that $\{U_\rho\}_{\rho>0}$ is a well-defined and uniformly bounded family of operators on X_c.

Theorem 6.19 *Let $\{U_\rho\}_{\rho>0}$ be a family of Mellin convolution operators generated by a local approximate identity $\{\kappa_\rho\}_{\rho>0} \subset X_c$. Then, for $p \geq 1$, we have*

$$\lim_{\rho\to\infty} \|U_\rho f - f\|_{X_c^p} = 0 \qquad (f \in X_c^p).$$

Proof For E_δ as introduced in Definition 6.17, let $\widetilde{E}_\delta := \mathbb{R}^+ \setminus E_\delta$. Then, according to Proposition 3.13, for every $\varepsilon > 0$, there exists a $\delta \in]0, 1[$ such that for all $h \in \widetilde{E}_\delta$, we have

$$\|\tau^c_{1/h} f - f\|_{X^p_c} < \varepsilon. \tag{6.21}$$

Theorem 4.4 shows us that $U_\rho f \in X^p_c$ for every $\rho > 0$. Since $[\kappa_\rho]^\wedge_{M_c}(c) = 1$, we may write

$$(U_\rho f)(x) - f(x) = \int_0^\infty \big((\tau^c_{1/u} f)(x) - f(x)\big)\kappa_\rho(u) u^{c-1} du.$$

Therefore, using Proposition 3.3, we find that

$$\begin{aligned}\|U_\rho f - f\|_{X^p_c} &\le \int_0^\infty \|\tau^c_{1/u} f - f\|_{X^p_c} |\kappa_\rho(u)| u^{c-1} du \\ &\le \left\{\int_{E_\delta} + \int_{\widetilde{E}_\delta}\right\} \|\tau^c_{1/u} f - f\|_{X^p_c} |\kappa_\rho(u)| u^{c-1} du.\end{aligned}$$

Now, by Proposition 3.12 we have $\|\tau^c_{1/u} f - f\|_{X^p_c} \le 2\|f\|_{X^p_c}$ and so, by the definition of an approximate identity and (6.21), we obtain

$$\limsup_{\rho\to\infty} \|U_\rho f - f\|_{X^p_c} \le \limsup_{\rho\to\infty} \left(\int_{E_\delta} 2\|f\|_{X^p_c} |\kappa_\rho(u)| u^{c-1} du + \varepsilon \|\kappa_\rho\|_{X_c}\right) \le K\varepsilon,$$

from which the assertion follows. □

For spaces the $X_{]a,b[}$ the notion of an approximate identity given before is not appropriate, because the condition $[\kappa_\rho]^\wedge_M(c) = 1$ cannot be satisfied for all $c \in]a, b[$. Indeed, the identity theorem for analytic functions would imply that $[\kappa_\rho]^\wedge_M \equiv 1$ on St$]a, b[$, contradicting Theorem 6.11. Therefore we modify the three previous definitions as follows.

Definition 6.20 Let $\widetilde{c} \in \mathbb{R}$. A family $\{\kappa_\rho\}_{\rho>0} \subset X_{]a,b[} \cap X_{\widetilde{c}}$, depending on a parameter $\rho > 0$, is said to be a *global kernel on* $X_{]a,b[}$, if for every $\rho > 0$,

$$\left[\kappa_\rho\right]^\wedge_M(\widetilde{c}) = 1 \tag{6.22}$$

$$\|\kappa_\rho\|_{X_{\widetilde{c}}} \le K \tag{6.23}$$

for a constant $K > 0$.

Definition 6.21 Let $\widetilde{c} \in \mathbb{R}$. A global kernel $\{\kappa_\rho\}_{\rho>0} \subset X_{]a,b[} \cap X_{\widetilde{c}}$ is said to be a *global approximate identity on* $X_{]a,b[}$ if for all $c \in]a, b[$ and each $\delta \in]0, 1[$, defining $E_\delta := \{x \in \mathbb{R}^+ : |1 - x| \ge \delta\}$, there holds

$$\lim_{\rho\to\infty} \int_{E_\delta} |\kappa_\rho(u)|(u^{c-1} + u^{\widetilde{c}-1}) du = 0. \tag{6.24}$$

Definition 6.22 Let $\widetilde{c} \in \mathbb{R}$ and let $\{\kappa_\rho\}_{\rho>0} \subset X_{]a,b[} \cap X_{\widetilde{c}}$ be a global approximate identity on $X_{]a,b[}$. Then the family of operators $\{U_\rho\}_{\rho>0}$, defined for $x \in \mathbb{R}^+$ by

$$U_\rho : X_{]a,b[} \longrightarrow X_{]a,b[}, \qquad (U_\rho f)(x) := (f * \kappa_\rho)(x),$$

is called a *global Mellin convolution operators on* $X_{]a,b[}$ generated by the family $\{\kappa_\rho\}_{\rho>0}$.

Remark 6.23 Note that a family $\{\kappa_\rho\}_{\rho>0}$ forms a global approximate identity on $X_{]a,b[}$ if it is a kernel on $X_{\widetilde{c}}$ for a particular value $\widetilde{c} \in \mathbb{R}$, whereas condition (6.20) has to be satisfied for all $c \in]a, b[$, not only for $\widetilde{c}$. One may choose $\widetilde{c} = 0$ or $\widetilde{c} \in]a, b[$.

The approximation by global Mellin convolution operators can be described as follows.

Theorem 6.24 *Let $\{U_\rho\}_{\rho>0}$ be a family of global Mellin convolution operators generated by a global approximate identity $\{\kappa_\rho\}_{\rho>0} \subset X_{]a,b[}$. Then, for $p \geq 1$, we have*

$$\lim_{\rho\to\infty} \|U_\rho f - f\|_{X_c^p} = 0, \tag{6.25}$$

for all $f \in X^p_{]a,b[}$ and all $c \in]a, b[$. Furthermore the convergence in (6.25) holds uniformly in $c \in [\alpha, \beta] \subset]a, b[$.

Proof In the following, let E_δ and $\widetilde{E}_\delta$ be as in the proof of Theorem 6.19. By Definition 6.21, $\{\kappa_\rho\}_{\rho>0}$ forms a kernel for some $\widetilde{c} \in \mathbb{R}$. Using Proposition 3.12, we see that

$$\|\tau^{\widetilde{c}}_{1/h} f - f\|_{X_c^p} \leq (h^{c-\widetilde{c}} + 1)\|f\|_{X_c^p} \qquad (h \in \mathbb{R}^+). \tag{6.26}$$

According to Proposition 3.13, for every $\varepsilon > 0$ there exists a $\delta \in]0, 1[$ such that for all $h \in \widetilde{E}_\delta$, we have

$$\|\tau^{\widetilde{c}}_{1/h} f - f\|_{X_c^p} < \varepsilon. \tag{6.27}$$

Proceeding as in the previous proof and using again Proposition 3.3, we find that

$$\|U_\rho f - f\|_{X_c^p} \leq \left\{\int_{E_\delta} + \int_{\widetilde{E}_\delta}\right\} \|\tau^{\widetilde{c}}_{1/u} f - f\|_{X_c^p} \, |\kappa_\rho(u)| u^{\widetilde{c}-1} du.$$

Now, by (6.26), (6.27) and the definition of a global approximate identity, we obtain

$$\limsup_{\rho\to\infty} \|U_\rho f - f\|_{X^p_c}$$
$$\leq \limsup_{\rho\to\infty} \left(\int_{E_\delta} \|f\|_{X^p_c} |\kappa_\rho(u)| \big(u^{\tilde{c}-1} + u^{c-1}\big) du + \varepsilon \|\kappa_\rho\|_{X_{\tilde{c}}} \right) \leq K\varepsilon,$$

which implies (6.25). The uniformity in $c \in [\alpha, \beta] \subset]a, b[$ follows by (3.1) and Proposition 3.13. □

Now we present two examples of approximate identities together with their basic properties.

Example 6.25 The *Mellin–Gauss–Weierstrass kernel*, denoted by $\{G_\rho\}_{\rho>0}$, is defined by (Fig. 6.1)

$$G_\rho(x) := \frac{\rho}{\sqrt{4\pi}} x^{-(\rho^2/4)\log x} \qquad (x \in \mathbb{R}^+, \rho > 0).$$

An obvious alternative representation is

$$G_\rho(x) := \frac{\rho}{\sqrt{4\pi}} \exp\left(-\left(\frac{\rho}{2}\log x\right)^2 \right) \qquad (x \in \mathbb{R}^+, \rho > 0). \tag{6.28}$$

Since for $a > 0$ the function $g(x) := x^{-a\log x}$, defined on $\mathbb{R}^+$, belongs to $X_{]-\infty,\infty[}$, we have $G_\rho \in X_{]-\infty,\infty[}$ for every $\rho > 0$. Moreover the following statement holds.

Proposition 6.26 *The Mellin–Gauss–Weierstrass kernel is local approximate identity on X_0 as well as a global approximate identity on $X_{]-\infty,\infty[}$ and*

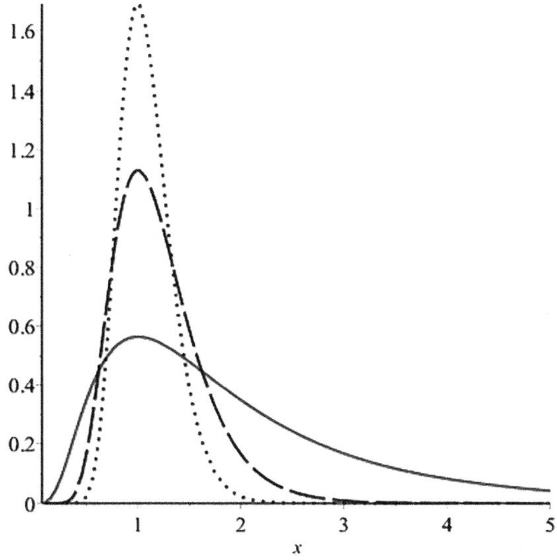

Fig. 6.1 The Mellin–Gauss–Weierstrass kernel with $\rho = 2$ (solid line), $\rho = 4$ (dashed line) and $\rho = 6$ (dotted line)

$$[G_\rho]^\wedge_M(s) = \exp\left(\left(\frac{s}{\rho}\right)^2\right) \qquad (s \in \mathbb{C},\ \rho > 0). \tag{6.29}$$

Proof First we turn to (6.29). In view of Theorem 6.5, both sides of the asserted equality are analytic for all $s \in \mathbb{C}$. Hence, according to the identity principle of analytic functions (Theorem 1.29), it suffices to prove (6.29) for $s = c \in \mathbb{R}$ only. For this, we note that

$$G_\rho(u)u^c = \frac{\rho}{2\sqrt{\pi}} \exp\left(\frac{c^2}{\rho^2}\right) \exp\left(-\left(\frac{\rho}{2}\log u - \frac{c}{\rho}\right)^2\right)$$

as can be verified by a standard calculation. Thus, substituting $x = \frac{\rho}{2}\log u - \frac{c}{\rho}$, we obtain

$$[G_\rho]^\wedge_M(c) = \int_0^\infty G_\rho(u)u^c \frac{du}{u} = \frac{1}{\sqrt{\pi}} \exp\left(\frac{c^2}{\rho^2}\right) \int_{-\infty}^\infty e^{-x^2} dx.$$

The integral on the right-hand side, also called the Gaussian integral, is known to be equal to $\sqrt{\pi}$ (see, e.g., [1, formula 7.4.1]). This proves (6.29) for all $c \in \mathbb{R}$.

Finally, we show that $\{G_\rho\}_{\rho>0}$ forms an approximate identity. Since G_ρ is positive, we have $\|G_\rho\|_{X_0} = [G_\rho]^\wedge_{M_0}(0) = 1$. Now let $\delta \in]0, 1[$ be fixed and $c \in \mathbb{R}$. Let $\rho > 0$ be such that $c - (\rho/2)^2 \log(1 - \delta) > 0$ and $c - (\rho/2)^2 \log(1 + \delta) < 0$, which is certainly satisfied for all sufficiently large ρ. Since G_ρ is positive, we have

$$\begin{aligned}
\int_{E_\delta} G_\rho(u)u^{c-1}du &= \frac{\rho}{\sqrt{4\pi}}\left\{\int_0^{1-\delta} + \int_{1+\delta}^\infty\right\} u^{c-1-(\rho/2)^2\log u}du \\
&\le \frac{\rho}{\sqrt{4\pi}} \int_0^{1-\delta} u^{c-1-(\rho/2)^2\log(1-\delta)}du \\
&\quad + \frac{\rho}{\sqrt{4\pi}} \int_{1+\delta}^\infty u^{c-1-(\rho/2)^2\log(1+\delta)}du \\
&\le \frac{\rho}{\sqrt{4\pi}}\left(\frac{(1-\delta)^{c-(\rho/2)^2\log(1-\delta)}}{c-(\rho/2)^2\log(1-\delta)} - \frac{(1+\delta)^{c-(\rho/2)^2\log(1+\delta)}}{c-(\rho/2)^2\log(1+\delta)}\right).
\end{aligned}$$

Taking the limit for $\rho \to \infty$, we obtain (6.20) as well as (6.24). □

Remark 6.27 The following interesting relations between the Mellin–Gauss–Weierstrass kernel and its Mellin transform can easily be verified:

$$G_\rho(x) = \frac{\rho}{\sqrt{4\pi}}[G_\rho]^\wedge_M\left(\frac{i\rho^2}{2}\log x\right) \qquad (x \in \mathbb{R}^+,\ \rho > 0),$$

$$G_\rho\left(\exp\left(\frac{2t}{\rho^2}\right)\right) = \frac{\rho}{\sqrt{4\pi}}[G_\rho]^\wedge_M(it) \qquad (t \in \mathbb{R},\ \rho > 0).$$

For the Mellin convolution operator generated by the Mellin–Gauss–Weierstrass kernel, we obtain the following statement.

Corollary 6.28 *For $f \in X_c^p$, where $p \geq 1$ and $c \in \mathbb{R}$, there holds*

$$\lim_{\rho\to\infty} \|f * G_\rho - f\|_{X_c^p} = 0. \tag{6.30}$$

For $f \in X_{]a,b[}$, the limit in (6.30) is uniform in $c \in [\alpha, \beta] \subset]a, b[$. Moreover, there exists a sequence $(\rho_n)_{n\in\mathbb{N}}$ such that

$$\lim_{n\to\infty} (f * G_{\rho_n})(x) = f(x) \qquad (a.e.\ for\ x \in \mathbb{R}^+). \tag{6.31}$$

Proof Formula (6.30) follows from Theorems 6.19 and 6.24, while (6.31) is a consequence of a classical result for L^p spaces which states that convergence in L^p of a sequence of functions implies the almost everywhere convergence of a subsequence (see, e.g., [171, p. 192]). □

Example 6.29 The *Mellin–Abel–Poisson kernel* $\{P_\rho\}_{\rho>0}$ is defined by (Fig. 6.2)

$$P_\rho(x) := \begin{cases} \frac{\rho}{2} x^\rho, & 0 < x \leq 1, \\ \frac{\rho}{2} x^{-\rho}, & x \geq 1. \end{cases}$$

In a more concise form it may be written as

$$P_\rho(x) = \frac{\rho}{2} \exp(-\rho|\log x|) \qquad (x \in \mathbb{R}^+,\ \rho > 0). \tag{6.32}$$

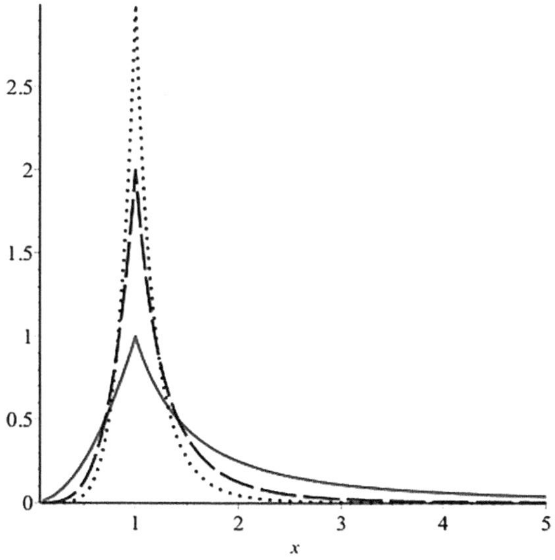

Fig. 6.2 The Mellin–Abel–Poisson kernel, with $\rho = 2$ (solid line), $\rho = 4$ (dashed line) and $\rho = 6$ (dotted line)

The basic properties of the Mellin–Abel–Poisson kernel may be stated as follows.

Proposition 6.30 *The Mellin transform of P_ρ is given by*

$$[P_\rho]^\wedge_M(s) = \left\{1 - \left(\frac{s}{\rho}\right)^2\right\}^{-1} \qquad \big(s = c + it \in \mathrm{St}]-\rho, \rho[,\ \rho > 0\big). \quad (6.33)$$

In particular,

$$\|P_\rho\|_{X_c} = \left\{1 - \left(\frac{c}{\rho}\right)^2\right\}^{-1} \qquad \big(c \in]-\rho, \rho[,\ \rho > 0\big). \qquad (6.34)$$

Furthermore, the kernel $\{P_\rho\}_{\rho>0}$ forms a local approximate identity on X_0.

Proof For $s \in \mathrm{St}]-\rho, \rho[$, we see that

$$[P_\rho]^\wedge_M(s) = \left\{\int_0^1 + \int_1^\infty\right\} P_\rho(u)u^{s-1}du = \frac{\rho}{2}\left(\frac{1}{s+\rho} - \frac{1}{s-\rho}\right)$$
$$= \left\{1 - \left(\frac{s}{\rho}\right)^2\right\}^{-1},$$

which is (6.33). Since P_ρ is positive, we have $\|P_\rho\|_{X_c} = [P_\rho]^\wedge_M(c)$ for $c \in]-\rho, \rho[$. In particular, taking $c = 0$, we obtain $\|P_\rho\|_{X_0} = 1$, and so (6.18) and (6.19) are satisfied. It is easy to show that for every $\delta \in]0, 1[$ we have

$$\lim_{\rho\to\infty} \int_{E_\delta} |P_\rho(u)|\frac{du}{u} = 0,$$

establishing (6.20). □

Remark 6.31 The Mellin–Abel–Poisson kernel cannot be a global kernel simply because $\|P_\rho\|_{X_c}$ does not exist as soon as $c \neq 0$ and $\rho \leq |c|$. But this is not really a problem. In all applications of kernels only the functions with a large index ρ are employed. Therefore we can simply shift the index. Indeed, for any $\rho_0 > 0$ and $\widetilde{c} = 0$ the family $\{P_{\rho_0+\rho}\}_{\rho>0}$ is an approximate identity on $X_{]-\rho_0,\rho_0[}$.

As a counterpart of Corollary 6.28 we have the following statement.

Corollary 6.32 *For $f \in X_0^p$, where $p \geq 1$, there holds*

$$\lim_{\rho\to\infty} \|f * P_\rho - f\|_{X_0^p} = 0.$$

Moreover, there exists a sequence $(\rho_n)_{n\in\mathbb{N}}$ *such that*

$$\lim_{n\to\infty}(f * P_{\rho_n})(x) = f(x) \qquad (a.e.\ for\ x \in \mathbb{R}^+). \tag{6.35}$$

6.4 Pointwise Convergence of Mellin–Fejér Type Operators

In this section we study some pointwise convergence properties of nets of Mellin convolution operators of Mellin–Fejér type, using the convolution defined in (4.3). Indeed, it seems that this approach may be more convenient from a technical point of view. On the other hand, corresponding results for the operators U_ρ are easily deduced (see Remark 6.41 below). First we establish a pointwise convergence theorem at the continuity points of a function $f \in X_0^p$ and then a uniform convergence theorem for log-uniformly continuous functions (see Definition 3.15).

6.4.1 *Pointwise and Uniform Convergence*

We begin with the following definition.

Definition 6.33 A function $K \in X_0$ is said to be a *kernel function* if

$$\int_0^\infty K(u)\frac{du}{u} = 1. \tag{6.36}$$

In this case, the family $\{K_\rho\}_{\rho>0}$, defined by $K_\rho(x) := \rho K(x^\rho)$, is called *the Mellin–Fejér type kernel* generated by K.

Note that the two examples of the previous section (Examples 6.25 and 6.29) are special cases of Mellin–Fejér type kernels. Another important example, known as *the* Mellin–Fejér kernel, will be discussed later within the theory of inversion of the Mellin transform (see Definition 6.83).

First we show that each Mellin–Fejér type kernel is really a kernel and even an approximate identity.

Proposition 6.34 *Let* $K \in X_0$ *be a kernel function. Then the family* $\{K_\rho\}_{\rho>0}$, *introduced by Definition 6.33 is a local approximate identity on* X_0. *In particular,* $\|K_\rho\|_{X_0} = \|K\|_{X_0}$ *for all* $\rho > 0$ *and*

$$\lim_{\rho\to\infty}\int_{|\log u|>\delta}|K_\rho(u)|\frac{du}{u} = 0 \tag{6.37}$$

for every $\delta > 0$.

Proof Clearly, for any $\rho > 0$, we find by substituting $u^\rho = v$ that

$$\int_0^\infty K_\rho(u)\frac{du}{u} = \rho\int_0^\infty K(u^\rho)\frac{du}{u} = \int_0^\infty K(v)\frac{dv}{v} = 1.$$

Analogously we obtain $\|K_\rho\|_{X_0} = \|K\|_{X_0}$. Next, for every $\delta \in]0,1[$, the same substitution and the assumption $K \in X_0$ yield

$$\lim_{\rho\to\infty}\int_{|1-u|>\delta}|K_\rho(u)|\frac{du}{u}$$

$$= \lim_{\rho\to\infty}\left\{\int_0^{(1-\delta)^\rho}|K(v)|\frac{dv}{v} + \int_{(1+\delta)^\rho}^\infty |K(v)|\frac{dv}{v}\right\} = 0. \tag{6.38}$$

Finally (6.37) is verified in the same way. □

Remark 6.35 The relations (6.37) and (6.38) are equivalent. However, in some of the subsequent considerations (6.37) will be more convenient.

Definition 6.36 For $\{K_\rho\}_{\rho>0}$ as introduced by Definition 6.33, consider the family of integral operators

$$(T_\rho f)(x) := \int_0^\infty K_\rho(u)f(xu)\frac{du}{u} \qquad (x \in \mathbb{R}^+).$$

Then we call $\{T_\rho\}_{\rho>0}$ the family of *Mellin–Fejér type operators* generated by the kernel function K.

By Proposition 4.5, we know that if $f \in X_0^p$ for $p \in [1,\infty]$, then $T_\rho f$ exists almost everywhere and $T_\rho f \in X_0^p$.

Remark 6.37 We mention that the Mellin–Fejér type operators are convolution operators for the modified convolution $\widetilde{*}$ defined in (4.3).

The results of the previous section about the norm-convergence on spaces X_c^p for the corresponding family of operators $\{U_\rho\}$ also hold for the family of operators $\{T_\rho\}$. Here we consider pointwise convergence. We begin with the following theorem.

Theorem 6.38 *Let $K \in X_0$ be a kernel function. Let $\{T_\rho\}_{\rho>0}$ be the family of Mellin–Fejér type operators generated by K (see Definition 6.36). Suppose that $f \in X_0^\infty$. Then the following statements hold:*

(i) If f is continuous at a point $x \in \mathbb{R}^+$, then

$$\lim_{\rho\to\infty}|T_\rho f(x) - f(x)| = 0.$$

(ii) If f is log-uniformly continuous, then

$$\lim_{\rho\to\infty} \|T_\rho f(x) - f(x)\|_{X_0^\infty} = 0.$$

Proof For a fixed x, we see that

$$|(T_\rho f)(x) - f(x)| \leq \int_0^\infty |K_\rho(u)||f(xu) - f(x)| \frac{du}{u}.$$

If f is continuous at x, then $f(\exp(\cdot))$ is continuous at $\log x$. Hence for every $\varepsilon > 0$ there exists a $\delta > 0$ such that for $|\log(xu) - \log x| = |\log u| \leq \delta$, we have $|f(xu) - f(x)| < \varepsilon$. Therefore we obtain

$$\begin{aligned} |(T_\rho f)(x) - f(x)| &\leq \int_{|\log u|\leq\delta} |K(u_\rho)||f(xu) - f(x)| \frac{du}{u} \\ &\quad + \int_{|\log u|>\delta} |K_\rho(u)||f(xu) - f(x)| \frac{du}{u} \\ &\leq \varepsilon\|K\|_{X_0} + 2\|f\|_{X_0^\infty} \int_{|\log u|>\delta} |K_\rho(u)| \frac{du}{u}. \end{aligned}$$

Now assertion (i) follows in view of (6.37). Assertion (ii) follows analogously, taking into account the fact that if f is log-uniformly continuous, then the involved $\delta > 0$ can be obtained independently of x. □

The following theorem extends Theorem 6.38 to the case $f \in X_0^p$ with $1 < p < \infty$. In this instance, we have to impose a further condition, namely for for any $\delta > 0$, we have

$$S_\delta := \sup_{\rho>0} \sup_{|\log x|>\delta} |K_\rho(x)| < \infty. \tag{6.39}$$

Theorem 6.39 *Let $K \in X_0$ be a kernel function satisfying (6.39). Let $\{T_\rho\}_{\rho>0}$ be the family of Mellin–Fejér type operators generated by K (see Definition 6.36). Suppose that $f \in X_0^p$ with $1 < p < \infty$. If the function f is continuous at the point $x \in \mathbb{R}^+$, then*

$$\lim_{\rho\to\infty} |T_\rho f(x) - f(x)| = 0.$$

Proof Let $f \in X_0^p$ with $1 < p < \infty$. Using the same arguments as in the proof of Theorem 6.38, we only have to estimate the term

$$\begin{aligned} I :&= \int_{|\log u|>\delta} |K_\rho(u)||f(xu) - f(x)| \frac{du}{u} \\ &\le \int_{|\log u|>\delta} |K_\rho(u)||f(xu)| \frac{du}{u} + |f(x)| \int_{|\log u|>\delta} |K_\rho(u)| \frac{du}{u} \\ &=: I_1 + I_2 \end{aligned}$$

differently. For the last term I_2, relation (6.37) applies. As to I_1, using Hölder's inequality and (6.39), we obtain, with p' being the conjugate exponent of p,

$$\begin{aligned} I_1 &\le \left\{ \int_{|\log u|>\delta} |K_\rho(u)|\, |f(xu)|^p \frac{du}{u} \right\}^{1/p} \left\{ \int_{|\log u|>\delta} |K_\rho(u)| \frac{du}{u} \right\}^{1/p'} \\ &\le S_\delta^{1/p} \|f\|_{X_0^p}^p \left\{ \int_{|\log u|>\delta} |K_\rho(u)| \frac{du}{u} \right\}^{1/p'}. \end{aligned}$$

Taking the limit for $\rho \to \infty$, the assertion follows again by (6.37). □

Remark 6.40 The case $p = 1$ is more delicate. For functions $f \in X_0$ we have to impose a condition stronger than (6.39), namely that for any $\delta > 0$, there holds

$$\lim_{\rho \to \infty} \sup_{|\log x|>\delta} |K_\rho(x)| = 0. \tag{6.40}$$

Details are left to the reader.

Remark 6.41 Corresponding results hold for the Mellin convolution operators $\{U_\rho\}$. Indeed, for every $x > 0$ and $\widetilde{f}(x) := f(1/x)$, we have

$$\begin{aligned} (U_\rho f)(x) &= \int_0^\infty K_\rho(u) f\left(\frac{x}{u}\right) \frac{du}{u} = \int_0^\infty K_\rho(u) \widetilde{f}\left(\frac{u}{x}\right) \frac{du}{u} \\ &= (T_\rho \widetilde{f})\left(\frac{1}{x}\right). \end{aligned} \tag{6.41}$$

Thus, since $\widetilde{f}$ satisfies the same assumptions as f, the convergence $(T_\rho \widetilde{f})(1/x) \to \widetilde{f}(1/x) = f(x)$ implies the convergence $(U_\rho f)(x) \to f(x)$.

6.4.2 *Asymptotic Behavior*

For a thorough study of the asymptotic behavior of the approximants $(T_\rho f)(x)$ by Mellin–Fejér type operators as $\rho \to \infty$, we shall establish a Taylor formula in the Mellin setting with local and global remainder, respectively. These results, corresponding to classical formulae of Peano and Lagrange, will be of independent interest. First we introduce a notion of local differentiability.

Definition 6.42 Let $f : \mathbb{R}^+ \to \mathbb{C}$, $x \in \mathbb{R}^+$ and $n \in \mathbb{N}$. We say that f is a C^n-*function locally at* x, and write $f \in C^n(x)$, if there exists a neighborhood V_x of x such that f is $(n-1)$-times continuously differentiable on V_x and the derivative of order n exists at x.

The following statement corresponds to Taylor's formula with Peano's remainder.

Proposition 6.43 *Let $f : \mathbb{R}^+ \to \mathbb{C}$ belong to $C^n(x)$, where $n \in \mathbb{N}$ and $x \in \mathbb{R}^+$. Then there exists $\delta > 1$ such that for $t \in]1/\delta, \delta[$, we have*

$$f(xt) = f(x) + \Theta f(x)\log t + \frac{\Theta^2 f(x)}{2!}\log^2 t + \cdots + \frac{\Theta^n f(x)}{n!}\log^n t + h(t)\log^n t$$

where h is a function depending on x and such that $h(t) \to 0$ as $t \to 1$. Moreover, if $f \in X_0^\infty$, then this formula holds for every $t \in \mathbb{R}^+$ and the function h is bounded on $\mathbb{R}^+$.

Proof First we consider the cases when $n = 1$ and $n = 2$. If f is differentiable at the point $x \in \mathbb{R}^+$, we have

$$\lim_{t\to 1}\frac{f(xt)-f(x)}{\log t} = \lim_{t\to 1}\frac{f(xt)-f(x)}{tx-x}\,\frac{x(t-1)}{\log t} = xf'(x) = \Theta f(x),$$

which implies the first assertion for $n = 1$.

Now let $f \in C^2(x)$. Then, putting $F(t) := f(x) + xf'(x)\log t$ and using L'Hospital's rule twice, we obtain

$$\begin{aligned}\lim_{t\to 1}\frac{f(xt)-F(x)}{\log^2 t} &= \frac{x}{2}\lim_{t\to 1}\frac{t(f'(xt)-f'(x))}{\log t} + \frac{xf'(x)}{2}\lim_{t\to 1}\frac{t-1}{\log t}\\ &= \frac{x^2}{2}f''(x) + \frac{x}{2}f'(x) = \frac{\Theta^2 f(x)}{2},\end{aligned}$$

which implies the first assertion for $n = 2$.

Now we turn to the general case. By introducing

$$F_{n-1}(x,t) := \sum_{j=0}^{n-1}\frac{(\Theta^j f)(x)}{j!}\log^j t,$$

it suffices to show that

$$\lim_{t\to 1}\frac{f(xt)-F_{n-1}(x,t)}{\log^n t} = \frac{(\Theta^n f)(x)}{n!}. \tag{6.42}$$

We want to calculate this limit by using L'Hospital's rule repeatedly. When we differentiate the numerator and the denominator with respect to t, we may always multiply them by t afterwards. This amounts to employing a modification of L'Hospital's rule in which the Mellin differential operator Θ is used instead of ordinary differentiation. For $k \in \{1, \dots, n-1\}$, we note that

$$\big(\Theta^k f(x\,\cdot)\big)(t) = \big(\Theta^k f\big)(xt),$$

$$\big(\Theta^k \log^n(\cdot)\big)(t) = \frac{n!}{(n-k)!} \log^{n-k} t,$$

$$\big(\Theta^k F_{n-1}(x,\cdot)\big)(t) = \sum_{j=k}^{n-1} \frac{(\Theta^j f)(x)}{(j-k)!} \log^{j-k} t.$$

Thus, applying the modified L'Hospital rule $(n-1)$-times, we find that

$$\lim_{t\to 1} \frac{f(xt) - F_{n-1}(x,t)}{\log^n t} = \lim_{t\to 1} \frac{(\Theta^{n-1} f)(xt) - (\Theta^{n-1} f)(x)}{n! \log t}.$$

By the hypothesis, f has a derivative of order n which need not be continuous. Therefore we have to proceed differently for calculating the limit on the right-hand side. We observe that

$$\begin{aligned}
&\lim_{t\to 1} \frac{(\Theta^{n-1} f)(xt) - (\Theta^{n-1} f)(x)}{n! \log t} \\
&\quad = \lim_{t\to 1} \frac{(\Theta^{n-1} f)(xt) - (\Theta^{n-1} f)(x)}{xt - x} \lim_{t\to 1} \frac{x(t-1)}{n! \log t} \\
&\quad = \big(\Theta^{n-1} f\big)'(x) \frac{x}{n!} = \frac{(\Theta^n f)(x)}{n!}.
\end{aligned}$$

Altogether, we have shown that (6.42) holds.

For the second assertion, we may write

$$|h(t)| = \left| \frac{f(xt) - F_n(x,t)}{\log^n t} \right|.$$

Since $h(t) \to 0$ as $t \to 1$, there is a $\delta > 1$ such that $|h(t)| \le 1$, for all $t \in [1/\delta, \delta]$. Moreover, for $t \in \mathbb{R}^+ \setminus [1/\delta, \delta]$, we have

$$|h(t)| \le \frac{|f(xt)|}{|\log^n t|} + \frac{|F_n(x,t)|}{|\log^n t|}.$$

Thus, since f is bounded, it is clear that $|h(t)|$ is also bounded outside the interval $[1/\delta, \delta]$ as a function of t. This completes the proof. □

The next statement corresponds to Taylor's formula with Lagrange's remainder.

Proposition 6.44 *Let $f \in C^{n+1}(\mathbb{R}^+)$ and let $x \in \mathbb{R}^+$ be fixed. Then for $t \in \mathbb{R}^+$ there exists θ between xt and x such that*

$$f(xt) = f(x) + \Theta f(x) \log t + \frac{\Theta^2 f(x)}{2!} \log^2 t + \cdots + \frac{\Theta^n f(x)}{n!} \log^n t + \frac{\Theta^{n+1} f(\theta)}{(n+1)!} \log^{n+1} t.$$

Proof First we prove the assertion for $n = 0$. Let $f \in C^1(\mathbb{R}^+)$. Setting

$$q(t) := \frac{f(xt) - f(x)}{\log t},$$

we consider the function

$$F(u) = f(xt) - f(u) - \log\left(\frac{xt}{u}\right) q(t)$$

and note that $F(x) = F(xt) = 0$. Hence, by Rolle's theorem, there exists a θ between x and xt such that $\Theta F(\theta) = 0$. Since $\Theta F(u) = -\Theta f(u) + q(t)$, we obtain

$$\Theta f(\theta) = \frac{f(xt) - f(x)}{\log t},$$

which yields the assertion for $n = 0$.

Now let $f \in C^{n+1}(\mathbb{R}^+)$. Using the notation of the previous proof, we define the function

$$\Phi(u) := f(xt) - F_n\left(u, \frac{xt}{u}\right) - \frac{\log^{n+1}(xt/u)}{\log^{n+1} t}\big(f(xt) - F_n(x, t)\big).$$

Then $\Phi(x) = \Phi(xt) = 0$ and so $(\Theta\Phi)(\theta) = 0$ for some θ between x and xt. Since

$$(\Theta\Phi)(u) = -\frac{(\Theta^{n+1} f(u)}{n!} \log^n\left(\frac{xt}{u}\right) + \frac{(n+1)\log^n(xt/u)}{\log^{n+1} t}\big(f(xt) - F_n(x, t)\big),$$

we readily arrive at the desired result. □

Remark 6.45 Note that there is a relationship between the expansion of Proposition 6.43 and that of Theorem 3.31. In fact, if in the latter we set $\theta = \theta_0 = 0,\ r_0 = x$ and $r = xt$ and truncate the resulting series, we obtain the approximation of Proposition 6.43.

Remark 6.46 The case $n = 0$ of Proposition 6.44 can be seen as a Mellin version of the classical Lagrange mean value theorem. Indeed, if f is a differentiable function on $\mathbb{R}^+$, then, given two points x_1, x_2 with $x_1 < x_2$, we deduce the existence of a point $\theta \in]x_1, x_2[$ such that

$$\frac{f(x_2) - f(x_1)}{\log(x_1/x_2)} = \Theta f(\theta).$$

Definition 6.47 For a function $K \in X_0$, we define the *logarithmic moment* of order $j \in \mathbb{N}$ by

$$m_j(K) := \int_0^\infty K(u) \log^j u \, \frac{du}{u}$$

and the *absolute logarithmic moment* of order $j \in \mathbb{N}$ by

$$M_j(K) := \int_0^\infty |K(u)||\log u|^j \, \frac{du}{u}.$$

It is easily seen that for $K_\rho(u) = \rho K(u^\rho)$, we obtain

$$m_j(K_\rho) = \frac{1}{\rho^j} m_j(K) \quad \text{and} \quad M_j(K_\rho) = \frac{1}{\rho^j} M_j(K). \tag{6.43}$$

We are now ready for studying the asymptotic behavior of $T_\rho f$ for $f \in X_0^p$, where $p \in [1, \infty]$. First we assume that the kernel function $K \in X_0$ is *Mellin-even* in the sense that $K(u) = K(1/u)$ for all $u \in \mathbb{R}^+$. Note that in this case the logarithmic moments of odd order are all zero while the moments of even order j may be expressed as

$$m_j(K) = 2 \int_0^1 K(u) \log^j u \, \frac{du}{u}.$$

We begin with the following statement.

Theorem 6.48 *Let $K \in X_0$ be a Mellin-even kernel function such that $M_1(K) < \infty$. Let $\{T_\rho\}_{\rho>0}$ be the family of Mellin–Fejér type operators generated by K (see Definition 6.36). Suppose that $f \in X_0^\infty \cap C^1(x)$ for an $x \in \mathbb{R}^+$. Then*

$$\lim_{\rho\to\infty} \rho\big((T_\rho f)(x) - f(x)\big) = 0.$$

Proof Under the hypotheses on f, it follows from Proposition 6.43 that

$$f(xt) - f(x) = \Theta f(x) \log t + h(t) \log t,$$

where h is a bounded function with $h(t) \to 0$ as $t \to 1$. Hence, given $\varepsilon > 0$, there exists a $\delta > 1$ be such that $|h(t)| < \varepsilon$ for $t \in]1/\delta, \delta[$. Since K is Mellin-even, we see with the help of (6.43) that $m_1(K_\rho) = 0$. Taking these facts into account, we conclude that

$$\begin{aligned}(T_\rho f)(x) - f(x) &= \int_0^\infty K_\rho(t)\big(f(xt) - f(x)\big)\frac{dt}{t} \\ &= \int_0^\infty K_\rho(t)\big(\Theta f(x)\log t + h(t)\log t\big)\frac{dt}{t} \\ &= \Theta f(x) m_1(K_\rho) + \int_0^\infty K_\rho(t)h(t)\log t\,\frac{dt}{t} \\ &= \int_{1/\delta}^{\delta} K_\rho(t)h(t)\log t\,\frac{dt}{t} + \int_{\mathbb{R}^+\setminus]1/\delta,\delta[} K_\rho(t)h(t)\log t\,\frac{dt}{t} \\ &=: I_1 + I_2.\end{aligned}$$

By our choice of δ, we may estimate I_1 as

$$|I_1| \le \rho\varepsilon \int_{1/\delta}^{\delta} |K(t^\rho)|\,|\log t|\frac{dt}{t} = \frac{\varepsilon}{\rho}\int_{1/\delta^\rho}^{\delta^\rho} |K(t)|\,|\log t|\frac{dt}{t} \le \frac{\varepsilon}{\rho} M_1(K).$$

Since h is known to be bounded on $\mathbb{R}^+$ and K is Mellin-even, we obtain

$$\begin{aligned}|I_2| &\le \|h\|_\infty^*\,\rho \int_{\mathbb{R}^+\setminus]1/\delta,\delta[} |K(t^\rho)|\,|\log t|\,\frac{dt}{t} \\ &= \frac{\|h\|_\infty^*}{\rho}\int_0^{1/\delta^\rho} |K(t)|\,|\log t|\,\frac{dt}{t} + \frac{\|h\|_\infty^*}{\rho}\int_{\delta^\rho}^\infty |K(t)|\,|\log t|\,\frac{dt}{t} \\ &= \frac{2\|h\|_\infty^*}{\rho}\int_0^{1/\delta^\rho} |K(t)|\,|\log t|\,\frac{dt}{t} \le \frac{2\|h\|_\infty^*}{\rho}\varepsilon\end{aligned}$$

for all sufficiently large ρ since $M_1(K) < \infty$. Combining the estimates of I_1 and I_2, we find that

$$\limsup_{\rho\to\infty} \rho\,\big|(T_\rho f)(x) - f(x)\big| \le \varepsilon\big(M_1(K) + 2\|h\|_\infty^*\big),$$

which implies the assertion. □

Under additional assumptions on the kernel function K, we can extend the previous result to functions $f \in X_0^p$ with $1 < p < \infty$. Note that for each p of this range the side conditions on K in the following theorem can be adjusted by an appropriate $\alpha > 1$. We always tacitly denote by p' the conjugate exponent of p.

Theorem 6.49 *Let $K \in X_0$ be a bounded, Mellin-even kernel function such that $M_1(K) < \infty$ and*

$$\int_{\delta\rho}^{\infty} |K(t)| \frac{dt}{t} = o(\rho^{-\alpha}) \qquad (\rho \to \infty)$$

for each $\delta > 1$ and some $\alpha > 1$. Let $\{T_\rho\}_{\rho>0}$ be the family of Mellin–Fejér type operators generated by K (see Definition 6.36). Suppose that $f \in X_0^p \cap C^1(x)$ for an $x \in \mathbb{R}^+$. If $(\alpha+1)/(\alpha-1) \le p < \infty$, then

$$\lim_{\rho\to\infty} \rho\big((T_\rho f)(x) - f(x)\big) = 0.$$

Proof Under the hypotheses on f, it follows from the first statement of Proposition 6.43 that

$$f(xt) - f(x) = \Theta f(x) \log t + h(t) \log t,$$

where $h(t) \to 0$ as $t \to 1$. Thus, for any given $\varepsilon > 0$, there exists $\delta > 1$ such that $|h(t)| < \varepsilon$ for $t \in]1/\delta, \delta[$. Now we can write

$$(T_\rho f)(x) - f(x) = \left\{ \int_{1/\delta}^{\delta} + \int_{\mathbb{R}^+ \setminus]1/\delta,\delta[} \right\} K_\rho(t)\big(f(xt) - f(x)\big)\frac{dt}{t} =: J_1 + J_1^*.$$

The integral J_1 can be represented as

$$J_1 = \int_{1/\delta}^{\delta} K_\rho(t)\big(\Theta f(x) + h(t)\big) \log t \,\frac{dt}{t} = \int_{1/\delta}^{\delta} K_\rho(t) h(t) \log t \,\frac{dt}{t},$$

where the second equation is a consequence of K being Mellin-even. The right-hand side is exactly the integral I_1 of the previous proof. With the former estimate we obtain

$$|J_1| \le \frac{\varepsilon}{\rho} M_1(K).$$

By splitting the integral J_1^*, we write

$$|J_1^*| \le \left\{ \int_0^{1/\delta} + \int_\delta^\infty \right\} |K_\rho(t)|\,|f(xt) - f(x)| \frac{dt}{t} =: J_2 + J_2^*.$$

First we continue with J_2^* which can obviously be estimated as

$$J_2^* \le \int_\delta^\infty |K_\rho(t)|\,|f(xt)| \frac{dt}{t} + |f(x)| \int_\delta^\infty |K_\rho(t)| \frac{dt}{t} =: J_3 + J_3^*.$$

Clearly, by the hypotheses on K,

$$J_3^* = |f(x)| \int_{\delta\rho}^{\infty} |K(t)| \frac{dt}{t} = o(\rho^{-\alpha}) \qquad (\rho \to \infty).$$

Concerning J_3, we find with the help of Hölder's inequality that

$$J_3 \leq \left(\int_{\delta}^{\infty} |K_\rho(t)|\,|f(xt)|^p \frac{dt}{t} \right)^{1/p} \left(\int_{\delta}^{\infty} |K_\rho(t)| \frac{dt}{t} \right)^{1/p'}.$$

Since K is bounded, there exists a constant $C > 0$ such that $|K(t)| \leq C$ for all $t \in \mathbb{R}^+$. Therefore

$$\begin{aligned} J_3 &\leq \rho^{1/p} C^{1/p} \|f\|_{X_0^p} \left(\int_{\delta\rho}^{\infty} |K(t)| \frac{dt}{t} \right)^{1/p'} \\ &= \rho^{1/p} C^{1/p} \|f\|_{X_0^p} \big(o(\rho^{-\alpha})\big)^{1/p'} \quad (\rho \to \infty). \end{aligned}$$

A short calculation shows that $1/p - \alpha/p' \leq -1$ if and only if $p \geq (\alpha+1)/(\alpha-1)$ and so, under our hypothesis on α and p, it follows that

$$J_2^* \leq J_3 + J_3^* = o(\rho^{-1}) \qquad (\rho \to \infty).$$

Keeping in mind that K is Mellin-even, we can argue for J_2 analogously and obtain the same asymptotic bound. Altogether we have shown that

$$\big|(T_\rho f)(x) - f(x)\big| \leq |J_1| + J_2 + J_2^* \leq \frac{\varepsilon}{\rho} M_1(K) + o(\rho^{-1}) \qquad (\rho \to \infty),$$

from which the assertion follows. □

The following theorem gives a more precise description of the asymptotic behavior under higher regularity of the function f.

Theorem 6.50 *Let $K \in X_0$ be a Mellin-even kernel function such that the logarithmic moments $m_2(K), m_4(K), \ldots, m_{2(k-1)}(K)$ and $M_{2k}(K)$ are finite. Let $\{T_\rho\}_{\rho>0}$ be the family of Mellin–Fejér type operators generated by K (see Definition 6.36). Suppose that $f \in X_0^\infty \cap C^{2k}(x)$ for an $x \in \mathbb{R}^+$. Then*

$$\lim_{\rho\to\infty} \rho^{2k} \left((T_\rho f)(x) - f(x) - \sum_{j=1}^{k-1} \frac{1}{\rho^{2j}} \frac{\Theta^{2j} f(x)}{(2j)!} m_{2j}(K) \right) = \frac{\Theta^{2k} f(x)}{(2k)!} m_{2k}(K).$$

Proof By Proposition 6.43 we have

$$f(xt) = f(x) + \sum_{j=1}^{2k} \frac{\Theta^j f(x)}{j!} \log^j t + h(t) \log^{2k} t, \tag{6.44}$$

where h is a bounded function such that $h(t) \to 0$ as $t \to 1$. Since K is Mellin-even, the error of the approximation by the Mellin–Fejér type operator can be represented as

$$(T_\rho f)(x) - f(x) = \int_0^1 K_\rho(t)\big(f(xt) + f(xt^{-1}) - 2f(x)\big)\frac{dt}{t}. \tag{6.45}$$

For the integrand, we find with the help of (6.44) that

$$f(xt) + f(xt^{-1}) - 2f(x) = 2\sum_{j=1}^{k} \frac{\Theta^{2j} f(x)}{(2j)!} \log^{2j} t + \big(h(t) + h(1/t)\big) \log^{2k} t. \tag{6.46}$$

Hence

$$\begin{aligned}(T_\rho f)(x) - f(x)&\\ &= 2\int_0^1 K_\rho(t) \sum_{j=1}^{k} \frac{\Theta^{2j} f(x)}{(2j)!} \log^{2j} t \,\frac{dt}{t} + \int_0^1 K_\rho(t)\big(h(t) + h(1/t)\big) \log^{2k} t \,\frac{dt}{t}\\ &=: J_1 + J_1^*.\end{aligned}$$

Clearly, J_1 may be expressed as

$$J_1 = 2\sum_{j=1}^{k} \frac{\Theta^{2j} f(x)}{(2j)!\rho^{2j}} \int_0^1 K(t) \log^{2j} t \,\frac{dt}{t} = \sum_{j=1}^{k} \frac{\Theta^{2j} f(x)}{(2j)!\rho^{2j}} m_{2j}(K).$$

Next we note that for any given $\varepsilon > 0$, there exists $\delta > 1$ with $|h(t) + h(1/t)| < \varepsilon$ for $t \in]1/\delta, 1[$. Therefore it is reasonable to split J_1^* as

$$J_1^* = \left(\int_0^{1/\delta} + \int_{1/\delta}^1\right) \rho K(t^\rho)(h(t) + h(1/t)) \log^{2k} t \,\frac{dt}{t} =: J_2 + J_2^*.$$

Since $M_{2k}(K) < \infty$, we see that for all sufficiently large ρ

$$|J_2| \le 2\frac{\|h\|_\infty^*}{\rho^{2k}} \int_0^{1/\delta^\rho} |K(t)| \log^{2k} t \,\frac{dt}{t} \le 2\|h\|_\infty^* \,\frac{\varepsilon}{\rho^{2k}}.$$

As for J_2^*, the following estimate holds:

$$|J_2^*| \le \varepsilon \int_{1/\delta}^{1} \rho |K(t^\rho)| \log^{2k} t \, \frac{dt}{t} = \frac{\varepsilon}{\rho^{2k}} \int_{1/\delta^\rho}^{1} |K(t)| \log^{2k} t \, \frac{dt}{t} \le \frac{\varepsilon}{\rho^{2k}} M_{2k}(K).$$

Combining the previous relations, we obtain

$$(T_\rho f)(x) - f(x) - \sum_{j=1}^{k-1} \frac{1}{\rho^{2j}} \frac{\Theta^{2j} f(x)}{(2j)!} m_{2j}(K) = \frac{1}{\rho^{2k}} \frac{\Theta^{2k} f(x)}{(2k)!} m_{2k}(K) + J_1^*,$$

where $J_1^* = o(\rho^{-2k})$ as $\rho \to \infty$. This implies the assertion. □

Now we give a version of Theorem 6.50 for functions $f \in X_0^p$.

Theorem 6.51 *Let $K \in X_0$ be a bounded, Mellin-even kernel function such that the logarithmic moments $m_2(K), m_4(K), \ldots, m_{2(k-1)}(K)$ and $M_{2k}(K)$ are finite and*

$$\int_{\delta\rho}^{\infty} |K(t)| \frac{dt}{t} = o(\rho^{-\alpha}), \quad \int_{\delta\rho}^{\infty} |K(t)| \log^{2j} t \, \frac{dt}{t} = o(\rho^{-2(k-j)}) \quad (j = 1, \ldots, k)$$

for each $\delta > 1$ and some $\alpha > 2k$ as $\rho \to \infty$. Let $\{T_\rho\}_{\rho>0}$ be the family of Mellin–Fejér type operators generated by K (see Definition 6.36). Suppose that $f \in X_0^p \cap C^{2k}(x)$ for an $x \in \mathbb{R}^+$. If $(\alpha + 1)/(\alpha - 2k) \le p < \infty$, then

$$\lim_{\rho \to \infty} \rho^{2k} \left((T_\rho f)(x) - f(x) - \sum_{j=1}^{k-1} \frac{1}{\rho^{2j}} \frac{\Theta^{2j} f(x)}{(2j)!} m_{2j}(K) \right) = \frac{\Theta^{2k} f(x)}{(2k)!} m_{2k}(K).$$

Proof We shall use essentially the same relations and estimates as in the proofs of the previous theorems. Beginning with the representation

$$(T_\rho f)(x) - f(x) = \int_0^\infty K_\rho(t) \big(f(xt) - f(x) \big) \frac{dt}{t}, \tag{6.47}$$

we deduce from Proposition 6.43 that

$$f(xt) - f(x) = \sum_{j=1}^{2k} \frac{\Theta^j f(x)}{j!} \log^j t + h(t) \log^{2k} t, \tag{6.48}$$

holding for t sufficiently close to 1. In particular, given any $\varepsilon > 0$, there exists $\delta > 1$ such that $|h(t)| < \varepsilon/2$ for all $t \in]1/\delta, \delta[$. With this δ, we split the integral in (6.47) as

$$(T_\rho f)(x) - f(x) = \left\{ \int_{1/\delta}^{\delta} + \int_{\mathbb{R}^+ \setminus]1/\delta, \delta[} \right\} K_\rho(t) \big(f(xt) - f(x) \big) \frac{dt}{t} =: J_1 + J_1^*.$$

Taking into account that K is Mellin-even and using (6.46), we can write the integral J_1 as

$$\begin{aligned} J_1 &= \int_{1/\delta}^{1} K_\rho(t)\big(f(xt) + f(xt^{-1}) - 2f(x)\big)\,\frac{dt}{t} \\ &= 2\sum_{j=1}^{k} \frac{\Theta^{2j} f(x)}{(2j)!} \int_{1/\delta}^{1} K_\rho(t) \log^{2j} t\, \frac{dt}{t} + \int_{1/\delta}^{1} K_\rho(t)\big(h(t) + h(1/t)\big) \log^{2k} t\, \frac{dt}{t} \\ &=: J_2 + J_2^*. \end{aligned}$$

Continuing with J_2, we find that

$$\begin{aligned} J_2 &= 2\sum_{j=1}^{k} \frac{\Theta^{2j} f(x)}{(2j)!} \int_{0}^{1} K_\rho(t) \log^{2j} t\, \frac{dt}{t} - 2\sum_{j=1}^{k} \frac{\Theta^{2j} f(x)}{(2j)!} \int_{0}^{1/\delta} K_\rho(t) \log^{2j} t\, \frac{dt}{t} \\ &= \sum_{j=1}^{k} \frac{\Theta^{2j} f(x)}{(2j)!} \frac{m_{2j}(K)}{\rho^{2j}} + J_3, \end{aligned}$$

where

$$J_3 := -2\sum_{j=1}^{k} \frac{\Theta^{2j} f(x)}{(2j)!} \int_{0}^{1/\delta} K_\rho(t) \log^{2j} t\, \frac{dt}{t} = o(\rho^{-2k}) \qquad (\rho \to \infty)$$

since, by the hypotheses on the kernel function K and the substitution $t = 1/u$, each term of this sum is seen to be of order $o(\rho^{-2k})$. As regards J_2^*, we readily obtain

$$\big|J_2^*\big| \le \varepsilon \int_{1/\delta}^{1} |K_\rho(t)| \log^{2k} t\, \frac{dt}{t} \le \frac{\varepsilon}{\rho^{2k}}\, M_{2k}(K).$$

It remains to estimate the integral J_1^*. It is exactly the same integral as J_1^* in the proof of Theorem 6.49. Proceeding step by step as in the former proof and noting that $1/p - \alpha/p' \le -2k$ if and only if $p \ge (\alpha + 1)/(\alpha - 2k)$, we find under the assumptions on α and p of the present theorem that $J_1^* = o(\rho^{-2k})$ as $\rho \to \infty$.

Altogether we have shown that

$$\begin{aligned} &\left| (T_\rho f)(x) - f(x) - \sum_{j=1}^{k} \frac{\Theta^{2j} f(x)}{(2j)!} \frac{m_{2j}(K)}{\rho^{2j}} \right| \\ &\qquad \le \frac{\varepsilon}{\rho^{2k}}\, M_{2k}(K) + o(\rho^{-2k}) \quad (\rho \to \infty). \end{aligned}$$

This implies the assertion. □

Now we introduce a modulus of continuity suitable for the Mellin setting.

Definition 6.52 The *logarithmic modulus of continuity* of a function $f : \mathbb{R}^+ \to \mathbb{C}$ is defined by

$$\omega(f, \eta) \equiv \omega(\eta) = \sup\{|f(x_1) - f(x_2)| : \ |\log x_1 - \log x_2| \le \eta\} \tag{6.49}$$

for $\eta > 0$. Furthermore, the logarithmic modulus of continuity of the derivatives $\Theta^j f(x)$ will be denoted by $\omega_j(f, \eta) \equiv \omega_j(\eta)$. We shall suppress f in the notation when there is no risk of confusion.

The function ω satisfies all the classical properties of a modulus of continuity. In particular, $\omega(\eta)$ is finite for every $\eta > 0$ and $\lim_{\eta\to 0+} \omega(\eta) = 0$ if and only if f is log-uniformly continuous on $\mathbb{R}^+$. Moreover, with similar reasonings as in Proposition 3.17, it can be shown that for every λ, $\eta \in \mathbb{R}^+$ and $j \in \mathbb{N}_0$, there holds

$$\omega_j(\lambda\eta) \le (\lambda + 1)\omega_j(\eta). \tag{6.50}$$

The following theorem presents two statements related to Theorems 6.48 and 6.50, respectively, but valid for all $x \in \mathbb{R}^+$.

Theorem 6.53 *Let $K \in X_0$ be a Mellin-even kernel function. Let $\{T_\rho\}_{\rho>0}$ be the family of Mellin–Fejér type operators generated by K (see Definition 6.36). Suppose that $f \in X_0^\infty$.*

(i) If the absolute logarithmic moments $M_1(K)$ and $M_2(K)$ are finite, Θf exists as a continuous function and $\omega_1(\eta)$ is finite for sufficiently small $\eta \in \mathbb{R}^+$, then for all $x \in \mathbb{R}^+$,

$$(T_\rho f)(x) = f(x) + \frac{\varepsilon_1(\rho)}{\rho},$$

where $|\varepsilon_1(\rho)| \le \big(M_1(K) + M_2(K)\big)\,\omega_1(1/\rho)$.

(ii) If the logarithmic moments $m_2(K), m_4(K), \ldots, m_{2(k-1)}(K)$ and $M_{2k}(K)$, $M_{2k+1}(K)$ are finite, $\Theta^{2k} f$ exists as a continuous function and $\omega_{2k}(\eta)$ is finite for sufficiently small $\eta \in \mathbb{R}^+$, then for all $x \in \mathbb{R}^+$,

$$(T_\rho f)(x) = f(x) + \sum_{j=1}^{k} \frac{1}{\rho^{2j}} \frac{\Theta^{2j} f(x)}{(2j)!} m_{2j}(K) + \frac{\varepsilon_{2k}(\rho)}{\rho^{2k}},$$

where $|\varepsilon_{2k}(\rho)| \le \dfrac{1}{(2k)!}\big(M_{2k}(K) + M_{2k+1}(K)\big)\,\omega_{2k}(1/\rho)$.

Proof We only prove part (ii). The proof of part (i) is similar and even simpler. Let $x \in \mathbb{R}^+$ be fixed. Since $\Theta^{2k} f$ exists and is continuous, we can employ

Proposition 6.44 for $n = 2k - 1$ and write

$$f(xt) - f(x) = \sum_{\ell=1}^{k} \frac{\Theta^{2\ell-1} f(x)}{(2\ell-1)!} \log^{2\ell-1} t + \sum_{j=1}^{k} \frac{\Theta^{2j} f(x)}{(2j)!} \log^{2j} t$$

$$+ \frac{\log^{2k} t}{(2k)!} \big(\Theta^{2k} f(\theta) - \Theta^{2k} f(x)\big)$$

with a suitable θ between xt and x. It follows that

$$(T_\rho f)(x) - f(x) = \int_0^\infty \sum_{\ell=1}^{k} \frac{\Theta^{2\ell-1} f(x)}{(2\ell-1)!} \log^{2\ell-1} t\, K_\rho(t) \frac{dt}{t}$$

$$+ \int_0^\infty \sum_{j=1}^{k} \frac{\Theta^{2j} f(x)}{(2j)!} \log^{2j} t\, K_\rho(t) \frac{dt}{t}$$

$$+ \int_0^\infty \frac{\log^{2k} t}{(2k)!} \big(\Theta^{2k} f(\theta) - \Theta^{2k} f(x)\big) K_\rho(t) \frac{dt}{t}$$

$$=: I_1 + I_2 + I_3.$$

Clearly, $I_1 = 0$ since K is Mellin-even. For I_2 we see that

$$I_2 = \sum_{j=1}^{k} \frac{\Theta^{2j} f(x)}{(2j)!} \frac{1}{\rho^{2j}} \int_0^\infty K(t) \log^{2j} t \,\frac{dt}{t} = \sum_{j=1}^{k} \frac{\Theta^{2j} f(x)}{(2j)!} \frac{m_{2j}(K)}{\rho^{2j}}.$$

Finally, by (6.50), we obtain

$$|\Theta^{2k} f(\theta) - \Theta^{2k} f(x)| \le \omega_{2k}(|\log t|) \le (1 + \rho|\log t|)\omega_{2k}(1/\rho),$$

and so

$$|I_3| \le \int_0^\infty \frac{\log^{2k} t}{(2k)!} (1 + \rho|\log t|)\omega_{2k}(1/\rho)|K_\rho(t)| \frac{dt}{t}$$

$$= \int_0^\infty \frac{\log^{2k} t}{(2k)!} \omega_{2k}(1/\rho)|K_\rho(t)| \frac{dt}{t} + \rho \int_0^\infty \frac{\log^{2k} t}{(2k)!} \omega_{2k}(1/\rho)|\log t|\,|K_\rho(t)| \frac{dt}{t}$$

$$= \frac{1}{\rho^{2k}} \frac{\omega_{2k}(1/\rho)}{(2k)!} \big(M_{2k}(K) + M_{2k+1}(K)\big).$$

From these relations the assertion follows immediately. □

As an easy consequence of Theorem 6.53, we can deduce the following result which gives the order of pointwise approximation by the integrals $T_\rho f$.

Corollary 6.54 *Let $K \in X_0$ be a Mellin-even kernel function. Let $\{T_\rho\}_{\rho>0}$ be the family of Mellin–Fejér type operators generated by K (see Definition 6.36). If under the hypotheses of Theorem 6.53 (i) the function Θf is log-uniformly continuous on $\mathbb{R}^+$, then for all $x \in \mathbb{R}^+$,*

$$\lim_{\rho\to\infty} \rho\big((T_\rho f)(x) - f(x)\big) = 0.$$

If under the hypotheses of Theorem 6.53 (ii) the function $\Theta^{2k} f$ is log-uniformly continuous, then for all $x \in \mathbb{R}^+$,

$$\lim_{\rho\to\infty} \rho^2\big((T_\rho f)(x) - f(x)\big) = \frac{\Theta^2 f(x)}{2} m_2(K). \tag{6.51}$$

Remark 6.55 If in the case $k = 1$ of Corollary 6.54 we have $m_2(K)\Theta^2 f(x) \neq 0$, then the difference $(T_\rho f)(x) - f(x)$ is exactly of order $\mathcal{O}(\rho^{-2})$ and this degree of approximation cannot be improved even if higher Mellin derivatives of f exist at x. In the next section, we shall consider a method which enables us to improve the order of approximation.

6.4.3 Linear Combinations of Mellin–Fejér Type Operators

For improving the order of approximation, we want to construct suitable linear combinations of $T_\rho f$. More precisely, we shall select real numbers $\alpha_1, \alpha_2, \ldots, \alpha_\nu$ such that $\sum_{\ell=1}^{\nu} \alpha_\ell = 1$, define a kernel

$$H_{\rho,\nu}(t) := \sum_{\ell=1}^{\nu} \alpha_\ell K_{\rho\ell}(t)$$

and consider the operator

$$(T_{\rho,\nu} f)(x) := \int_0^\infty H_{\rho,\nu}(u) f(xu)\,\frac{du}{u} = \sum_{\ell=1}^{\nu} \alpha_\ell (T_{\rho\ell} f)(x) \qquad (x \in \mathbb{R}^+).$$

A general result in the spirit of Theorem 6.53 (ii) can now be stated as follows.

Theorem 6.56 *Let $K \in X_0$ be a Mellin-even kernel function. Then, under the hypotheses of Theorem 6.53 (ii), there holds, for $2 \le \nu \le k$,*

$$(T_{\rho,\nu}f)(x) = f(x) + \sum_{j=1}^{k} \frac{1}{\rho^{2j}} \frac{\Theta^{2j}f(x)}{(2j)!} m_{2j}(K) \Big(\sum_{\ell=1}^{\nu} \frac{\alpha_\ell}{\ell^{2j}} \Big) \\ + \mathcal{O}\big(\rho^{-2k}\omega_{2k}(\rho^{-1})\big) \quad (\rho \to \infty).$$

Proof The conclusion easily follows from Theorem 6.53 (ii) by noting that the moments of the kernel $H_{\rho,\nu}$ are given by

$$m_{2j}(H_{\rho,\nu}) = \frac{m_{2j}(K)}{\rho^{2j}} \sum_{\ell=1}^{\nu} \frac{\alpha_\ell}{\ell^{2j}}.$$

□

If we choose the coefficients $\alpha_1, \alpha_2, \ldots, \alpha_\nu$ such that

$$\sum_{\ell=1}^{\nu} \alpha_\ell = 1, \quad \sum_{\ell=1}^{\nu} \frac{\alpha_\ell}{\ell^2} = 0, \quad \ldots, \quad \sum_{\ell=1}^{\nu} \frac{\alpha_\ell}{\ell^{2\nu-2}} = 0$$

and $\Theta^{2\nu} f$ is log-uniformly continuous, we obtain

$$\lim_{\rho\to\infty} \rho^{2\nu}\big((T_{\rho,\nu}f)(x) - f(x)\big) = \frac{m_{2\nu}(K)}{(2\nu)!} \sum_{\ell=1}^{\nu} \frac{\alpha_\ell}{\ell^{2\nu}}\, \Theta^{2\nu} f(x).$$

Let us consider some particular cases. For $\nu = 2$ we have

$$(T_{\rho,2}f)(x) = \alpha_1 (T_\rho f)(x) + \alpha_2 (T_{2\rho} f)(x).$$

Thus, for $k \geq 2$, we obtain under the hypotheses of Theorem 6.53 (ii) that

$$(T_{\rho,2}f)(x) = f(x) + \sum_{j=1}^{k} \frac{\Theta^{2j}f(x)}{(2j)!} \frac{m_{2j}(K)}{\rho^{2j}} \Big(\alpha_1 + \frac{\alpha_2}{2^{2j}}\Big) + \mathcal{O}\big(\rho^{-2k}\omega_{2k}(\rho^{-1})\big)$$

as $\rho \to \infty$. This suggests to determine the coefficients α_1, α_2 such that the linear system

$$\alpha_1 + \alpha_2 = 1$$
$$\alpha_1 + \frac{\alpha_2}{4} = 0$$

is satisfied. Its unique solution is given by $\alpha_1 = -\dfrac{1}{3}$ and $\alpha_2 = \dfrac{4}{3}$. If $\Theta^4 f$ is log-uniformly continuous, this choice of the coefficients yields

$$\lim_{\rho\to\infty} \rho^4\big((T_{\rho,2}f)(x) - f(x)\big) = -\frac{m_4(K)}{96}\Theta^4 f(x). \tag{6.52}$$

For $\nu = 3$ we have

$$(T_{\rho,3}f)(x) = \alpha_1(T_\rho f)(x) + \alpha_2(T_{2\rho}f)(x) + \alpha_3(T_{3\rho}f)(x).$$

Thus, for $k \geq 3$, we obtain under the hypotheses of Theorem 6.53 (ii) that

$$(T_{\rho,3}f)(x) = f(x) + \sum_{j=1}^{k} \frac{\Theta^{2j} f(x)}{(2j)!} \frac{m_{2j}(K)}{\rho^{2j}} \left(\alpha_1 + \frac{\alpha_2}{2^{2j}} + \frac{\alpha_3}{3^{2j}}\right)$$
$$+\mathcal{O}\big(\rho^{-2k}\omega_{2k}(\rho^{-1})\big)$$

as $\rho \to \infty$. This suggests to determine the coefficients α_1, α_2, α_3 such that the linear system

$$\begin{aligned} \alpha_1 + \alpha_2 + \alpha_3 &= 1 \\ \alpha_1 + \frac{\alpha_2}{4} + \frac{\alpha_3}{9} &= 0 \\ \alpha_1 + \frac{\alpha_2}{16} + \frac{\alpha_3}{81} &= 0 \end{aligned}$$

is satisfied. Its unique solution is given by $\alpha_1 = \dfrac{1}{24}$, $\alpha_2 = -\dfrac{16}{15}$ and $\alpha_3 = \dfrac{81}{40}$. If $\Theta^6 f$ is log-uniformly continuous, this choice of the coefficients yields

$$\lim_{\rho\to\infty} \rho^6\big((T_{\rho,3}f)(x) - f(x)\big) = \frac{m_6(K)}{25920}\Theta^6 f(x). \tag{6.53}$$

6.4.4 Mellin–Fejér Type Operators for General Kernel Functions

In this subsection the kernel function K is supposed to be not necessarily Mellin-even. The results of the previous two subsections can be generalized to this situation by slight modifications of the former proofs. We limit ourselves to functions $f \in X_0^\infty$.

The following statement generalizes Theorem 6.48.

Theorem 6.57 *Let $K \in X_0$ be a kernel function such that $M_1(K) < \infty$. Let $\{T_\rho\}_{\rho>0}$ be the family of Mellin–Fejér type operators generated by K (see Definition 6.36). Suppose that $f \in X_0^\infty \cap C^1(x)$ for an $x \in \mathbb{R}^+$. Then*

$$\lim_{\rho\to\infty} \rho\big((T_\rho f)(x) - f(x)\big) = m_1(K)\Theta f(x).$$

Proof As in the proof of Theorem 6.48, we find that

$$(T_\rho f)(x) - f(x) = \frac{m_1(K)}{\rho}\Theta f(x) + \int_0^\infty K_\rho(t)h(t)\log t\,\frac{dt}{t}.$$

Now it suffices to observe that for sufficiently large ρ the former estimate for the integral can also be achieved when K is not necessarily Mellin-even. □

As regards Theorem 6.50, the following generalization holds.

Theorem 6.58 *Let $K \in X_0$ be a kernel function such that the logarithmic moments $m_1(K), m_2(K), \ldots, m_{k-1}(K)$ and $M_k(K)$ are finite. Let $\{T_\rho\}_{\rho>0}$ be the family of Mellin–Fejér type operators generated by K (see Definition 6.36). Suppose that $f \in X_0^\infty \cap C^k(x)$ for an $x \in \mathbb{R}^+$. Then*

$$\lim_{\rho\to\infty} \rho^k\left((T_\rho f)(x) - f(x) - \sum_{j=1}^{k-1}\frac{1}{\rho^j}\frac{\Theta^j f(x)}{j!}m_j(K)\right) = \frac{\Theta^k f(x)}{k!}m_k(K).$$

Proof We know from Proposition 6.43 that

$$f(xt) = f(x) + \sum_{j=1}^{k}\frac{\Theta^j f(x)}{j!}\log^j t + h(t)\log^k t,$$

where $h(t)$ is a bounded function such that $h(t) \to 0$ as $t \to 1$. Hence

$$\begin{aligned}
&(T_\rho f)(x) - f(x)\\
&\quad = \sum_{j=1}^{k}\frac{\Theta^j f(x)}{j!}\int_0^\infty K_\rho(t)\log^j t\,\frac{dt}{t} + \int_0^\infty K_\rho(t)h(t)\log^k t\,\frac{dt}{t}\\
&\quad =: J_1 + J_1^*.
\end{aligned}$$

Clearly, J_1 may be expressed as

$$J_1 = \sum_{j=1}^{k}\frac{\Theta^j f(x)}{j!\,\rho^j}\,m_j(K).$$

Choosing an appropriate $\delta > 1$ and writing

$$J_1^* = \left(\int_0^{1/\delta} + \int_{1/\delta}^{\delta} + \int_\delta^\infty\right)K_\rho(t)h(t)\log^k t\,\frac{dt}{t},$$

we conclude by arguments similar to those in the proof of Theorem 6.50 that $J_1^* = o(\rho^{-k})$ as $k \to \infty$. From these relations the assertion follows immediately. □

As to uniform convergence, the following generalization of Theorem 6.53 holds.

Theorem 6.59 *Let $K \in X_0$ be a kernel function. Let $\{T_\rho\}_{\rho>0}$ be the family of Mellin–Fejér type operators generated by K (see Definition 6.36). Suppose that $f \in X_0^\infty$.*

(i) If the absolute logarithmic moments $M_1(K)$ and $M_2(K)$ are finite, Θf exists as a continuous function and $\omega_1(\eta)$ is finite for sufficiently small $\eta \in \mathbb{R}^+$, then for all $x \in \mathbb{R}^+$,

$$(T_\rho f)(x) = f(x) + \frac{1}{\rho}\,\Theta f(x) m_1(K) + \frac{\varepsilon_1(\rho)}{\rho},$$

where $|\varepsilon_1(\rho)| \le \big(M_1(K) + M_2(K)\big)\,\omega_1(1/\rho)$.

(ii) If the logarithmic moments $m_1(K), m_2(K), \ldots, m_{k-1}(K)$ and $M_k(K)$, $M_{k+1}(K)$ are finite, $\Theta^k f$ exists as a continuous function and $\omega_k(\eta)$ is finite for sufficiently small $\eta \in \mathbb{R}^+$, then for all $x \in \mathbb{R}^+$,

$$(T_\rho f)(x) = f(x) + \sum_{j=1}^{k} \frac{1}{\rho^j} \frac{\Theta^j f(x)}{j!} m_j(K) + \frac{\varepsilon_k(\rho)}{\rho^k},$$

where $|\varepsilon_k(\rho)| \le \dfrac{1}{k!}\big(M_k(K) + M_{k+1}(K)\big)\,\omega_k(1/\rho)$.

Proof The proof is similar to that of Theorem 6.53. □

Theorem 6.57 shows us that if $m_1(K)\Theta f(x) \neq 0$, then the difference $(T_\rho f)(x) - f(x)$ is exactly of order $\mathcal{O}(\rho^{-1})$, and the rate of approximation cannot be improved even if higher Mellin derivatives of f exist at x. However, as in the symmetric case, we can increase the order of approximation by appropriate linear combinations of $T_\rho f$. In the notation of the previous subsection, we deduce from Theorem 6.59 the following statement.

Theorem 6.60 *Let $K \in X_0$ be a kernel function. Then, under the hypotheses of Theorem 6.59 (ii), there holds, for $2 \le \nu \le k$,*

$$(T_{\rho,\nu} f)(x) = f(x) + \sum_{j=1}^{k} \frac{1}{\rho^j} \frac{\Theta^j f(x)}{j!} m_j(K) \Big(\sum_{\ell=1}^{\nu} \frac{\alpha_\ell}{\ell^j} \Big)$$
$$+ \mathcal{O}\big(\rho^{-k}\omega_k(\rho^{-1})\big) \quad (\rho \to \infty).$$

If we choose the coefficients $\alpha_1, \alpha_2, \ldots, \alpha_\nu$ such that

$$\sum_{\ell=1}^{\nu} \alpha_\ell = 1, \quad \sum_{\ell=1}^{\nu} \frac{\alpha_\ell}{\ell} = 0, \quad \ldots, \quad \sum_{\ell=1}^{\nu} \frac{\alpha_\ell}{\ell^{\nu-1}} = 0$$

and $\Theta^\nu f$ is log-uniformly continuous, we obtain

$$\lim_{\rho\to\infty} \rho^\nu\big((T_{\rho,\nu} f)(x) - f(x)\big) = \frac{m_\nu(K)}{\nu!} \sum_{\ell=1}^{\nu} \frac{\alpha_\ell}{\ell^\nu}\, \Theta^\nu f(x).$$

Let us consider some particular cases. For $\nu = 2$ we have

$$(T_{\rho,2} f)(x) = \alpha_1 (T_\rho f)(x) + \alpha_2 (T_{2\rho} f)(x).$$

Thus, for $k \geq 2$, we obtain under the hypotheses of Theorem 6.59 (ii) that

$$\begin{aligned}(T_{\rho,2} f)(x) = f(x) + \sum_{j=1}^{k} \frac{\Theta^j f(x)}{j!} \frac{m_j(K)}{\rho^j} \left(\alpha_1 + \frac{\alpha_2}{2^j}\right) \\ + \mathcal{O}\big(\rho^{-k}\omega_k(\rho^{-1})\big) \quad (\rho \to \infty).\end{aligned}$$

This suggests to determine the coefficients α_1, α_2 such that the linear system

$$\begin{aligned}\alpha_1 + \alpha_2 &= 1 \\ \alpha_1 + \frac{\alpha_2}{2} &= 0\end{aligned}$$

is satisfied. Its unique solution is given by $\alpha_1 = -1$ and $\alpha_2 = 2$. If $\Theta^2 f$ is log-uniformly continuous, this choice of the coefficients yields

$$\lim_{\rho\to\infty} \rho^2\big((T_{\rho,2} f)(x) - f(x)\big) = -\frac{m_2(K)}{4}\, \Theta^2 f(x). \tag{6.54}$$

For $\nu = 3$ we have

$$(T_{\rho,3} f)(x) = \alpha_1 (T_\rho f)(x) + \alpha_2 (T_{2\rho} f)(x) + \alpha_3 (T_{3\rho} f)(x).$$

Thus, for $k \geq 3$, we obtain under the hypothesis of Theorem 6.59 (ii) that

$$(T_{\rho,3} f)(x) = f(x) + \sum_{j=1}^{k} \frac{\Theta^j f(x)}{j!} \frac{m_j(K)}{\rho^j} \left(\alpha_1 + \frac{\alpha_2}{2^j} + \frac{\alpha_3}{3^j}\right) + \mathcal{O}\big(\rho^{-k}\omega_k(\rho^{-1})\big)$$

as $\rho \to \infty$. This suggests to determine the coefficients α_1, α_2, α_3 such that the linear system

$$\begin{aligned}\alpha_1 + \alpha_2 + \alpha_3 &= 1\\ \alpha_1 + \frac{\alpha_2}{2} + \frac{\alpha_3}{3} &= 0\\ \alpha_1 + \frac{\alpha_2}{4} + \frac{\alpha_3}{9} &= 0\end{aligned}$$

is satisfied. Its unique solution is given by $\alpha_1 = \frac{1}{2}$, $\alpha_2 = -4$ and $\alpha_3 = \frac{9}{2}$. If $\Theta^3 f$ is log-uniformly continuous, this choice of the coefficients yields

$$\lim_{\rho\to\infty} \rho^3\big((T_{\rho,3}f)(x) - f(x)\big) = \frac{m_3(K)}{36}\Theta^3 f(x). \tag{6.55}$$

6.4.5 Some Examples

Here we illustrate the results on linear combinations by specific examples of kernels. Both the Mellin–Gauss–Weierstrass kernel and the Mellin–Abel–Poisson kernel satisfy the required assumptions. We give details for the Mellin–Gauss–Weierstrass kernel only. Analogous considerations for the other kernel are left to the reader.

Example 6.61 The Mellin–Gauss–Weierstrass kernel $G_\rho(x)$, introduced as Example 6.25, is a Mellin–Fejér type kernel generated by the kernel function

$$G(x) := \frac{1}{\sqrt{4\pi}} \exp\left(-\left(\frac{1}{2}\log x\right)^2\right) \qquad (x \in \mathbb{R}^+).$$

Clearly G is Mellin-even and satisfies the hypotheses of the theorems in Sects. 6.4.2 and 6.4.3. The corresponding integral operator is given by

$$(\mathcal{G}_\rho f)(x) := \int_0^\infty G_\rho(t) f(xt)\,\frac{dt}{t} \qquad (x \in \mathbb{R}^+).$$

Now we consider linear combinations. The following calculation of logarithmic moments will end up with integrals which could be expressed by the gamma function. For the direct evaluation of these integrals given here see, e.g., Gradshteyn–Ryzhik [160, formula 3.371]. For $\nu = 2$, we choose $\alpha_1 = -\frac{1}{3}$, $\alpha_2 = \frac{4}{3}$ and calculate

$$m_4(G) = \frac{1}{\sqrt{4\pi}} \int_0^\infty \exp\left(-\left(\frac{1}{2}\log t\right)^2\right) \log^4 t\,\frac{dt}{t} = \frac{2^5}{\sqrt{4\pi}} \int_0^\infty e^{-t} t^{3/2} dt = 12.$$

Then, by (6.52), we obtain

$$\lim_{\rho\to\infty} \rho^4\big((\mathcal{G}_{\rho,2}f)(x) - f(x)\big) = -\frac{1}{8}\,\Theta^4 f(x)$$

provided that $\Theta^4 f$ is log-uniformly continuous.

For $\nu = 3$, we choose $\alpha_1 = \frac{1}{24}$, $\alpha_2 = -\frac{16}{15}$, $\alpha_3 = \frac{81}{40}$ and calculate

$$m_6(G) = \frac{1}{\sqrt{4\pi}} \int_0^\infty \exp\left(-\left(\frac{1}{2}\log t\right)^2\right) \log^6 t\,\frac{dt}{t} = \frac{2^7}{\sqrt{4\pi}} \int_0^\infty e^{-t} t^{5/2} dt = 120.$$

Then, by (6.53), we obtain

$$\lim_{\rho\to\infty} \rho^6\big((\mathcal{G}_{\rho,3}f)(x) - f(x)\big) = \frac{1}{216}\,\Theta^6 f(x)$$

provided that $\Theta^6 f$ is log-uniformly continuous.

Example 6.62 As a second example, we consider the following *Mellin–Poisson–Cauchy kernel* which is a Mellin–Fejér type kernel generated by the function

$$C(x) := \frac{2^{n-1}(n-1)!}{\pi(2n-3)!!}\,\frac{1}{(1+\log^2 x)^n} \qquad (x \in \mathbb{R}^+)$$

for $n \in \mathbb{N}\setminus\{1\}$, where $(2n-3)!! := 1\cdot 3\cdots(2n-3)$. The corresponding integral operator is given by

$$(\mathcal{C}_\rho f)(x) := \frac{2^{n-1}(n-1)!}{\pi(2n-3)!!} \int_0^\infty \frac{\rho}{(1+\rho^2\log^2 t)^n} f(xt)\,\frac{dt}{t}.$$

Again C is a Mellin-even kernel function. Moreover, it can be shown that all the previous hypotheses on the involved kernel functions are satisfied for sufficiently large values of n. Thus the theory developed in the previous subsections can be applied to $\mathcal{C}_\rho$. Now we consider linear combinations of the Mellin–Poisson–Cauchy operator. The following calculation of logarithmic moments will end up with integrals that could be expressed by the Euler beta function which in turn could be represented in terms of the gamma function. For the direct evaluation of these integrals given here see, e.g., Gradshteyn–Ryzhik [160, formula 3.251(4)]. For $\nu = 2$, we choose α_1 and α_2 as before and calculate

$$\begin{aligned} m_4(C) &= \frac{2^{n-1}(n-1)!}{\pi(2n-3)!!} \int_0^\infty \frac{1}{(1+\log^2 t)^n} \log^4 t\,\frac{dt}{t} \\ &= \frac{2^n(n-1)!}{\pi(2n-3)!!} \int_0^\infty \frac{u^4}{(1+u^2)^n} du = \frac{3}{(2n-3)(2n-5)}. \end{aligned}$$

Thus, by (6.52), we obtain

$$\lim_{\rho\to\infty} \rho^4\big((\mathcal{C}_{\rho,2}f)(x) - f(x)\big) = -\frac{1}{32(2n-3)(2n-5)}\,\Theta^4 f(x)$$

provided that $\Theta^4 f$ is log-uniformly continuous.

For $\nu = 3$, we choose $\alpha_1, \alpha_2, \alpha_3$ as in the previous example and calculate

$$\begin{aligned} m_6(C) &= \frac{2^{n-1}(n-1)!}{\pi(2n-3)!!}\int_0^\infty \frac{1}{(1+\log^2 t)^n}\log^6 t\,\frac{dt}{t} \\ &= \frac{2^n(n-1)!}{\pi(2n-3)!!}\int_0^\infty \frac{u^6}{(1+u^2)^n}du = \frac{15}{(2n-3)(2n-5)(2n-7)}. \end{aligned}$$

Then, by (6.53), we obtain

$$\lim_{\rho\to\infty} \rho^6\big((\mathcal{C}_{\rho,3}f)(x) - f(x)\big) = \frac{1}{1728(2n-3)(2n-5)(2n-7)}\,\Theta^6 f(x)$$

provided $\Theta^6 f$ is log-uniformly continuous. Note that the approximation can be improved by increasing n.

Example 6.63 As a third example, we consider the so-called *moment kernel* defined by the generating function

$$K(x) = x\chi_{]0,1[}(x) \qquad (x \in \mathbb{R}^+),$$

where $\chi_{]0,1[}(x)$ is the characteristic function of the interval $]0,1[$. The corresponding integral operator is given by

$$(\mathcal{K}_\rho f)(x) := \rho\int_0^1 t^\rho f(xt)\,\frac{dt}{t}.$$

Obviously K satisfies (6.36) and so it is a kernel function. However K is not Mellin-even but fulfills the hypotheses of the theorems in Sect. 6.4.4. Again we consider linear combinations. For $\nu = 2$, we choose $\alpha_1 = -1$, $\alpha_2 = 2$ and calculate

$$m_2(K) = \int_0^1 \log^2 t\,dt = 2.$$

Then, by (6.54), we obtain

$$\lim_{\rho\to\infty} \rho^2\big((\mathcal{K}_{\rho,2}f)(x) - f(x)\big) = -\frac{1}{2}\,\Theta^2 f(x)$$

provided that $\Theta^2 f$ is log-uniformly continuous.

For $\nu = 3$, we choose $\alpha_1 = \frac{1}{2}, \alpha_2 = -4, \alpha_3 = \frac{9}{2}$ and calculate

$$m_3(K) = \int_0^1 \log^3 t \, dt = -6.$$

Then, by (6.55), we obtain

$$\lim_{\rho\to\infty} \rho^3\big((\mathcal{K}_{\rho,3} f)(x) - f(x)\big) = -\frac{1}{6}\,\Theta^3 f(x)$$

provided that $\Theta^3 f$ is log-uniformly continuous.

6.4.6 Mellin–Lebesgue Points and Convergence Almost Everywhere

Let $f \in X_{0,\mathrm{loc}}$. We say that a point $x \in \mathbb{R}^+$ is a *Mellin–Lebesgue point* of f if

$$\lim_{v\to 1} \left| \frac{1}{\log v} \int_1^v |f(xt) - f(x)| \, \frac{dt}{t} \right| = 0.$$

Note that this condition is equivalent to

$$\lim_{v\to 1-} \frac{1}{-\log v} \int_v^1 |f(xt) - f(x)| \, \frac{dt}{t} + \lim_{v\to 1+} \frac{1}{\log v} \int_1^v |f(xt) - f(x)| \, \frac{dt}{t} = 0.$$

The following proposition shows that Mellin–Lebesgue points occur very frequently.

Proposition 6.64 *Let $f \in X_{0,\mathrm{loc}}$. Then Mellin–Lebesgue points of f exist almost everywhere on $\mathbb{R}^+$.*

Proof For $x \in \mathbb{R}^+$ and $v \in \,]0,1[$, we can write

$$\begin{aligned}\frac{1}{-\log v} \int_v^1 |f(xt) - f(x)| \, \frac{dt}{t} &= \frac{1}{-\log v} \int_{xv}^x |f(t) - f(x)| \, \frac{dt}{t} \\ &\le \frac{v-1}{v \log v} \, \frac{1}{x(1-v)} \int_{xv}^x |f(t) - f(x)| \, dt.\end{aligned}$$

It is easily seen that

$$\lim_{v\to 1} \frac{v-1}{v \log v} = 1$$

while by Proposition 1.15,

$$\lim_{v\to 1-} \frac{1}{x(1-v)} \int_{xv}^{x} |f(t) - f(x)| dt = 0$$

for $x \in \mathbb{R}^+$ almost everywhere. Thus

$$\lim_{v\to 1-} \frac{1}{-\log v} \int_{v}^{1} |f(xt) - f(x)| \frac{dt}{t} = 0$$

almost everywhere on $\mathbb{R}^+$. The limit $v \to 1+$ is treated analogously. □

Now we show that under suitable assumptions on the Mellin–Fejér type kernel $\{K_\rho\}_{\rho>0}$ we obtain convergence of $T_\rho f$ at each Mellin–Lebesgue point of $f \in X_0^p$ with $1 \le p \le \infty$.

We begin with the following statement.

Theorem 6.65 *Let $K \in X_0$ be a kernel function such that $|K|$ is increasing on $]0, 1[$ and decreasing on $]1, \infty[$, and $S_t := \sup_{\rho>0} |\rho K(t^\rho)| < \infty$ for each $t \in \mathbb{R}^+ \setminus \{1\}$. Let $\{T_\rho\}_{\rho>0}$ be the family of Mellin–Fejér type operators generated by K (see Definition 6.36). Suppose that $f \in X_0^p$, where $1 < p \le \infty$. Then*

$$\lim_{\rho\to\infty} |(T_\rho f)(x) - f(x)| = 0$$

at each Mellin–Lebesgue point $x \in \mathbb{R}^+$ of f.

Proof Let x be a Mellin–Lebesgue point. For $\delta > 1$ and $U_\delta := [1/\delta, \delta]$, we can write

$$\begin{aligned} |(T_\rho f)(x) - f(x)| &\le \int_0^\infty |K_\rho(t)||f(xt) - f(x)| \frac{dt}{t} \\ &= \int_{U_\delta} |K_\rho(t)||f(xt) - f(x)| \frac{dt}{t} \\ &\quad + \int_{\mathbb{R}^+\setminus U_\delta} |K_\rho(t)||f(xt) - f(x)| \frac{dt}{t} \\ &=: I_1 + I_2. \end{aligned}$$

Turning first to I_1, we split it as

$$I_1 = \left(\int_{1/\delta}^{1} + \int_{1}^{\delta} \right) |K_\rho(t)||f(xt) - f(x)| \frac{dt}{t} =: I_1^1 + I_1^2.$$

For estimating I_1^1, we consider the absolutely continuous function

$$\Lambda^- f(t) := \int_t^1 |f(x\xi) - f(x)| \frac{d\xi}{\xi}$$

for $t \in]0, 1[$. Now let $\varepsilon > 0$ be given. Since $x \in \mathbb{R}^+$ is a Mellin–Lebesgue point, we can choose $\delta > 1$ such that

$$\Lambda^- f(t) \leq -\varepsilon \log t$$

for every $t \in]1/\delta, 1[$. Thus, taking into account that $|K_\rho|$ is increasing on $]0, 1[$, we find by two partial integrations that

$$\begin{aligned} I_1^1 &= \int_{1/\delta}^1 |K_\rho(t)||f(xt) - f(x)| \frac{dt}{t} \\ &= -\int_{1/\delta}^1 |K_\rho(t)|\, d\Lambda^- f(t) = |K_\rho(1/\delta)|\Lambda^- f(1/\delta) + \int_{1/\delta}^1 \Lambda^- f(t)\, d|K_\rho(t)| \\ &\leq |K_\rho(1/\delta)|\Lambda^- f(1/\delta) - \int_{1/\delta}^1 \varepsilon \log t\, d|K_\rho(t)| \\ &\leq |K_\rho(1/\delta)|\varepsilon \log \delta - \varepsilon\left(|K_\rho|(1/\delta) \log \delta - \int_{1/\delta}^1 |K_\rho(t)| \frac{dt}{t}\right) \leq \varepsilon \|K\|_{X_0}. \end{aligned}$$

Introducing the function

$$\Lambda^+ f(t) := \int_1^t |f(x\xi) - f(x)| \frac{d\xi}{\xi}$$

and taking into account that $|K_\rho|$ is decreasing on $]1, \infty[$, we can argue analogously for estimating I_1^2 and arrive at the same bound. Hence $I_1 \leq 2\varepsilon \|K\|_{X_0}$.

Now we estimate I_2. If $f \in X_0^\infty$, we readily see that

$$I_2 \leq 2\|f\|_{X_0^\infty} \int_{\mathbb{R}^+ \setminus U_\delta} |K_\rho(t)| \frac{dt}{t} = 2\|f\|_{X_0^\infty}\left(\int_0^{1/\delta^\rho} + \int_{\delta^\rho}^\infty \right)|K(t)| \frac{du}{u},$$

which tends to zero as $\rho \to \infty$.

In the case $f \in X_0^p$ with $1 < p < \infty$, we conclude that

$$\begin{aligned} I_2 \leq \int_0^{1/\delta} |K_\rho(t)||f(xt)| \frac{dt}{t} + |f(x)| \int_0^{1/\delta} |K_\rho(t)| \frac{dt}{t} \\ + \int_\delta^\infty |K_\rho(t)||f(xt)| \frac{dt}{t} + |f(x)| \int_\delta^\infty |K_\rho(t)| \frac{dt}{t}. \end{aligned}$$

As we have noted already, the integrals of the second and the fourth term tend to zero as $\rho \to \infty$. For the first term we use Hölder's inequality with p' being the conjugate exponent of p and obtain by involving S_t,

$$\int_0^{1/\delta} |K_\rho(t)||f(xt)| \frac{dt}{t} \le \left(\int_0^{1/\delta} |K_\rho(t)||f(xt)|^p \frac{dt}{t} \right)^{1/p} \left(\int_0^{1/\delta} |K_\rho(t)| \frac{dt}{t} \right)^{1/p'}$$
$$\le S_{1/\delta} \|f\|_{X_0^p} \left(\int_0^{1/\delta} |K_\rho(t)| \frac{dt}{t} \right)^{1/p'},$$

which tends to zero as $\rho \to \infty$. The third term is treated analogously. Hence in any case, $I_2 \to 0$ as $\rho \to \infty$. Since ε can be an arbitrary positive number, we arrive at the desired result. □

In order to obtain a convergence result for functions $f \in X_0$, we have to impose a further condition on the kernel function.

Theorem 6.66 *Let $K \in X_0$ be a kernel function satisfying the hypotheses of Theorem 6.65 and assume in addition that*

$$\lim_{\rho \to \infty} |K_\rho(t)| = 0 \tag{6.56}$$

for each $t \in \mathbb{R}^+ \setminus \{1\}$. Suppose that $f \in X_0$. Then

$$\lim_{\rho \to \infty} |(T_\rho f)(x) - f(x)| = 0$$

at each Mellin–Lebesgue point $x \in \mathbb{R}^+$ of f.

Proof We proceed as in the previous proof except that we use the estimate

$$\int_0^{1/\delta} |K_\rho(t)||f(xt)| \frac{dt}{t} \le |K_\rho(1/\delta)| \|f\|_{X_0}$$

instead of Hölder's inequality and employ condition (6.56). □

Remark 6.67 Theorem 6.65 allows some flexibility. If $f \in X_c^p$, then f_c, defined by $f_c(x) := x^c f(x)$, belongs to X_0^p, and so the assertion holds for f_c in place of f.

If the *monotonicity assumption*, requiring that $|K|$ is increasing on $]0, 1[$ and decreasing on $]1, \infty[$, is violated but there exists a majorant K^* of $|K|$ which satisfies this assumption, then the assertion is still valid. This is seen by analyzing the proof of Theorem 6.65. An interesting example is the Mellin–Fejér kernel $\{F_\rho^c\}_{\rho>0}$ of Definition 6.83. For $c \ge 0$ the function F_1^c satisfies the hypotheses of Theorem 6.65 on K except for the monotonicity assumption. But the latter holds for a majorant.

If, in addition to the assumptions of Theorem 6.65, the function K is Mellin-even, then, in view of Remark 4.10, the assertion also holds for the operator U_ρ generated by $\{K_\rho\}_{\rho>0}$ (see Definition 6.18). In particular, this assertion is valid for $K = F_1^0$.

Similar comments apply to Theorem 6.66.

6.5 Inversion Theory

The aim of the inversion theory of an integral transform is to reconstruct the original function from its transformation. This is particularly useful in the theory of partial differential equations as we shall see later. In the Mellin instance one looks for an operator M_c^{-1} such that $M_c^{-1}[M_c[f]] = f$ for $f \in X_c$ with $c \in \mathbb{R}$.

In the following, for a given $c \in \mathbb{R}$ and $p \in [1, \infty]$, we denote by $L^p(c + i\mathbb{R})$ the set of all functions $g : c + i\mathbb{R} \to \mathbb{C}$ such that $g(c + i\cdot) \in L^p(\mathbb{R})$. Analogously, we denote by $L^p(\mathrm{St}(]a, b[)$ the set of all functions $g : \mathrm{St}]a, b[\to \mathbb{C}$ such that $g \in L^p(c + i\mathbb{R})$ for each $c \in]a, b[$.

Definition 6.68 The *inverse Mellin transform* $M_c^{-1}[g]$ of a function $g \in L^1(c+i\mathbb{R})$ is defined by

$$M_c^{-1}[g](x) \equiv M_c^{-1}[g(c + i(\cdot))](x) := \frac{x^{-c}}{2\pi} \int_{-\infty}^{\infty} g(c + it)x^{-it}dt \quad (x \in \mathbb{R}^+). \tag{6.57}$$

The name "inverse Mellin transform" is justified by Theorem 6.75 below, in which we show that M_c^{-1} acts as an inverse operation.

Concerning the global Mellin transform as a mapping defined on the space $X_{]a,b[}$ into the space $H(\mathrm{St}]a, b[)$ of analytic functions on the strip $\mathrm{St}]a, b[$, it is not surjective. For example, by Theorem 6.11, there does not exist a function $f \in X_{]a,b[}$ such that $[f]^\wedge_M = 1$, although this constant function is analytic. In order that a function $g \in H(\mathrm{St}]a, b[)$ can be expressed as the Mellin transform of a function $f \in X_{]a,b[}$, it is at least necessary that

$$\lim_{|t|\to\infty} g(c + it) = 0 \quad \text{uniformly in } c \in [\alpha, \beta] \text{ for all } [\alpha, \beta] \subset]a, b[. \tag{6.58}$$

Subsequently we shall see (cf. Theorem 6.77 (d)) that (6.58) is also sufficient provided that $g \in L^1(\mathrm{St}]a, b[)$. In statement (b) of the following proposition these conditions ensure that $M_c^{-1}[g]$ produces a function which does not depend on c.

Proposition 6.69

(a) Let $g \in L^1(c + i\mathbb{R})$ for some $c \in \mathbb{R}$. Then the inverse Mellin transform M_c^{-1} is well-defined with

$$|M_c^{-1}[g](x)| \le \frac{x^{-c}}{2\pi} \int_{-\infty}^{\infty} |g(c+it)|dt < \infty \qquad (x \in \mathbb{R}^+). \tag{6.59}$$

(b) Let $g \in H(\mathrm{St}]a,b[) \cap L^1(\mathrm{St}]a,b[)$. If condition (6.58) is satisfied, then the inverse Mellin transform $M_c^{-1}[g]$ is independent of the choice of $c \in]a,b[$, that is, if $c_1 \in]a,b[$ is arbitrary, then

$$M_c^{-1}[g](x) = M_{c_1}^{-1}[g](x) \qquad (x \in \mathbb{R}^+). \tag{6.60}$$

Since in (6.60) the function g is analytic, the inverse Mellin transform can be expressed as a line integral with the complex variable $s \in \mathbb{C}$, namely

$$M_c^{-1}[g](x) = \lim_{R_1,R_2\to\infty} \frac{1}{2\pi i} \int_{c-iR_1}^{c+iR_2} g(s)x^{-s}ds. \tag{6.61}$$

Proof (of Proposition 6.69) The estimate in (a) is obvious. As to (b), by the continuity of g, the inverse transform $M_c^{-1}[g]$ may be understood in the sense of (6.61), so that $M_c^{-1}[g]$ is a well-defined, complex valued function on $\mathbb{R}^+$. Now we show that the inverse transform is independent of $c \in]a,b[$. Let $c_1 \in]a,b[$ be arbitrary and choose $[\alpha,\beta] \subset]a,b[$ with $c, c_1 \in]\alpha,\beta[$. Without loss of generality, we may assume that $c_1 < c$. Then, for $R_1, R_2 > 0$, consider the rectangle $\mathcal{R}$ with vertices $c - iR_1$, $c + iR_2$, $c_1 + iR_2$, $c_1 - iR_1$ and let $\partial\mathcal{R}$ be its positively oriented boundary. Since $g(\cdot)x^{(\cdot)} \in H(\mathrm{St}]a,b[)$ for each $x \in \mathbb{R}^+$, we conclude with the help of Cauchy's integral theorem (Theorem 1.24) that

$$\int_{\partial\mathcal{R}} g(s)x^{-s}ds = 0,$$

and so

$$\int_{c_1-iR_1}^{c_1+iR_2} g(s)x^{-s}ds = \left\{ \int_{c-iR_1}^{c+iR_2} + \int_{c_1-iR_1}^{c-iR_1} - \int_{c_1+iR_2}^{c+iR_2} \right\} g(s)x^{-s}ds. \tag{6.62}$$

Now let $\varepsilon > 0$ be given. By condition (6.58) there exists a $T = T(\alpha,\beta) > 0$ such that $|g(\widetilde{c}+it)| < \varepsilon$ for all $|t| \ge T$ and all $\widetilde{c} \in [\alpha,\beta]$. Hence for $R \in \{R_1, R_2\}$ with $|R| \ge T$, we obtain

$$\left| \int_{c_1+iR}^{c+iR} g(s)x^{-s}ds \right| \le \left| \int_{c_1}^{c} \varepsilon x^{-u}du \right| \le K_x \varepsilon |\beta - \alpha|,$$

where $K_x := \max_{u\in[\alpha,\beta]} x^{-u}$. With this we readily conclude by letting $R_1, R_2 \to \infty$ that (6.62) implies (6.60). □

Remark 6.70 In Definition 6.68 the inverse Mellin transform is defined for $c \in \mathbb{R}$. If a given function $g \in L^1(\mathrm{St}]a,b[)$ is analytic on a strip $\mathrm{St}]a,b[$ and (6.58) is satisfied, then $M_c^{-1}[g]$ is well defined for all $c \in]a,b[$ and is independent of c. Therefore we may drop the subscript c.

The following lemma presents important basic properties of the inverse Mellin transform.

Lemma 6.71

(a) Let $g \in L^1(c+i\mathbb{R})$ for some $c \in \mathbb{R}$. Then the inverse Mellin transform $M_c^{-1}[g]$ is continuous on $\mathbb{R}^+$.

(b) Let $g \in H(\mathrm{St}]a,b[) \cap L^1(\mathrm{St}]a,b[)$ such that (6.58) is satisfied. Then $M^{-1}[g] \in X_{]a,b[}$.

Proof

(a) Obviously it suffices to show that the function G, defined by

$$G(x) := 2\pi x^c M_c^{-1}[g](x) = \int_{-\infty}^{\infty} g(c+it)x^{-it}dt,$$

is continuous on $\mathbb{R}^+$. For an arbitrary $x \in \mathbb{R}^+$, let $(x_n)_{n\in\mathbb{N}}$ be a sequence in $\mathbb{R}^+$ such that $x_n \to x$ as $n \to \infty$. Set $f_n(t) := g(c+it)x_n^{-it}$. Then $|f_n(t)| = |g(c+it)|$, and the right-hand side is an $L^1(\mathbb{R})$ majorant as a function of t. Hence by Lebesgue's dominated convergence theorem, we obtain

$$\lim_{n\to\infty} G(x_n) = \lim_{n\to\infty} \int_{-\infty}^{\infty} f_n(t)dt = \int_{-\infty}^{\infty} g(c+it)x^{-it}dt = G(x).$$

(b) Given $c \in]a,b[$, we choose α and β such that $c \in [\alpha,\beta] \subset]a,b[$. Then, applying (6.59) of Proposition 6.69 and taking also (6.60) into account, we conclude that

$$\begin{aligned}
\int_0^{\infty} |M^{-1}[g](x)|x^{c-1}dx &= \left\{\int_0^1 + \int_1^{\infty}\right\} |M^{-1}[g](x)|x^{c-1}dx \\
&\le \frac{1}{2\pi}\|g(\alpha+i\cdot)\|_{L^1(\mathbb{R})} \int_0^1 x^{c-\alpha-1}dx \\
&\quad + \frac{1}{2\pi}\|g(\beta+i\cdot)\|_{L^1(\mathbb{R})} \int_1^{\infty} x^{c-\beta-1}dx \\
&= \frac{1}{2\pi(c-\alpha)}\|g(\alpha+i\cdot)\|_{L^1(\mathbb{R})} \\
&\quad + \frac{1}{2\pi(\beta-c)}\|g(\beta+i\cdot)\|_{L^1(\mathbb{R})} < \infty.
\end{aligned}$$

Hence $M^{-1}[g] \in X_c$ for each $c \in]a,b[$, as was to be shown. □

The next lemma provides a reconstruction of the Mellin–Gauss–Weierstrass kernel (see Example 6.25) from its Mellin transform. This inversion formula will be useful in order to state a general inversion formula for a large class of functions, using the fact that $\{G_\rho\}_{\rho>0}$ is a global approximate identity, which implies that

$$\lim_{n\to\infty} (f * G_{\rho_n})(x) = f(x) \qquad \text{(a.e. on } \mathbb{R}^+)$$

for any function $f \in X_c$ and a suitable sequence (ρ_n) such that $\rho_n \to \infty$ (see Corollary 6.28).

Lemma 6.72 *For the Mellin–Gauss–Weierstrass kernel* $\{G_\rho\}_{\rho>0}$ *the following inversion formula holds*

$$G_\rho(x) = M^{-1}\big[M[G_\rho]\big](x) = \frac{1}{2\pi i}\int_{c-i\infty}^{c+i\infty} [G_\rho]^\wedge_M(s)x^{-s}ds \tag{6.63}$$

for all $x \in \mathbb{R}^+$ *and all* $c \in \mathbb{R}$.

Proof By Remark 6.27 we can write

$$G_\rho(x) = \frac{\rho}{\sqrt{4\pi}}[G_\rho]^\wedge_M\Big(\frac{i\rho^2}{2}\log x\Big) = \frac{\rho}{\sqrt{4\pi}}\int_0^\infty G_\rho(u)u^{i(\rho^2/2)\log x}\frac{du}{u}$$

$$= \frac{\rho^2}{4\pi}\int_0^\infty [G_\rho]^\wedge_M\Big(\frac{i\rho^2}{2}\log u\Big)u^{i(\rho^2/2)\log x}\frac{du}{u}.$$

Now substituting $(i\rho^2/2)\log u = -c - it$, which implies that $u^{i(\rho^2/2)\log x} = x^{-c-it}$, and noting that by (6.29) the Mellin transform $[G_\rho]^\wedge_M(s)$ is even, we obtain

$$G_\rho(x) = \frac{2\rho^2}{4\pi\rho^2}\int_{-\infty}^{\infty} [G_\rho]^\wedge_M(-c-it)x^{-c-it}dt = \frac{1}{2\pi i}\int_{c-i\infty}^{c+i\infty} [G_\rho]^\wedge_M(s)x^{-s}ds.$$

Thus (6.63) verified. □

We shall need the following representation for the convolution $f * G_\rho$.

Lemma 6.73 *Let* $f \in X_c$ *and let* $\{G_\rho\}_{\rho>0}$ *be the Mellin–Gauss–Weierstrass kernel, then*

$$(f * G_\rho)(x) = \frac{1}{2\pi}\int_{-\infty}^{\infty} [f]^\wedge_{M_c}(c+it)[G_\rho]^\wedge_{M_c}(c+it)x^{-c-it}dt \quad (x \in \mathbb{R}^+). \tag{6.64}$$

Proof Since $G_\rho \in X_{]-\infty,\infty[}$ is uniformly continuous and bounded, the convolution $(f * G_\rho)(x)$ exists for all $x \in \mathbb{R}^+$, according to Theorem 4.3. Employing Lemma 6.72 and Fubini's theorem, we find that for $x \in \mathbb{R}^+$,

$$\begin{aligned}(f * G_\rho)(x) &= \int_0^\infty f(u) G_\rho\left(\frac{x}{u}\right)\frac{du}{u} \\ &= \int_0^\infty f(u)\left\{\frac{1}{2\pi}\int_{-\infty}^\infty [G_\rho]^\wedge_{M_c}(c+it)\left(\frac{x}{u}\right)^{-c-it} dt\right\}\frac{du}{u} \\ &= \frac{1}{2\pi}\int_{-\infty}^\infty [G_\rho]^\wedge_{M_c}(c+it)x^{-c-it}\left\{\int_0^\infty f(u)u^{c+it-1}du\right\}dt.\end{aligned}$$

This is (6.64). We supplement that Fubini's theorem is applicable since by (6.29) and the proof of Proposition 6.26, we have

$$\int_{-\infty}^\infty \int_0^\infty |[G_\rho]^\wedge_{M_c}(c+it)x^{-c-it}f(u)u^{c+it-1}|du dt$$

$$= x^{-c}\|f\|_{X_c}\int_{-\infty}^\infty \left|\exp\left(\left(\frac{c+it}{\rho}\right)^2\right)\right|dt = x^{-c}\|f\|_{X_c}\rho\exp\left(\frac{c^2}{\rho^2}\right)\sqrt{\pi} < \infty.$$

□

In conjunction with Corollary 6.28 and Proposition 6.26, Lemma 6.73 yields the following statement.

Corollary 6.74 *For $f \in X_c$ there exists a sequence $(\rho_n)_{n\in\mathbb{N}}$ with $\lim_{n\to\infty}\rho_n = \infty$ such that*

$$\lim_{n\to\infty}\frac{1}{2\pi}\int_{-\infty}^\infty [f]^\wedge_{M_c}(c+it)|\exp\left(\left(\frac{c+it}{\rho_n}\right)^2\right)x^{-c-it}dt = f(x) \qquad (6.65)$$

almost everywhere on $\mathbb{R}^+$.

We are now ready for presenting the following general inversion theorem.

Theorem 6.75 *Let $f \in X_c$ with $[f]^\wedge_{M_c} \in L^1(c+i\mathbb{R})$ for some $c \in \mathbb{R}$. Then*

$$\begin{aligned}f(x) &= M_c^{-1}\big[M_c[f]\big](x) \\ &= \frac{1}{2\pi}\int_{-\infty}^\infty [f]^\wedge_{M_c}(c+it)x^{-c-it}dt \quad (a.e.\ on\ \mathbb{R}^+).\end{aligned} \qquad (6.66)$$

If f is continuous on $\mathbb{R}^+$, then this equation is valid for all $x \in \mathbb{R}^+$.

Proof Under the hypotheses on f, the conclusion of Corollary 6.74 holds. When $\rho_n \geq 1$, the integrand in (6.65) may be estimated as

$$\left|[f]^{\wedge}_{M_c}(c+it)\exp\left(\left(\frac{c+it}{\rho_n}\right)^2\right)x^{-c-it}\right| \leq e^{c^2}x^{-c}|[f]^{\wedge}_{M_c}(c+it)|.$$

As a function of t, the right-hand side is an $L^1(\mathbb{R})$ majorant. Thus, by Lebesgue's dominated convergence theorem, we obtain

$$\begin{aligned}\lim_{n\to\infty}\frac{1}{2\pi}&\int_{-\infty}^{\infty}[f]^{\wedge}_{M_c}(c+it)|\exp\left(\left(\frac{c+it}{\rho_n}\right)^2\right)x^{-c-it}dt\\ &=\frac{1}{2\pi}\int_{-\infty}^{\infty}[f]^{\wedge}_{M_c}(c+it)x^{-c-it}dt,\end{aligned}$$

and so (6.66) is verified.

Finally, by Lemma 6.71, the inverse Mellin transform defines a continuous function so that the equation holds for all $x \in \mathbb{R}^+$ provided that f is continuous. □

By suitable substitutions, one can express the Mellin transform by its inverse and vice versa.

Lemma 6.76 *Let $f \in X_c$ for $c \in \mathbb{R}$ and $g(\widetilde{c}+it) := f(e^{-t})e^{-ct}$ for $\widetilde{c}, t \in \mathbb{R}$. Then $g \in L^1(\widetilde{c}+i\mathbb{R})$ and*

$$M_c[f](c+it) = 2\pi e^{\widetilde{c}t} M^{-1}_{\widetilde{c}}[g](e^t) \qquad (t \in \mathbb{R}^+). \tag{6.67}$$

Conversely, let $g \in L^1(\widetilde{c}+i\mathbb{R})$ and $f(x) := x^{-c}g(\widetilde{c}-i\log x)$ for $x \in \mathbb{R}^+$. Then $f \in X_c$ and

$$M^{-1}_{\widetilde{c}}[g](x) = \frac{x^{-\widetilde{c}}}{2\pi}M_c[f](c+i\log x) \qquad (x \in \mathbb{R}^+). \tag{6.68}$$

Proof As to (6.67), the substitution $u = e^{-v}$ yields for any $t \in \mathbb{R}$,

$$\begin{aligned}M_c[f](c+it) &= \int_0^{\infty} f(u)u^{c+it}\frac{du}{u} = \int_{-\infty}^{\infty}\left(f(e^{-v})e^{-cv}\right)e^{-itv}dv\\ &= e^{\widetilde{c}t}\int_{-\infty}^{\infty} g(\widetilde{c}+iv)e^{-(\widetilde{c}+iv)t}dv = 2\pi e^{\widetilde{c}t}M^{-1}_{\widetilde{c}}[g](e^t).\end{aligned}$$

Equation (6.68) is verified similarly. □

We summarize the above results in the following theorem.

Theorem 6.77

(a) If $f \in X_c$ such that $[f]^{\wedge}_{M_c} \in L^1(c+i\mathbb{R})$ for some $c \in \mathbb{R}$, then

$$M_c^{-1}\big[M_c[f]\big](x) = \frac{1}{2\pi}\int_{-\infty}^{\infty} [f]^{\wedge}_{M_c}(c+it)x^{-c-it}dt = f(x) \quad (a.e.\ on\ \mathbb{R}^+). \tag{6.69}$$

(b) If $f \in X_{]a,b[}$ such that $[f]^{\wedge}_{M} \in L^1(\mathrm{St}]a,b[)$, then for each $c \in]a,b[$,

$$M_c^{-1}\big[M[f]\big](x) = \frac{1}{2\pi i}\int_{c-i\infty}^{c+i\infty} [f]^{\wedge}_{M}(s)x^{-s}ds = f(x) \quad (a.e.\ on\ \mathbb{R}^+). \tag{6.70}$$

In particular $M_{c_1}^{-1} \equiv M_{c_2}^{-1}$ for all $c_1, c_2 \in]a,b[$.

(c) If $g \in L^1(c+i\mathbb{R})$ such that $M_c^{-1}[g] \in X_c$, then

$$M_c\big[M_c^{-1}[g]\big](c+it) = g(c+it) \qquad (a.e.\ for\ t \in \mathbb{R}). \tag{6.71}$$

(d) Let $g \in H(\mathrm{St}]a,b[) \cap L^1(\mathrm{St}]a,b[)$ and suppose that (6.58) is satisfied. Then $M^{-1}[g] \in X_{]a,b[}$, and for any $c \in]a,b[$, we have

$$M\big[M^{-1}[g]\big](s) = g(s) \qquad (s \in \mathrm{St}]a,b[).$$

If f or g are continuous on $\mathbb{R}^+$ or $c+i\mathbb{R}$, respectively, then all the results (a)–(c) hold for all $x \in \mathbb{R}^+$ and $t \in \mathbb{R}$, respectively.

Proof Parts (a) and (b) follow from Theorem 6.75 and Proposition 6.69. As to (6.71), setting $B(it) := e^{-ct}M_c^{-1}[g](e^{-t})$ with $t \in \mathbb{R}$, then, according to Lemma 6.76, the function B belongs to $L^1(i\mathbb{R})$ and

$$M_c\big[M_c^{-1}[g]\big](c+it) = 2\pi M_0^{-1}[B](e^t). \tag{6.72}$$

Using (6.68) of Lemma 6.76 with the roles of the pair $(\widetilde{c}, c)$ being taken by $(c, 0)$, we find that

$$e^{-ct}M_c^{-1}[g](e^{-t}) = e^{-ct}\frac{e^{ct}}{2\pi}M_0[g(c-i\log(\cdot))](-it) = \frac{1}{2\pi}M_0[g(c+i\log(\cdot)](it),$$

where for the last step we simply have to substitute $u = 1/v$ in the integral of the Mellin transform. Thus $B(\cdot) = (2\pi)^{-1}M_0[g(c+i\log(\cdot)](\cdot)$. Inserting this in (6.72) and applying part (a), we obtain

$$\begin{aligned} M_c\big[M_c^{-1}[g]\big](c+it) &= 2\pi M_0^{-1}[B](e^t) \\ &= M_0^{-1}\big[M_0[g(c+i\log(\cdot)](\cdot)\big](e^t) = g(c+it), \end{aligned}$$

almost everywhere for $t \in \mathbb{R}$, proving (c). Finally, (d) follows from (b), since the assumptions in (d) and Lemma 6.71 already ensure that $M^{-1}[g] \in X_{]a,b[}$.

Finally, the statement on continuous functions f or g should be obvious. □

Remark 6.78 For $f \in X_c$, the inversion formula (6.66) of Theorem 6.75 is valid under the sole assumption that $[f]^{\wedge}_{M_c}$ is integrable along the line $c + i\mathbb{R}$. Note that this is necessary for the existence of the inverse Mellin transform. Indeed, take the function $f(x) = \chi_{[1,2]}(x)$. Then $f \in X_{]-\infty,\infty[}$ but $[f]^{\wedge}_M \notin L^1(c + i\mathbb{R})$ for all $c \in \mathbb{R}$. Otherwise $M^{-1}\big[[f]^{\wedge}_M\big](x)$ would be continuous and by Theorem 6.77 (a), $M^{-1}\big[[f]^{\wedge}_M\big](x) = f(x)$ almost everywhere on $\mathbb{R}^+$, which is not possible since f is the characteristic function of an interval.

6.6 The Uniqueness Theorem and the Parseval Equation for Convolutions

We begin with a fundamental consequence of the inversion formula, which shows the uniqueness of the Mellin transform.

Theorem 6.79

(a) Let $f, g \in X_c$ for some $c \in \mathbb{R}$ and suppose that $[f]^{\wedge}_{M_c}(c + it) = [g]^{\wedge}_{M_c}(c + it)$ for all $t \in \mathbb{R}$. Then $f = g$ almost everywhere on $\mathbb{R}^+$.

(b) Let $f, g \in X_{]a,b[}$ and suppose that $[f]^{\wedge}_M = [g]^{\wedge}_M$ on some subset $E \subset \mathrm{St}]a, b[$ having a finite cluster point in the strip $\mathrm{St}]a, b[$. Then $f = g$ almost everywhere on $\mathbb{R}^+$.

Proof (a) By the hypotheses and the linearity of the Mellin transform we have $[f - g]^{\wedge}_{M_c}(c + it) = 0$ for all $t \in \mathbb{R}$. Therefore Theorem 6.75 yields

$$f(x) - g(x) = \frac{1}{2\pi} \int_{-\infty}^{\infty} [f - g]^{\wedge}_{M_c}(c + it)x^{-c-it}dt = 0 \qquad \text{(a.e. on } \mathbb{R}^+\text{)}.$$

As to (b), since by Theorem 6.5, $[f]^{\wedge}_M, [g]^{\wedge}_M \in H(\mathrm{St}]a, b[)$, the assertion follows by Theorem 1.29 and part (a). □

The previous uniqueness theorem has the following consequence.

Corollary 6.80 *The Mellin transforms, understood as mappings $M_c : X_c \to C(c + i\mathbb{R})$ and $M : X_{]a,b[} \to H(\mathrm{St}]a, b[)$, are injective operators.*

Our next aim is a generalization of Lemma 6.73, which gives a Mellin version of the Parseval equation for convolutions.

Theorem 6.81 *Let $f, g \in X_c$ and suppose that $[f]^{\wedge}_{M_c} \in L^1(c + i\mathbb{R})$ for some $c \in \mathbb{R}$. Then*

$$(f * g)(x) = M_c^{-1}\big[[f]^{\wedge}_{M_c}\,[g]^{\wedge}_{M_c}\big](x),$$

that is,

$$\int_0^\infty f\left(\frac{x}{u}\right)g(u)\frac{du}{u} = \frac{1}{2\pi}\int_{-\infty}^{\infty}[f]^\wedge_{M_c}(c+it)[g]^\wedge_{M_c}(c+it)x^{-c-it}dt.$$

A similar result holds for $f, g \in X_{]a,b[}$.

Proof Lemma 6.4 in conjunction with Theorem 6.11 ensure that $|[g]^\wedge_{M_c}(c+it)|$ is bounded. Hence the product $[f]^\wedge_{M_c}[g]^\wedge_{M_c}$ belongs to $L^1(c+i\mathbb{R})$ and so all the integrals exist in the sense of Lebesgue. Now Theorem 6.75 together with Fubini's theorem yield

$$\begin{aligned}(f*g)(x) &= \int_0^\infty \left\{\frac{1}{2\pi}\int_{-\infty}^{\infty}[f]^\wedge_{M_c}(c+it)\left(\frac{x}{u}\right)^{-c-it}dt\right\}g(u)\frac{du}{u} \\ &= \frac{1}{2\pi}\int_{-\infty}^{\infty}[f]^\wedge_{M_c}(c+it)x^{-c-it}\left\{\int_0^\infty g(u)u^{c+it-1}du\right\}dt \quad (x\in\mathbb{R}^+),\end{aligned}$$

which gives the desired result. □

Example 6.82 Let $f(x) := e^{-x}$. Then $f \in X_{]0,\infty[}$ and the Mellin transform of f is the Euler gamma function Γ defined by

$$\Gamma(s) = \int_0^\infty e^{-x}x^{s-1}dx \qquad (s = c+it,\ c\in\mathbb{R}^+, t\in\mathbb{R}).$$

It is known that Γ is analytic in the right half-plane and

$$|\Gamma(c+it)|^2 = |\Gamma(c)|^2\prod_{k=0}^{\infty}\left(1+\frac{t^2}{(c+k)^2}\right)^{-1} \qquad (c\in\mathbb{R}^+,\ t\in\mathbb{R});$$

see, e.g., Abramowitz and Stegun [1, formula 6.1.25] or Gradsteyn and Ryzhik [160, formula 8.326]. Since

$$\prod_{k=0}^{\infty}\left(1+\frac{t^2}{(c+k)^2}\right)^{-1} \le \left(1+\frac{t^2}{c^2}\right)^{-1}\left(1+\frac{t^2}{(c+1)^2}\right)^{-1} \le \left(1+\frac{t^2}{(c+1)^2}\right)^{-2},$$

we see that

$$|\Gamma(c+it)| \le \frac{|\Gamma(c)|}{1+(c+1)^{-2}t^2} \qquad (c\in\mathbb{R}^+,\ t\in\mathbb{R}).$$

Hence $[f]^\wedge_M \in L^1(c+i\mathbb{R})$ for $c\in\mathbb{R}^+$ and condition (6.58) is satisfied. Therefore Theorem 6.81 is applicable to $f(x) = g(x) = e^{-x}$ and yields

$$(f * g)(x) = (f * f)(x) = \int_0^\infty e^{-x/u} e^{-u} \frac{du}{u} = \frac{1}{2\pi i} \int_{c-i\infty}^{c+i\infty} \Gamma^2(s) x^{-s} ds.$$

Another interesting application is related to the calculation of the Mellin transform of the Mellin–Fejér kernel which was announced immediately after Definition 6.33. It may be introduced as follows.

Definition 6.83 The *Mellin–Fejér kernel* $\{F_\rho^c\}_{\rho>0}$ for $c \in \mathbb{R}$ is defined by (Fig. 6.3)

$$F_\rho^c(x) := -\frac{x^{-c}}{2\pi\rho}\Big(\frac{x^{i\rho/2} - x^{-i\rho/2}}{\log x}\Big)^2 \qquad (x \in \mathbb{R}^+ \setminus \{1\})$$

with the continuous extension $F_\rho^c(1) = \rho/(2\pi)$. An alternative representation in terms of the sinc function is

$$F_\rho^c(x) = \frac{\rho}{2\pi} x^{-c} \mathrm{sinc}^2\big(\frac{\rho}{2\pi}\log x\big) \qquad (x \in \mathbb{R}^+ \setminus \{1\}), \tag{6.73}$$

where $\mathrm{sinc}(x) := \sin(\pi x)/(\pi x)$ for $x \in \mathbb{R} \setminus \{0\}$ and $\mathrm{sinc}(0) = 1$.

Lemma 6.84 *Let $c \in \mathbb{R}$. The Mellin–Fejér kernel $\{F_\rho^c\}_{\rho>0}$ belongs to X_c and is nonnegative with Mellin transform*

$$[F_\rho^c]^\wedge_{M_c}(c + it) = \begin{cases} 1 - \dfrac{|t|}{\rho}, & 0 \le |t| \le \rho, \\ 0, & |t| > \rho. \end{cases} \tag{6.74}$$

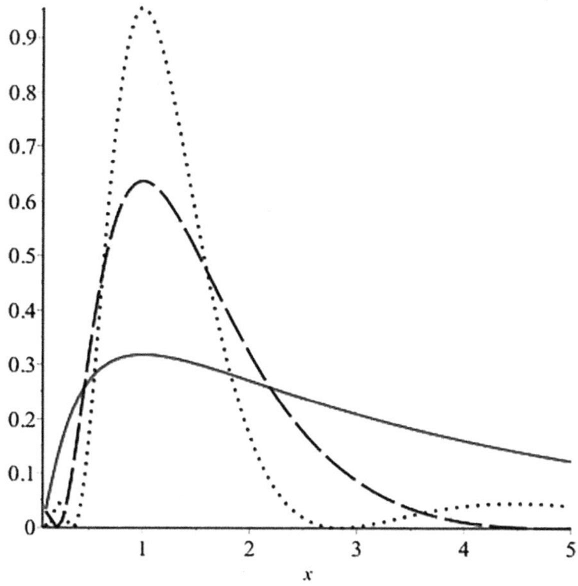

Fig. 6.3 The Mellin–Fejér kernel with $c = 0$, $\rho = 2$ (solid line), $\rho = 4$, (dashed line) and $\rho = 6$ (dotted line)

Furthermore, the Mellin–Fejér kernel forms a local approximate identity on X_c (Fig. 6.3).

Proof The nonnegativity follows immediately from (6.73). Moreover, using the representation (6.73) and the well-known formula

$$\int_{-\infty}^{\infty} \left(\frac{\sin u}{u}\right)^2 du = \pi,$$

we easily see that $F_\rho^c \in X_c$ and $\|F_\rho^c\|_{X_c} = [F_\rho^c]^\wedge_{M_c}(c) = 1$, which shows that we have indeed a local kernel in the sense of Definition 6.16. Now, setting $g(c+it) := 1 - |t|/\rho$ for $0 \le |t| \le \rho$ and $g(c+it) := 0$ for $|t| > \rho$, then obviously $g \in L^1(c + i\mathbb{R})$. Therefore applying the inverse Mellin transform, we obtain

$$\begin{aligned} M_c^{-1}[g](x) &= \frac{x^{-c}}{2\pi} \int_{-\rho}^{\rho} \left(1 - \frac{|t|}{\rho}\right) x^{-it} dt \\ &= \frac{x^{-c}}{2\pi} \int_0^{\rho} \left(1 - \frac{t}{\rho}\right) (x^{-it} + x^{it}) dt = \frac{x^{-c}}{2\pi} \frac{2 - x^{-i\rho} - x^{i\rho}}{\rho \log^2 x} \\ &= -\frac{x^{-c}}{2\pi\rho} \left(\frac{x^{i\rho/2} - x^{-i\rho/2}}{\log x}\right)^2. \end{aligned}$$

Thus, $M_c^{-1}[g](x) = F_\rho^c(x)$ for all $x \in \mathbb{R}^+$, and so (6.74) follows from Theorem 6.77 (c).

It remains to prove that the Mellin–Fejér kernel is a local approximate identity. For this, we take any $\widetilde{\delta} > 0$ and introduce the set $\widetilde{E_{\widetilde{\delta}}} := \{x \in \mathbb{R}^+ : |\log x| \ge \widetilde{\delta}\}$. Then, using the substitution $v = u^\rho$, we find that

$$\begin{aligned} \int_{\widetilde{E_{\widetilde{\delta}}}} F_\rho^c(u) u^{c-1} du &= -\frac{1}{2\pi} \int_{\widetilde{E}_{\rho\widetilde{\delta}}} \left(\frac{v^{i/2} - v^{-i/2}}{\log v}\right)^2 \frac{dv}{v} \\ &= \int_{\widetilde{E}_{\rho\widetilde{\delta}}} F_1^0(v) \frac{dv}{v}. \end{aligned}$$

Finally, for an arbitrary $\delta \in]0, 1[$ and E_δ as in Definition 6.17, we choose $\widetilde{\delta} > 0$ such that $E_\delta \subset \widetilde{E_{\widetilde{\delta}}}$. Then, by the previous equations,

$$\lim_{\rho\to\infty} \int_{E_\delta} F_\rho^c(u) u^{c-1} du \le \lim_{\rho\to\infty} \int_{\widetilde{E_{\widetilde{\delta}}}} F_\rho^c(u) u^{c-1} du = \lim_{\rho\to\infty} \int_{\widetilde{E}_{\rho\widetilde{\delta}}} F_1^0(v) \frac{dv}{v} = 0.$$

This completes the proof. □

6.7 An Equivalence Theorem for Mellin–Sobolev Spaces

We begin with the following proposition.

Proposition 6.85 *Let $f \in X_c$, where $c \in \mathbb{R}$. Suppose that there exist $g \in X_c$ and $r \in \mathbb{N}$ such that*

$$(-it)^r [f]^{\wedge}_{M_c}(s) = [g]^{\wedge}_{M_c}(s) \qquad (s = c + it,\ t \in \mathbb{R}). \tag{6.75}$$

Then

$$f(x) = x^{-c} \int_0^x \int_0^{u_1} \cdots \int_0^{u_{r-1}} g(u_r) u_r^c \frac{du_r}{u_r} \cdots \frac{du_2}{u_2} \frac{du_1}{u_1} \qquad (a.e.\ x \in \mathbb{R}^+). \tag{6.76}$$

Moreover, each of the iterated integrals

$$x^{-c} \int_0^x \int_0^{u_1} \cdots \int_0^{u_{j-1}} g(u_j) u_j^c \frac{du_j}{u_j} \cdots \frac{du_2}{u_2} \frac{du_1}{u_1}$$

belongs to X_c for $1 \le j \le r$.

Proof Let $h \in]0, 1[$ and $r = 1$. By the hypotheses, Theorem 6.10 and (6.14), we find that for $s = c + it$,

$$[\tau_h^c f - f]^{\wedge}_{M_c}(s) = (h^{-it} - 1)[f]^{\wedge}_{M_c}(s) = \frac{h^{-it} - 1}{-it} [g]^{\wedge}_{M_c}(s) = [m_h^c * g]^{\wedge}_{M_c}(s).$$

Since

$$\left(m_h^c * g\right)(x) = -\int_1^{1/h} g\left(\frac{x}{u}\right) u^{-c} \frac{du}{u} = -x^{-c} \int_{xh}^x g(v) v^{c-1} dv,$$

Theorem 6.79 yields

$$(\tau_h^c f)(x) - f(x) = -x^{-c} \int_{xh}^x g(u) u^{c-1} du \qquad (\text{a.e. } x \in \mathbb{R}^+),$$

and so

$$\begin{aligned} &\int_1^y \left((\tau_h^c f)(x) - f(x)\right) x^{c-1} dx \\ &= -\int_1^y \int_{xh}^x g(u) u^{c-1} du \frac{dx}{x} \qquad (y \in \mathbb{R}^+). \end{aligned} \tag{6.77}$$

In view of $f \in X_c$, we have

$$\lim_{h\to 0+} \int_1^y (\tau_h^c f)(x) x^{c-1} dx = \lim_{h\to 0+} \int_h^{yh} f(u) u^{c-1} du = 0. \tag{6.78}$$

As regards the right-hand side of (6.77), Lebesgue's dominated convergence theorem gives

$$\lim_{h\to 0+} \int_1^y \int_{xh}^x g(u) u^{c-1} du \, \frac{dx}{x} = \int_1^y \int_0^x g(u) u^{c-1} du \, \frac{dx}{x},$$

where we have taken into account that $\int_0^x |g(u)| u^{c-1} du$ is an integrable majorant for the integration over $[1, y]$. This limit relation together with that of (6.78) allow us to deduce from (6.77) that

$$\int_1^y f(x) x^{c-1} dx = \int_1^y \int_0^x g(u) u^{c-1} du \, \frac{dx}{x} \qquad (y \in \mathbb{R}^+).$$

Finally, differentiating with respect to y yields

$$f(y) = y^{-c} \int_0^y g(u) u^{c-1} du \qquad (\text{a.e. } y \in \mathbb{R}^+), \tag{6.79}$$

which proves the proposition for $r = 1$.

Now we proceed by induction on r. Consider any $r \geq 2$, suppose that (6.75) holds and assume that the proposition is true for $r - 1$. Using the binomial formula, we may express $(it)^r$ as

$$(it)^r = \big((1 + it) - 1\big)^r = \sum_{j=0}^{r} (-1)^{r-j} \binom{r}{j} (1 + it)^j.$$

Multiplying by $(1 + it)^{1-r} [f]^\wedge_{M_c}(s)$ and writing $(it)^r = (-1)^r (-it)^r$, we obtain

$$(-1)^r \frac{(-it)^r [f]^\wedge_{M_c}(s)}{(1 + it)^{r-1}}$$
$$= \sum_{j=0}^{r-1} (-1)^{r-j} \binom{r}{j} (1 + it)^{j+1-r} \, [f]^\wedge_{M_c}(s) + (1 + it) \, [f]^\wedge_{M_c}(s). \tag{6.80}$$

Several of these terms can be interpreted as a Mellin transform of a convolution. Indeed, introducing the function $\nu_c(x) := x^{-c} \chi_{]0,1]}(x)$ for $x \in \mathbb{R}^+$ and noting that $\nu_{c-1} \in X_c$, with Mellin transform

$$[\nu_{c-1}]^\wedge_{M_c}(s) = \frac{1}{1+it} \qquad (s = c+it,\ t \in \mathbb{R}),$$

we have by Theorem 6.12,

$$(1+it)^{j+1-r}\,[f]^\wedge_{M_c}(s) = \big[\big((\nu_{c-1}*)^{r-j-1} * f\big)\big]^\wedge_{M_c}(s).$$

For the left-hand side, we may employ (6.75) to conclude that

$$\frac{(-it)^r[f]^\wedge_{M_c}(s)}{(1+it)^{r-1}} = \big[\big((\nu_{c-1}*)^{r-1} * g\big)\big]^\wedge_{M_c}(s).$$

Thus, using the linearity of the Mellin transform and rearranging terms, we may rewrite (6.80) as

$$(-it)[f]^\wedge_{M_c}(s) = \Bigg[f + (-1)^{r+1}\big((\nu_{c-1}*)^{r-1} * g\big) + \sum_{j=0}^{r-1}(-1)^{r-j}\binom{r}{j}\big((\nu_{c-1}*)^{r-j-1} * f\big)\Bigg]^\wedge_{M_c}(s).$$

Since $\nu_{c-1} \in X_c$, we conclude with the help of Theorem 4.4 that the function inside the square brackets also belongs to X_c. In other words, there exists a function $\widetilde{g} \in X_c$ such that

$$(-it)\,[f]^\wedge_{M_c}(s) = [\widetilde{g}]^\wedge_{M_c}(s) \qquad (s = c+it,\ t \in \mathbb{R}). \tag{6.81}$$

Therefore the case $r = 1$ implies that

$$f(x) = x^{-c}\int_0^x \widetilde{g}(u)u^c\,\frac{du}{u} \qquad (\text{a.e. } x \in \mathbb{R}^+). \tag{6.82}$$

Furthermore, by (6.81) and (6.75),

$$(-it)^{r-1}[\widetilde{g}]^\wedge_{M_c}(s) = (-it)^r\,[f]^\wedge_{M_c}(s) = [g]^\wedge_{M_c}(s).$$

Hence the induction hypothesis applies with $\widetilde{g}$ taking the role of f, and so

$$\widetilde{g}(x) = x^{-c}\int_0^x\int_0^{u_1}\cdots\int_0^{u_{r-2}} g(u_{r-1})u_{r-1}^c\,\frac{du_{r-1}}{u_{r-1}}\cdots\frac{du_1}{u_1} \qquad (\text{a.e. } x \in \mathbb{R}^+). \tag{6.83}$$

Substituting (6.83) into (6.82) we obtain (6.76). Consequently, the iterated integral on the right-hand side of (6.76) also belongs to X_c.

Concerning the last statement about the iterated integrals for $1 \leq j \leq r$, we know from the formulae (6.81) and (6.82) that under the hypotheses of the proposition, there exists a function $g_1 := \widetilde{g} \in X_c$ such that

$$(-it)[f]^{\wedge}_{M_c}(s) = [g_1]^{\wedge}_{M_c}(s) \quad \text{and} \quad f(x) = x^{-c}\int_0^x g_1(u_1)u_1^c \frac{du_1}{u_1} \qquad (\text{a.e. } x \in \mathbb{R}^+).$$

Hence the right-hand side belongs to X_c as a function of x. Now for $r \geq 2$ the hypothesis (6.75) can be rewritten as

$$(-it)^{r-1}[g_1]^{\wedge}_{M_c}(s) = [g]^{\wedge}_{M_c}(s)$$

and so the role of f can be taken by g_1. This allows us to conclude that there exists a function $g_2 \in X_c$ such that

$$(-it)[g_1]^{\wedge}_{M_c}(s) = [g_2]^{\wedge}_{M_c}(s),$$

$$g_1(x) = x^{-c}\int_0^x g_2(u_2)u_2^c \frac{du_2}{u_2} \qquad (\text{a.e. } x \in \mathbb{R}^+).$$

Therefore

$$f(x) = x^{-c}\int_0^x \int_0^{u_1} g_2(u_2)u_2^c \frac{du_2}{u_2}\frac{du_1}{u_1} \qquad (\text{a.e. } x \in \mathbb{R}^+),$$

which implies that the second iterated integral belongs to X_c. Repeating this procedure successively, we verify the second assertion. □

Now we show that the iterated integrals of the previous proposition can be expressed as a single integral.

Proposition 6.86 *Let $g \in X_c$ be such that the iterated integrals*

$$I := x^{-c}\int_0^x \int_0^{u_1} \cdots \int_0^{u_r} g(u_{r+1})u_{r+1}^c \frac{du_{r+1}}{u_{r+1}} \cdots \frac{du_2}{u_2}\frac{du_1}{u_1}$$

exist. Then

$$I = \frac{1}{r!}\int_0^x \left(\frac{u}{x}\right)^c \log^r\left(\frac{x}{u}\right) g(u)\frac{du}{u}. \tag{6.84}$$

Proof For $0 < \varepsilon < x$ we introduce

$$J_\varepsilon := \int_\varepsilon^x \int_\varepsilon^{u_1} \cdots \int_\varepsilon^{u_r} g(u_{r+1}) u_{r+1}^c \frac{du_{r+1}}{u_{r+1}} \cdots \frac{du_2}{u_2} \frac{du_1}{u_1}.$$

Then it suffices to prove that

$$J_\varepsilon = \frac{1}{r!} \int_\varepsilon^x u^c \log^r \left(\frac{x}{u}\right) g(u) \frac{du}{u}$$

since the desired result will follow by letting $\varepsilon \to 0+$. We proceed by induction on r. The case $r = 0$ is obvious. For $r \geq 1$ we introduce

$$G(u) := \int_\varepsilon^u g(t) t^c \frac{dt}{t}.$$

Then

$$J_\varepsilon = \int_\varepsilon^x \int_\varepsilon^{u_1} \cdots \int_\varepsilon^{u_{r-1}} G(u_r) \frac{du_r}{u_r} \cdots \frac{du_2}{u_2} \frac{du_1}{u_1}.$$

For the right-hand side the induction hypothesis with $c = 0$ applies. It yields that

$$J_\varepsilon = \frac{1}{(r-1)!} \int_\varepsilon^x \log^{r-1} \left(\frac{x}{u}\right) G(u) \frac{du}{u}.$$

Noting that $u^{-1} \log^{r-1}(x/u)$ is the derivative of

$$H(u) := -\frac{1}{r} \log^r \left(\frac{x}{u}\right),$$

we obtain, using integration by parts,

$$\begin{aligned}
J_\varepsilon &= \frac{1}{(r-1)!} \int_\varepsilon^x G(u) H'(u) du \\
&= \frac{1}{(r-1)!} \left\{ \left[G(u) H(u) \right]_\varepsilon^x - \int_\varepsilon^x G'(u) H(u) du \right\} \\
&= \frac{1}{r!} \int_\varepsilon^x u^c \log^r \left(\frac{x}{u}\right) g(u) \frac{du}{u},
\end{aligned}$$

as was to be shown. □

Remark 6.87 By Proposition 6.86, we may rewrite the conclusion (6.76) of Proposition 6.85 as

$$f(x) = \frac{1}{(r-1)!}\int_0^x \left(\frac{u}{x}\right)^c \left(\log\left(\frac{x}{u}\right)\right)^{r-1} g(u)\frac{du}{u} \qquad (\text{a.e. } x \in \mathbb{R}^+). \quad (6.85)$$

Thus the integral on the right-hand side belongs to X_c as a function of x.

Remark 6.88 Under the assumption (6.75), we cannot have $g \geq 0$, unless g is zero almost everywhere. For example, in the case $r = 1$, Proposition 6.85 yields

$$f(x) = x^{-c}\int_0^x g(u)u^{c-1}du =: F(x)$$

almost everywhere on $\mathbb{R}^+$. Therefore we deduce that $F \in X_c$. Now assume that $g > 0$ almost everywhere on $\mathbb{R}^+$. Then, taking into account that $\chi_{]0,x[}(u) = \chi_{]u,\infty[}(x)$, we obtain by Fubini's theorem,

$$\|F\|_{X_c} = \int_0^\infty F(x)x^{c-1}dx = \int_0^\infty g(u)u^{c-1}\left\{\int_u^\infty \frac{dx}{x}\right\}du = \infty.$$

Therefore the only possibility for F to belong to X_c is that the function g is zero almost everywhere or has different signs on subsets of positive measure. Also note that for $c = 0$ the function F would be positive and increasing if $g > 0$ almost everywhere.

Now we are ready to state the following equivalence theorem.

Theorem 6.89 *For $f \in X_c$ and $r \in \mathbb{N}$, the following four assertions are equivalent:*

(i) $f \in W_c^{r,1}(\mathbb{R}^+)$.
(ii) There exists $g_1 \in X_c$ such that

$$(-it)^r[f]^\wedge_{M_c}(s) = [g_1]^\wedge_{M_c}(s) \qquad (s = c + it,\ t \in \mathbb{R}). \qquad (6.86)$$

(iii) There exists $g_2 \in X_c$ such that

$$f(x) = x^{-c}\int_0^x\int_0^{u_1}\cdots\int_0^{u_{r-1}} g_2(u_r)u_r^c\frac{du_r}{u_r}\cdots\frac{du_2}{u_2}\frac{du_1}{u_1} \qquad (6.87)$$

almost everywhere on $\mathbb{R}^+$.
(iv) There exists $g_3 \in X_c$ such that

$$\lim_{h\to 1}\left\|\sum_{k=0}^r(-1)^{r-k}\binom{r}{k}\frac{\tau_{h^k}^c f}{(h-1)^r} - g_3\right\|_{X_c} = 0. \qquad (6.88)$$

Moreover, if one of these statements holds, then $\Theta_c^j f \in X_c$ *for all* $1 \le j \le r$ *and*

$$\Theta_c^r f = g_1 = g_2 = g_3 \qquad (a.e.\ on\ \mathbb{R}^+).$$

Proof For proving these equivalences, it is enough to verify the following implications:

$$(i) \Longrightarrow (ii) \Longrightarrow (iii) \Longrightarrow (i) \quad \text{and} \quad (iv) \Longrightarrow (ii), \quad (i) \Longrightarrow (iv).$$

$(i) \Rightarrow (ii)$: This follows from Theorem 6.14 with $g_1 := \Theta_c^r f$.
$(ii) \Rightarrow (iii)$: This follows from Proposition 6.85.
$(iii) \Rightarrow (i)$: If (iii) holds, then $f \in C^{r-1}(\mathbb{R}^+)$. Successive Mellin differentiation yields

$$\Theta_c^j f(x) = x^{-c} \int_0^x \int_0^{u_{j+1}} \cdots \int_0^{u_{r-1}} g_2(u_r) u_r^c \frac{du_r}{u_r} \cdots \frac{du_{j+1}}{u_{j+1}} \qquad (x \in \mathbb{R}^+)$$

for $j = 1, 2, \dots, r-1$ and $\Theta_c^r f(x) = g_2(x)$ almost everywhere on $\mathbb{R}^+$. Now, considering the formula for $j = r-1$ and expressing $\Theta_c^{r-1} f(x)$ with the help of Proposition 3.21 by ordinary derivatives, we find that

$$f^{(r-1)}(x) = x^{1-r-c} \int_0^x g_2(u) u^{c-1} du - \sum_{k=0}^{r-2} S_c(r-1, k) x^{k+1-r} f^{(k)}(x).$$

The functions on the right-hand side belong to $AC_{\text{loc}}(\mathbb{R}^+)$. Hence $f^{(r-1)} \in AC_{\text{loc}}(\mathbb{R}^+)$, and since $\Theta_c^r f = g_2 \in X_c$, we have shown that $f \in W_c^{r,1}(\mathbb{R}^+)$. Furthermore, the second statement of Proposition 6.85 implies that $\Theta_c^j f \in X_c$ for $1 \le j \le r$.

With the equivalence of the statements (i) to (iii) we have also obtained that $\Theta_c^r f = g_1 = g_2$ almost everywhere on $\mathbb{R}^+$.
$(iv) \Rightarrow (ii)$: Suppose that (iv) holds. By Theorem 6.10 we have

$$\left[\sum_{k=0}^{r} (-1)^{r-k} \binom{r}{k} \frac{\tau_{h^k}^c f}{(h-1)^r} \right]_{M_c}^{\wedge} (s) = \left(\frac{h^{-it} - 1}{h-1} \right)^r [f]_{M_c}^{\wedge}(s).$$

Taking into account that

$$\lim_{h \to 1} \left(\frac{h^{-it} - 1}{h-1} \right)^r = (-it)^r,$$

and employing (6.2), we obtain

$$\left|(-it)^r [f]^\wedge_{M_c}(s) - [g_3]^\wedge_{M_c}(s)\right| = \lim_{h\to 1} \left|\left[\left(\frac{h^{-it}-1}{h-1}\right)^r f - g_3\right]^\wedge_{M_c}(s)\right|$$

$$\leq \lim_{h\to 1} \left\| \sum_{k=0}^{r} (-1)^{r-k}\binom{r}{k} \frac{\tau^c_{h^k} f}{(h-1)^r} - g_3 \right\|_{X_c} = 0,$$

and so (ii) holds. Moreover, Theorem 6.14 together with Theorem 6.79 show us that $\Theta^r_c f = g_3$ almost everywhere on $\mathbb{R}^+$.

$(i) \Rightarrow (iv)$: Suppose that (i) holds. Then, given $\varepsilon > 0$, there exists by Proposition 3.13 a $\delta \in]0, 1[$ such that

$$\|\tau^{c-1}_h \Theta^r_c f - \Theta^r_c f\|_{X_c} < \varepsilon$$

for all $h \in]1-\delta, 1+\delta[$. Employing formula (3.17), we find by a change of variables that

$$\sum_{k=0}^{r} (-1)^{r-k}\binom{r}{k} (\tau^c_{h^k} f)(x) = \int_1^h \cdots \int_1^h (\tau^c_{u_1\cdots u_r} \Theta^r_c f)(x) \frac{du_r}{u_r} \cdots \frac{du_1}{u_1}.$$

Hence

$$\left\| \sum_{k=0}^{r} (-1)^{r-k}\binom{r}{k} \frac{\tau^c_{h^k} f}{(h-1)^r} - \Theta^r_c f \right\|_{X_c}$$

$$= \left\| \frac{1}{(h-1)^r} \int_1^h \cdots \int_1^h (\tau^{c-1}_{u_1\cdots u_r} \Theta^r_c f - \Theta^r_c f) du_r \cdots du_1 \right\|_{X_c}$$

$$\leq \left| \frac{1}{(h-1)^r} \int_1^h \cdots \int_1^h \|\tau^{c-1}_{u_1\cdots u_r} \Theta^r_c f - \Theta^r_c f\|_{X_c} du_r \cdots du_1 \right| < \varepsilon,$$

where the estimate is obtained by applying the generalized Minkowski inequality (Proposition 1.9) repeatedly. This implies (6.88) with $g_3 = \Theta^r_c f$. □

Remark 6.90 The Mellin differential operator may be defined in the strong topology as follows. Let $f \in X_c$ and suppose that there exists a function $g \in X_c$ such that

$$\lim_{h\to 1} \left\| \frac{\tau^c_h f - f}{h-1} - g \right\|_{X_c} = 0.$$

Then set $\Theta_c f := g$. Theorem 6.89 shows that the pointwise definition of the Mellin derivative, given by $\Theta_c f(x) := x f'(x) + c f(x)$, is equivalent to the strong Mellin derivative.

In case of the global Mellin transform on the space $X_{]a,b[}$, it would be desirable to have a Mellin differential operator that is independent of $c \in]a,b[$. In the following we show that we can use just one fixed $\widetilde{c} \in \mathbb{R}$ instead of working with all $c \in]a,b[$.

Corollary 6.91 *Let $f \in W_c^{r,1}(\mathbb{R}^+)$, where $c \in \mathbb{R}$ and $r \in \mathbb{N}$. Then, for an arbitrary $\widetilde{c} \in \mathbb{R}$, there holds $\Theta_{\widetilde{c}}^k f \in X_c$ for all $1 \le k \le r$ with*

$$\Theta_{\widetilde{c}}^r f = \sum_{k=0}^{r} \binom{r}{k} (\widetilde{c} - c)^{r-k} \Theta_c^k f \tag{6.89}$$

and

$$[\Theta_{\widetilde{c}}^r f]_{M_c}^{\wedge}(s) = (\widetilde{c} - s)^r [f]_{M_c}^{\wedge}(s) \qquad (s = c + it,\ t \in \mathbb{R}). \tag{6.90}$$

Furthermore, $(\cdot)^{c-\widetilde{c}} f(\cdot) \in W_{\widetilde{c}}^{r,1}(\mathbb{R}^+)$ and

$$f(x) = x^{-c} \int_0^x \int_0^{u_1} \cdots \int_0^{u_{r-1}} \Theta_{\widetilde{c}}^r \big(u_r^{c-\widetilde{c}} f(u_r)\big) u_r^{\widetilde{c}} \frac{du_r}{u_r} \cdots \frac{du_2}{u_2} \frac{du_1}{u_1}. \tag{6.91}$$

almost everywhere on $\mathbb{R}^+$.

Proof By Theorem 6.89 the derivatives $\Theta_c^k f$ belong to X_c for $1 \le k \le r$, and by Definition 3.19 we have $\Theta_{\widetilde{c}}^j = (\Theta_c + (\widetilde{c} - c)I_d)^j$. For $j = r$ we obtain (6.89) by using the binomial formula on the right-hand side. We also conclude that $\Theta_{\widetilde{c}}^k f \in W_c^{r-k,1}(\mathbb{R}^+)$ for $0 \le k \le r$. From (6.89) we deduce with the help of Theorem 6.14 that

$$[\Theta_{\widetilde{c}}^r f]_{M_c}^{\wedge}(s) = \sum_{k=0}^{r} \binom{r}{k} (\widetilde{c} - c)^{r-k} (-it)^k [f]_{M_c}^{\wedge}(s) = (\widetilde{c} - s)^r [f]_{M_c}^{\wedge}(s).$$

Finally, noting that $[f]_{M_c}^{\wedge}(c+it) = [(\cdot)^{c-\widetilde{c}} f]_{M_{\widetilde{c}}}^{\wedge}(\widetilde{c}+it)$, we can use the equivalence of statements (i) to (iii) of Theorem 6.89 for obtaining $(\cdot)^{c-\widetilde{c}} f(\cdot) \in W_{\widetilde{c}}^{r,1}(\mathbb{R}^+)$ and formula (6.91). □

As a consequence of Corollary 6.91, we can acquire counterparts of the results of Theorem 6.89 for the spaces $X_{]a,b[}$. Taking for example $\widetilde{c} = 0$, we obtain the following equivalence.

Corollary 6.92 *For $f \in X_{]a,b[}$ and $r \in \mathbb{N}$ the following statements are equivalent:*

(i) $f \in W_{]a,b[}^{r,1}(\mathbb{R}^+)$.
(ii) *There exists $g \in X_{]a,b[}$ such that*

$$(-s)^r [f]_M^{\wedge}(s) = [g]_M^{\wedge}(s) \qquad (s \in \mathrm{St}]a,b[).$$

In this case we have $\Theta^r f = g$ almost everywhere on $\mathbb{R}^+$.

6.8 The Fundamental Theorem of the Differential and Integral Calculus in the Mellin Setting

As for the notion of Mellin differential operator, we also consider a notion of Mellin integral, which was already used in the proof of Proposition 6.85.

Definition 6.93 For $f \in X_c$, where $c \in \mathbb{R}$, the *Mellin integral* $J_c^r f$ of order $r \in \mathbb{N}$ is defined by

$$J_c^r f(x) := \big(f * (\mu_c *)^r\big)(x) = x^{-c} \int_0^x \int_0^{u_1} \cdots \int_0^{u_{r-1}} f(u_r) u_r^c \frac{du_r}{u_r} \cdots \frac{du_2}{u_2} \frac{du_1}{u_1}$$

for $x \in \mathbb{R}^+$ with $\mu_c(x) := x^{-c} \chi_{[1,\infty[}(x)$. We set $J_c := J_c^1$ and $J^r := J_0^r$.

Remark 6.94 Note that μ_c does not belong to X_c. However, since $f \in X_c$, the integral

$$x^c (f * \mu_c)(x) = \int_0^x f(u) u^c \frac{du}{u}$$

exists for $x \in \mathbb{R}^+$ as a continuous function and vanishes for $x \to 0+$. Thus, $J_c f(x)$ is well defined, and iteration shows that $J_c^r f(x)$ is also well defined on $\mathbb{R}^+$ but is not necessarily an element of X_c. For example, the function $f(x) := (\log x)^{-2} \chi_{[e,\infty[}(x)$ belongs to X_0, but the function

$$J_0 f(x) = \int_0^x f(u) \frac{du}{u} = \begin{cases} 1 - \dfrac{1}{\log x}, & x \geq e, \\ 0, & x \in]0, e[\end{cases}$$

does not belong to X_0.

The fundamental theorem of the differential and integral calculus in the Mellin setting may be stated as follows.

Theorem 6.95 *Let $f \in W_c^{r,1}(\mathbb{R}^+)$ for $c \in \mathbb{R}$ and $r \in \mathbb{N}$. Then*

$$J_c^r \Theta_c^r f(x) = f(x) \qquad (a.e.\ x \in \mathbb{R}^+).$$

Conversely, let $f \in X_c$ and suppose that $J_c^r f \in X_c$. Then $J_c^r f \in W_c^{r,1}(\mathbb{R}^+)$ and

$$\Theta_c^r J_c^r f(x) = f(x) \qquad (a.e.\ x \in \mathbb{R}^+).$$

In particular, in this case,

$$[J_c^r f]^{\wedge}_{M_c}(s) = (-it)^{-r}[f]^{\wedge}_{M_c}(s) \qquad (s = c + it,\ t \in \mathbb{R} \setminus \{0\}).$$

The same results hold for $f \in X_{]a,b[}$, *that is, for each* $c \in\,]a, b[$.

Proof These assertions are immediate consequences of Theorem 6.89. □

6.9 The Mellin–Poisson Summation Formula

In Fourier analysis the famous Poisson summation formula is a fundamental relation which connects the finite and the continuous Fourier transform. It has numerous applications, from quadrature formulae to signal analysis. Here we discuss a version of the Poisson summation formula, called *Mellin–Poisson summation formula*, in the setting of the Mellin transform.

We begin with the following lemma.

Lemma 6.96 *Let* $A \subset \mathbb{R}$ *be a measurable set, and let* $(g_k)_{k\in\mathbb{Z}}$ *be a sequence of measurable functions* $g_k : \mathbb{R} \to \mathbb{C}$. *If*

$$\sum_{k=-\infty}^{\infty} \int_A |g_k(x)|dx < \infty, \tag{6.92}$$

then the series of functions $\sum_{k=-\infty}^{\infty} g_k(x)$ *converges absolutely almost everywhere on* A *and*

$$\int_A \sum_{k=-\infty}^{\infty} g_k(x)dx = \sum_{k=-\infty}^{\infty} \int_A g_k(x)dx.$$

Proof First we remark that, by Beppo-Levi's Theorem, assumption (6.92) implies that

$$\int_A \sum_{k=-\infty}^{\infty} |g_k(x)|dx < \infty.$$

Therefore the series $G(x) := \sum_{k=-\infty}^{\infty} g_k(x)$ converges absolutely almost everywhere on A and it is summable on A. Setting $S_N(x) := \sum_{k=-N}^{N} g_k(x)$, we have $\lim_{N\to\infty} S_N(x) = G(x)$ almost everywhere on A. Moreover,

$$|S_N(x)| = \left|\sum_{k=-N}^{N} g_k(x)\right| \leq \sum_{k=-\infty}^{\infty} |g_k(x)| < \infty.$$

Thus, applying Lebesgue's dominated convergence theorem, we obtain

$$\lim_{n\to\infty} \sum_{k=-N}^{N} \int_A g_k(x)dx = \int_A G(x)dx,$$

as remained to be shown. □

The following theorem provides an important relation between the finite Mellin transform and that on X_c.

Theorem 6.97 *Let $f \in X_c$, where $c \in \mathbb{R}$. Set*

$$f^c(x) := \sum_{j=-\infty}^{\infty} f(e^{2\pi j}x)e^{2\pi jc} \qquad (x \in \mathbb{R}^+). \tag{6.93}$$

Then the series in (6.93) converges absolutely almost everywhere on the interval $[e^{-\pi}, e^{\pi}]$ and $f^c \in Y_c$ with

$$\|f^c\|_{Y_c} \le \|f\|_{X_c}. \tag{6.94}$$

Moreover, the finite Mellin transform of $f^c \in Y_c$ is equal to the restriction to the set $c + i\mathbb{Z}$ of the Mellin transform $[f]^\wedge_{M_c}$, that is,

$$[f^c]^\wedge_{\mathscr{M}_c}(k) = [f]^\wedge_{M_c}(c + ik) \qquad (k \in \mathbb{Z}). \tag{6.95}$$

Proof First, it is immediately seen that the function f^c defined by (6.93) is c-recurrent. Next, for $x \in [e^{-\pi}, e^{\pi}]$ and any fixed $k \in \mathbb{Z}$ set

$$g_j(x) := f(e^{2\pi j}x)e^{2\pi jc}x^{c+ik-1} \qquad (j \in \mathbb{Z}).$$

Then, by the change of variable $u = e^{2\pi j}x$, we obtain

$$\sum_{j=-\infty}^{\infty} \int_{e^{-\pi}}^{e^{\pi}} |g_j(x)|dx = \sum_{j=-\infty}^{\infty} \int_{e^{(2j-1)\pi}}^{e^{(2j+1)\pi}} |f(u)|u^{c-1}du = \|f\|_{X_c},$$

which shows that the hypotheses of Lemma 6.96 are satisfied. The conclusion of this lemma, used for the sequence $(|g_j|)_{j\in\mathbb{Z}}$, yields

$$\int_{e^{-\pi}}^{e^{\pi}} |f^c(x)|x^{c-1}dx \le \int_{e^{-\pi}}^{e^{\pi}} \sum_{j=-\infty}^{\infty} |g_j(x)|dx = \sum_{j=-\infty}^{\infty} \int_{e^{-\pi}}^{e^{\pi}} |g_j(x)|dx = \|f\|_{X_c},$$

which is (6.94). Using again Lemma 6.96 and the substitution $u = e^{2\pi j}x$, we conclude that

$$[f^c]^\wedge_{\mathscr{M}_c}(k) = \int_{e^{-\pi}}^{e^{\pi}} \sum_{j=-\infty}^{\infty} g_j(x)dx = \sum_{j=-\infty}^{\infty} \int_{e^{-\pi}}^{e^{\pi}} g_j(x)dx$$

$$= \sum_{j=-\infty}^{\infty} \int_{e^{-\pi}}^{e^{\pi}} f(e^{2\pi j}x)e^{2\pi jc}x^{c+ik-1}dx$$

$$= \sum_{j=-\infty}^{\infty} \int_{e^{(2j-1)\pi}}^{e^{(2j+1)\pi}} f(u)u^{c+ik-1}du = [f]^\wedge_{M_c}(c+ik),$$

and so (6.95) holds. □

We are now ready for stating the Poisson summation formula in the Mellin setting.

Theorem 6.98 *Let $f \in X_c$ for $c \in \mathbb{R}$ be a continuous function on $\mathbb{R}^+$ and suppose that*

$$\sum_{k=-\infty}^{\infty} |[f]^\wedge_{M_c}(c+ik)| < \infty. \tag{6.96}$$

If the series (6.93) defining the function f^c converges uniformly on $[e^{-\pi}, e^{\pi}]$, then

$$f^c(x) := \sum_{k=-\infty}^{\infty} f(e^{2\pi k}x)e^{2\pi kc}$$

$$= \frac{x^{-c}}{2\pi} \sum_{k=-\infty}^{\infty} [f]^\wedge_{M_c}(c+ik)x^{-ik} \quad (x \in \mathbb{R}^+). \tag{6.97}$$

Proof For $x \in [e^{-\pi}, e^{\pi}]$ we see with the help of (6.96) that

$$x^{-c} \sum_{k=-\infty}^{\infty} |[f]^\wedge_{M_c}(c+ik)x^{-ik}| \le e^{\pi|c|} \sum_{k=-\infty}^{\infty} |[f]^\wedge_{M_c}(c+ik)| < \infty.$$

Therefore, the series on the right-hand side of (6.97) converges uniformly on $[e^{-\pi}, e^{\pi}]$ and defines a continuous function. By the hypotheses, f is continuous and the series defining f^c converges uniformly on $[e^{-\pi}, e^{\pi}]$. Hence f^c is itself continuous on $[e^{-\pi}, e^{\pi}]$. Moreover, by the c-recurrency , the continuity of f^c extends to the whole $\mathbb{R}^+$. Finally, using (6.95), we see that the right-hand side of (6.97) is the Mellin series of f^c and so Proposition 5.17 (c) implies that

$$f^c(x) = \frac{x^{-c}}{2\pi} \sum_{k=-\infty}^{\infty} [f^c]^\wedge_{\mathscr{M}^c}(k)x^{-ik} = \frac{x^{-c}}{2\pi} \sum_{k=-\infty}^{\infty} [f]^\wedge_{M_c}(c+ik)x^{-ik} \qquad (x \in \mathbb{R}^+).$$

This completes the proof. □

In the following we shall present a convenient version of the Mellin–Poisson summation formula for functions belonging to some Mellin–Sobolev spaces. This special case is based on the following proposition.

Proposition 6.99 *Let $f \in W_c^{1,1}$ be a continuous function on $\mathbb{R}^+$. Then the series (6.93) defining f^c converges absolutely and uniformly on $[e^{-\pi}, e^{\pi}]$. In particular, f^c is continuous on $\mathbb{R}^+$.*

Proof Introducing

$$F_k(x) := \frac{1}{2\pi} \int_{e^{2\pi k}x}^{e^{2\pi(k+1)}x} f(u)u^c \frac{du}{u} \qquad (x \in \mathbb{R}^+,\ k \in \mathbb{Z}),$$

we may write

$$\sum_{k=-\infty}^{\infty} |f(e^{2\pi k}x)e^{2\pi kc}| \leq x^{-c} \sum_{k=-\infty}^{\infty} |f(e^{2\pi k}x)(e^{2\pi k}x)^c - F_k(x)| + x^{-c} \sum_{k=-\infty}^{\infty} |F_k(x)|. \tag{6.98}$$

Now we turn to the first term on the right-hand side of (6.98). The hypotheses on f allows us to use formula (6.87) of Theorem 6.89 for $r = 1$, which implies that

$$f(e^{2\pi k}x)(e^{2\pi k}x)^c - f(u)u^c = \int_u^{e^{2\pi k}x} \Theta_c f(v)v^c \frac{dv}{v}.$$

Seeing this equation as a relation between two functions of u, we apply the functional

$$I_{k,x} : \varphi \longmapsto \frac{1}{2\pi} \int_{e^{2\pi k}x}^{e^{2\pi(k+1)}x} \varphi(u)\frac{du}{u}$$

on both sides. Since $I_{k,x}(1) = 1$ and $I_{k,x}(f(\cdot)(\cdot)^c) = F_k(x)$, we conclude that

$$\begin{aligned}
\left|f(e^{2\pi k}x)(e^{2\pi k}x)^c - F_k(x)\right| &= \left|\frac{1}{2\pi} \int_{e^{2\pi k}x}^{e^{2\pi(k+1)}x} \left\{\int_u^{e^{2\pi k}x} \Theta_c f(v)v^c\frac{dv}{v}\right\} \frac{du}{u}\right| \\
&\leq \frac{1}{2\pi} \int_{e^{2\pi k}x}^{e^{2\pi(k+1)}x} \frac{du}{u} \int_{e^{2\pi k}x}^{e^{2\pi(k+1)}x} |\Theta_c f(v)|v^c \frac{dv}{v} \\
&= \int_{e^{2\pi k}x}^{e^{2\pi(k+1)}x} |\Theta_c f(v)|v^c \frac{dv}{v}.
\end{aligned}$$

Thus the first series on the right-hand side of (6.98) is bounded by $\|\Theta_c f\|_{X_c}$. For the second series on the right-hand side of (6.98) we have by the definition of F_k,

$$\sum_{k=-\infty}^{\infty} |F_k(x)| \leq \frac{1}{2\pi} \sum_{k=-\infty}^{\infty} \int_{e^{2\pi k}x}^{e^{2\pi(k+1)}x} |f(u)|u^c \frac{du}{u} = \frac{1}{2\pi}\|f\|_{X_c}.$$

Altogether (6.98) implies that

$$\sum_{k=-\infty}^{\infty} |f(e^{2\pi k}x)|e^{2\pi kc} \leq e^{\pi|c|}\left(\|\Theta_c f\|_{X_c} + \frac{1}{2\pi}\|f\|_{X_c}\right) \qquad (x \in [e^{-\pi}, e^{\pi}]).$$

Hence the series f^c converges absolutely and uniformly on $[e^{-\pi}, e^{\pi}]$. The continuity of f^c is seen as in the previous proof. □

Proposition 6.99 allows us to deduce from Theorem 6.98 the following useful result.

Corollary 6.100 *Let $f \in W_c^{1,1}(\mathbb{R}^+) \cap C(\mathbb{R}^+)$, where $c \in \mathbb{R}$, and suppose that*

$$\sum_{k=-\infty}^{\infty} \left|[f]^{\wedge}_{M_c}(c + 2\pi i T k)\right| < \infty$$

for some $T > 0$. Then

$$\sum_{k=-\infty}^{\infty} f(e^{k/T})e^{kc/T} = T \sum_{k=-\infty}^{\infty} [f]^{\wedge}_{M_c}(c + 2\pi i T k),$$

both series being absolutely convergent.

Proof For $g(x) := f(x^{1/(2\pi T)})$, we have by Proposition 6.8 (d) that $g \in W_{\widetilde{c}}^{1,1}(\mathbb{R}^+)$ with $\widetilde{c} := c/(2\pi T)$ and $[g]^{\wedge}_{M_c}(\widetilde{c} + ik) = 2\pi T[f]^{\wedge}_{M_c}(c + 2\pi i T k)$. Now Proposition 6.99 applies to the function g and shows that the hypotheses of Theorem 6.98 are satisfied for g. Setting $x = 1$, we obtain

$$\sum_{k=-\infty}^{\infty} g(e^{2\pi k})e^{2\pi k\widetilde{c}} = \frac{1}{2\pi} \sum_{k=-\infty}^{\infty} [g]^{\wedge}_{M_c}(\widetilde{c} + ik).$$

When rewritten in terms of f, we have the assertion. □

6.10 Notes

The Mellin transform is treated in several books on integral transforms and operational calculus, often as a marginal appendix to the Fourier or Laplace transform, since it is formally equivalent to the latter; see, e.g., Doetsch [137–139], Titchmarsh [286], Widder [302, 303], Sneddon [275, 276], Prudnikov et al. [247], Brychkov et al. [81], Glaeske et al. [157], Szpankowski [285, Chap. 9].

Let us briefly point out the relations of the Mellin transform with the integral transforms of Fourier and Laplace (see also Butzer and Jansche [89]). First we recall that the normalized Fourier transform of a function $f \in L^1(\mathbb{R})$ is defined as

$$\widehat{f}(v) := \frac{1}{\sqrt{2\pi}} \int_{-\infty}^{\infty} f(x)e^{-ivx}dx \qquad (v \in \mathbb{R}),$$

where the integral exists as a Lebesgue integral (see Sect. 1.6). For $f \in X_c$, we find by substituting $x = e^v$ that

$$[f]^{\wedge}_{M_c}(t) = \int_0^{\infty} f(x)x^{c+it}\frac{dx}{x} = \int_{-\infty}^{\infty} f(e^v)e^{cv}e^{ivt}dv = \sqrt{2\pi}\,\widehat{g}(-t),$$

where $g(x) := f(e^x)e^{cx}$ for $x \in \mathbb{R}$. This shows that the Mellin transform can be expressed in terms of the Fourier transform and vice versa.

Next we turn to the Laplace transforms. Let $f : \mathbb{R}^+ \to \mathbb{C}$ be such that the function $g_s : u \mapsto f(u)e^{-su}$ belongs to $L^1(\mathbb{R}^+)$ for $s \in \mathbb{C}$. Then the *Laplace transform* of f is defined by

$$\mathcal{L}[f](s) := \int_0^{\infty} f(u)e^{-su}du.$$

If $g_\sigma \in L^1(\mathbb{R}^+)$ for some $\sigma \in \mathbb{R}$, then it is easily seen that the Laplace transform $\mathcal{L}[f](s)$ exists as a Lebesgue integral whenever $s = c + it$ with $c \geq \sigma$ and $t \in \mathbb{R}$.

Now let $f : \mathbb{R} \to \mathbb{C}$ and let g_s be defined as before. Then the *bilateral Laplace transform* of f, denoted by $\mathcal{L}_B$, is defined by

$$\mathcal{L}_B[f](s) := \int_{-\infty}^{\infty} f(u)e^{-su}du.$$

The two kinds of Laplace transforms are connected by

$$\mathcal{L}_B[f(\cdot)](s) = \mathcal{L}[f(\cdot)](s) + \mathcal{L}[f(-\,\cdot)](-s).$$

If $\sigma_1, \sigma_2 \in \mathbb{R}$ are such that $\sigma_1 < \sigma_2$, $g_{\sigma_1}(\cdot) \in L^1(\mathbb{R}^+)$ and $g_{\sigma_2}(-\,\cdot) \in L^1(\mathbb{R}^+)$, then the bilateral Laplace transform is well defined as a Lebesgue integral for every $s = c+it$ with $c \in]\sigma_1, \sigma_2[$ and $t \in \mathbb{R}$. For further convergence results on the Laplace transforms see, e.g., Doetsch [137, pp. 60, 403 ff], Widder [302, p. 237 ff], Zayed [309] and Glaeske et al. [157].

Relations of the Mellin transform with the Laplace transforms can be found in Butzer and Jansche [89]. If the functions $f : \mathbb{R}^+ \to \mathbb{C}$ and $F : \mathbb{R} \to \mathbb{C}$ are related by

$$F(u) = f(e^{-u}) \quad (u \in \mathbb{R}) \qquad \text{or} \qquad f(x) = F(-\log x) \quad (x \in \mathbb{R}^+),$$

then $f \in X_{]a,b[}$ if and only if $\mathcal{L}_B[F(\cdot)](s)$ exists for all $s = c + it \in \text{St}]a, b[$. In this case we have

$$\mathcal{L}_B[F](s) = [f]^{\wedge}_M(s) \qquad (s \in \text{St}]a, b[).$$

As regards the inverse transforms, it is shown in [89] that if $g \in L^1(c + i\mathbb{R})$ with $c \in \mathbb{R}$, then

$$\mathcal{L}_B^{-1}[g](u) = M_c^{-1}[g](e^{-u}) \quad (u \in \mathbb{R}),$$
$$M_c^{-1}[g](x) = \mathcal{L}_B^{-1}[g](-\log x) \quad (x \in \mathbb{R}^+).$$

Although the integral transforms under consideration can be expressed by each other, they have their individual properties. In particular, the Fourier transform has several interesting and useful properties which are not shared by the Mellin transform in an analogous way. We mention some of them:

(i) $f \in L^1(\mathbb{R})$ and $\widehat{f}$ have the same domain;
(ii) there exist subspaces of $L^1(\mathbb{R})$ which are mapped by the Fourier transform into itself;
(iii) for $f, g \in L^1(\mathbb{R})$ there holds the Parseval formula

$$\int_{-\infty}^{\infty} \widehat{f}(u)\, g(u) du = \int_{-\infty}^{\infty} f(u)\, \widehat{g}(u) du$$

(see Butzer and Nessel [95, Proposition 5.1.4]);
(iv) the inverse of the Fourier transform is the Fourier transform itself apart from a change of the sign on its domain;
(v) the Fourier transform has fixpoints, that is, there exist functions $g \in L^1(\mathbb{R})$ such that $\widehat{g} = g$.

Properties (ii) and (iii) are very useful for a distributional version of the Fourier transform and its extension to distributions (see the notes of Chap. 7 and Folland [151, Sect. 9.4]). As a consequence of property (iv), characterizations of (Fourier) bandlimited functions yield characterizations of duration-limited functions and vice versa. Property (v) provides useful auxiliary functions for computations in which simultaneous control over a truncation error and an aliasing error is desired.

Clearly, $f \in X_c$ and $[f]^{\wedge}_{M_c}$ cannot have the same domain and this is already the reason why properties (ii)–(v) cannot remain true when we simply replace $L^1(\mathbb{R})$ with X_c and the Fourier transform with the Mellin transform. Of course, we can express in formulae the involved Fourier transforms in terms of Mellin transforms but this will in some cases produce cumbersome relations of little use. Therefore it is desirable to study the Mellin transform as an independent subject and unveil its true nature, as we attempt to do in some of the subsequent chapters.

Moreover, the Mellin transform is not an artificial operator but has some interesting applications. It is a useful tool for certain boundary value problems (see, e.g., Sneddon, [276, Chap. 4], Colombo [128, Chap. II], Colombo and Lavoine [129], Uflyand [294], Zemanian [310], Szmydt and Zemian [284]) and plays a role in sampling theory (see, e.g., Ostrowsky et al. [233], Bertero and Pike [70], Gori [159], Butzer and Jansche [90], Bardaro et al. [52, 54] and Chaps. 11–13 of this book).

A first approach to an independent and self-contained theory of the Mellin transform and Mellin analysis was initiated by Mamedov and Orudhzev in a series of papers [209–211] beginning in the late seventieth of the last century. In 1991, the basic ideas were included in the book of Mamedov [208]. One of the author's aims was to produce a theory of the Mellin transform in X_0 parallel to the Fourier transform by taking inspiration from the book of Butzer and Nessel [95]. Only a few years after this pioneering work, in 1997, the Mellin transform was studied in a short monograph by Brychkov [79]. In 1996, Butzer and Jansche turned to the Mellin transform and developed an extensive theory for the spaces X_c in [88] and [89]. For further contributions to Mellin analysis by Butzer and Jansche, see the bibliography. A wide and useful compendium of various applications of the Mellin transform to problems of pure and applied mathematics can be found in Marichev [213] and Brychkov et al. [82].

The material contained in Sects. 6.1–6.3 and 6.5–6.8 of the present chapter is mainly based on [89]. Sections 6.4.2–6.4.5 originate from Bardaro and Mantellini [33]. Section 6.4.6 is inspired by Bardaro and Mantellini [30, Sect. 3]. In this paper the notion of Mellin–Lebesgue points was introduced for the first time. Section 6.9 is borrowed from Butzer and Jansche [91, Sect. 7]. As an application of the Mellin–Poisson summation formula, we mention that it gives another proof of the transformation formula of the Jacobi theta function (see [91, Sect. 8]). This in turn is known to be equivalent to the famous functional equation of the Riemann zeta function (see, e.g., Rademacher [251, p. 88]). Moreover, the Mellin–Poisson summation formula is very useful in sampling theory and can be interpreted as a quadrature formula on the positive real line with various favorable properties (see Chap. 14).

As research related to Sect. 6.4, we mention that the asymptotic behavior of certain Mellin convolution operators is also studied in Bardaro and Mantellini [31, 32, 34, 40], Mantellini [212] and Popa [245]. The idea of considering linear combinations of integral operators, presented in Sect. 6.4.3, was originally inspired by a classical approach of P.L. Butzer for Bernstein polynomial (see [84]), which gives a way of improving the order of approximation; for further contributions, see Bardaro and Mantellini [35–37]. For convergence results with respect to variation properties and in more general function spaces such as Orlicz, Musielak–Orlicz spaces and modular function spaces, see, e.g., Angeloni and Vinti [8–12], Bardaro and Mantellini [29, 30] and Bardaro et al. [49, 51]). An abstract approach for nonlinear integral operators was introduced in Bardaro et al. [46]. Special attention may be payed to the paper by Tyr and Radouan [292], in which the authors study the absolute convergence of the Mellin transform of a function f belonging to generalized Mellin–Lipschitz or Zygmund classes.

Next, we refer to special applications of the inverse Mellin transform. Certain integrals, named Mellin–Barnes integrals (see Paris and Kaminski [236]), play an important role in convolution theory (Nguyen and Yakubovich [228]), the theory of higher transcendental functions (Erdélyi [147]), probability (Mainardi and Pagnini [206]), quantum theory (Suzuki et al. [281]) and special functions (Katsurada [186]). These are complex integrals of the form

$$\mathrm{MB}(z) :=$$

$$\frac{1}{2\pi i}\int_{\gamma-i\infty}^{\gamma+i\infty} \frac{\Gamma(a_1 + A_1 s)\cdots\Gamma(a_m + A_m s)}{\Gamma(c_1 + C_1 s)\cdots\Gamma(c_p + C_p s)}\,\frac{\Gamma(b_1 - B_1 s)\cdots\Gamma(b_n - B_n s)}{\Gamma(d_1 - D_1 s)\cdots\Gamma(d_q - D_q s)}\, z^{-s}\,ds,$$

where $\gamma \in \mathbb{R}$, A_j, B_j, C_j, D_j are positive constants and a_j, b_j, c_j, d_j are complex numbers. Thus the Mellin–Barnes integrals are particular inverse Mellin transforms of ratios of gamma functions in the integrand (see formula (6.61)).

A different approach to the study of the various integral transforms and their applications, in particular to Mellin–Barnes integrals, can be found in Brychkov et al. [80], which contains a survey of methods of representing integral transforms of convolution type as direct and inverse compositions of other integral transforms (see also Tuan et al. [291], and Yakubovich and Luchko [306] for results in terms of hypergeometric functions).

For an efficient computation of the Mellin transform, one discretizes the equation

$$[f]^{\wedge}_{M_c}(s) = \int_0^{\infty} f(u)u^s\,\frac{du}{u}$$

such that a linear relation between N geometrically spaced samples of f and N arithmetically spaced samples of its Mellin transform is obtained approximately, where N is a sufficiently large integer. Then the algorithm of the fast Fourier transform (FFT algorithm) can be used for computing approximations of the samples of the Mellin transform. For details, see Bertrand et al. [71, Chap. 11, Subsect. 11.3.2].

Chapter 7
The Mellin Transform in Spaces X_c^p for $1 < p \leq 2$

In this chapter, we first define the Mellin transform in the Hilbert space X_c^2. Then, using the Riesz–Thorin interpolation theorem (Theorem 1.22), we extend the definition to the spaces X_c^p for $1 < p < 2$. The case $p > 2$ can be treated within the theory of distributions. However, we will not cover this topic in the present book but give a brief outline in the notes.

The extension from X_c^p to the spaces $X_{]a,b[}^p$ can be performed analogously to the case $p = 1$.

7.1 The Mellin Transform in X_c^2

We first remark that the integral defining the classical Mellin transform in X_c,

$$[f]^{\wedge}_{M_c}(s) = \int_0^{\infty} f(u)u^s \frac{du}{u} \qquad (s = c + it;\ c, t \in \mathbb{R})$$

is well-defined as a Lebesgue integral. But for functions $f \in X_c^2$ it may not exist as an absolutely convergent integral. Therefore, in this setting, we have to change the notion of Mellin transform considerably, but in a such a way that the new definition is consistent with the classical one, that is, the two definitions are equivalent for $f \in X_c \cap X_c^2$.

We begin with the following basic result (for the notation $L^2(c + i\mathbb{R})$ see Sect. 6.5).

Proposition 7.1 *Let $f \in X_c \cap X_c^2$, where $c \in \mathbb{R}$. Then $[f]^{\wedge}_{M_c} \in L^2(c + i\mathbb{R})$ and*

$$\int_0^{\infty} |x^c f(x)|^2 \frac{dx}{x} = \frac{1}{2\pi} \int_{-\infty}^{\infty} |[f]^{\wedge}_{M_c}(c + it)|^2 dt, \tag{7.1}$$

C. Bardaro et al., *Mellin Analysis, Transform Theory, and Applications*,
https://doi.org/10.1007/978-3-031-96672-9_7

that is,

$$\|f\|_{X_c^2} = \frac{1}{\sqrt{2\pi}} \|[f]^\wedge_{M_c}\|_{L^2(c+i\mathbb{R})}. \tag{7.2}$$

Proof Define $f^*(x) := x^{-2c}\overline{f(1/x)}$ and $h(x) := (f * f^*)(x)$. Now $f \in X_c$ implies that $f^* \in X_c$ since

$$\int_0^\infty u^c \left(\frac{1}{u^{2c}} \left| \overline{f\left(\frac{1}{u}\right)} \right| \right) \frac{du}{u} = \int_0^\infty u^c |\overline{f(u)}| \frac{du}{u} < \infty.$$

Analogously, $f \in X_c^2$ implies that $f^* \in X_c^2$. Hence together with f also $f^* \in X_c \cap X_c^2$. With the help of Theorems 4.3 and 4.4, we deduce that h is a continuous function on $\mathbb{R}^+$ which belongs to $X_c \cap X_c^2$ as well. Since

$$h(x) = \int_0^\infty \left(\frac{u}{x}\right)^{2c} \overline{f\left(\frac{u}{x}\right)} f(u) \frac{du}{u},$$

we observe that

$$h(1) = \int_0^\infty |u^c f(u)|^2 \frac{du}{u} = \|f\|^2_{X_c^2}. \tag{7.3}$$

Next we consider the convolution

$$(U_\rho h)(x) := (h * G_\rho^c)(x) \qquad (x, \rho \in \mathbb{R}^+),$$

where $G_\rho^c(x) := x^{-c} G_\rho(x)$ with $\{G_\rho\}_{\rho>0}$ being the Mellin–Gauss–Weierstrass kernel (see Example 6.25). Proposition 6.8(c) applied to the function G_ρ shows us that $[G_\rho^c]^\wedge_{M_c}(c+it) = [G_\rho]^\wedge_{M_c}(it)$. Employing Proposition 6.26 and Lemma 6.73, we conclude that

$$(U_\rho h)(1) = \frac{1}{2\pi} \int_{-\infty}^\infty [h]^\wedge_{M_c}(c+it) \exp\left(-\frac{t^2}{\rho^2}\right) dt. \tag{7.4}$$

In view of Remark 6.41, we may write $(U_\rho h)(x) = (T_\rho \widetilde{h})(1/x)$, where T_ρ is the Mellin–Fejér type operator associated with U_ρ and $\widetilde{h}(x) := h(1/x)$ for $x \in \mathbb{R}^+$. Since the continuity of h on $\mathbb{R}^+$ entails that of $\widetilde{h}$, we deduce from statement (i) of Theorem 6.38 that

$$\lim_{\rho\to\infty} (U_\rho h)(1) = h(1). \tag{7.5}$$

By Theorem 6.12 and the relations

$$[f^*]^\wedge_{M_c}(c+it) = [\overline{f}]^\wedge_{M_c}(c-it) = \overline{[f]^\wedge_{M_c}(c+it)},$$

we obtain

$$[h]^{\wedge}_{M_c}(c+it) = [f]^{\wedge}_{M_c}(c+it)\,[f^*]^{\wedge}_{M_c}(c+it)$$
$$= [f]^{\wedge}_{M_c}(c+it)\,[\overline{f}]^{\wedge}_{M_c}(c-it) = |[f]^{\wedge}_{M_c}(c+it)|^2. \tag{7.6}$$

Thus, the integrand in (7.4) is a nonnegative function depending on ρ which builds for each fixed $t \in \mathbb{R}$ a non-decreasing sequence that converges to $[h]^{\wedge}_{M_c}(c+it)$ as $\rho \to \infty$. In this situation Beppo Levi's theorem (see Proposition 1.3) yields that $[h]^{\wedge}_{M_c} \in L^1(c+i\mathbb{R})$ and, together with (7.5),

$$\int_{-\infty}^{\infty} [h]^{\wedge}_{M_c}(c+it)dt = 2\pi h(1).$$

Finally, combining this equation with (7.3) and substituting (7.6), we arrive at (7.2). □

Definition 7.2 For $f \in X_c^2$, where $c \in \mathbb{R}$, the Mellin transform $M_c^2[f] \equiv [f]^{\wedge}_{M_c^2} : c+i\mathbb{R} \to \mathbb{C}$ is defined by

$$\lim_{\varrho\to\infty} \left\| [f]^{\wedge}_{M_c^2}(\cdot) - \int_{1/\varrho}^{\varrho} f(u)u^{(\cdot)}\frac{du}{u} \right\|_{L^2(c+i\mathbb{R})} = 0, \tag{7.7}$$

i.e., by the limit in mean

$$[f]^{\wedge}_{M_c^2}(c+it) := \overset{(2)}{\text{l.i.m.}}_{\varrho\to\infty} \int_{1/\varrho}^{\varrho} f(u)u^{c+it}\frac{du}{u}. \tag{7.8}$$

One can show that (7.7) is equivalent to

$$\lim_{\substack{\varrho_1\to 0\\ \varrho_2\to\infty}} \left\| [f]^{\wedge}_{M_c^2}(\cdot) - \int_{\varrho_1}^{\varrho_2} f(u)u^{(\cdot)}\frac{du}{u} \right\|_{L^2(c+i\mathbb{R})} = 0. \tag{7.9}$$

Proposition 7.3 *The Mellin transform M_c^2 for $c \in \mathbb{R}$ is well-defined as a linear operator*

$$M_c^2 : X_c^2 \longrightarrow L^2(c+i\mathbb{R}), \quad f \longmapsto M_c^2[f]. \tag{7.10}$$

In particular, for $f \in X_c^2$ the function $M_c^2[f]$ is uniquely determined as a function in $L^2(c+i\mathbb{R})$.

Proof For $f \in X_c^2$ and $\varrho > 0$, we consider the truncated function $f_\varrho := f\chi_{[1/\varrho,\varrho]}$. Clearly, $f_\varrho \in X_c^2 \cap X_c$,

$$[f_\varrho]^{\wedge}_{M_c}(c+it) = \int_{1/\varrho}^{\varrho} f(u)u^{c+it}\frac{du}{u}, \tag{7.11}$$

and Proposition 7.1 implies that $[f_\varrho]^\wedge_{M_c} \in L^2(c + i\mathbb{R})$ for all $\varrho > 0$. Next we show that $([f_\varrho]^\wedge_{M_c})_{\varrho>0}$ is a Cauchy sequence in $L^2(c + i\mathbb{R})$. For this, take any numbers ϱ_1 and ϱ_2 such that $1 < \varrho_1 < \varrho_2$. Then, by Proposition 7.1,

$$\frac{1}{2\pi} \left\|[f_{\varrho_1}]^\wedge_{M_c} - [f_{\varrho_2}]^\wedge_{M_c}\right\|^2_{L^2(c+i\mathbb{R})} = \left\|f_{\varrho_1} - f_{\varrho_2}\right\|^2_{X_c^2}$$
$$= \int_{\varrho_1}^{\varrho_2} |f(u)u^c|^2 \frac{du}{u} + \int_{1/\varrho_2}^{1/\varrho_1} |f(u)u^c|^2 \frac{du}{u},$$

and since $f \in X_c^2$, these integrals approach zero as $\varrho_1 \to \infty$. Finally, the completeness of $L^2(c + i\mathbb{R})$ guarantees the existence of a uniquely determined function $g \in L^2(c + i\mathbb{R})$ such that

$$\lim_{\varrho\to\infty} \|[f_\varrho]^\wedge_{M_c} - g\|_{L^2(c+i\mathbb{R})} = 0.$$

This completes the proof by showing that $M_c^2[f] \equiv [f]^\wedge_{M_c^2} = g$. □

By Proposition 7.1, it follows that for $f \in X_c \cap X_c^2$,

$$\left\|[f]^\wedge_{M_c} - [f]^\wedge_{M_c^2}\right\|_{L^2(c+i\mathbb{R})} \leq \limsup_{\varrho\to\infty} \left\|[f]^\wedge_{M_c^2}(\cdot) - \int_{1/\varrho}^{\varrho} f(u)u^{(\cdot)} \frac{du}{u}\right\|_{L^2(c+i\mathbb{R})}$$
$$+ \limsup_{\varrho\to\infty} \sqrt{2\pi}\, \|f_\varrho - f\|_{X_c^2} = 0. \tag{7.12}$$

Hence the following consistency statement holds.

Lemma 7.4 *For $f \in X_c \cap X_c^2$, the Mellin transforms $M_c[f]$ and $M_c^2[f]$ coincide, that is, $M_c[f](c + it) = M_c^2[f](c + it)$ almost everywhere for $t \in \mathbb{R}$.*

Some authors use the same notation for both types of Mellin transforms. Then the type of the Mellin transform has to be chosen according to the hypotheses on the involved function f. In formulae which are valid for both types, we shall always prefer the notation $[f]^\wedge_{M_c}$ or equivalently $M_c[f]$ when the Mellin transform is mainly considered as an operator.

Next we show that the conclusion of Proposition 7.1 also holds under the weaker hypothesis that $f \in X_c^2$ only.

Lemma 7.5 *The operator $M_c^2 : X_c^2 \to L^2(c+i\mathbb{R})$ is bounded and norm preserving in the sense that for $f \in X_c^2$,*

$$\|f\|_{X_c^2} = \frac{1}{\sqrt{2\pi}} \left\|M_c^2[f]\right\|_{L^2(c+i\mathbb{R})}. \tag{7.13}$$

Proof Since $f_\varrho := f\chi_{[1/\varrho,\varrho]} \in X_c \cap X_c^2$, Proposition 7.1 in conjunction with Lemma 7.4 yields

$$\|[f_\varrho]^\wedge_{M_c^2}\|_{L^2(c+i\mathbb{R})} = \sqrt{2\pi}\|f_\varrho\|_{X_c^2} \qquad (\varrho \in \mathbb{R}^+).$$

By Definition 7.2, we have $\lim_{\varrho\to\infty} \|[f]^\wedge_{M_c^2} - [f_\varrho]^\wedge_{M_c^2}\|^2_{L^2(c+i\mathbb{R})} = 0$, which allows us to conclude that

$$\lim_{\varrho\to\infty} \|[f_\varrho]^\wedge_{M_c^2}\|_{L^2(c+i\mathbb{R})} = \|[f]^\wedge_{M_c^2}\|_{L^2(c+i\mathbb{R})}.$$

Furthermore, $\lim_{\varrho\to\infty} \|f_\varrho\|_{X_c^2} = \|f\|_{X_c^2}$. Now the proof is completed by combining these relations. □

The basic properties of the Mellin transform in X_c can be easily transferred to that in the space X_c^2 just by using Definition 7.2 and observing that $f_\varrho \in X_c \cap X_c^2$.

For example, the convolution theorem (Theorem 6.12) takes the following form.

Proposition 7.6 *Let $f \in X_c^2$ and $g \in X_c$. Then $f * g \in X_c^2$ and*

$$[f * g]^\wedge_{M_c^2}(s) = [f]^\wedge_{M_c^2}(s)[g]^\wedge_{M_c}(s) \qquad (s \in c + i\mathbb{R}). \tag{7.14}$$

In Sect. 7.2 we shall present a general form of the convolution theorem for the spaces X_c^p (see Theorem 7.15). Now we introduce the inverse Mellin transform for functions in $L^2(c + i\mathbb{R})$.

Definition 7.7 For $g \in L^2(c + i\mathbb{R})$ with $c \in \mathbb{R}$, the inverse Mellin transform $M_c^{2,-1}[g] : \mathbb{R}^+ \to \mathbb{C}$ is defined by

$$\lim_{\varrho\to\infty} \left\| M_c^{2,-1}[g](\cdot) - \frac{1}{2\pi}\int_{-\varrho}^{\varrho} g(c+it)(\cdot)^{-c-it}dt \right\|_{X_c^2} = 0. \tag{7.15}$$

Lemma 7.8 *The inverse Mellin transform $M_c^{2,-1}$ with $c \in \mathbb{R}$, is well-defined as a linear, bounded operator*

$$M_c^{2,-1} : L^2(c + i\mathbb{R}) \longrightarrow X_c^2, \quad g \longmapsto M_c^{2,-1}[g]$$

and

$$\|g\|_{L^2(c+i\mathbb{R})} = \sqrt{2\pi}\|M_c^{2,-1}[g]\|_{X_c^2}. \tag{7.16}$$

Proof For $g \in L^2(c + i\mathbb{R})$, it is easily verified that the function

$$f : x \longmapsto x^{-c}g(c - i\log x)$$

belongs to X_c^2 and $\|f\|_{X_c^2} = \|g\|_{L^2(c+i\mathbb{R})}$. Furthermore,

$$2\pi x^c M_c^{2,-1}[g](x) = M_c^2[f](c + i\log x) \qquad (\text{a.e. } x \in \mathbb{R}^+). \tag{7.17}$$

Indeed, for a given $\varrho > 0$, we obtain for the truncated integrals, by using the substitution $t = -\log u$, that

$$2\pi x^c \frac{1}{2\pi}\int_{-\varrho}^{\varrho} g(c+it)x^{-c-it}dt = \int_{e^{-\varrho}}^{e^{\varrho}} g(c - i\log u)x^{i\log u}\frac{du}{u}$$
$$= \int_{e^{-\varrho}}^{e^{\varrho}} u^c f(u) u^{i\log x}\frac{du}{u}$$

and (7.17) follows by letting $\varrho \to \infty$. From (7.17) we deduce that

$$\sqrt{2\pi}\,\|M_c^{2,-1}[g]\|_{X_c^2} = \frac{1}{\sqrt{2\pi}}\left\{\int_0^\infty \left|M_c^2[f](c+i\log x)\right|^2 \frac{dx}{x}\right\}^{1/2}$$
$$= \frac{1}{\sqrt{2\pi}}\left\|M_c^2[f]\right\|_{L^2(c+i\mathbb{R})}.$$

Finally, using Lemma 7.5, we arrive at

$$\sqrt{2\pi}\,\|M_c^{2,-1}[g]\|_{X_c^2} = \|f\|_{X_c^2} = \|g\|_{L^2(c+i\mathbb{R})},$$

which completes the proof. □

Remark 7.9 With reference to Lemma 7.4, we mention that a consistency statement also holds for the inverse Mellin transform. If $g \in L^1(c+i\mathbb{R}) \cap L^2(c+i\mathbb{R})$, then $M_c^{2,-1}[g](x) = M_c^{-1}[g](x)$ almost everywhere for $x \in \mathbb{R}^+$.

The studies of this section can be extended to a Mellin–Plancherel type theorem and stated as follows.

Theorem 7.10 *The Mellin transform for X_c^2 with $c \in \mathbb{R}$, as introduced by Definition 7.2, is a linear isomorphism from X_c^2 onto $L^2(c+i\mathbb{R})$ which preserves inner products in the sense that for $f, g \in X_c^2$,*

$$\langle f, g\rangle_{X_c^2} = \int_0^\infty f(u)\overline{g(u)}u^{2c-1}du$$
$$= \frac{1}{2\pi}\int_{-\infty}^{\infty} [f]^\wedge_{M_c^2}(c+it)\overline{[g]^\wedge_{M_c^2}(c+it)}dt = \frac{1}{2\pi}\big\langle [f]^\wedge_{M_c^2}, [g]^\wedge_{M_c^2}\big\rangle_{L^2(c+i\mathbb{R})}. \tag{7.18}$$

Furthermore, the inverse operator of the Mellin transform is given by $M_c^{2,-1}$, that is,

$$M_c^{2,-1}[M_c^2[f]](x) = f(x) \qquad (a.e.\ x \in \mathbb{R}^+). \tag{7.19}$$

Proof For proving (7.19), we have to verify that $\|f - M_c^{2,-1}[M_c^2[f]]\|_{X_c^2} = 0$. In view of the definitions of M_c^2 and $M_c^{2,-1}$ in conjunction with the consistency statements, it suffices to show that

$$\lim_{\varrho\to\infty} \|f_\varrho - M_c^{-1}[[f_\varrho]^\wedge_{M_c}\chi_{[-\varrho,\varrho]}]\|^2_{X_c^2} = 0, \tag{7.20}$$

where again $f_\varrho := f\chi_{[1/\varrho,\varrho]}$. Using (7.16), we obtain

$$\begin{aligned}
&\|f_\varrho - M_c^{-1}[[f_\varrho]^\wedge_{M_c}\chi_{[-\varrho,\varrho]}]\|^2_{X_c^2} \\
&\qquad = \|f_\varrho\|^2_{X_c^2} + \frac{1}{2\pi}\|[f_\varrho]^\wedge_{M_c}\chi_{[-\varrho,\varrho]}\|^2_{L^2(c+i\mathbb{R})} \\
&\qquad\quad - 2\Re\int_0^\infty \overline{f_\varrho(u)}\,\frac{1}{2\pi}\int_{-\varrho}^{\varrho}[f_\varrho]^\wedge_{M_c}(c+it)u^{-c-it}dt\,u^{2c}\frac{du}{u}.
\end{aligned} \tag{7.21}$$

Employing Fubini's theorem, we can calculate the double integral as

$$\begin{aligned}
&-2\Re\int_0^\infty \overline{f_\varrho(u)}\,\frac{1}{2\pi}\int_{-\varrho}^{\varrho}[f_\varrho]^\wedge_{M_c}(c+it)u^{-c-it}dt\,u^{2c}\frac{du}{u} \\
&\qquad = -\frac{1}{\pi}\Re\int_{-\varrho}^{\varrho}[f_\varrho]^\wedge_{M_c}(c+it)\int_{1/\varrho}^{\varrho}\overline{f_\varrho(u)}u^{c-it}\frac{du}{u}dt \\
&\qquad = -\frac{1}{\pi}\Re\int_{-\varrho}^{\varrho}\left|[f_\varrho]^\wedge_{M_c}(c+it)\right|^2 dt = -\frac{1}{\pi}\|[f_\varrho]^\wedge_{M_c}\chi_{[-\varrho,\varrho]}\|^2_{L^2(c+i\mathbb{R})}.
\end{aligned} \tag{7.22}$$

Combining (7.21) and (7.22), we obtain

$$\begin{aligned}
&\lim_{\varrho\to\infty}\left\| f_\varrho - M_c^{-1}[[f_\varrho]^\wedge_{M_c}\chi_{[-\varrho,\varrho]}]\right\|^2_{X_c^2} \\
&\qquad = \lim_{\varrho\to\infty}\left[\frac{1}{2\pi}\|[f_\varrho]^\wedge_{M_c}\|^2_{L^2(c+i\mathbb{R})} - \frac{1}{2\pi}\|[f_\varrho]^\wedge_{M_c}\chi_{[-\varrho,\varrho]}\|^2_{L^2(c+i\mathbb{R})}\right] = 0,
\end{aligned}$$

where again Lemma 7.5 has been used. Hence (7.20) holds and so (7.19) is verified. The latter relation in connection with Lemmas 7.5 and 7.8 yields that M_c^2 is injective and surjective.

It remains to verify (7.18) for $f, g \in X_c^2$. But this is an immediate consequence of Lemma 7.5 by observing that any inner product $\langle f, g\rangle$ can be represented by its corresponding norm as

$$4\langle f, g\rangle = \|f+g\|^2 - \|f-g\|^2 - i\|if+g\|^2 + i\|if-g\|^2.$$

□

Remark 7.11 By Theorem 7.10, formula (7.15) can also be stated as

$$\lim_{\varrho\to\infty} \left\| f(\cdot) - \frac{1}{2\pi}\int_{-\varrho}^{\varrho} [f_\varrho]^\wedge_{M_c}(c+it)(\cdot)^{-c-it}dt \right\|_{X_c^2} = 0.$$

Indeed, this norm can be written and estimated as

$$\left\| f - M_c^{2,-1}\big[[f_\varrho]^\wedge_{M_c}\chi_{[-\varrho,\varrho]}\big] \right\|_{X_c^2}$$

$$\le \|f - f_\varrho\|_{X_c^2} + \left\| f_\varrho - M_c^{2,-1}\big[[f_\varrho]^\wedge_{M_c}\chi_{[-\varrho,\varrho]}\big] \right\|_{X_c^2},$$

and by (7.20) the right-hand side approaches zero as $\varrho \to \infty$.

7.2 The Mellin Transform in X_c^p for $1 < p < 2$

By virtue of the Riesz–Thorin interpolation theorem (Proposition 1.22), it is possible to define a Mellin transform as an ordinary function defined almost everywhere for functions belonging to X_c^p, when $p \in]1, 2[$. As it happens for Fourier transforms, this is not possible without employing distributions if $p > 2$.

By Proposition 3.4 we know that $S := C^\infty_{\text{comp}}(\mathbb{R}^+)$ is dense in every spaces X_c^p. Therefore, from inequality (6.2), Lemmas 6.4 and 7.5, we know that the operator $T_0 f := [f]^\wedge_{M_c}$ defined on S is such that for every $f \in S$,

$$\|T_0 f\|_{L^\infty(c+i\mathbb{R})} \le \|f\|_{X_c} \tag{7.23}$$

and

$$\|T_0 f\|_{L^2(c+i\mathbb{R})} = \sqrt{2\pi}\,\|f\|_{X_c^2}. \tag{7.24}$$

This means that T_0 is of strong type $(1, \infty)$ between the spaces S and $L^\infty(c + i\mathbb{R})$ and also of strong type $(2, 2)$ between the spaces S and $L^2(c + i\mathbb{R})$. Thus we can apply the Riesz–Thorin interpolation theorem (Proposition 1.22) with the measure spaces $R_1 := \mathbb{R}^+$ and $R_2 := c + i\mathbb{R}$, endowed with the logarithmic measure dt/t and the Lebesgue measure on the line $c + i\mathbb{R}$, respectively. Therefore, the operator T_0 is also of strong type (p, p') for any $p \in]1, 2[$ with p' being the conjugate exponent of p and

$$\|T_0 f\|_{L^{p'}(c+i\mathbb{R})} \le (2\pi)^{1/p'}\|f\|_{X_c^p}. \tag{7.25}$$

By the Hahn–Banach theorem (Proposition 1.18), the operator T_0 can be extended to a (unique) linear and bounded operator T on the space X_c^p with values in $L^{p'}(c+i\mathbb{R})$ such that

$$\|Tf\|_{L^{p'}(c+i\mathbb{R})} \le (2\pi)^{1/p'}\|f\|_{X_c^p} \qquad (f \in X_c^p). \tag{7.26}$$

Now for $\varrho > 1$ and $f \in X_c^p$, we consider again the truncated function $f_\varrho(x) = f(x)\chi_{[1/\varrho,\varrho]}(x)$. By Hölder's inequality, it is easily seen that $f_\varrho \in X_c \cap X_c^p$. The following result will be a crucial tool for defining a Mellin transform for X_c^p.

Theorem 7.12 *Let $f \in X_c^p$ with $1 < p < 2$. Then*

$$\lim_{\varrho\to\infty} \|Tf - [f_\varrho]^\wedge_{M_c}\|_{L^{p'}(c+i\mathbb{R})} = 0. \tag{7.27}$$

Proof Since $f_\varrho \in X_c \cap X_c^p$, the Mellin transform $[f_\varrho]^\wedge_{M_c}$ is well-defined as a Lebesgue integral. Moreover, it is obvious that

$$\lim_{\varrho\to\infty} \|f_\varrho - f\|_{X_c^p} = 0. \tag{7.28}$$

Let us now fix $\varrho > 1$. Since S is dense in X_c^p there exists a sequence $(h_k)_{k\in\mathbb{N}} \subset S$ such that

$$\lim_{k\to\infty} \|f_\varrho - h_k\|_{X_c^p} = 0. \tag{7.29}$$

Using again Hölder's inequality, we conclude that (7.29) holds for $p = 1$ as well and so

$$\lim_{k\to\infty} \|[f_\varrho]^\wedge_{M_c} - [h_k]^\wedge_{M_c}\|_{L^\infty(c+i\mathbb{R})} = 0.$$

Hence $\lim_{k\to\infty}[h_k]^\wedge_{M_c}(c+it) = [f_\varrho]^\wedge_{M_c}(c+it)$ for every $t \in \mathbb{R}$. By (7.26) we have

$$\|Tf_\varrho - [h_k]^\wedge_{M_c}\|_{L^{p'}(c+i\mathbb{R})} \leq (2\pi)^{1/p'}\|f_\varrho - h_k\|_{X_c^p},$$

and then

$$\lim_{k\to\infty} \|Tf_\varrho - [h_k]^\wedge_{M_c}\|_{L^{p'}(c+i\mathbb{R})} = 0.$$

Thus there exists a subsequence $(h_{k_j})_{j\in\mathbb{N}}$ such that

$$\lim_{j\to\infty} [h_{k_j}]^\wedge_{M_c}(c+it) = (Tf_\varrho)(c+it) \qquad (\text{a.e. } t \in \mathbb{R}).$$

Consequently, $(Tf_\varrho)(c+it) = [f_\varrho]^\wedge_{M_c}(c+it)$ almost everywhere. This extends the operator T to truncated functions as their Mellin transform. Using again (7.26), we obtain by (7.28)

$$\lim_{\varrho\to\infty} \|Tf - [f_\varrho]^\wedge_{M_c}\|_{L^{p'}(c+i\mathbb{R})} \leq \lim_{\varrho\to\infty} (2\pi)^{1/p'}\|f - f_\varrho\|_{X_c^p} = 0.$$

From this the assertion follows. □

We are now ready for defining the Mellin transform of functions in X_c^p when $1 < p < 2$.

Definition 7.13 For $f \in X_c^p$, where $1 < p < 2$, the Mellin transform of f is the unique function $M_c^p[f] \equiv [f]^\wedge_{M_c^p} \in L^{p'}(c + i\mathbb{R})$ such that

$$\lim_{\rho\to\infty} \left\| M_c^p[f](\cdot) - \int_{1/\rho}^{\rho} f(u)u^{(\cdot)} \frac{du}{u} \right\|_{L^{p'}(c+i\mathbb{R})} = 0. \tag{7.30}$$

From the above theory, we can deduce a Mellin version of the classical Titchmarsh inequality of Fourier analysis.

Proposition 7.14 *Let $f \in X_c^p$, where $1 < p < 2$. Then*

$$\|[f]^\wedge_{M_c^p}\|_{L^{p'}(c+i\mathbb{R})} \le (2\pi)^{1/p'} \|f\|_{X_c^p}.$$

Moreover, it is easy to show the consistency property, namely if $f \in X_c \cap X_c^p$, then

$$[f]^\wedge_{M_c} = [f]^\wedge_{M_c^p} \tag{7.31}$$

almost everywhere. The proof is left to the reader as an exercise.

As in the case $p = 2$, the basic properties of the Mellin transform in X_c can be transferred to the general case $1 < p < 2$ by using the consistency property applied to the truncated functions f_ϱ.

For more information, see the notes. Here we will give two important properties concerning the convolution product of a function $f \in X_c^p$ and a function $g \in X_c$, and the formula for the Mellin transform of Mellin derivatives.

The following theorem extends Theorem 6.12 to $1 \le p \le 2$.

Theorem 7.15 *Let $f \in X_c^p$ and $g \in X_c$, where $c \in \mathbb{R}$ and $p \in [1, 2]$. Then $f * g \in X_c^p$ and*

$$[f * g]^\wedge_{M_c^p}(s) = [f]^\wedge_{M_c^p}(s)\,[g]^\wedge_{M_c}(s) \qquad (s = c + it, \text{ a.e. } t \in \mathbb{R}).$$

Proof For $f \in X_c^p \cap X_c$, the result follows immediately from the convolution theorem in X_c and the consistency property. Now let $f \in X_c^p$ only. Then, for the truncated functions f_ϱ, we have $f_\varrho \in X_c^p \cap X_c$ and so

$$[f_\varrho * g]^\wedge_{M_c^p}(s) = [f_\varrho]^\wedge_{M_c^p}(s)\,[g]^\wedge_{M_c}(s)$$

for every $s = c + it$. Using now Minkowski's inequality, Proposition 7.14 and Theorem 4.4, we conclude that

$$\begin{aligned}
&\|[f * g]^{\wedge}_{M_c^p} - [f]^{\wedge}_{M_c^p}\,[g]^{\wedge}_{M_c}\|_{L^{p'}(c+i\mathbb{R})} \\
&\leq \|[f * g]^{\wedge}_{M_c^p} - [f_\varrho * g]^{\wedge}_{M_c^p}\|_{L^{p'}(c+i\mathbb{R})} \\
&\quad + \|[f_\varrho]^{\wedge}_{M_c^p}\,[g]^{\wedge}_{M_c} - [f]^{\wedge}_{M_c^p}\,[g]^{\wedge}_{M_c}\|_{L^{p'}(c+i\mathbb{R})} \\
&\leq (2\pi)^{1/p'}\|(f - f_\varrho) * g\|_{X_c^p} \\
&\quad + \|[g]^{\wedge}_{M_c}\|_{L^\infty(c+i\mathbb{R})}\|[f_\varrho - f]^{\wedge}_{M_c^p}\|_{L^{p'}(c+i\mathbb{R})} \\
&\leq 2(2\pi)^{1/p'}\|g\|_{X_c}\|f - f_\varrho\|_{X_c^p}.
\end{aligned}$$

Since $\lim_{\varrho\to\infty}\|f - f_\varrho\|_{X_c^p} = 0$, the assertion follows. □

The following theorem gives an extension of Theorem 6.14 for the Mellin–Sobolev spaces $W^{r,p}(\mathbb{R}^+)$.

Theorem 7.16 *Let $f \in W_c^{r,p}(\mathbb{R}^+)$, where $c \in \mathbb{R}$, $r \in \mathbb{N}$ and $p \in [1, 2]$. Then*

$$[\Theta_c^r f]^{\wedge}_{M_c^p}(s) = (-it)^r [f]^{\wedge}_{M_c^p}(s) \qquad (s = c + it,\ t \in \mathbb{R}). \tag{7.32}$$

Proof Let $r = 1$. As in the proof of Proposition 3.24, we have

$$(\tau_h^c f)(x) - f(x) = x^{-c}\int_x^{hx} u^c\Theta_c f(u)\frac{du}{u} = \int_{1/h}^1 \Theta_c f\left(\frac{x}{v}\right)v^{-c}\frac{dv}{v}.$$

The last integral is the convolution $\Theta_c f * m_h^c$, where the function $m_h^c \in X^p_{]-\infty,\infty[}$ is defined by (6.13). Therefore, taking the Mellin transforms on both sides, we obtain by Proposition 7.15 and the Mellin transform of the translation operator that

$$(h^{-it} - 1)[f]^{\wedge}_{M_c^p}(s) = [\Theta_c f]^{\wedge}_{M_c^p}(s)\,[m_h^c]^{\wedge}_{M_c}(s) = [\Theta_c f]^{\wedge}_{M_c^p}(s)\frac{h^{-it} - 1}{-it},$$

and so

$$[\Theta_c f]^{\wedge}_{M_c^p}(s) = (-it)[f]^{\wedge}_{M_c^p}(s),$$

which is (7.32) for $r = 1$. The general case can be proved by induction. □

Concerning the inversion theory for the Mellin transform in X_c^p with $1 < p \leq 2$, we only present the following result.

Theorem 7.17 *Let $f \in X_c^p$, where $c \in \mathbb{R}$ and $1 < p \leq 2$. Suppose that $[f]^{\wedge}_{M_c^p} \in L^1(c + i\mathbb{R})$. Then the following inversion formula holds:*

$$f(x) = \frac{1}{2\pi}\int_{-\infty}^{\infty} [f]^{\wedge}_{M_c^p}(c + it)x^{-c-it}dt \qquad (a.e.\ on\ \mathbb{R}^+). \tag{7.33}$$

Moreover, if f is continuous, then this equation is valid for all $x \in \mathbb{R}^+$.

Proof The proof is similar to that of Theorem 6.75 with Theorems 4.4 and 7.15 as auxiliary results. Details are left to the reader. □

For more information about the inversion theory for the Mellin transform in X_c^p spaces, see the subsequent notes.

7.3 Notes

The theory of the Mellin transform in the space X_c^2 was developed by Butzer and Jansche [92]. The material described in Sect. 7.1 is based on [92, Sect. 2]. The definition of a Mellin transform for the spaces X_c^p with $1 < p < 2$ seems to have not been considered in literature in its full generality. For $c = 0$ a treatment with a different approach can be found in Titchmarsh [286, Chap. 4, Sect. 4.14]. The present approach is based on the Riesz–Thorin interpolation theorem (see Zygmund [312, Chap. XII], Butzer and Nessel [95, Sect. 4.4, p. 186]). However, in the Mellin setting the use of the space $C^{\infty}_{\mathrm{comp}}(\mathbb{R}^+)$ instead of simple functions turns out to be more convenient.

In the spaces X_c^2, Tyr and Radouan [293] gained Mellin versions of some classical inequalities, like Jackson type inequalities.

The inversion theory of the Fourier transform in L^p spaces with $1 < p < 2$ appears to be quite intricate since it is based on the theory of the Hilbert transform; see Butzer and Nessel [95, Theorem 8.3.8]. By a standard substitution which leads from the Fourier to the Mellin instance, we can deduce the following result:

Let $f \in X_c^p$, where $1 < p < 2$. Then

$$\lim_{\rho\to\infty} \left\| f(\cdot) - \frac{1}{2\pi}\int_{-\rho}^{\rho} [f]^{\wedge}_{M_c^p}(c+it)(\cdot)^{-c-it}dt \right\|_{X_c^p} = 0.$$

When $p > 2$, the previous approach cannot be used for defining the Mellin transform as a function. In the Fourier theory, this has been shown in the book of Zygmund [312, Chap. XVI, Sect. 3] by appropriately constructed examples. As an efficient alternative, distributions are used in place of functions. Subsequently we sketch this approach.

Distributions

An extension of the integral transformations of Fourier or Mellin to distributions is very useful. It gives access to larger classes of functions, for example to the spaces $L^p(\mathbb{R})$ or X_c^p when $p > 2$, and also to classical singular distributions used in various applications such as the Dirac distribution.

In Fourier analysis a suitable class of test functions on which one can define a (one-dimensional) distributional version of the Fourier transform is the Schwartz class $\mathcal{S}$ consisting of all C^{∞} functions φ on $\mathbb{R}$ such that

$$(1+|x|)^n|\varphi^{(k)}(x)| \le C_{n,k} \qquad (n, k \in \mathbb{N}),$$

where $C_{n,k}$ are constants. In $\mathcal{S}$ it is possible to introduce a notion of convergence in the following way: A sequence $(\varphi_j)_{j\in\mathbb{N}}$ in $\mathcal{S}$ converges to a function $\psi \in \mathcal{S}$ if for every $n, k \in \mathbb{N}$ we have

$$\lim_{j\to\infty} \max_{x\in\mathbb{R}} (1+|x|)^n|\varphi_j^{(k)}(x) - \psi^{(k)}(x)| = 0.$$

A (Fourier) *tempered distribution* is a continuous linear functional T on $\mathcal{S}$. We denote by $\mathcal{S}^*$ the space of all tempered distributions on $\mathcal{S}$, namely the dual space of $\mathcal{S}$. The classical Fourier transform $\widehat{\varphi}$ of $\varphi \in \mathcal{S}$ belongs to $\mathcal{S}$ and the Fourier transform is a continuous bijection of $\mathcal{S}$ on itself. Therefore, we can define the Fourier transform of a distribution as

$$\widehat{T}[\varphi] = T[\widehat{\varphi}] \qquad (\varphi \in \mathcal{S}). \tag{7.34}$$

Now let f be a *tempered function*, that is a Lebesgue measurable function such that

$$|f(x)| \le C(1+|x|)^m \qquad (x \in \mathbb{R})$$

for some $C > 0$ and $m \in \mathbb{N}_0$ or a function $f \in L^p(\mathbb{R})$, where $1 \le p \le \infty$. Then f defines a distribution $T_f[\varphi]$ given by

$$T_f[\varphi] = \int_{\mathbb{R}} f(x)\varphi(x)dx \qquad (\varphi \in \mathcal{S}).$$

Note that the integral exists as a Lebesgue integral. This is obvious in case of tempered functions while for functions in L^p it can be shown with the help of Hölder's inequality (see, e.g., Triebel [288, Sect. 2.3.2, Example 1]). These distributions T_f are called *regular distributions* and they are identified with the generating function f. For this we note that T_f is uniquely determined by f and write $T_f = f$.

The definition of the Fourier transform of a distribution takes inspiration from the classical Parseval equation for Fourier transforms in $L^1(\mathbb{R})$ (see Butzer and Nessel [95, Chap. 5, Proposition 5.1.4]). Indeed, for $f \in L^1(\mathbb{R})$ we have

$$\int_{\mathbb{R}} f(x)\widehat{\varphi}(x)dx = \int_{\mathbb{R}} \widehat{f}(x)\varphi(x)dx$$

for every $\varphi \in \mathcal{S}$. This also suggests the definition (7.34). Thus we can assign a Fourier transform also to all tempered functions f or functions belonging to L^p spaces with $p > 2$, namely the distribution $\widehat{f}$ defined by

$$\widehat{f}[\varphi] = f[\widehat{\varphi}] \qquad (\varphi \in \mathcal{S}).$$

In the Mellin setting the situation is quite different. We cannot imitate the Fourier approach directly since a corresponding form of the Parseval equation with Mellin transforms instead of Fourier transforms is not valid.

A distributional setting of the Mellin transform has been studied mainly by Fung Kang [185], Zemanian [310], Misra and Lavoine [227], (see also Lysik [204], Ziemian [311]). We report on two treatments that can be found in Bertrand et al. [71, Chap. 11, Para. 11.1.3]) and Szmydt and Zemian [284, Chap. 2]).

In the first approach in [71] a class of test functions is defined, which is suitable when one considers Mellin transforms with a strip of type $a_1 < \operatorname{Re} s < a_2$ as analyticity domain. Here we model the definition for the line $c + it$ of the complex plane.

For $c \in \mathbb{R}$, the space of test functions $\mathcal{T}_c$ comprises all C^∞ functions on $\mathbb{R}^+$ for which exists two real numbers ζ_1 and ζ_2 such that for every $k \in \mathbb{N}_0$, the following conditions hold:

$$x^{k+1-c+\zeta_1}\varphi^{(k)}(x) \to 0, \quad x \to 0,$$

$$x^{k+1-c+\zeta_2}\varphi^{(k)}(x) \to 0, \quad x \to \infty.$$

In particular, the functions g_s, defined by $g_s(x) := x^{s-1}$, where $s = c + it$, belong to $\mathcal{T}_c$. Note that this class is quite different from the Schwartz class. Now, every continuous, linear functional on $\mathcal{T}_c$ is called a *Mellin distribution*. If f is such a Mellin distribution, then we denote by $f[\varphi]$ the evaluation of f at $\varphi \in \mathcal{T}_c$, or we may say the action of f on φ. In this notation, the Mellin transform of a Mellin distribution f is simply defined as

$$\mathcal{M}[f](s) := f[g_s].$$

Thus, the Mellin transform of f is a conventional function on the line $c + it$ with $t \in \mathbb{R}$. An interesting example of a Mellin distribution is the Dirac distribution δ_a, where $a > 0$. It is defined by $\delta_a[\varphi] := \varphi(a)$ for $\varphi \in \mathcal{T}_c$. Obviously, it is a Mellin distribution and its Mellin transform is given by

$$\mathcal{M}[\delta_a](s) = \delta_a[g_s] = g_s(a) = a^{s-1}.$$

However, a locally integrable function f can, in general, not be associated with a Mellin distribution. Therefore the notion of regular distribution similar to the Fourier case is not available in this approach.

In the approach by Szmydt and Zemian, the matter is transferred to the Fourier case by using suitable substitutions. The class of test functions is related to the Schwartz class $\mathcal{S}$ of Fourier analysis. Indeed, it is defined by

$$\mathbb{M}_c = \{\varphi \in C^\infty(\mathbb{R}^+) \,:\, \big((\cdot)^{c+1}\varphi(\cdot)\big) \circ \sigma \in \mathcal{S}\},$$

where $c \in \mathbb{R}$ and $\sigma(t) = e^{-t}$ for $t \in \mathbb{R}$. The space of the Mellin distributions is the dual space of $\mathbb{M}_c$ and it is denoted by $\mathbb{M}_c^*$. In [284, Chap. 2, Remark 1, p. 30] it is observed that T is a Mellin distribution if and only if $e^{c(\cdot)}(T \circ \sigma(\cdot))$ is a tempered Fourier distribution. This implies that a Mellin distribution T is regular if and only if the distribution $e^{c(\cdot)}(T \circ \sigma(\cdot))$ on $\mathcal{S}$ is regular.

The construction of the Mellin transform of a Mellin distribution $T \in \mathbb{M}_c^*$, considered in [284], is as follows. Given $c \in \mathbb{R}$, we first define the operators $\mathcal{M}_c : \mathcal{S} \to \mathbb{M}_c$ and $\mathcal{M}_c^{-1} : \mathbb{M}_c \to \mathcal{S}$ by

$$(\mathcal{M}_c\psi)(x) := \int_{-\infty}^{\infty} x^{-c-it-1}\psi(t)dt \qquad (\psi \in \mathcal{S}),$$

$$(\mathcal{M}_c^{-1}\varphi)(t) := \int_{0}^{\infty} x^{c+it}\varphi(x)dx \qquad (\varphi \in \mathbb{M}_c).$$

Both transformations are continuous and mutually inverse. Now the Mellin transform of a Mellin distribution T is the linear functional defined by

$$(M_cT)[\psi] := T[\mathcal{M}_c\psi] \qquad (\psi \in \mathcal{S}).$$

The Mellin transform $M_c : \mathbb{M}_c^* \to \mathcal{S}^*$ is an isomorphism with inverse M_c^{-1} defined by

$$(M_c)^{-1}R[\varphi] := R[(\mathcal{M}_c)^{-1}\varphi] \qquad (R \in \mathcal{S}^*, \ \varphi \in \mathbb{M}_c).$$

The basic properties of the classical Mellin transform can be transferred to distributions. For details see [284].

Chapter 8
Mellin Bandlimited Functions and Paley–Wiener Theorems in Mellin Setting

In this chapter we shall study the structure of the space which comprises all functions $f : \mathbb{R}^+ \to \mathbb{C}$ belonging to some weighted Lebesgue spaces and having a compactly supported Mellin transform.

8.1 Paley–Wiener Spaces in Mellin Setting

The considered spaces can be introduced as follows.

Definition 8.1 Let $c \in \mathbb{R}$ and $T \in \mathbb{R}^+$. The *Mellin–Paley–Wiener space*, denoted by $B^2_{c,T}$, is the space of all functions in $X^2_c \cap C(\mathbb{R}^+)$ such that $[f]^\wedge_{M^2_c}(c+it) = 0$ almost everywhere for $|t| > T$. The functions of $B^2_{c,T}$ are called *Mellin bandlimited* to the interval $[-T, T]$. Analogously, by $B^1_{c,T}$ we denote the Mellin–Paley–Wiener space of all functions $f \in X_c \cap C(\mathbb{R}^+)$ such that $[f]^\wedge_{M_c}(c+it) = 0$ for all $|t| > T$.

Remark 8.2 Note that the slight difference between the above definitions of the Mellin–Paley–Wiener spaces $B^2_{c,T}$ and $B^1_{c,T}$ is due to the fact that the Mellin transform of a function in X^2_c is defined only almost everywhere while for the space X_c it is defined everywhere as a continuous function (see Lemma 6.4). We can also define the Mellin–Paley–Wiener space $B^p_{c,T}$ for $1 < p < 2$ simply by employing the Mellin transform in X^p_c. Here we shall study the classical setting of X^2_c spaces only. By using properties of the Mellin transform in X^p_c, the achieved results can be extended to general p.

We begin with the following inclusions.

Lemma 8.3 *For $c \in \mathbb{R}$ and $T \in \mathbb{R}^+$ we have $B^1_{c,T} \subset B^2_{c,T} \subset X^2_c$.*

C. Bardaro et al., *Mellin Analysis, Transform Theory, and Applications*,
https://doi.org/10.1007/978-3-031-96672-9_8

Proof If $f \in B^1_{c,T}$, then $[f]^\wedge_{M_c} \in L^1(c+i\mathbb{R})$. Therefore, by Theorem 6.75,

$$f(x) = \frac{x^{-c}}{2\pi}\int_{-T}^{T}[f]^\wedge_{M_c}(c+it)x^{-it}dt \qquad (x>0). \tag{8.1}$$

Thus

$$|x^c f(x)| \le \frac{1}{2\pi}\int_{-\infty}^{\infty}|[f]^\wedge_{M_c}(c+it)|dt \le (2\pi)^{-1}\|[f]^\wedge_{M_c}\|_{L^1(c+i\mathbb{R})} =: Q$$

and so

$$\int_0^\infty |x^c f(x)|^2\frac{dx}{x} \le Q\int_0^\infty x^c|f(x)|\frac{dx}{x} = Q\|f\|_{X_c}.$$

Hence $B^1_{c,T} \subset X^2_c$. As an easy consequence of the consistency property of the Mellin transform in X^2_c (see Lemma 7.4), we deduce that $B^1_{c,T} \subset B^2_{c,T}$. The rest is obvious. □

The following result presents the behaviour of a Mellin bandlimited function at the origin and at infinity.

Theorem 8.4 *Let $f \in B^2_{c,T}$, where $c \in \mathbb{R}$ and $T \in \mathbb{R}^+$. Then $f \in C^\infty(\mathbb{R}^+)$ and*

$$\lim_{x\to 0+} x^c\Theta^k_c f(x) = \lim_{x\to\infty} x^c\Theta^k_c f(x) = 0 \qquad (k \in \mathbb{N}_0).$$

Proof Set $F(t) := [f]^\wedge_{M^2_c}(c+it)$. Then $F \in L^1(c+i\mathbb{R})$, as a consequence of f being Mellin bandlimited. Therefore the Mellin inversion formula (see Theorem 7.17) applies and gives

$$f(x) = \frac{1}{2\pi}\int_{-T}^{T}F(t)x^{-c-it}dt = \frac{x^{-c}}{2\pi}\int_{-T}^{T}F(t)e^{-it\log x}dt,$$

or

$$g(x) := x^c f(x) = \frac{1}{2\pi}\int_{-T}^{T}F(t)e^{-it\log x}dt \qquad (x>0). \tag{8.2}$$

Hence by the classical Riemann–Lebesgue lemma (Proposition 1.1), we obtain

$$\lim_{x\to 0+} x^c f(x) = \lim_{x\to\infty} x^c f(x) = 0.$$

Now differentiation of (8.2) yields

$$g'(x) = \frac{-i}{2\pi} \int_{-T}^{T} tF(t)e^{-it\log x} \frac{dt}{x},$$

or

$$xg'(x) = \frac{-i}{2\pi} \int_{-T}^{T} tF(t)e^{-it\log x} dt.$$

Since the function $tF(t)$ is integrable over $[-T, T]$, we may use again the Riemann–Lebesgue lemma to conclude that

$$\lim_{x\to 0+} \Theta g(x) = \lim_{x\to\infty} \Theta g(x) = 0.$$

As $x^c \Theta_c f(x) = \Theta g(x)$, we obtain the assertion for $k = 1$. The general case follows by the same method together with the formula

$$x^c \Theta_c^k f(x) = \Theta^k g(x),$$

which can be verified by induction on k. □

In particular, for $c = 0$, we have $g = f$ and

$$\lim_{x\to 0+} \Theta^k f(x) = \lim_{x\to\infty} \Theta^k f(x) = 0 \qquad (k \in \mathbb{N}_0),$$

which implies that

$$\lim_{x\to 0+} x^k f^{(k)}(x) = \lim_{x\to\infty} x^k f^{(k)}(x) = 0.$$

Now we extend the definition of the integral in (8.2) with $F(t) = [f]^{\wedge}_{M_c^2}(c + it)$ to $z \in \Omega$, where

$$\Omega := \{\zeta \in \mathbb{C} : \zeta \notin \mathbb{R}_0^+\},$$

that is, we define

$$z^c f(z) := \frac{1}{2\pi} \int_{-T}^{T} [f]^{\wedge}_{M_c^2}(c + it)e^{-it\log z} dt \qquad (z \in \Omega), \tag{8.3}$$

which is an analytic function on Ω, as may be proved by using the classical theorem of Morera (see Theorem 1.25). The complex logarithm may be written as $\log z = \log |z| + i \arg(z)$, where $0 \leq \arg(z) < 2\pi$. Therefore $e^{-it\log z} = e^{-it\log|z|}e^{t\arg(z)}$ and so

$$
\begin{aligned}
|z^c f(z)| &\leq \frac{1}{2\pi}\int_{-T}^{T} |[f]^{\wedge}_{M^2_c}(c+it)| e^{t\arg(z)} dt \\
&\leq \frac{e^{T|\arg(z)|}}{2\pi}\int_{-T}^{T} |[f]^{\wedge}_{M^2_c}(c+it)| dt \;=\; Ce^{T|\Im(\log z)|},
\end{aligned}
$$

where C is a constant depending on f and c but not on T and z.

As regards the extension of the function f to the whole complex plane, the following restrictive statement holds.

Theorem 8.5 *Let $f \in B^2_{c,T}$ be not identically zero. Then the function $g(z) := z^c f(z)$ cannot be extended to the whole complex plane as an entire function.*

Proof By (8.3) the function g is analytic on Ω. For $x \in]0, \infty[$, we define

$$
g(x_\pm) := \lim_{\varepsilon\to 0+} \frac{1}{2\pi}\int_{-T}^{T} [f]^{\wedge}_{M^2_c}(c+it) e^{-it\log(x\pm i\varepsilon)} dt
$$

and note that

$$
\lim_{\varepsilon\to 0+} e^{-it\log(x\pm i\varepsilon)} = \begin{cases} e^{-it\log x} & \text{for the upper sign,} \\ e^{-it\log x} e^{2\pi t} & \text{for the lower sign.} \end{cases}
$$

Now assume that the right-hand side of (8.3) has an analytic continuation to the whole of $\mathbb{C}$. Then, interchanging the limit and the integration, we must have $g(x_+) = g(x_-)$, or equivalently,

$$
\frac{1}{2\pi}\int_{-T}^{T} [f]^{\wedge}_{M^2_c}(c+it) e^{-it\log x} e^{\pi t} \sinh(\pi t) dt = 0.
$$

If x runs through $\mathbb{R}^+$, then $u := \log x$ runs through the whole of $\mathbb{R}$. Since f is Mellin bandlimited to $[-T, T]$, we may extend the range of integration to $\mathbb{R}$, i.e.,

$$
\frac{1}{2\pi}\int_{-\infty}^{\infty} [f]^{\wedge}_{M^2_c}(c+it) e^{-itu} e^{\pi t} \sinh(\pi t) dt = 0
$$

for all $u \in \mathbb{R}$. This means that the Fourier transform of $[f]^{\wedge}_{M^2_c}(c+it)e^{\pi t}\sinh(\pi t)$ is identically zero. As a consequence, $[f]^{\wedge}_{M^2_c}(c+it)$ must be zero almost everywhere. Now Lemma 7.5 and the continuity of f imply that f is identically zero. This is a contradiction. □

Remark 8.6 As a consequence of the previous result, the restriction to $\mathbb{R}^+$ of an entire function f cannot be Mellin bandlimited unless f is identically zero.

8.2 Mellin–Paley–Wiener Theorem for Mellin–Bernstein Spaces

We have seen that a Mellin bandlimited function cannot be extended to the whole complex plane as an entire function. However it has an analytic extension on the Riemann surface $S_{\log}$ of the logarithm. Indeed, we shall show that every Mellin bandlimited function has an extension as a function belonging to a Mellin–Bernstein space as introduced in Definition 3.25. Moreover, this property characterizes the Mellin bandlimited functions. The following Mellin–Paley–Wiener theorem holds.

Theorem 8.7 *Every function $f \in B^2_{c,T}$ has a unique extension to a function in $\widetilde{B}^2_{c,T}$ and conversely every function $f \in \widetilde{B}^2_{c,T}$ belongs to the space $B^2_{c,T}$ when it is restricted to $\mathbb{R}^+$.*

Proof Let f be any function from $B^2_{c,T}$. Then, by the inversion formula for the Mellin transform (see Theorem 7.17), we have

$$h(x) := e^{cx} f(e^x) = \frac{1}{2\pi} \int_{-T}^{T} [f]^{\wedge}_{M^2_c}(c+it)e^{-itx}\, dt \qquad (x > 0).$$

Now it follows from the classical Paley–Wiener theorem (see Theorem 1.61) that h has an extension to an entire function such that

$$|h(x+iy)| \le Ce^{T|y|} \qquad (x, y \in \mathbb{R}) \tag{8.4}$$

with some constant $C > 0$ and

$$\lim_{x\to\pm\infty} h(x+iy) = 0 \tag{8.5}$$

uniformly with respect to y on bounded intervals.

Formally, we have

$$g(z) := z^c f(z) = h(\log z).$$

This relation certainly holds for $z \in \mathbb{R}^+$. The logarithm can be extended to its Riemann surface $S_{\log}$, and since h is an entire function, this entails an extension of g on $S_{\log}$. On $\Omega := \{z \in \mathbb{C} : z \notin \mathbb{R}^+_0\}$, the extension of the logarithm has analytic branches

$$\log_k(re^{i\theta}) := \log r + i(2\pi k + \theta) \qquad (r \in \mathbb{R}^+,\ \theta \in \,]0, 2\pi[\,,\ k \in \mathbb{Z}).$$

They induce branches g_k of g, defined by

$$g_k(re^{i\theta}) := h(\log_k(re^{i\theta})) = h(\log r + i(2\pi k + \theta)),$$

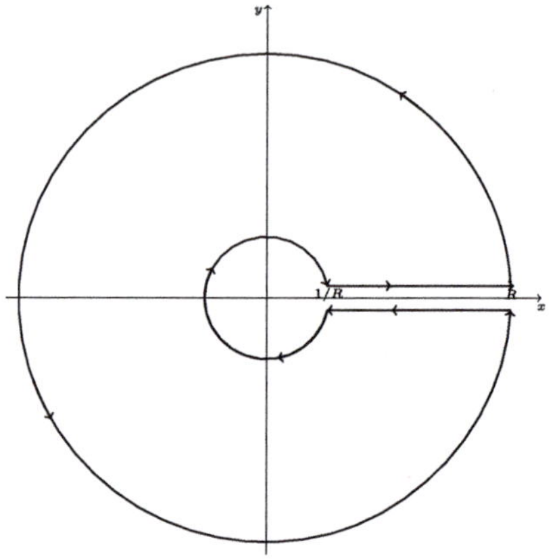

Fig. 8.1 The closed curve $\gamma(R)$

which have the properties (i) and (ii) of Sect. 3.6. As consequences of (8.4) and (8.5), we also see that conditions (3.19) and (3.20) hold. Altogether we have shown that $f \in \widetilde{B}^2_{c,T}$ and so $B^2_{c,T} \subset \widetilde{B}^2_{c,T}$ under these extensions.

Now we prove the converse inclusion. Let $f \in \widetilde{B}^2_{c,T}$. We have to show that $[f]^{\wedge}_{M^2_c}(c+it) = 0$ almost everywhere for $|t| > T$. For any $k \in \mathbb{Z}$ and $z = re^{i\theta}$ with $0 \le \theta < 2\pi$, we define

$$F_k(z) := \frac{g_k(z)}{z} \exp\bigl(it(\log|z| + i(2\pi k + \theta))\bigr),$$

which is analytic in Ω since the argument of the exponential function is an analytic branch of $it \log z$. Hence the integral of F_k along the closed curve $\gamma(R)$ as shown in Fig. 8.1 vanishes.

We may even let the two line segments of $\gamma(R)$ coincide on $\mathbb{R}$ provided that we replace the integrand by

$$F_k^+(x) := \frac{g_k^+(x)}{x} \exp(it(\log x + i2\pi k)),$$

for the integration from $1/R$ to R, and by

$$\begin{aligned} F_k^-(x) &:= \frac{g_k^-(x)}{z} \exp(it(\log x + i2\pi(k+1))) \\ &= \frac{g_{k+1}^+(x)}{z} \exp(it(\log x + i2\pi(k+1))) = F_{k+1}^+(x), \end{aligned}$$

for the integration from R to $1/R$. Now, for any $\rho > 0$, we consider the circle $\lambda(\rho)$ given by $z = \rho e^{i\theta}$, for $0 \le \theta \le 2\pi$, and define

$$I_k(\rho) := \int_{\lambda(\rho)} F_k(z)dz = i \int_0^{2\pi} g_k(\rho e^{i\theta}) e^{it\log\rho} e^{-t(2\pi k+\theta)} d\theta$$

and

$$J_k(\rho) := \int_{1/\rho}^{\rho} F_k^+(u)du = \int_{1/\rho}^{\rho} g_k^+(u)e^{it\log u}e^{-t2\pi k}\frac{du}{u}.$$

Then

$$0 = \int_{\gamma(R)} F_k(z)dz = J_k(R) + I_k(R) - J_{k+1}(R) - I_k(1/R),$$

and so

$$J_k(R) - J_{k+1}(R) = I_k(1/R) - I_k(R). \tag{8.6}$$

Summing over $k = 0, 1, \ldots n-1$, we obtain by (i) of Sect. 3.6

$$\int_{1/R}^{R} g(u)e^{it\log u}\frac{du}{u} = J_n(R) + \sum_{k=0}^{n-1}\big(I_k(1/R) - I_k(R)\big). \tag{8.7}$$

First let $t > T$ and let $\varepsilon > 0$ be fixed. For sufficiently large R, we choose $n := \lfloor \log R \rfloor$, and estimate the terms on the right-hand side of (8.7). Employing (3.19), we find that

$$|J_n(R)| \leq Ce^{-(t-T)2\pi n}\int_{1/R}^{R}\frac{du}{u} = Ce^{-(t-T)2\pi n}\log R^2,$$

and so by our choice of n the term $J_n(R)$ approaches 0 as $R \to \infty$.

Next we estimate $I_k(\rho)$. Using again (3.19), we conclude that

$$|I_k(\rho)| \leq C\int_0^{2\pi} e^{-(t-T)(2\pi k+\theta)}d\theta = \frac{C}{t-T}e^{-2\pi k(t-T)}\left(1 - e^{-2\pi(t-T)}\right).$$

Thus for an integer $n_0 \in \,]0, n-1[$, we obtain

$$\begin{aligned}\sum_{k=n_0}^{n-1}|I_k(1/R) - I_k(R)| &\leq \frac{2C}{t-T}\left(1 - e^{-2\pi(t-T)}\right)\sum_{k=n_0}^{\infty}e^{-2\pi k(t-T)}\\ &= \frac{2C}{t-T}e^{-2\pi n_0(t-T)}.\end{aligned}$$

Now, the right-hand side can be made smaller than ε by fixing n_0 as a sufficiently large integer, which is possible since R and so n shall tend to infinity. Finally, by (3.20) we have

$$\lim_{R\to\infty}\sum_{k=0}^{n_0-1}(I_k(1/R) - I_k(R)) = 0.$$

Altogether, we have proved that for $t > T$,

$$\lim_{R\to\infty}\left|\int_{1/R}^{R} g(u)e^{it\log u}\,\frac{du}{u}\right| < \varepsilon,$$

for any $\varepsilon > 0$. This implies that $[f]^\wedge_{M^2_c}(c+it) = 0$ almost everywhere for $t > T$. When $t < -T$, we choose $n = -\lfloor \log R \rfloor$, sum in (8.7) from n to -1 and proceed analogously. □

Employing the Mellin–Bernstein inequality (Proposition 3.28), we can give an alternative, shorter proof for the second part of the Mellin–Paley–Wiener theorem by using real variable arguments only. We begin with the following remark.

Remark 8.8 The support of the Mellin transform of functions in X^2_c can be defined as the set of all points $t \in \mathbb{R}$ that possess no neighbourhood throughout which $[f]^\wedge_{M^2_c}(c+it) = 0$ almost everywhere. We will denote it by $\operatorname{supp}[f]^\wedge_{M^2_c}$. The support is always a closed set.

The inclusion $\widetilde{B}^2_{c,T} \subset B^2_{c,T}$ for the restrictions of functions to $\mathbb{R}^+$ can now be stated as follows and be verified by a shorter proof.

Theorem 8.9 *If $f \in \widetilde{B}^2_{c,T}$, then* $\operatorname{supp}[f]^\wedge_{M^2_c} \subset [-T, T]$.

Proof Let $f \in \widetilde{B}^2_{c,T}$. By Proposition 3.28, we have $\Theta^r_c f \in X^2_c$ for every $r \in \mathbb{N}$ and

$$\limsup_{r\to\infty} \|\Theta^r_c f\|^{1/r}_{X^2_c} \le T \limsup_{r\to\infty} \|f\|^{1/r}_{X^2_c} = T. \tag{8.8}$$

Now let $t_0 \in \operatorname{supp}[f]^\wedge_{M^2_c}$, where $t_0 \neq 0$, and let $\varepsilon > 0$ be such that $|t_0| - \varepsilon > 0$. By Lemma 7.5 and Theorem 7.16, we obtain

$$\begin{aligned}
\|\Theta^r_c f\|^2_{X^2_c} &= \frac{1}{2\pi}\|[\Theta^r_c f]^\wedge_{M^2_c}(c+i\cdot)\|^2_{L^2(c+i\mathbb{R})}\\
&= \frac{1}{2\pi}\|(-i\cdot)^r[f]^\wedge_{M^2_c}(c+i\cdot)\|^2_{L^2(c+i\mathbb{R})}\\
&= \frac{1}{2\pi}\int_{-\infty}^{\infty} |t|^{2r}|[f]^\wedge_{M^2_c}(c+it)|^2 dt\\
&\ge \frac{(|t_0|-\varepsilon)^{2r}}{2\pi}\int_{|t|>|t_0|-\varepsilon} |[f]^\wedge_{M^2_c}(c+it)|^2 dt.
\end{aligned}$$

Therefore, setting $H_\varepsilon := \int_{|t|>|t_0|-\varepsilon} |[f]^\wedge_{M^2_c}(c+it)|^2 dt$, we have $H_\varepsilon \neq 0$ and

$$(|t_0|-\varepsilon)^{2r} \le \frac{2\pi}{H_\varepsilon}\|\Theta^r_c f\|^2_{X^2_c},$$

and then

$$|t_0| - \varepsilon \le \left(\frac{2\pi}{H_\varepsilon}\right)^{1/2r} \|\Theta_c^r f\|_{X_c^2}^{1/r}.$$

Finally, passing to the lim sup and employing (8.8), we deduce that $|t_0| \le T$. □

8.3 Polar Form of the Mellin–Paley–Wiener Theorem

In this section we present and prove a version of the Mellin–Paley–Wiener theorem by employing the Mellin–Bernstein spaces $\mathscr{B}^2_{c,T}$ of polar-analytic functions (see Sect. 3.8). This approach is simpler than the previous one since we do not need Riemann surfaces and analytic branches.

Theorem 8.10 (Mellin–Paley–Wiener) *A function $\varphi : \mathbb{R}^+ \to \mathbb{C}$ belongs to the Mellin–Paley–Wiener space $B^2_{c,T}$ if and only if there exists a function $f \in \mathscr{B}^2_{c,T}$ such that $f(\cdot, 0) = \varphi(\cdot)$.*

Proof First suppose that $\varphi \in B^2_{c,T}$. Then, by the inversion formula,

$$\varphi(r) = \frac{1}{2\pi}\int_{-T}^{T} [\varphi]^\wedge_{M_c^2}(c+it) r^{-c-it} dt \qquad (r > 0).$$

Now we extend φ to the (r, θ)-plane $\mathbb{H}$ with $r > 0$ and $\theta \in \mathbb{R}$ by replacing r with $re^{i\theta}$ on the right-hand side. Denoting this extension by $f(r, \theta)$, we have

$$f(r,\theta) := \frac{1}{2\pi}\int_{-T}^{T} [\varphi]^\wedge_{M_c^2}(c+it)(re^{i\theta})^{-c-it} dt. \tag{8.9}$$

This implies that

$$|f(r,\theta)| \le \frac{r^{-c}e^{T|\theta|}}{2\pi}\int_{-T}^{T} |[\varphi]^\wedge_{M_c^2}(c+it)| dt.$$

Hence (iii) of Definition 3.41 holds with

$$C_f := \frac{1}{2\pi}\int_{-T}^{T} |[\varphi]^\wedge_{M_c^2}(c+it)| dt = \frac{1}{2\pi}\int_{-\infty}^{\infty} |[\varphi]^\wedge_{M_c^2}(c+it)| dt.$$

Concerning polar analyticity of f, we choose an arbitrary point $(r_0, \theta_0) \in \mathbb{H}$ and consider the difference quotient

$$\frac{f(r,\theta) - f(r_0,\theta_0)}{re^{i\theta} - r_0e^{i\theta_0}} = \frac{1}{2\pi}\int_{-T}^{T} [\varphi]^\wedge_{M_c^2}(c+it)\frac{(re^{i\theta})^{-c-it} - (r_0e^{i\theta_0})^{-c-it}}{re^{i\theta} - r_0e^{i\theta_0}} dt.$$

The limit $(r, \theta) \to (r_0, \theta_0)$ carried out inside the integral leads to an ordinary differentiation of z^{-c-it} with respect to z at the point $z_0 := r_0 e^{i\theta_0}$. However, we have to justify that the limit and the integration can be interchanged. For this, we take a closed rectangle Q centered at the point (r_0, θ_0) such that $Q \subset \mathbb{H}$. Now the function $h : Q \times [-T, T] \to \mathbb{C}$, defined by

$$h(r, \theta, t) := \frac{(re^{i\theta})^{-c-it} - (r_0 e^{i\theta_0})^{-c-it}}{re^{i\theta} - r_0 e^{i\theta_0}} \qquad \text{for } (r, \theta) \neq (r_0, \theta_0)$$

and

$$h(r_0, \theta_0, t) := -(c + it)(r_0 e^{i\theta_0})^{-c-1-it},$$

is continuous and therefore its absolute value is bounded on $Q \times [-T, T]$. Hence the desired interchange is guaranteed by Lebesgue's dominated convergence theorem (see Proposition 1.4). Thus we have proved that also (i) of Definition 3.41 holds while (ii) is obvious. Altogether we have shown that f belongs to $\mathscr{B}^2_{c,T}$.

Now we prove the reverse implication. Let $f \in \mathscr{B}^2_{c,T}$. For any fixed $t \in \mathbb{R}$, define the function

$$g(r, \theta) := e^{(c-1+it)(\log r + i\theta)} f(r, \theta),$$

which is polar-analytic on $\mathbb{H}$.

Let us first assume that $t > T$. For $R > 1$, we define $n := \lfloor \log R \rfloor$ and consider the piecewise smooth curve γ which is the boundary of the rectangle in $\mathbb{H}$ defined by $[1/R, R] \times [0, 2n\pi]$. Then, using Theorem 2.6, we have

$$\int_\gamma g(r, \theta) e^{i\theta} (dr + ird\theta) = 0.$$

This easily implies that

$$\begin{aligned}\int_{1/R}^{R} r^{c+it-1} \varphi(r) dr &= \int_{1/R}^{R} e^{(c-1+it)(\log r + 2n\pi i)} f(r, 2n\pi) dr \\ &\quad + \frac{i}{R} \int_0^{2n\pi} e^{(c-1+it)(-\log R + i\theta) + i\theta} f\left(\frac{1}{R}, \theta\right) d\theta \\ &\quad - iR \int_0^{2n\pi} e^{(c-1+it)(\log R + i\theta) + i\theta} f(R, \theta) d\theta \\ &=: I_1 + I_2 - I_3,\end{aligned}$$

where φ is the restriction of f to $\mathbb{R}^+$.

We now estimate the integrals on the right-hand side. Using (iii) of Definition 3.41, we find that

$$|I_1| \le \int_{1/R}^{R} e^{(c-1)\log r} e^{-2n\pi t} |f(r, 2\pi n)| dr$$

$$\le C_f \int_{1/R}^{R} e^{-(t-T)2n\pi} \frac{dr}{r} = C_f e^{-(t-T)2n\pi} \log R^2.$$

Thus, according to our choice of n, we obtain $|I_1| \to 0$ as $R \to \infty$.

As to I_2, we easily see that

$$|I_2| \le \int_0^{2n\pi} \frac{1}{R^c} \left| f\left(\frac{1}{R}, \theta\right) \right| e^{-t\theta} d\theta \le \int_0^{\infty} \frac{1}{R^c} \left| f\left(\frac{1}{R}, \theta\right) \right| e^{-t\theta} d\theta.$$

Now, since the integrand is dominated by $C_f e^{-(t-T)\theta}$, which is integrable as a function of θ, we conclude with the help of statement (b) of Theorem 3.42 and the theorem of dominated convergence that $|I_2| \to 0$ as $R \to \infty$. The same result is obtained for $|I_3|$. Altogether we have

$$\lim_{R\to\infty} \left| \int_{1/R}^{R} r^{c-1+it} \varphi(r) dr \right| = 0.$$

This implies that $[\varphi]^{\wedge}_{M_c^2}(c + it) = 0$ almost everywhere for $t > T$.

When $t < -T$, we choose $n := -\lfloor \log R \rfloor$ and proceed analogously using the boundary of the rectangle $[1/R, R] \times [-2n\pi, 0]$. Hence $\varphi(\cdot) := f(\cdot, 0)$ belongs to $B^2_{c,T}$. □

8.4 Mellin–Paley–Wiener Theorem for Exponentially Decaying Functions

In this section we aim at a generalization of the Paley–Wiener theorem for the Mellin setting which will later be applied for characterizing function spaces in terms of their distance from a Mellin–Bernstein space. We shall need the following space $H_c^*(\mathbb{H}_a)$, which lies between two Mellin–Hardy spaces (see Definition 3.49).

Definition 8.11 Let $a, c \in \mathbb{R}$ with $a > 0$ be fixed numbers. The class $H_c^*(\mathbb{H}_a)$ comprises all functions $f : \mathbb{H}_a \to \mathbb{C}$ with the following properties:

(a) f is polar-analytic on $\mathbb{H}_a$;
(b) $\lim_{r\to 0+} r^c f(r, 0) = \lim_{r\to\infty} r^c f(r, 0) = 0$;

(c) for every $\varepsilon \in]0, a[$ there exists a constant $K(f, \varepsilon)$ such that

$$|f(r, \theta)| \leq r^{-c} K(f, \varepsilon)$$

for all $(r, \theta) \in \mathbb{H}_{a-\varepsilon}$;

(d) for every $\theta \in]-a, a[$ and all $t \in \mathbb{R}$,

$$I_c(f, \theta, t) := \lim_{R \to \infty} \int_{1/R}^{R} f(r, \theta) r^{c+it-1} dr$$

exists and $|I_c(f, \theta, t)| \leq K(f)$ with a constant depending on f only.

For the subsequent considerations, we need some lemmas concerning the $(C, 1)$-summability of the Mellin inversion in L^2 and a result on the inversion of the Cauchy principal value of a Mellin transform.

Lemma 8.12 *Let $G(c + i\cdot)$ be a function in $L^2(c + i\mathbb{R})$. Then, denoting by $g \in X_c^2$ the inverse Mellin transform of G, one has*

$$g(r) = \lim_{\rho \to \infty} \frac{1}{2\pi} \int_{-\rho}^{\rho} \left(1 - \frac{|t|}{\rho}\right) G(c + it) r^{-c-it} dt \qquad (a.e.\ r \in \mathbb{R}^+).$$

Proof First, by Theorem 7.10, the inverse Mellin transform $g \in X_c^2$ exists. Next we want to employ the Mellin–Fejér kernel, defined by

$$F_\rho^c(r) := -\frac{1}{2\pi\rho} r^{-c} \left(\frac{r^{\rho i/2} - r^{-\rho i/2}}{\log r} \right)^2 \quad (r \neq 1) \quad \text{and} \quad F_\rho^c(1) = \frac{\rho}{2\pi}$$

(see Definition 6.83). Introducing the function $g_c : r \mapsto r^c g(r)$, we obtain by Mellin convolution

$$\begin{aligned}(F_\rho^c * g)(r) &= \int_0^\infty F_\rho^c\left(\frac{r}{u}\right) g(u) \frac{du}{u} = r^{-c} \int_0^\infty F_\rho^0\left(\frac{r}{u}\right) g_c(u) \frac{du}{u} \\ &= r^{-c} \int_0^\infty F_\rho^0(t) g_c(rt) \frac{dt}{t},\end{aligned}$$

where we have used that $F_\rho^0(r/u) \equiv F_\rho^0(u/r)$ for all $r, u \in \mathbb{R}^+$. It is easily seen that $g_c \in X_0^2 \cap X_{0,\mathrm{loc}}$. Therefore, by Theorem 6.65 in conjunction with Proposition 6.64 and Remark 6.67, we conclude that the last integral on the right-hand side converges to $g_c(r)$ almost everywhere for $r \in \mathbb{R}^+$ as $\rho \to \infty$. Hence

$$\lim_{\rho \to \infty} (F_\rho^c * g)(r) = g(r) \qquad (\text{a.e. } r \in \mathbb{R}^+). \tag{8.10}$$

Since the Mellin transform of F_ρ^c is given by

$$[F_\rho^c]^\wedge_{M_c}(c+it) = \begin{cases} \left(1 - \dfrac{|t|}{\rho}\right) & \text{if } 0 \le |t| \le \rho, \\ 0 & \text{if } |t| > \rho, \end{cases}$$

we find with the help of Theorems 7.15 and 7.17 that

$$(F_\rho^c * g)(r) = \frac{1}{2\pi} \int_{-\rho}^{\rho} \left(1 - \frac{|t|}{\rho}\right) G(c+it) r^{-c-it} dt \qquad (\text{a.e. } r \in \mathbb{R}^+).$$

Combined with (8.10), the assertion follows. □

Lemma 8.13 *Suppose that $\phi : \mathbb{R}^+ \to \mathbb{C}$ is integrable on every compact subinterval and*

$$G(c+it) := \lim_{R\to\infty} \int_{1/R}^{R} \phi(r) r^{c-1+it} dr$$

exists for each $t \in \mathbb{R}$ and is integrable on compact subintervals. Then,

$$\phi(r) = \lim_{\rho\to\infty} \frac{1}{2\pi} \int_{-\rho}^{\rho} \left(1 - \frac{|t|}{\rho}\right) G(c+it) r^{-c-it} dt$$

almost everywhere for $r \in \mathbb{R}^+$.

We do not give a proof of this lemma but refer to accompanying comments in Notes 8.5. Combining Lemmas 8.12 and 8.13, we obtain the following basic lemma.

Lemma 8.14 *Let $\phi : \mathbb{R}^+ \to \mathbb{C}$ be integrable on every compact subinterval of $\mathbb{R}^+$, and suppose that*

$$M_c^*[\phi](c+it) := \lim_{R\to\infty} \int_{1/R}^{R} \phi(r) r^{c+it-1} dr$$

exists for all $t \in \mathbb{R}$. If $M_c^*[\phi] \in L^2(c+i\mathbb{R})$, then $\phi \in X_c^2$ and $M_c^*[\phi](c+it) = M_c^2[\phi](c+it) \equiv [\phi]^\wedge_{M_c^2}(c+it)$ almost everywhere for $t \in \mathbb{R}$.

Proof If $M_c^*[\phi] \in L^2(c+i\mathbb{R})$, then by Theorem 7.10, there exists a function $\psi \in X_c^2$ such that $M_c^2[\psi] = M_c^*[\phi]$. Employing Lemma 8.12, we conclude that

$$\psi(r) = \lim_{\rho\to\infty} \frac{1}{2\pi} \int_{-\rho}^{\rho} \left(1 - \frac{|t|}{\rho}\right) M_c^*[\phi](c+it) r^{-c-it} dt \qquad (\text{a.e. } r \in \mathbb{R}^+).$$

On the other hand, by Lemma 8.13, we have

$$\phi(r) = \lim_{\rho\to\infty} \frac{1}{2\pi} \int_{-\rho}^{\rho} \left(1 - \frac{|t|}{\rho}\right) M_c^*[\phi](c+it) r^{-c-it} dt \qquad (\text{a.e. } r \in \mathbb{R}^+).$$

Hence $\phi(r) = \psi(r)$ almost everywhere for $r \in \mathbb{R}^+$. Now the assertion follows immediately. □

Remark 8.15 Using the notation of Lemma 8.14, we may express the quantity figuring in statement (d) of Definition 8.11 as

$$I_c(f, \theta, t) = M_c^*[f(\cdot, \theta)](c + it).$$

With the help of Lemma 8.14, we can easily see that $H_c^*(\mathbb{H}_a)$ constitutes a normed linear space.

Proposition 8.16 *Let $f \in H_c^*(\mathbb{H}_a)$. By introducing*

$$\| f \|_{H_c^*(\mathbb{H}_a)} := \sup\{|I_c(f, \theta, t)| \,;\, \theta \in]-a, a[\,, t \in \mathbb{R}\},$$

the class $H_c^*(\mathbb{H}_a)$ becomes a normed linear space.

Proof As the only non-trivial assertion, we have to show that if $f \in H_c^*(\mathbb{H}_a)$ and $\|f\|_{H_c^*(\mathbb{H}_a)} = 0$, then $f(r, \theta) = 0$ for all $(r, \theta) \in \mathbb{H}_a$. Now, if $\|f\|_{H_c^*(\mathbb{H}_a)} = 0$, then, clearly, $I_c(f, \theta, t) = 0$ for all $t \in \mathbb{R}$. In view of Lemma 8.14, we obtain that $M_c^2[f(\cdot, \theta)](c + it) = 0$ almost everywhere for $t \in \mathbb{R}$, and application of Lemma 7.5 shows that $f(r, \theta) = 0$ almost everywhere for $r \in \mathbb{R}^+$. Since a polar-analytic function is continuous, we conclude that $f(r, \theta) = 0$ for all $(r, \theta) \in \mathbb{H}_a$. □

Next we want to show that in Definition 8.11, statement (b) can be generalized by employing (a) and (c).

Proposition 8.17 *Let $f : \mathbb{H}_a \to \mathbb{C}$ satisfy conditions (a)–(c) of Definition 8.11. Then*

$$\lim_{r\to 0+} r^c f(r, \theta) = \lim_{r\to\infty} r^c f(r, \theta) = 0$$

uniformly with respect to θ on all compact subintervals of $]-a, a[$.

Proof For $x \in \mathbb{R}$ and $y \in]-a, a[$, we define

$$F(x + iy) := e^{c(x+iy)} f(e^x, y).$$

Then (a) of Definition 8.11 implies that F is analytic in the strip $S_a := \{z \in \mathbb{C} : |\Im z| < a\}$ and (c) shows that for every $\varepsilon \in]0, a[$ it is bounded on $S_{a-\varepsilon}$.

Furthermore, by (b) we have $F(x) \to 0$ as $x \to \pm\infty$. Then, by Proposition 1.31, it follows that $F(x+iy) \to 0$ as $x \to \pm\infty$ uniformly with respect to y on compact subintervals of $]-a, a[$. Expressing this conclusion in terms of f, we arrive at the desired result. □

Now we want to compare $H_c^*(\mathbb{H}_a)$ with $H_c^1(\mathbb{H}_a)$.

Proposition 8.18 *We have*

$$H_c^1(\mathbb{H}_a) \subsetneqq H_c^*(\mathbb{H}_a) \quad \text{and} \quad \|f\|_{H_c^*(\mathbb{H}_a)} \le 2\|f\|_{H_c^1(\mathbb{H}_a)} \quad (f \in H_c^1(\mathbb{H}_a)). \tag{8.11}$$

Proof Let $f \in H_c^1(\mathbb{H}_a)$. First we have to show that f satisfies conditions (a)–(d) of Definition 8.11. Clearly, (a) is guaranteed by the definition of $H_c^1(\mathbb{H}_a)$. Furthermore, (b) and (c) follow from Proposition 3.52 and Proposition 3.50, respectively. As regards (d), we note that $I_c(f, \theta, t)$ exists by the hypotheses of $H_c^1(\mathbb{H}_a)$ and

$$|I_c(f, \theta, t)| \le \lim_{R\to\infty} \int_{1/R}^{R} |f(r, \theta)| r^{c-1} dr = \|f(\cdot, \theta)\|_{X_c} \le 2\|f\|_{H_c^1(\mathbb{H}_a)}.$$

This not only verifies (d) but proves (8.11) as well.

The inclusion in (8.11) is strict since it can be shown that

$$f(r, \theta) := \frac{r^{-c} e^{-ic\theta}}{a + i(\log r + i\theta)}$$

belongs to $H_c^*(\mathbb{H}_a)$ but $f \notin H_c^1(\mathbb{H}_a)$. □

We can also compare $H_c^*(\mathbb{H}_a)$ with a corresponding Hardy type space for L^2 norms. The following proposition implies that $H_c^*(\mathbb{H}_a)$ is a subspace of $H_c^2(\mathbb{H}_{a-\varepsilon})$ for every $\varepsilon \in]0, a[$.

Proposition 8.19 *Let $f \in H_c^*(\mathbb{H}_a)$. Then, for all $\alpha \in]-a, a[$, we have*

$$|M_c^*[f(\cdot, \alpha)](c + it)| \le e^{-|t|(a-|\alpha|)} \|f\|_{H_c^*(\mathbb{H}_a)}, \tag{8.12}$$

$f(\cdot, \alpha) \in X_c^2$ and

$$\|f(\cdot, \alpha)\|_{X_c^2} \le \frac{1}{\sqrt{2\pi(a - |\alpha|)}} \|f\|_{H_c^*(\mathbb{H}_a)}. \tag{8.13}$$

Proof We proceed in a way similar to the proof of Theorem 8.10. For $f \in H_c^*(\mathbb{H}_a)$ and $t \in \mathbb{R}$, we consider the function g defined by

$$g(r, \theta) := e^{(c-1+it)(\log r + i\theta)} f(r, \theta),$$

which is polar-analytic on $\mathbb{H}_a$. For $R > 1$ and $\alpha \in]-a, a[\setminus\{0\}$, let γ be the positively oriented rectangular curve with vertices at $(1/R, 0)$, $(R, 0)$, $(1/R, \alpha)$ and (R, α). By Theorem 2.6, we have

$$\int_\gamma g(r,\theta)e^{i\theta}(dr + ird\theta) = 0.$$

This equation may be written in terms of ordinary integrals as

$$\int_{1/R}^{R} g(r,0)dr = \int_{1/R}^{R} g(r,\alpha)e^{i\alpha}dr + I(1/R,t) - I(R,t), \tag{8.14}$$

where, for any $r > 0$,

$$I(r,t) := \int_0^\alpha g(r,\theta)ire^{i\theta}d\theta = ir^{c+it}\int_0^\alpha e^{-(t-ic)\theta}f(r,\theta)d\theta.$$

Using statement (d) of Definition 8.11, we find that the first two integrals in (8.14) converge as $R \to \infty$ while the last two expressions approach 0 as a consequence of Proposition 8.17. More precisely, in the notation of Definition 8.11, the limit $R \to \infty$ in (8.14) leads to

$$I_c(f,0,t) = e^{i\alpha(c+it)}I_c(f,\alpha,t)$$

and so

$$|I_c(f,0,t)| = e^{-\alpha t}|I_c(f,\alpha,t)|. \tag{8.15}$$

This holds for any $\alpha \in]-a, a[\setminus\{0\}$. Hence, given $t \neq 0$, there exist a β with $|\beta| = |\alpha|$ and $\beta t > 0$ such that (8.15) holds with β in place of α. Since $|I_c(f,\beta,t)| \leq \|f\|_{H_c^*(\mathbb{H}_a)}$, we conclude that

$$|I_c(f,0,t)| \leq e^{-|\beta t|}\|f\|_{H_c^*(\mathbb{H}_a)}.$$

Letting $|\beta|$ approach a, we obtain

$$|I_c(f,0,t)| \leq e^{-a|t|}\|f\|_{H_c^*(\mathbb{H}_a)}.$$

Combining this with (8.15), we arrive at

$$|I_c(f,\alpha,t)| \leq e^{-|t|(a-|\alpha|)}\|f\|_{H_c^*(\mathbb{H}_a)}. \tag{8.16}$$

Since $I_c(f,\alpha,t) = M_c^*[f(\cdot,\alpha)](c+it)$, we have proved (8.12).

Using Lebesgue's dominated convergence theorem, we conclude from (8.16) that $I_c(f,\alpha,\cdot) \in L^2(\mathbb{R})$ and

$$\|I_c(f,\alpha,\cdot)\|_{L^2(\mathbb{R})} \le \frac{1}{\sqrt{a-|\alpha|}}\|f\|_{H_c^*(\mathbb{H}_a)}.$$

Now Lemma 8.14 implies that $f(\cdot,\alpha) \in X_c^2$ for $|\alpha| < a$. Finally, the same lemma and Lemma 7.5 lead us to inequality (8.13). □

We are ready for the announced generalization of the Paley–Wiener theorem in the Mellin setting.

Theorem 8.20 *A continuous function $\phi : \mathbb{R}^+ \to \mathbb{C}$ is the restriction to $\mathbb{R}^+$ of a function $f \in H_c^*(\mathbb{H}_a)$ if and only if $M_c^*[\phi]$, as defined in Lemma 8.14, exists and*

$$|M_c^*[\phi](c+it)| \le Ce^{-a|t|} \qquad (t \in \mathbb{R}) \tag{8.17}$$

with a constant C that may be taken to be $\|f\|_{H_c^*(\mathbb{H}_a)}$.

Proof If ϕ is the restriction to $\mathbb{R}^+$ of a function $f \in H_c^*(\mathbb{H}_a)$, then Proposition 8.19 applies and (8.12) for $\alpha = 0$ shows that (8.17) holds with $C = \|f\|_{H_c^*(\mathbb{H}_a)}$.

Conversely, suppose that $M_c^*[\phi]$ as defined in Lemma 8.14 exists and (8.17) holds. Then, as a function of t, the expression $M_c^*[\phi](c+it)e^{i\alpha t}$ belongs to $L^1(\mathbb{R}) \cap L^2(\mathbb{R})$ whenever $|\alpha| < a$. Therefore,

$$f(r,\theta) := \frac{1}{2\pi}\int_{-\infty}^{\infty} M_c^*[\phi](c+it)(re^{i\theta})^{-c-it}dt$$

exists for $(r,\theta) \in \mathbb{H}_a$. By Lemma 8.14 and the Mellin inversion theorem in X_c^2, we obtain that $f(r,0) = \phi(r)$ almost everywhere on $\mathbb{R}^+$. Since $M_c^*[\phi](c+i\cdot) \in L^1(\mathbb{R})$, we conclude that $f(\cdot,0)$ is continuous. Hence $f(r,0) = \phi(r)$ for every $r \in \mathbb{R}^+$.

Now we want to show that f satisfies (a)–(d) of Definition 8.11. By a refinement of the proof of Theorem 8.10, one can verify that f is polar-analytic in $\mathbb{H}_a$. For the reader's convenience we give details. Let (r_0,θ_0) be any point in $\mathbb{H}_a$. Denote by Q a sufficiently small closed rectangle centered at (r_0,θ_0) such that $Q \subset \mathbb{H}_a$. We consider the difference quotient

$$\frac{f(r,\theta)-f(r_0,\theta_0)}{re^{i\theta}-r_0e^{i\theta_0}} = \frac{1}{2\pi}\int_{-\infty}^{\infty} M_c^*[\phi](c+it)\frac{(re^{i\theta})^{-c-it}-(r_0e^{i\theta_0})^{-c-it}}{re^{i\theta}-r_0e^{i\theta_0}}dt. \tag{8.18}$$

For $(r,\theta) \in Q$, the limit $(r,\theta) \to (r_0,\theta_0)$ carried out inside the integral leads to an ordinary differentiation of z^{-c-it} with respect to z at the point $z_0 := r_0e^{i\theta_0}$ and the integral with the derivative in place of the difference quotient exists. However, we have to justify that the limit and the integration can be interchanged. For this purpose, we first note that the difference quotient on the right-hand side of (8.18) can be expressed as

$$\frac{(re^{i\theta})^{-c-it}-(r_0e^{i\theta_0})^{-c-it}}{re^{i\theta}-r_0e^{i\theta_0}} = -(c+it)\int_0^1 (r_0e^{i\theta_0}+s(re^{i\theta}-r_0e^{i\theta_0}))^{-c-1-it}ds,$$

which implies that

$$\left|\frac{(re^{i\theta})^{-c-it} - (r_0e^{i\theta_0})^{-c-it}}{re^{i\theta} - r_0e^{i\theta_0}}\right| \le \sqrt{c^2+t^2} \sup_{(r,\theta)\in Q} \left|\left(re^{i\theta}\right)^{-c-1+it}\right|$$
$$\le \frac{\sqrt{c^2+t^2}}{r_1^{c+1}} e^{\theta_1|t|}$$

for some $r_1 > 0$ and $\theta_1 \in]0, a[$. Thus, for $(r, \theta) \in Q$ the absolute value of the integrand in (8.18) is bounded by

$$\frac{C\sqrt{c^2+t^2}}{r_1^{c+1}} e^{-(a-\theta_1)|t|},$$

which is integrable with respect to t over $\mathbb{R}$. Hence the desired interchange is guaranteed by Lebesgue's dominated convergence theorem. This completes the proof of property (a).

Next we note that (b) is a consequence of the Riemann–Lebesgue lemma (see Proposition 1.1). Furthermore, using (8.17), we see that

$$|f(r,\theta)| \le \frac{1}{2\pi}\int_{-\infty}^{\infty} |M_c^*[\phi](c+it)| r^{-c} e^{\theta t} dt$$
$$\le \frac{Cr^{-c}}{2\pi}\int_{-\infty}^{\infty} e^{-|t|(a-|\theta|)} dt = r^{-c}\frac{C}{\pi(a-|\theta|)},$$

which shows that (c) also holds.

It remains to verify (d). By the hypothesis $I_c(f, 0, t) = M_c^*[\phi](c + it)$ exists. Now performing a contour integration as in the proof of Proposition 8.19 and noting that Proposition 8.17 is applicable to f, we conclude that $I_c(f, \alpha, t)$ also exists for $|\alpha| < a$ and (8.15) holds. Combined with (8.17), this yields

$$|I_c(f,\alpha,t)| \le Ce^{-|t|(a-|\alpha|)},$$

which shows that (d) holds with $K(f) = C$. □

Remark 8.21 The hard part of Theorem 8.10 (i.e., a function $f \in \mathscr{B}^2_{c,T}$ is such that $f(\cdot, 0) \in B^2_{c,T}$) can now be deduced from Theorem 8.20. Indeed, let $f \in \mathscr{B}^2_{c,T}$. First assume in addition that $f(\cdot, 0) \in X_c$. Then $M_c^*[f(\cdot, 0)] = M_c[f(\cdot, 0)]$ and $f \in H_c^*(\mathbb{H}_a)$ for every $a > 0$. By Theorem 3.42, we conclude that

$$\|f\|_{H_c^*(\mathbb{H}_a)} \le \sup_{|\theta|<a}\int_0^{\infty} |f(r,\theta)| r^{c-1} dr \le \sup_{|\theta|<a} \|f(\cdot,\theta)\|_{X_c}$$
$$\le e^{aT}\|f(\cdot,0)\|_{X_c}.$$

Now by Theorem 8.20 we have

$$|M_c[f(\cdot,0)](c+it)| \le e^{-a|t|}e^{aT}\,\|f(\cdot,0)\|_{X_c}$$

for every $a > 0$. Letting $a \to \infty$, we find that

$$M_c[f(\cdot,0)](c+it) = 0 \qquad \text{whenever} \quad |t| > T,$$

that is, $f(\cdot,0)$ is Mellin bandlimited to $[-T,T]$.

We still have to get rid of the additional assumption that $f(\cdot,0) \in X_c$. This can be achieved as follows. For $f \in \mathscr{B}^2_{c,T}$, consider the functions

$$f_n(r,\theta) := r^{-c/n}e^{-ic\theta/n} f\left(r^{1-1/n}, (1-1/n)\theta\right)$$
$$\operatorname{sinc}\left(\frac{T}{n\pi}(\log r + i\theta)\right) \qquad (n \in \mathbb{N}\setminus\{1\}).$$

It can be verified by straightforward calculations that $f_n \in \mathscr{B}^2_{c,T}$, $f_n(\cdot,0) \in X^2_c \cap X_c$ and

$$\|f_n(\cdot,0) - f(\cdot,0)\|_{X^2_c} \longrightarrow 0 \qquad (n \to \infty).$$

Using the Mellin–Plancherel theorem (Theorem 7.10), we also have

$$\|M^2_c[f_n(\cdot,0)](c+i\cdot) - M^2_c[f(\cdot,0)](c+i\cdot)\|_{L^2(c+i\mathbb{R})} \longrightarrow 0 \qquad (n \to \infty).$$

Since the first part of the proof applies to f_n, we may write

$$\begin{aligned}
&\|M^2_c[f_n(\cdot,0)](c+i\cdot) - M^2_c[f(\cdot,0)](c+i\cdot)\|^2_{L^2(c+i\mathbb{R})} \\
&\quad = \|\big(M^2_c[f_n(\cdot,0)](c+i\cdot) - M^2_c[f(\cdot,0)](c+i\cdot)\big)\chi_{[-T,T]}\|^2_{L^2(c+i\mathbb{R})} \\
&\qquad + \|M^2_c[f(\cdot,0)](c+i\cdot)\chi_{\{|t|>T\}}\|^2_{L^2(c+i\mathbb{R})}.
\end{aligned}$$

As this holds for every $n \in \mathbb{N}\setminus\{1\}$, we deduce that

$$\|M^2_c[f(\cdot,0)](c+i\cdot)\chi_{\{|t|>T\}}\|^2_{L^2(c+i\mathbb{R})} = 0,$$

and so $M^2_c[f(\cdot,0)](c+it) = 0$ almost everywhere for $|t| > T$.

8.5 Notes

The classical version of the Paley–Wiener theorem of Fourier analysis with a proof involving complex analysis first appeared in 1934 in the book by Paley and Wiener [234]. It can also be found in various standard works, e.g., Boas [73, Chap. 6, Theorem 6.8.1], Rudin [265, Chap. 19], [266, Chap. 7], and Yosida [307, Chap. VI, Sect. 4]. A version for distributions, known as the Paley–Wiener–Schwartz theorem, is presented in the book of Higgins [173, Chap. 7].

Real versions of the Paley–Wiener theorem for Fourier transforms were acquired by Andersen [5] and Andersen and de Jeu [6], also in a multidimensional form by using an approach which involves certain differential operators.

For a Mellin version of the Paley–Wiener theorem, one has to achieve a characterization of a Mellin bandlimited function f by regularity and growth properties of an extension of f to a domain that contains $\mathbb{R}^+$. In Sects. 8.1 and 8.2, which are based on Bardaro et al. [55, Sects. 3–5], this has been done in the setting of classical complex analysis using the Riemann surface of the logarithm as an extension of $\mathbb{R}^+$. Although this approach rests on familiar notions, it can become very cumbersome in applications when appropriate analytic branches have to be selected.

The polar form, presented in Sect. 8.3, is considerably more comfortable, using simply the right half-plane as an extension of $\mathbb{R}^+$ and avoiding a splitting into branches. It originates from Bardaro et al. [58, Sect. 5].

Prior to the contributions by Bardaro et al., another approach to a Paley–Wiener theorem for Mellin transforms was established by Z. Szmydt and her school in Cracow (see Szmydt [283] and Szmydt and Ziemian [284, Chap. II, Sect. 9]). Related studies were performed in Lysik [204]; also see the review paper on Z. Szmydt's life by Lysik [205]. However, these authors start with functions (distributions) of compact support and characterize the resulting Mellin transform. This is a quite different problem since, other than in Fourier analysis, the inverse of a Mellin transform is not a Mellin transform. Therefore when one characterizes compact support, it makes an essential difference if it is done for the original function or for its Mellin transform.

A real version of a Paley–Wiener type theorem for the multidimensional Mellin transform, based on certain inequalities involving the transmutation (differential) operators, was obtained by Tuan [290]. For a different approach to the multidimensional case, see Pesenson [241, Theorem 8.3].

The contents of Sect. 8.4 originates from Bardaro et al. [59, Sect. 3.2]. The inspiration for this study came from Dryanov et al. [142, Theorem 2.1] who acquired a generalization of the classical Paley–Wiener theorem by characterizing functions whose Fourier transform exists as a Cauchy principal value and has an exponentially decaying bound as its argument approaches $\pm\infty$. This includes compact support as a special case. The notion of polar-analytic functions allows us to establish a Mellin analogue in the concise form of Theorem 8.20. A formulation in terms of classical complex analysis would be a torture for the reader.

An important auxiliary result is Lemma 8.13. Its Fourier analogue, employed in [142], is stated and proved in Titchmarsh [286, Theorem 120] and in Zygmund [312, Theorem 10.3]. Both proofs are quite long and need several intermediate results. For this reason, we made no attempts to design an intrinsic Mellin type proof, but deduced Lemma 8.13 by using the classical substitution from the Fourier to the Mellin instance. The modern literature on integral transforms prefers an L^p setting. However, as soon as contour integration of analytic functions is involved, Cauchy principal values occur in a natural way. It would be desirable to find a short direct approach to Lemma 8.13 independent of the Fourier case.

Chapter 9
Mellin Transforms and Fractional Analysis

In this chapter we shall introduce and study fractional integral and differential operators in the setting of Mellin analysis.

9.1 Fractional Mellin Integrals

In terms of Mellin analysis, the natural operator of fractional integration is not the classical Riemann–Liouville fractional integral of order $\alpha > 0$ on $\mathbb{R}^+$, namely (see [99, 100, 215, 267, Chap. 1, Sect. 2])

$$(I^{\alpha}_{0+}f)(x) = \frac{1}{\Gamma(\alpha)}\int_0^x (x-u)^{\alpha-1}f(u)du \qquad (x>0)$$

but the Hadamard fractional integral, first introduced by J. Hadamard [163] in 1892 in connection with Taylor expansions and defined by

$$(J^{\alpha}_{0+}f)(x) = \frac{1}{\Gamma(\alpha)}\int_0^x \left(\log\frac{x}{u}\right)^{\alpha-1} f(u)\frac{du}{u} \qquad (x>0).$$

However, for the development of the theory, it is important to consider the following generalization.

Definition 9.1 Let $f : \mathbb{R}^+ \to \mathbb{C}$ be a measurable function. For $\mu \in \mathbb{R}$, we define the *(left-sided) Mellin fractional integral* of order $\alpha > 0$ of f as

$$(J^{\alpha}_{0+,\mu}f)(x) = \frac{1}{\Gamma(\alpha)}\int_0^x \left(\frac{u}{x}\right)^{\mu}\left(\log\frac{x}{u}\right)^{\alpha-1} f(u)\frac{du}{u} \qquad (x>0). \tag{9.1}$$

C. Bardaro et al., *Mellin Analysis, Transform Theory, and Applications*,
https://doi.org/10.1007/978-3-031-96672-9_9

It is convenient to admit $\alpha = 0$ by agreeing that $J^0_{0+,\mu}$ is the identity operator, i.e., $J^0_{0+,\mu} f = f$.

For $\mu = 0$ we obtain the original definition of Hadamard so that $J^\alpha_{0+,0} = J^\alpha_{0+}$. We mention that also a right-sided Mellin fractional integral has been introduced and studied (see Sect. 9.5 Notes).

Remark 9.2 The notation $J^\alpha_{0+,\mu}$ with the subscript 0+ has been used in the literature in order to distinguish the fractional integral from the one in which the interval $[0, x]$ is replaced by $[x, \infty[$, the latter being denoted by $J^\alpha_{-,\mu}$ (see [105–107]).

Note that for $\alpha = r \in \mathbb{N}$ and $\mu = c$ formula (9.1) turns into the iterated integral representation

$$(J^r_{0+,c} f)(x) = x^{-c} \int_0^x \int_0^{u_1} \cdots \int_0^{u_{r-1}} f(u_r) u_r^c \frac{du_r}{u_r} \cdots \frac{du_2}{u_2} \frac{du_1}{u_1} \qquad (x > 0) \tag{9.2}$$

(see Proposition 6.86, and Remark 6.87).

For $\mu > c$ we can view $J^\alpha_{0+,\mu} f$ as a Mellin convolution between two functions $f, g^*_\mu \in X_c$, where

$$g^*_\mu\left(\frac{x}{u}\right) := \left(\frac{x}{u}\right)^{-\mu} \frac{\chi_{]0,x]}(u)}{\Gamma(\alpha)} \left(\log \frac{x}{u}\right)^{\alpha-1}.$$

Indeed, we have

$$\begin{aligned}(J_{0+,\mu} f)(x) &= \frac{1}{\Gamma(\alpha)} \int_0^x \left(\frac{u}{x}\right)^\mu \left(\log \frac{x}{u}\right)^{\alpha-1} f(u) \frac{du}{u} \\ &= \frac{1}{\Gamma(\alpha)} \int_0^\infty \left(\frac{u}{x}\right)^\mu \chi_{]0,x]}(u) \left(\log \frac{x}{u}\right)^{\alpha-1} f(u) \frac{du}{u} \\ &= \int_0^\infty g^*_\mu\left(\frac{x}{u}\right) f(u) \frac{du}{u} = (f * g^*_\mu)(x).\end{aligned}$$

Noting that $\chi_{]0,x]}(u) = \chi_{[1,\infty[}(x/u)$, we obtain

$$g^*_\mu(u) = u^{-\mu} \frac{\chi_{[1,\infty]}(u)}{\Gamma(\alpha)} (\log u)^{\alpha-1},$$

which is easily seen to belong to X_c if $\mu > c$.

However, we are mainly interested in properties of $J^\alpha_{0+,\mu} f$ when $\mu = c$ since in the subsequent considerations fractional integrals with parameter c will appear in connection with the space X_c. Therefore we have to study the domain and the range of these fractional operators. For example, we shall show that for any non-trivial and nonnegative function f in the domain of $J^\alpha_{0+,c}$, the image $J^\alpha_{0+,c} f$ cannot belong to

X_c. This stems from the fact that $g_c^* \notin X_c$ and so its Mellin transform in the space X_c does not exist.

Now, for $c \in \mathbb{R}$ and $\alpha \in \mathbb{R}^+$, we define the *domain* of $J^\alpha_{0+,c}$ as the class of all the functions $f : \mathbb{R}^+ \to \mathbb{C}$ such that

$$\int_0^x u^c \left(\log \frac{x}{u}\right)^{\alpha-1} |f(u)| \frac{du}{u} < \infty \tag{9.3}$$

almost everywhere for $x \in \mathbb{R}^+$. In the following we shall denote the domain of $J^\alpha_{0+,c}$ by dom $J^\alpha_{0+,c}$.

Example 9.3 Let $\alpha > 0$ and let $b, c \in \mathbb{R}$ be such that $b + c > 0$. Then the function $g(t) := t^b$, where $t \in \mathbb{R}^+$, belongs to dom $J^\alpha_{0+,c}$ and there holds

$$(J^\alpha_{0+,c} g)(x) = (b + c)^{-\alpha} x^b.$$

Indeed, using the substitutions $\log(x/u) = v$ and $(b + c)v = y$, we find that

$$\begin{aligned}(J^\alpha_{0+,c} g)(x) &= \frac{1}{\Gamma(\alpha)} \int_0^x \left(\frac{u}{x}\right)^c \left(\log \frac{x}{u}\right)^{\alpha-1} u^{b-1} du \\ &= \frac{x^b}{\Gamma(\alpha)} \int_0^\infty e^{-(b+c)v} v^{\alpha-1} dv = (b + c)^{-\alpha} x^b,\end{aligned}$$

as was to be shown.

We recall (see Sect. 3.1) that $X_{c,\mathrm{loc}}$ denotes the space of all the functions $f : \mathbb{R}^+ \to \mathbb{C}$ such that $(\cdot)^{c-1} f(\cdot) \in L^1(]0, a[)$ for every $a > 0$. We state two simple properties of these spaces.

Proposition 9.4 *The following properties hold:*

(i) If $f \in X_{c,\mathrm{loc}}$, then the function $(\cdot)^c f(\cdot) \in X_{1,\mathrm{loc}}$.
(ii) If $c < c'$, then $X_{c,\mathrm{loc}} \subset X_{c',\mathrm{loc}}$.

Proof Let $f \in X_{c,\mathrm{loc}}$. Then, for any $a > 0$,

$$\int_0^a x^c |f(x)| dx = \int_0^a x x^{c-1} |f(x)| dx \le a \int_0^a x^{c-1} |f(x)| dx,$$

which shows that (i) holds.

Likewise, setting $\alpha = c' - c$, we can write

$$\int_0^a x^{c'-1} |f(x)| dx \le a^\alpha \int_0^a x^{c-1} |f(x)| dx,$$

which implies (ii). □

Note that the inclusion in (ii) does not hold for the spaces X_c. Concerning the domain of the operator $J^{\alpha}_{0+,c}$, we begin with the following proposition.

Proposition 9.5 *Let $c \in \mathbb{R}$ and $\alpha > 1$. Then* dom $J^{\alpha}_{0+,c} \subset X_{c,\mathrm{loc}}$.

Proof Assume that the integral $(J^{\alpha}_{0+,c}|f|)(x)$, exists almost everywhere for $x \in \mathbb{R}^+$ and set $F(u) := u^{c-1}f(u)$. We have to show that F is integrable over $]0, a[$, for any $a > 0$. Let $a > 0$ be fixed and let $x > a$ be such that $(J^{\alpha}_{0+,c}|f|)(x)$ exists. Then, since $|\log(x/u)| \geq |\log(x/a)|$ for $u \in]0, a[$, we have

$$|F(u)| \leq \left|\log \frac{x}{u}\right|^{\alpha-1} |F(u)| \left|\log \frac{x}{a}\right|^{1-\alpha},$$

and the right-hand side of the inequality is integrable as a function of u. □

Note that for $\alpha = 1$ we find immediately that dom $J^{1}_{0+,c} = X_{c,\mathrm{loc}}$. The case of $0 < \alpha < 1$ is more delicate. We shall show that in this instance $X_{c,\mathrm{loc}} \subset$ dom $J^{\alpha}_{0+,c}$.

In order to give a more precise description of the domain of $J^{\alpha}_{0+,c}$, we now present a direct proof of the semigroup property in the domain of fractional integrals.

Theorem 9.6 *Let $c \in \mathbb{R}$ and $\alpha, \beta \in \mathbb{R}^+$. Suppose that $f \in$ dom $J^{\alpha+\beta}_{0+,c}$. Then*

(i) $f \in$ dom $J^{\alpha}_{0+,c} \cap$ dom $J^{\beta}_{0+,c}$,
(ii) $J^{\alpha}_{0+,c}f \in$ dom $J^{\beta}_{0+,c}$ *and* $J^{\beta}_{0+,c}f \in$ dom $J^{\alpha}_{0+,c}$,
(iii) $(J^{\alpha+\beta}_{0+,c}f)(x) = (J^{\alpha}_{0+,c}(J^{\beta}_{0+,c}f))(x) \qquad (a.e.\ x \in \mathbb{R}^+)$.
(iv) If $\alpha < \beta$, then dom $J^{\beta}_{0+,c} \subset$ dom $J^{\alpha}_{0+,c}$.

Proof First we assume in addition that $f \in$ dom $J^{\alpha+\beta}_{0+,c}$ is a nonnegative function. Then the integral

$$(J^{\alpha+\beta}_{0+,c}f)(x) = \frac{1}{\Gamma(\alpha+\beta)} \int_0^x \left(\frac{v}{x}\right)^c \left(\log \frac{x}{v}\right)^{\alpha+\beta-1} f(v)\, \frac{dv}{v}$$

is finite and nonnegative almost everywhere for $x \in \mathbb{R}^+$. Now for positive α and β we use the formula

$$\int_v^x \left(\log \frac{x}{u}\right)^{\alpha-1} \left(\log \frac{u}{v}\right)^{\beta-1} \frac{du}{u} = B(\alpha, \beta) \left(\log \frac{x}{v}\right)^{\alpha+\beta-1}, \tag{9.4}$$

where B is the beta function. It can be verified by substituting first $\log(x/u) = s$ in the integral on the left-hand side, then $s = \log(x/v)t$ in the resulting integral and recalling the definition of the beta function (see Sect. 1.5.1). Now, by using (9.4) and Fubini's theorem for iterated integrals of nonnegative functions, we conclude that

$$
\begin{aligned}
(J^{\alpha+\beta}_{0+,c}f)(x) &= \frac{1}{\Gamma(\alpha)\Gamma(\beta)}\frac{\Gamma(\alpha)\Gamma(\beta)}{\Gamma(\alpha+\beta)}\int_0^x\left(\frac{v}{x}\right)^c\left(\log\frac{x}{v}\right)^{\alpha+\beta-1}f(v)\,\frac{dv}{v}\\
&= \frac{x^{-c}}{\Gamma(\alpha)\Gamma(\beta)}\int_0^x v^c f(v)\left[B(\alpha,\beta)\left(\log\frac{x}{v}\right)^{\alpha+\beta-1}\right]\frac{dv}{v}\\
&= \frac{x^{-c}}{\Gamma(\alpha)\Gamma(\beta)}\int_0^x v^c f(v)\left[\int_v^x\left(\log\frac{x}{u}\right)^{\alpha-1}\left(\log\frac{u}{v}\right)^{\beta-1}\frac{du}{u}\right]\frac{dv}{v}\\
&= \frac{x^{-c}}{\Gamma(\alpha)\Gamma(\beta)}\int_0^x\int_0^x v^c\chi_{]v,x[}(u)\left(\log\frac{x}{u}\right)^{\alpha-1}\left(\log\frac{u}{v}\right)^{\beta-1}f(v)\,\frac{dv}{v}\frac{du}{u}\\
&= \frac{x^{-c}}{\Gamma(\alpha)\Gamma(\beta)}\int_0^x\int_0^x v^c\chi_{]0,u[}(v)\left(\log\frac{x}{u}\right)^{\alpha-1}\left(\log\frac{u}{v}\right)^{\beta-1}f(v)\,\frac{dv}{v}\frac{du}{u}\\
&= \frac{1}{\Gamma(\alpha)}\int_0^x\left(\frac{u}{x}\right)^c\left(\log\frac{x}{u}\right)^{\alpha-1}\\
&\quad\times\left[\frac{1}{\Gamma(\beta)}\int_0^u\left(\frac{v}{u}\right)^c\left(\log\frac{u}{v}\right)^{\beta-1}f(v)\,\frac{dv}{v}\right]\frac{du}{u}\\
&= \big(J^{\alpha}_{0+,c}(J^{\beta}_{0+,c}f)\big)(x).
\end{aligned}
$$

This proves the assertions (i)–(iii) for nonnegative functions. In the general case, we can apply this reasoning to the nonnegative functions f^+ and f^- using the linearity of integrals for extending the results to $f = f^+ - f^-$. Property (iv) follows from (i) by writing $\beta = \alpha + (\beta - \alpha)$. □

By combining some of the previous statements, we obtain the following inclusions.

Proposition 9.7 *Let $c \in \mathbb{R}$ and $0 < \alpha < 1 < \beta$. Then*

$$
\operatorname{dom} J^{\beta}_{0+,c} \subset \operatorname{dom} J^{1}_{0+,c} = X_{c,\mathrm{loc}} \subset \operatorname{dom} J^{\alpha}_{0+,c}. \tag{9.5}
$$

The inclusions are strict.

Proof The equation in (9.5) is obvious and has been mentioned already on page 242. The inclusions follow from Proposition 9.5 and statement (iv) of Theorem 9.6. It remains to show that they are strict.

Consider the function

$$
f(x) = \frac{x^{-c}}{|\log x|^{\beta}}\,\chi_{]0,1/2[}(x).
$$

Then $f \in X_{c,\mathrm{loc}}$, but for any $x > 1$,

$$\Gamma(\beta)(J^{\beta}_{0+,c}f)(x) = x^{-c}\int_0^x u^c \left(\log\frac{x}{u}\right)^{\beta-1} f(u)\,\frac{du}{u} = x^{-c}\int_0^{1/2} \frac{\left(\log\frac{x}{u}\right)^{\beta-1}}{u|\log u|^{\beta}}\,du$$

$$\geq x^{-c}\int_0^{1/2} \frac{\left(\log\frac{1}{u}\right)^{\beta-1}}{u|\log u|^{\beta}}\,du = x^{-c}\int_0^{1/2} \frac{1}{u|\log u|}\,du = \infty.$$

Hence $f \notin \operatorname{dom} J^{\beta}_{0+,c}$.

Next consider

$$f(x) = \frac{x^{-c}}{|\log x|^{\gamma}}\,\chi_{]0,1/2[}(x), \tag{9.6}$$

where $\alpha < \gamma < 1$. Then $f \notin X_{c,\mathrm{loc}}$, but $(J^{\alpha}_{0+,c}f)(x)$ is finite almost everywhere for $x \in \mathbb{R}^+$. Indeed, if $0 < x < 1/2$, then

$$\Gamma(\alpha)(J^{\alpha}_{0+,c}f)(x) = x^{-c}\int_0^x \frac{1}{u\left(\log\frac{1}{u}\right)^{\gamma-\alpha+1}}\,\frac{\left(\log\frac{1}{u}\right)^{1-\alpha}}{\left(\log\frac{x}{u}\right)^{1-\alpha}}\,du.$$

The integrand, denoted by $F(x, u)$, is continuous on $]0, x[$ as a function of u but approaches ∞ as $u \to 0+$ and $u \to x-$. We conclude that there exist positive constants $C_{1,x}$ and $C_{2,x}$ such that

$$F(x,u) \leq \begin{cases} \dfrac{C_{1,x}}{u\,|\log u|^{\gamma-\alpha+1}} & \text{for } 0 < u \leq x/2, \\[2ex] \dfrac{C_{2,x}}{\left|\log\frac{x}{u}\right|^{1-\alpha}} & \text{for } x/2 < u < x. \end{cases}$$

The right-hand side forms a majorant which is integrable on $[0, x]$ with respect to u. For $x > 1/2$, we have

$$\Gamma(\alpha)(J^{\alpha}_{0+,c}f)(x) = x^{-c}\int_0^{1/2} \frac{1}{u\left(\log\frac{1}{u}\right)^{\gamma-\alpha+1}}\,\frac{\left(\log\frac{1}{u}\right)^{1-\alpha}}{\left(\log\frac{x}{u}\right)^{1-\alpha}}\,du$$

$$\leq \frac{C_x}{x^c}\int_0^{1/2} \frac{1}{u|\log u|^{\gamma-\alpha+1}}\,du < \infty$$

with a positive constant C_x since the second quotient in the first integral remains bounded for $u \in [0, 1/2]$. □

The function (9.6) with $\alpha < \gamma < \beta$ can also be used to show that the inclusion of statement (iv) of Theorem 9.6 is strict.

We now give a sufficient condition for a function f to belong to the domain of the fractional integrals of order $\alpha > 1$.

Proposition 9.8 *Let $c \in \mathbb{R}$, $\alpha > 1$ and suppose that $f \in X_{c,\text{loc}}$. If $r \in]0, 1[$ and $f(u) = \mathcal{O}(u^{-(r+c-1)})$ as $u \to 0+$, then $f \in \operatorname{dom} J^{\alpha}_{0+,c}$.*

Proof For fixed $x > 0$ define

$$I_a^b := \int_a^b u^{c-1}|f(u)| \left(\log \frac{x}{u}\right)^{\alpha-1} du \qquad (0 \le a < b).$$

We have to show that $I_0^x < \infty$. The prescribed asymptotic behavior of f at the point 0 implies that there exists an $\varepsilon > 0$ such that $I_0^{\varepsilon} < \infty$ and so $I_0^x < \infty$ if $0 < x \le \varepsilon$. If $x > \varepsilon$, then $\left(\log \frac{x}{u}\right)^{\alpha-1} \le \left(\log \frac{x}{\varepsilon}\right)^{\alpha-1}$ for $u \in [\varepsilon, x]$. Hence

$$I_0^x \le I_0^{\varepsilon} + I_{\varepsilon}^x \le I_0^{\varepsilon} + \left(\log \frac{x}{\varepsilon}\right)^{\alpha-1} \int_0^x u^{c-1}\,|f(u)|\,du < \infty$$

since $f \in X_{c,\text{loc}}$. □

The functions figuring in the previous proposition form the class

$$\widetilde{X}_{c,\text{loc}} := \left\{ f \in X_{c,\text{loc}} : \exists r \in]0, 1[\text{ such that } f(u) = \mathcal{O}(u^{-(r+c-1)}) \text{ as } u \to 0+ \right\}.$$

With this notation, Proposition 9.7 in conjunction with Proposition 9.8 imply the following result.

Corollary 9.9 *Let $c \in \mathbb{R}$. Then*

$$\widetilde{X}_{c,\text{loc}} \subset \bigcap_{\alpha>0} \operatorname{dom} J^{\alpha}_{0+,c}.$$

Now let f be given as a uniformly convergent power series

$$f(x) = \sum_{k=0}^{\infty} a_k x^k \qquad (a_k \in \mathbb{C},\ k \in \mathbb{N}_0). \tag{9.7}$$

For these functions the following series representation for $J^{\alpha}_{0+,c} f$ holds provided that c is positive.

Lemma 9.10 *Let $c, \alpha \in \mathbb{R}^+$. Suppose that f is given by the power series (9.7) converging uniformly on an interval $[0, a]$ with $a > 0$. Then*

$$(J^{\alpha}_{0+,c} f)(x) = \sum_{k=0}^{\infty} (c+k)^{-\alpha} a_k x^k \qquad \big(x \in [0, a]\big).$$

Proof Denoting by $S_m(x)$ the partial sums of the first $m+1$ terms of the series defining the function f, we see by Example 9.3 that

$$(J^{\alpha}_{0+,c}S_m)(x) = \sum_{k=0}^{m}(c+k)^{-\alpha}a_k x^k.$$

It remains to prove convergence to $(J^{\alpha}_{0+,c}f)(x)$ as $m \to \infty$. By the uniform convergence of the power series defining f, there exists for given $\varepsilon > 0$ an integer $N > 0$ such that for any $m \geq N$ and every $x \in [0,a]$, we have $|f(x) - S_m(x)| < \varepsilon$. Thus, using Example 9.3 with $b = 0$, we obtain

$$\begin{aligned}
&|(J^{\alpha}_{0+,c}f)(x) - (J^{\alpha}_{0+,c}S_m)(x)| \\
&\leq \frac{1}{\Gamma(\alpha)}\int_0^x \left(\frac{u}{x}\right)^c \left(\log\left(\frac{x}{u}\right)\right)^{\alpha-1} |f(u) - S_m(u)|\,\frac{du}{u} \\
&\leq \frac{\varepsilon}{\Gamma(\alpha)}\int_0^x \left(\frac{u}{x}\right)^c \left(\log\left(\frac{x}{u}\right)\right)^{\alpha-1} \frac{du}{u} \\
&= \varepsilon (J^{\alpha}_{0+,c}1)(x) = \varepsilon c^{-\alpha}.
\end{aligned}$$

This completes the proof. □

In Lemma 9.10, the assumption that $c > 0$ cannot be relaxed. For $c = 0$, corresponding to the classical Hadamard integrals, we state the following result.

Proposition 9.11 *Let $\alpha \in \mathbb{R}^+$. Suppose that f is given as $f(x) = \sum_{k=0}^{\infty} a_k x^k$. Then $f \in \operatorname{dom} J^{\alpha}_{0+}$ if and only if*

$$a_0 = 0 \quad \textit{and} \quad \limsup_{k\to\infty} |a_k|^{1/k} = 0. \tag{9.8}$$

In this case,

$$(J^{\alpha}_{0+}f)(x) = \sum_{k=1}^{\infty} a_k k^{-\alpha} x^k \qquad (x \in \mathbb{R}^+). \tag{9.9}$$

Proof Assume that $f \in \operatorname{dom} J^{\alpha}_{0+}$. Then, by the definition of domains, f exists on the whole of $\mathbb{R}^+$. This implies that the power series defining f converges everywhere on $\mathbb{R}$ or, equivalently, $\limsup_{k\to\infty} |a_k|^{1/k} = 0$. Furthermore, the integral (9.3) is finite for $c = 0$ and all $x \in \mathbb{R}^+$. Consequently, the following integral must also be finite:

$$\begin{aligned}
\int_0^x \left(\log\frac{x}{u}\right)^{\alpha-1} f(u)\frac{du}{u} &= \int_0^x \left(\log\frac{x}{u}\right)^{\alpha-1} \sum_{k=1}^{\infty} a_k u^k \frac{du}{u} + a_0 \int_0^x \left(\log\frac{x}{u}\right)^{\alpha-1} \frac{du}{u} \\
&=: I_1 + I_2.
\end{aligned}$$

Since the power series converges uniformly on compact subsets of $\mathbb{R}$ and absolutely everywhere, we conclude that

$$|I_1| \leq \int_0^x \left(\log \frac{x}{u}\right)^{\alpha-1} \sum_{k=1}^{\infty} |a_k| u^{k-1} du \leq \sum_{k=1}^{\infty} |a_k| x^{k-1} \int_0^x \left(\log \frac{x}{u}\right)^{\alpha-1} du.$$

By the change of variables $\log(x/u) = t$, we find that

$$\int_0^x \left(\log \frac{x}{u}\right)^{\alpha-1} du = x\Gamma(\alpha).$$

Thus

$$|I_1| \leq \Gamma(\alpha) \sum_{k=1}^{\infty} |a_k| x^k < \infty \qquad (x \in \mathbb{R}^+).$$

Since

$$\int_0^x \left(\log \frac{x}{u}\right)^{\alpha-1} \frac{du}{u} = \infty,$$

we obtain $|I_2| < \infty$ if and only if $a_0 = 0$. We have shown that the conditions (9.8) are necessary.

Conversely, assume that (9.8) holds. Then the previous estimates can be used to show that the integral (9.3) is finite for $c = 0$ and all $x \in \mathbb{R}^+$. Hence $f \in \text{dom}\, J_{0+}^{\alpha}$. As to formula (9.9), we find by interchanging integration and summation that

$$\begin{aligned}
(J_{0+}^{\alpha} f)(x) &= \sum_{k=1}^{\infty} \frac{a_k}{\Gamma(\alpha)} \int_0^x \left(\log \frac{x}{u}\right)^{\alpha-1} u^{k-1} du \\
&= \sum_{k=1}^{\infty} a_k \big(J_{0+}^{\alpha} (\cdot)^k\big)(x) = \sum_{k=1}^{\infty} a_k k^{-\alpha} x^k,
\end{aligned}$$

where Example 9.3 has been used in the last step. □

As regards the range of the operators $J_{0+,c}^{\alpha}$, we state the following propositions.

Proposition 9.12 *Let $c \in \mathbb{R}$ and $\alpha \in \mathbb{R}^+$. If $f \in \text{dom}\, J_{0+,c}^{\alpha+1}$, then $J_{0+,c}^{\alpha} f \in X_{c,\text{loc}}$.*

Proof We can assume that f is nonnegative, which implies that $J_{0+,c}^{\alpha} f$ is nonnegative. Then, for any $a > 0$,

$$\Gamma(\alpha)\int_0^a x^{c-1}(J^\alpha_{0+,c}f)(x)dx = \int_0^a u^{c-1}f(u)\Big[\int_u^a \frac{1}{x}\Big(\log\frac{x}{u}\Big)^{\alpha-1}dx\Big]du$$
$$= \frac{1}{\alpha}\int_0^a u^{c-1}f(u)\Big(\log\frac{a}{u}\Big)^\alpha du < \infty$$

if $f \in \operatorname{dom} J^{\alpha+1}_{0+,c}$. This shows that $J^\alpha_{0+,c}f \in X_{c,\mathrm{loc}}$. □

In view of Proposition 9.12, we mention that if $f \in \operatorname{dom} J^\alpha_{0+,c}$, then $J^\alpha_{0+,c}f$ may not belong to $X_{c,\mathrm{loc}}$. For example, when f is the function of (9.6) with $\alpha < \gamma < \alpha + 1$, then $f \in \operatorname{dom} J^\alpha_{0+,c}$ but $f \notin \operatorname{dom} J^{\alpha+1}_{0+,c}$ and $J^\alpha_{0+,c}f \notin X_{c,loc}$. In particular, $\operatorname{dom} J^{\alpha+1}_{0+,c}$ is a proper subspace of $\operatorname{dom} J^\alpha_{0+,c}$.

For the spaces X_c, we state the following result.

Proposition 9.13 *Let $c \in \mathbb{R}$ and $\alpha \in \mathbb{R}^+$. If $f \in \operatorname{dom} J^\alpha_{0+,c}$ is a nonnegative function, then $J^\alpha_{0+,c}f \notin X_c$, unless $f = 0$ almost everywhere on $\mathbb{R}^+$.*

Proof Suppose that $f \geq 0$. Then

$$\int_0^\infty x^{c-1}(J^\alpha_{0+,c}f)(x)dx$$
$$= \int_0^\infty x^{-1}\Big(\frac{1}{\Gamma(\alpha)}\int_0^\infty u^{c-1}\chi_{]0,x[}(u)\Big(\log\frac{x}{u}\Big)^{\alpha-1}f(u)du\Big)dx$$
$$= \int_0^\infty \frac{1}{\Gamma(\alpha)}\Big(\int_0^\infty x^{-1}u^{c-1}\chi_{]u,\infty[}(x)\Big(\log\frac{x}{u}\Big)^{\alpha-1}f(u)dx\Big)du$$
$$= \frac{1}{\Gamma(\alpha)}\int_0^\infty\Big(\int_u^\infty x^{-1}\Big(\log\frac{x}{u}\Big)^{\alpha-1}dx\Big)u^{c-1}f(u)du.$$

Since

$$\int_u^\infty \frac{1}{(\log\frac{x}{u})^{1-\alpha}}\frac{dx}{x} = \infty$$

for every $u > 0$, the assertion follows immediately. □

The above result implies that a function $f \in \operatorname{dom} J^\alpha_{0+,c}$ such that $J^\alpha_{0+,c}f \in X_c$ must necessarily change its sign. However, a change of sign is not sufficient for $J^\alpha_{0+,c}f \in X_c$ even if in addition $f \in X_c$, as is seen by the following example. For a given $a > 1$, take

$$f(x) = -\chi_{[1/a,1]}(x) + \chi_{]1,a]}(x).$$

Then $f \in \operatorname{dom} J^\alpha_{0+,c} \cap X_c$, but an estimate of $(J^\alpha_{0+,c}f)(x)$ for $x > a$ shows that $J^\alpha_{0+,c}f \notin X_c$ for any $c \in \mathbb{R}$.

Next we establish a formula for the Mellin transform of the fractional integral $J^{\alpha}_{0+,c}f$.

Proposition 9.14 *Let $c \in \mathbb{R}$ and $\alpha \in \mathbb{R}^+$. Suppose that $f \in \operatorname{dom} J^{\alpha}_{0+,c} \cap X_c$, $J^{\alpha}_{0+,c} f \in X_c$ and $J^{\alpha-\lfloor\alpha\rfloor}_{0+,c} f \in X_c$. Then*

$$[J^{\alpha}_{0+,c} f]^{\wedge}_{M_c}(c+it) = (-it)^{-\alpha}[f]^{\wedge}_{M_c}(c+it) \qquad (t \in \mathbb{R}\setminus\{0\}).$$

The power of exponent $-\alpha$ has to be taken as the principal value, that is,

$$(-it)^{-\alpha} = |t|^{-\alpha}\, e^{i\,\mathrm{sgn}(t)\,\pi\alpha/2}. \tag{9.10}$$

Proof First let $0 < \alpha < 1$ which implies that $\lfloor\alpha\rfloor = 0$. By the hypotheses, we can apply Fubini's theorem in the following calculations. Taking into account that $\chi_{]0,x[}(u) = \chi_{]u,\infty[}(x)$ and using the change of variable $\log(x/u) = y$, we obtain

$$\begin{aligned}
&[J^{\alpha}_{0+,c} f]^{\wedge}_{M_c}(c+it)\\
&= \frac{1}{\Gamma(\alpha)}\int_0^{\infty} x^{c+it}\left\{\int_0^{\infty}\left(\frac{u}{x}\right)^{c}\left(\log\frac{x}{u}\right)^{\alpha-1} f(u)\chi_{]0,x[}(u)\,\frac{du}{u}\right\}\frac{dx}{x}\\
&= \frac{1}{\Gamma(\alpha)}\int_0^{\infty} u^{c} f(u)\left\{\int_u^{\infty} x^{it}\left(\log\frac{x}{u}\right)^{\alpha-1}\frac{dx}{x}\right\}\frac{du}{u}\\
&= \frac{1}{\Gamma(\alpha)}\int_0^{\infty} u^{c+it} f(u)\,\frac{du}{u}\int_0^{\infty} y^{\alpha-1}e^{ity}\,dy.
\end{aligned} \tag{9.11}$$

The last integral on the right-hand side can be calculated with the help of Cauchy's Theorem 1.24. Let $0 < r < R$ and suppose that $t > 0$. Denote by γ the positively oriented, closed, quadrilateral curve lying in the first quadrant of the complex plane and being composed of the interval $[r, R]$, the circular arcs $re^{i\theta}$, $Re^{i\theta}$, where $\theta \in [0, \pi/2]$ and the line segment that connects iR with ir. Then, for $g(z) := z^{\alpha-1}e^{itz}$, we obtain $\int_{\gamma} g(z)dz = 0$. By standard estimates one can show that the contributions coming from the circular arcs converge to zero as $r \to 0+$ and $R \to \infty$. Therefore

$$\begin{aligned}
\int_0^{\infty} y^{\alpha-1}e^{ity}\,dy &= \int_0^{\infty} (i\xi)^{\alpha-1}e^{-t\xi}\, i\, d\xi\\
&= \left(\frac{i}{t}\right)^{\alpha}\int_0^{\infty} x^{\alpha-1}e^{-x}\,dx = (-it)^{-\alpha}\Gamma(\alpha).
\end{aligned}$$

For $t < 0$, we use the curve γ' obtained by reflecting γ on the real line and arrive at the same result. Combined with (9.11), it completes the proof for $\alpha \in]0, 1[$.

Now let $\alpha \geq 1$. Then $\beta := \alpha - \lfloor\alpha\rfloor \in [0, 1[$. Using the semigroup property of fractional integrals (Theorem 9.6 (iii)), formula (9.2), Theorem 6.89 and, in case of $\beta \neq 0$, the previous part of the proof, we conclude that

$$[J^{\alpha}_{0+,c}f]^{\wedge}_{M_c}(c+it) = [J^{\lfloor\alpha\rfloor}_{0+,c}(J^{\beta}_{0+,c}f)]^{\wedge}_{M_c}(c+it) = (-it)^{-\lfloor\alpha\rfloor}[J^{\beta}_{0+,c}f)]^{\wedge}_{M_c}(c+it)$$
$$= (-it)^{-\lfloor\alpha\rfloor}(-it)^{-\beta}[f]^{\wedge}_{M_c}(c+it) = (-it)^{-\alpha}[f]^{\wedge}_{M_c}(c+it).$$

Finally, the two parts of the proof reveal that the power $(-it)^{-\alpha}$ has to be specified as in (9.10). □

The following corollary shows an interesting property of the functions f figuring in Proposition 9.14.

Corollary 9.15 *Under the hypotheses of Proposition 9.14, it follows that*

$$\int_0^{\infty} x^{c-1} f(x)dx = 0.$$

Proof The previous proposition yields

$$[J^{\alpha}_{0+,c}f]^{\wedge}_{M_c}(s) = (-it)^{-\alpha}[f]^{\wedge}_{M_c}(s) \qquad (s = c+it,\ t \in \mathbb{R}\setminus\{0\}).$$

The transforms of f and $J^{\alpha}_{0+,c}f$ are continuous and bounded functions on the line $s = c + it$. Therefore, letting $t \to 0$, we must have $[f]^{\wedge}_{M_c}(c) = 0$, as was to be shown. □

Functions $f \in \text{dom}\, J^{\alpha}_{0+,c} \cap X_c$ for which $J^{\alpha}_{0+,c}f \in X_c$ can be constructed within the class of functions having non-constant sign and compact support in $\mathbb{R}^+$.

For the spaces $X_{[\nu,c]}$, we state the following results.

Proposition 9.16 *Let $c, \nu \in \mathbb{R}$ with $\nu < c$ and let $\alpha \in \mathbb{R}^+$. If $f \in \text{dom}\, J^{\alpha}_{0+,c} \cap X_{[\nu,c]}$, then $J^{\alpha}_{0+,c}f \in X_\nu$ and*

$$\|J^{\alpha}_{0+,c}f\|_{X_\nu} \le \frac{\|f\|_{X_\nu}}{(c-\nu)^{\alpha}}.$$

Moreover,

$$[J^{\alpha}_{0+,c}f]^{\wedge}_{M_\nu}(\nu+it) = (c-\nu-it)^{-\alpha}[f]^{\wedge}_{M_\nu}(\nu+it) \qquad (t \in \mathbb{R}) \qquad (9.12)$$

with the power of exponent $-\alpha$ being defined by its principal value and

$$|[J^{\alpha}_{0+,c}f]^{\wedge}_{M_\nu}(\nu+it)| \le \frac{\|f\|_{X_\nu}}{(c-\nu)^{\alpha}} \qquad (t \in \mathbb{R}).$$

Proof By Fubini's theorem,

$$\Gamma(\alpha)\|J^{\alpha}_{0+,c}f\|_{X_\nu} = \Gamma(\alpha)\int_0^\infty u^{\nu-1}|(J^{\alpha}_{0+,c}f)(u)|du$$

$$\leq \int_0^\infty u^{\nu-1}\left[\int_0^u \left(\frac{y}{u}\right)^c \left(\log\frac{u}{y}\right)^{\alpha-1} |f(y)|\,\frac{dy}{y}\right]du$$

$$= \int_0^\infty \left[\int_y^\infty u^{\nu-1-c}\left(\log\frac{u}{y}\right)^{\alpha-1} du\right] y^{c-1}|f(y)|dy.$$

Setting $\log(u/y) = w$, we may write the inner integral as

$$\int_y^\infty u^{\nu-1-c}\left(\log\frac{u}{y}\right)^{\alpha-1} du = \int_0^\infty y^{\nu-c}e^{-(c-\nu)w}w^{\alpha-1}dw = \frac{y^{\nu-c}}{(c-\nu)^\alpha}\Gamma(\alpha),$$

and so

$$\Gamma(\alpha)\|J^{\alpha}_{0+,c}f\|_{X_\nu} \leq \frac{\Gamma(\alpha)}{(c-\nu)^\alpha}\|f\|_{X_\nu}.$$

Formula (9.12) can be proved by recalling that for $c > \nu$, we can write $J^{\alpha}_{0+,c}f$ as a Mellin convolution of f and g^*_c given by

$$g^*_c(u) := u^{-c}(\log u)^{\alpha-1}\chi_{]1,\infty[}(u)(\Gamma(\alpha))^{-1} \qquad (u \in \mathbb{R}^+)$$

(see the discussion on page 240). Since $[g^*_c]^\wedge_{M_\nu}(\nu + it) = (c - \nu - it)^{-\alpha}$, we obtain (9.12) by employing the convolution theorem for Mellin transforms (Theorem 6.12). Finally, the last inequality follows from the first by writing

$$|[J^{\alpha}_{0+,c}f]^\wedge_{M_\nu}(\nu + it)| \leq \|J^{\alpha}_{0+,c}f\|_{X_\nu} \leq \frac{\|f\|_{X_\nu}}{(c-\nu)^\alpha}.$$

□

For $0 < \alpha < 1$ the assumption $f \in \text{dom}\, J^{\alpha}_{0+,c} \cap X_{[\nu,c]}$ can be replaced by $f \in X_{[\nu,c]}$ since, by Propositions 9.7 and 9.4, it follows that $X_{[\nu,c]} \subset \text{dom}\, J^{\alpha}_{0+,c}$.

9.2 Pointwise Mellin–Hadamard Fractional Derivatives

Using the Mellin fractional integrals, we now introduce a notion of fractional derivative in the Mellin setting.

Definition 9.17 Let $c \in \mathbb{R}$, $\alpha \in \mathbb{R}^+$ and set $m := \lceil\alpha\rceil$. For $f \in \text{dom}\, J^{m-\alpha}_{0+,c}$, the *(left-sided) pointwise Mellin–Hadamard fractional derivative of order* α at the point $x \in \mathbb{R}^+$ is defined and denoted by

$$(D^{\alpha}_{0+,c}f)(x) := \Theta^m_c\left(J^{m-\alpha}_{0+,c}f\right)(x)$$

provided that the right-hand side exists. For $c = 0$, we agree that $(D^\alpha_{0+} f)(x) := (D^\alpha_{0+,0} f)(x)$.

Also a definition of right-sided pointwise Mellin–Hadamard derivative was considered in literature (see Sect. 9.5 Notes).

Now, let us explain how the operator $D^\alpha_{0+,c}$ is designed. When $\alpha \in \mathbb{N}$, then $m = \alpha$. Recalling that $J^0_{0+,c}$ was defined as the identity operator, we obtain $(D^\alpha_{0+,c} f)(x) = \Theta^\alpha_c f(x)$. This is an ordinary Mellin differentiation of integer order α. When $\alpha \in \mathbb{R}^+ \setminus \mathbb{N}$, then the deviation of α from the closest bigger integer m, which is $m - \alpha$, is first compensated by a fractional integration of order $m - \alpha$ in the spirit of the fundamental theorem of calculus and then an ordinary Mellin differentiation of integer order m is performed.

Note that for $m > \alpha$, the hypothesis that $f \in \operatorname{dom} J^{m-\alpha}_{0+,c}$ cannot be relaxed. For example, let $g(x) \equiv 1$ and $c = 0$. Then $\Theta^m g(x) = 0$ on $\mathbb{R}^+$ for all $m \in \mathbb{N}$, but $g \notin \operatorname{dom} J^{m-\alpha}_{0+}$ and $(D^\alpha_{0+} g)(x)$ does not exist. Another example is the function $h(x) = \log x$ for $x \in \mathbb{R}^+$. Then $\Theta h(x) = 1$ and $\Theta^m h(x) = 0$ for all integers $m > 1$ but $(D^\alpha_{0+} h)(x)$ does not exist when $\alpha \in \mathbb{R}^+ \setminus \mathbb{N}$.

For the calculation of Mellin–Hadamard fractional derivatives the following proposition is sometimes useful.

Proposition 9.18 *Let $c \in \mathbb{R}$ and $m \in \mathbb{N}$. Suppose that $\Theta^m_c f(x)$ exists at the point $x \in \mathbb{R}^+$. Then $x^c \Theta^m_c f(x) = \Theta^m (x^c f(x))$.*

Proof For $m = 1$ we have

$$x^c \Theta_c f(x) = x^c (cf(x) + xf'(x)) = x(cx^{c-1} f(x) + x^c f'(x)) = \Theta(x^c f(x)).$$

Proceeding by induction, we assume that the assertion holds for m and have to verify it for $m + 1$. This way together with the argumentation for $m = 1$, we conclude that

$$\begin{aligned} x^c \Theta^{m+1}_c f(x) &= x^c \Theta_c (\Theta^m_c f(x)) \\ &= \Theta(x^c \Theta^m_c f(x)) = \Theta(\Theta^m (x^c f(x)) = \Theta^{m+1}(x^c f(x)), \end{aligned}$$

as was to be shown. □

Employing Proposition 9.18, we may rewrite the definition of the Mellin–Hadamard derivative of order α with $m := \lceil \alpha \rceil$ as

$$(D^\alpha_{0+,c} f)(x) := x^{-c} \Theta^m \big(x^c (J^{m-\alpha}_{0+,c} f)(x)\big).$$

For $r = m \in \mathbb{N}$, formula (3.16) of Proposition 3.21 may be used to express the Mellin–Hadamard derivative of order m as

$$(D^m_{0+,c} f)(x) = \sum_{k=0}^{m} S_c(m,k) x^k f^{(k)}(x), \tag{9.13}$$

where $S_c(m,k)$ for $0 \le k \le m$ denote the generalized Stirling numbers of the second kind. We now state an extension of (9.13) for the fractional instance. For an alternative extension of (9.13) in case of infinitely differentiable functions f involving the Stirling functions of the second kind, see Sect. 9.5 Notes.

Proposition 9.19 *Let $c \in \mathbb{R}$, let $\alpha \in \mathbb{R}^+$ and set $m := \lceil \alpha \rceil$. Suppose that $f \in X_{c,\text{loc}}$ is m-times differentiable such that $f^{(k)} \in X_{c,\text{loc}}$ for $k = 1, \ldots, m$. Then*

$$(D^{\alpha}_{0+,c} f)(x) = \sum_{k=0}^{m} S_c(m,k) x^k \big(J^{m-\alpha}_{0+,c+k} f^{(k)}\big)(x).$$

Proof Obviously, for $\alpha \in \mathbb{N}$ this formula reduces to (9.13). Therefore it suffices to consider the case of $\alpha \in \mathbb{R}^+ \setminus \mathbb{N}$ where $m - \alpha \in \,]0,1[$. In view of Propositions 9.4 and 9.7, the hypotheses on f imply that $f^{(k)} \in \operatorname{dom} J^{\gamma}_{0+,c+k}$ for $k = 0, \ldots, m$ and any $\gamma \in \,]0,1[$. By a simple change of variable we find that

$$(J^{\gamma}_{0+,c} f)(x) = \frac{1}{\Gamma(\gamma)} \int_1^{\infty} \frac{1}{v^{c+1}} (\log v)^{\gamma-1} f\left(\frac{x}{v}\right) dv.$$

Differentiation with respect to x, carried out inside the integral on the right-hand side, leads us to

$$(J^{\gamma}_{0+,c} f)'(x) = \frac{1}{\Gamma(\gamma)} \int_1^{\infty} \frac{1}{v^{c+2}} (\log v)^{\gamma-1} f'\left(\frac{x}{v}\right) dv = (J^{\gamma}_{0+,c+1} f')(x).$$

Since $f' \in \operatorname{dom} J^{\gamma}_{0+,c+1}$, we know that the integrand is absolutely integrable, which justifies the interchange of differentiation and integration. Repeating this procedure successively, we obtain

$$(J^{\gamma}_{0+,c} f)^{(k)}(x) = (J^{\gamma}_{0+,c+k} f^{(k)})(x) \qquad (k = 1, \ldots, m).$$

Hence Proposition 3.21 yields

$$(D^{\alpha}_{0+,c} f)(x) = \Theta^m_c (J^{m-\alpha}_{0+,c} f)(x) = \sum_{k=0}^{m} S_c(m,k) x^k (J^{m-\alpha}_{0+,c+k} f^{(k)})(x),$$

as was to be shown. □

There exists a simple formula for $D^{\alpha}_{0+} f$ when f is given by a power series in accordance with Proposition 9.11.

Proposition 9.20 *Let $\alpha \in \mathbb{R}^+$. Suppose that f is given by $f(x) = \sum_{k=1}^{\infty} a_k x^k$ with the series converging on $\mathbb{R}^+$. Then*

$$(D^{\alpha}_{0+} f)(x) = \sum_{k=1}^{\infty} a_k k^{\alpha} x^k \qquad (x \in \mathbb{R}^+).$$

Proof Setting $m = \lceil \alpha \rceil$ and employing Proposition 9.11, we find that

$$(D^{\alpha}_{0+} f)(x) = \Theta^m (J^{m-\alpha}_{0+} f)(x) = \Theta^m \sum_{k=1}^{\infty} k^{-(m-\alpha)} a_k x^k$$

$$= \sum_{k=1}^{\infty} a_k k^{-(m-\alpha)} \Theta^m x^k = \sum_{k=1}^{\infty} a_k k^{\alpha} x^k,$$

where the interchange of the operator Θ^m with the summation can be justified by the absolute convergence of the power series defining f. □

Now we introduce certain Mellin–Sobolev type spaces in a way slightly different from that of Sect. 3.5, but more suitable for the subsequent considerations. First we recall the space AC_{loc} and redefine it in the following equivalent form

$$AC_{\text{loc}} = \left\{ f : \mathbb{R}^+ \to \mathbb{C} : f(x) = \int_0^x g(t)dt \text{ for } g \in L^1_{\text{loc}}(\mathbb{R}^+) \right\}.$$

Note that for any $f \in AC_{\text{loc}}$ there exists the usual derivative f' satisfying $f' = g$ almost everywhere. For $c \in \mathbb{R}$, we say that $f \in AC_{c,\text{loc}}$ if $(\cdot)^c f(\cdot) \in AC_{\text{loc}}$. For $m \in \mathbb{N}$ these spaces are generalized by defining $AC^1_{c,\text{loc}} := AC_{c,\text{loc}}$ and

$$AC^m_{c,\text{loc}} := \Big\{ f \in AC_{c,\text{loc}} : \Theta^j((\cdot)^c f(\cdot)) \in X_{0,\text{loc}} \; (j = 1, 2, \ldots m-1) \text{ and}$$

$$\Theta^{m-1}((\cdot)^c f(\cdot)) \in AC_{\text{loc}} \Big\}.$$

The following result is of basic interest.

Lemma 9.21 *Let $c \in \mathbb{R}$ and let $m \geq 2$ be an integer. Suppose that $f \in AC^m_{c,\text{loc}}$. Then*

$$\Theta^j(x^c f(x)) = \int_0^x \Theta^{j+1}(u^c f(u)) \frac{du}{u} \qquad (j = 0, 1, \ldots, m-2)$$

and $\lim_{x \to 0+} \Theta^j(x^c f(x)) = 0$ *for* $j = 0, 1, \ldots, m-1$.

Proof Since by the hypotheses $\Theta^{m-1}((\cdot)^c f(\cdot)) \in AC_{\text{loc}}$, we immediately see that $\lim_{x \to 0+} \Theta^{m-1}(x^c f(x)) = 0$.

Now, turning to $j = m-2$, we note that

$$\frac{d}{dx}\Big(\Theta^{m-2}(x^c f(x))\Big) = x^{-1}(x\frac{d}{dx})\Theta^{m-2}(x^c f(x)) = x^{-1}\Theta^{m-1}(x^c f(x))$$

and so $\Theta^{m-2}((\cdot)^c f(\cdot) \in C^1(\mathbb{R}^+)$. Integrating over an interval $[t, x]$ with $0 < t < x$, we obtain

$$\Theta^{m-2}(x^c f(x)) = \Theta^{m-2}(t^c f(t)) + \int_t^x \Theta^{m-1}(u^c f(u))\frac{du}{u}.$$

Since the integrand belongs to $X_{0,\mathrm{loc}}$, we conclude that, for fixed x, the limit

$$\lim_{t\to 0+} \Theta^{m-2}(t^c f(t)) =: A$$

exists. If A were positive, then there would exist a $\nu > 0$ such that $\Theta^{m-2}(t^c f(t)) > A/2$ for every $t \in]0, \nu[$. Hence

$$\int_0^\nu \Theta^{m-2}(t^c f(t))\frac{dt}{t} > \frac{A}{2}\int_0^\nu \frac{dt}{t} = \infty,$$

which is impossible since $\Theta^{m-2}((\cdot)^c f(\cdot)) \in X_{0,loc}$. Similarly one can show that A cannot be negative. Thus $A = 0$ and therefore

$$\Theta^{m-2}(x^c f(x)) = \int_0^x \Theta^{m-1}(u^c f(u))\frac{du}{u},$$

that is, $\Theta^{m-2}((\cdot)^c f(\cdot)) \in AC_{loc}$ with $g(t) = t^{-1}\Theta^{m-1}(t^c f(t))$ in the above definition. This shows that the assertion holds for $j = m - 2$.

By repeating this reasoning for $j = m-3, m-4, \ldots, 0$, we complete the proof. □

Now, for $c \in \mathbb{R}$ and $m \in \mathbb{N}$, we introduce a local version of the Mellin–Sobolev space $W_c^{m,1}(\mathbb{R}^+)$ by

$$W^{m,1}_{c,\mathrm{loc}} := \{f \in X_{c,\mathrm{loc}} \ : \ f = g \text{ a.e. on } \mathbb{R}^+, \text{ where } g \in AC^m_{c,\mathrm{loc}}\}.$$

By Theorem 6.95, we recall that if $f \in X_c$ and $J^m_{0+,c} f \in X_c$, then $J^m_{0+,c} f \in W_c^{m,1}(\mathbb{R}^+)$.

The fact that $f \in W_c^{m,1}(\mathbb{R}^+)$ implies $f \in X_c$ and $\Theta_c^m f \in X_c$, suggests a way how to introduce fractional versions of the above spaces. For a given $\alpha > 0$, we define

$$W(X_c, D^\alpha_{0+,c}) := \{f \in X_c \ : \ D^\alpha_{0+,c} f \text{ exists as a function in } X_c\}$$

and its local version

$$W(X_{c,\mathrm{loc}}, D^\alpha_{0+,c}) := \{f \in X_{c,\mathrm{loc}} \ : \ D^\alpha_{0+,c} f \text{ exists as a function in } X_{c,\mathrm{loc}}\}.$$

Analogously, if J is an interval, we define the space $W(X_J, D^\alpha_{0+})$ as

$$W(X_J, D^\alpha_{0+}) := \left\{ f \in X_J \ : \ D^\alpha_{0+,c} f \ \text{ exists for each } \ c \in J \text{ as a function in } X_J \right\}$$

and its local version $W(X_{J,\mathrm{loc}}, D^\alpha_{0+})$.

The following proposition gives a representation of the Mellin–Hadamard fractional derivative $D^\alpha_{0+,c} f$ under local assumptions on f.

Proposition 9.22 *Let $c \in \mathbb{R}$, $\alpha \in \mathbb{R}^+ \setminus \mathbb{N}$ and set $m := \lceil \alpha \rceil$. Suppose that $f \in W(X_{c,\mathrm{loc}}, D^\alpha_{0+,c})$ and $\Theta^m_c f \in X_{c,\mathrm{loc}}$. Then*

$$(D^\alpha_{0+,c} f)(x) = \Theta^m_c (J^{m-\alpha}_{0+,c} f)(x) = J^{m-\alpha}_{0+,c} (\Theta^m_c f)(x).$$

Proof By Proposition 9.7, the hypotheses on f imply that $f \in \operatorname{dom} J^{m-\alpha}_{0+,c}$ and $\Theta^m_c f \in \operatorname{dom} J^{m-\alpha}_{0+,c}$. Hence the fractional integrals exist. The first equation holds by the definition of the Mellin–Hadamard fractional derivative. We only have to prove the second one. Recalling the definition of the involved operators, substituting $v = x/t$ and employing Proposition 3.21, we conclude that

$$\begin{aligned}
(D^\alpha_{0+,c} f)(x) &= \Theta^m_c (J^{m-\alpha}_{0+,c} f)(x) \\
&= \Theta^m_c \left[\frac{1}{\Gamma(m-\alpha)} \int_0^x \left(\frac{v}{x}\right)^c \left(\log \frac{x}{v}\right)^{m-\alpha-1} f(v) \frac{dv}{v} \right] \\
&= \Theta^m_c \left[\frac{1}{\Gamma(m-\alpha)} \int_1^\infty (\log t)^{m-\alpha-1} f\left(\frac{x}{t}\right) \frac{dt}{t^{c+1}} \right] \\
&= \frac{1}{\Gamma(m-\alpha)} \int_1^\infty \sum_{k=0}^m S_c(m,k) x^k \frac{d^k}{dx^k} f\left(\frac{x}{t}\right) (\log t)^{m-\alpha-1} \frac{dt}{t^{c+1}}.
\end{aligned}$$

Now, resubstituting $t = x/v$ and using again Proposition 3.21, we obtain

$$\begin{aligned}
(D^\alpha_{0+,c} f)(x) &= \frac{1}{\Gamma(m-\alpha)} \int_0^x \sum_{k=0}^m S_c(m,k) v^k f^{(k)}(v) \left(\log \frac{x}{v}\right)^{m-\alpha-1} \left(\frac{v}{x}\right)^c \frac{dv}{v} \\
&= \frac{1}{\Gamma(m-\alpha)} \int_0^x \left(\frac{v}{x}\right)^c \left(\log \frac{x}{v}\right)^{m-\alpha-1} \Theta^m_c f(v) \frac{dv}{v} \\
&= J^{m-\alpha}_{0+,c} (\Theta^m_c f)(x).
\end{aligned}$$

This completes the proof. □

Next we extend the formula for the Mellin transform of $\Theta^m_c f$ to Mellin–Hadamard fractional derivatives. Here the power of a complex number z with exponent α is defined by its principal value that is, the argument of z is taken in the interval $]-\pi, \pi]$.

Proposition 9.23 *Let $\alpha \in \mathbb{R}^+$ and set $m := \lceil \alpha \rceil$.*

(i) Let $c \in \mathbb{R}$. Suppose that $f \in W(X_c, D^\alpha_{0+,c})$ and in addition that $J^{m-\alpha}_{0+,c} f \in W^{m,1}_c$ if $m > \alpha$. Then

$$[D^\alpha_{0+,c} f]^\wedge_{M_c}(c+it) = (-it)^\alpha \, [f]^\wedge_{M_c}(c+it) \qquad (t \in \mathbb{R}).$$

(ii) Let $a, b \in \mathbb{R}$ with $a < b$. Suppose that $f \in W(X_{[a,b]}, D^\alpha_{0+})$ and that $J^{m-\alpha}_{0+,c} f \in W^{m,1}_\nu$ for $c, \nu \in [a, b]$. Then

$$\left[D^\alpha_{0+,c} f\right]^\wedge_M(\nu+it) = (c-\nu-it)^\alpha [f]^\wedge_M(\nu+it) \qquad (t \in \mathbb{R},\ c, \nu \in [a, b]).$$

Proof When $\alpha \in \mathbb{N}$, the results are exactly those of Theorems 6.14 and 6.15. We may therefore assume that $\alpha \in \mathbb{R}^+ \setminus \mathbb{N}$.

As to (i), we first recall that $f \in \operatorname{dom} J^{m-\alpha}_{0+,c}$; see Proposition 9.7. Now, using Definition 9.17, Theorem 6.14 and Proposition 9.14, we obtain

$$\begin{aligned}
[D^\alpha_{0+,c} f]^\wedge_{M_c}(c+it) &= \left[\Theta^m_c \left(J^{m-\alpha}_{0+,c} f\right)\right]^\wedge_{M_c}(c+it) = (-it)^m \left[J^{m-\alpha}_{0+,c} f\right]^\wedge_{M_c}(c+it) \\
&= (-it)^m \, (-it)^{\alpha-m} \, [f]^\wedge_{M_c}(c+it) = (-it)^\alpha [f]^\wedge_{M_c}(c+it)
\end{aligned}$$

with $(-it)^\alpha = |t|^\alpha e^{-i\,\mathrm{sgn}(t)\pi\alpha/2}$.

As to (ii), let $c, \nu \in [a, b]$ be fixed. Then by the hypotheses in conjunction with Theorem 6.14 and formula (9.12), we conclude that

$$\begin{aligned}
\left[D^\alpha_{0+,c} f\right]^\wedge_M(\nu+it) &= \left[\Theta^m_c \left(J^{m-\alpha}_{0+,c} f\right)\right]^\wedge_M(\nu+it) \\
&= (c-\nu-it)^m \left[J^{m-\alpha}_{0+,c} f\right]^\wedge_M(\nu+it) \\
&= (c-\nu-it)^m (c-\nu-it)^{\alpha-m} [f]^\wedge_M(\nu+it) \\
&= (c-\nu-it)^\alpha [f]^\wedge_M(\nu+it),
\end{aligned}$$

as was to be shown. □

Our next aim is to establish a fractional version of the fundamental theorem of differential and integral calculus in the Mellin setting. First we consider again the case $\alpha = m \in \mathbb{N}$ and recall that $J^m_{0+,c} f$, represented by iterated integrals, is m-times differentiable whenever $f \in \operatorname{dom} J^m_{0+,c}$. Now we restate the fundamental theorem of differential and integral calculus for Mellin integrals of integer order (see Theorem 6.95) for the spaces $X_{c,\mathrm{loc}}$.

Proposition 9.24 *For $c \in \mathbb{R}$, the following statements hold:*

(i) Let $f \in W^{1,1}_{c,\mathrm{loc}}$. Then

$$(J^1_{0+,c}(\Theta_c f))(x) = f(x) \qquad (a.e.\ x \in \mathbb{R}^+).$$

(ii) Let $m \in \mathbb{N}$ with $m > 1$, and let $f \in W^{m,1}_{c,\mathrm{loc}}$ be such that $\Theta_c^m f \in \operatorname{dom} J^m_{0+,c}$. Then

$$(J^m_{0+,c}(\Theta_c^m f))(x) = f(x) \qquad (a.e.\ x \in \mathbb{R}^+).$$

(iii) Let $f \in X_{c,\mathrm{loc}}$. Then

$$\Theta_c(J^1_{0+,c} f)(x) = f(x) \qquad (a.e.\ x \in \mathbb{R}^+).$$

(iv) Let $m \in \mathbb{N}$ and let $f \in \operatorname{dom} J^m_{0+,c}$. Then

$$\Theta_c^m(J^m_{0+,c} f)(x) = f(x) \qquad (a.e.\ x \in \mathbb{R}^+).$$

Proof As to (i), we have by the absolute continuity,

$$(J^1_{0+,c}(\Theta_c f))(x) = \int_0^x \left(\frac{u}{x}\right)^c (\Theta_c f)(u)\,\frac{du}{u} = x^{-c}\int_0^x \frac{d}{du}\big(u^c f(u)\big)du = f(x)$$

almost everywhere on $\mathbb{R}^+$.

Concerning (ii), we may use Theorem 9.6 (iii) for representing $J^m_{0+,c} f$ by m iterated integrals as in formula (9.2). Then, employing Lemma 9.21 and Proposition 9.18, we can perform m steps similar to the proof of (i) for arriving at the desired result.

As to (iii), we note that

$$\begin{aligned}\Theta_c(J^1_{0+,c} f)(x) &= x\frac{d}{dx}(J^1_{0+,c} f)(x) + c(J^1_{0+,c} f)(x)\\ &= -cx^{-c}\int_0^x u^{c-1} f(u)du + f(x) + c\int_0^x \left(\frac{u}{x}\right)^c f(u)\,\frac{du}{u} = f(x)\end{aligned}$$

almost everywhere on $\mathbb{R}^+$.

Finally we prove (iv) by induction. Assuming that (iv) holds for $m-1$, we conclude by using Theorem 9.6 (iii) that

$$\Theta_c^m(J^m_{0+,c} f)(x) = \Theta_c(\Theta_c^{m-1}(J^{m-1}_{0+,c}(J^1_{0+,c} f)))(x) = f(x)$$

almost everywhere on $\mathbb{R}^+$. □

Proposition 9.24 gives a version of Theorem 6.95 for the spaces $X_{c,\mathrm{loc}}$ without the use of Mellin transforms and under minimal assumptions. As a consequence for the spaces X_c, we can deduce again the formula for the Mellin transform of $J^m_{0+,c}f$ (see Proposition 9.14) whenever $J^m_{0+,c}f \in X_c$, namely

$$[J^m_{0+,c}f]^\wedge_{M_c}(c+it) = (-it)^{-m}[f]^\wedge_{M_c}(c+it).$$

Now we are ready to prove the fundamental theorem of the fractional differential and integral calculus in the Mellin setting.

Theorem 9.25 *For $c \in \mathbb{R}$, $\alpha \in \mathbb{R}^+$ and $m := \lceil \alpha \rceil$ the following statements hold:*

(a) Let $f \in W(X_{c,\mathrm{loc}}, D^\alpha_{0+,c}) \cap W^{m,1}_{c,\mathrm{loc}}$ and suppose that $D^\alpha_{0+,c}f, \Theta^m_c f \in$ dom $J^m_{0+,c}$. Then

$$(J^\alpha_{0+,c}(D^\alpha_{0+,c}f))(x) = f(x) \qquad (a.e.\ x \in \mathbb{R}^+).$$

(b) Let $f \in$ dom $J^m_{0+,c}$. Then

$$(D^\alpha_{0+,c}(J^\alpha_{0+,c}f))(x) = f(x) \qquad (a.e.\ x \in \mathbb{R}^+).$$

Proof We may assume that $m > \alpha$ since the case $m = \alpha$ is already settled by Proposition 9.24.

As to part (a), we conclude by using Proposition 9.22, Theorem 9.6 and Proposition 9.24 that

$$\begin{aligned}(J^\alpha_{0+,c}(D^\alpha_{0+,c}f))(x) &= \left(J^\alpha_{0+,c}(\Theta^m_c(J^{m-\alpha}_{0+,c}f))\right)(x)\\ &= \left(J^\alpha_{0+,c}(J^{m-\alpha}_{0+,c}(\Theta^m_c f))\right)(x) = (J^m_{0+,c}(\Theta^m_c f))(x) = f(x)\end{aligned}$$

almost everywhere on $\mathbb{R}^+$.

As to part (b), we obtain by employing Theorem 9.6 and Proposition 9.24 that

$$(D^\alpha_{0+,c}(J^\alpha_{0+,c}f))(x) = \Theta^m_c\left(J^{m-\alpha}_{0+,c}(J^\alpha_{0+,c}f)\right)(x) = \Theta^m_c(J^m_{0+,c}f)(x) = f(x)$$

almost everywhere on $\mathbb{R}^+$. □

Now we consider compositions between the operators of Hadamard type fractional integrals and derivatives in local spaces.

Theorem 9.26 *Let $c \in \mathbb{R}$, let $\alpha, \beta \in \mathbb{R}^+$ with $\beta > \alpha$, and set $m := \lceil \alpha \rceil$.*

(a') If $f \in W(X_{c,\mathrm{loc}}, D^\alpha_{0+,c}) \cap W^{m,1}_{c,\mathrm{loc}}$, $D^\alpha_{0+,c}f \in$ dom $J^\beta_{0+,c}$ and $\Theta^m_c f \in$ dom $J^{m+\beta-\alpha}_{0+,c}$, then

$$(J^\beta_{0+,c}(D^\alpha_{0+,c}f))(x) = (J^{\beta-\alpha}_{0+,c}f)(x) \qquad (a.e.\ x \in \mathbb{R}^+).$$

(b') If $f \in \operatorname{dom} J_{0+,c}^{m+\beta-\alpha}$*, then*

$$(D_{0+,c}^{\alpha}(J_{0+,c}^{\beta} f))(x) = (J_{0+,c}^{\beta-\alpha} f)(x) \qquad (a.e.\ x \in \mathbb{R}^+).$$

Proof Regarding (a'), we first assume that $m > \alpha$ and find by arguing as in the proof of the previous theorem that

$$\begin{aligned}(J_{0+,c}^{\beta}(D_{0+,c}^{\alpha} f))(x) &= \big(J_{0+,c}^{\beta}(\Theta_c^m(J_{0+,c}^{m-\alpha} f))\big)(x) = \big(J_{0+,c}^{\beta}(J_{0+,c}^{m-\alpha}(\Theta_c^m f))\big)(x)\\ &= \big(J_{0+,c}^{\beta-\alpha}(J_{0+,c}^{m}(\Theta_c^m f))\big)(x) = (J_{0+,c}^{\beta-\alpha} f)(x)\end{aligned}$$

almost everywhere on $\mathbb{R}^+$. For $m = \alpha$, we simply have to replace $J_{0+,c}^{m-\alpha}$ with the identity operator.

Regarding (b'), we have

$$(D_{0+,c}^{\alpha}(J_{0+,c}^{\beta} f))(x) = \Theta_c^m(J_{0+,c}^{m-\alpha+\beta} f)(x) = (J_{0+,c}^{\beta-\alpha} f)(x)$$

almost everywhere on $\mathbb{R}^+$. □

Employing Proposition 9.23, we now state the fundamental formulae of integral and differential calculus in the fractional Mellin setting for the spaces $W(X_c, D_{0+,c}^{\alpha})$. In view of the uniqueness theorem (Theorem 6.79), it always suffices to show that the asserted equations hold for their Mellin transforms. We begin with the following statement.

Theorem 9.27 *Let* $c \in \mathbb{R}$*, let* $\alpha \in \mathbb{R}^+$ *and set* $m := \lceil \alpha \rceil$*.*

(a) Suppose that $f \in W(X_c, D_{0+,c}^{\alpha})$*,* $D_{0+,c}^{\alpha} f \in \operatorname{dom} J_{0+,c}^{\alpha}$*,* $J_{0+,c}^{\alpha}(D_{0+,c}^{\alpha} f) \in X_c$*,* $J_{0+,c}^{\alpha-\lfloor\alpha\rfloor}(D_{0+,c}^{\alpha} f) \in X_c$ *and* $J_{0+,c}^{m-\alpha} f \in W_c^{m,1}$*. Then*

$$(J_{0+,c}^{\alpha}(D_{0+,c}^{\alpha} f))(x) = f(x) \qquad (a.e.\ x \in \mathbb{R}^+).$$

(b) Suppose that $f \in \operatorname{dom} J_{0+,c}^{m} \cap X_c$*,* $J_{0+,c}^{m} f \in W_c^{m,1}$*,* $J_{0+,c}^{\alpha} f \in W(X_c, D_{0+,c}^{\alpha})$ *and* $J_{0+,c}^{\alpha-\lfloor\alpha\rfloor} f \in X_c$*. Then*

$$(D_{0+,c}^{\alpha} J_{0+,c}^{\alpha} f)(x) = f(x) \qquad (a.e.\ x \in \mathbb{R}^+).$$

Proof As to part (a), we use Propositions 9.14 and 9.23 for calculating the Mellin transforms and obtain

$$\begin{aligned}[J_{0+,c}^{\alpha}(D_{0+,c}^{\alpha} f)]_{M_c}^{\wedge}(c+it) &= (-it)^{-\alpha}[D_{0+,c}^{\alpha} f]_{M_c}^{\wedge}(c+it)\\ &= (-it)^{-\alpha}(-it)^{\alpha}[f]_{M_c}^{\wedge}(c+it) = [f]_{M_c}^{\wedge}(c+it),\end{aligned}$$

which implies the assertion.

As to part (b), we see with the help of Theorem 9.6 that $f \in \operatorname{dom} J^{\alpha}_{0+,c} \cap \operatorname{dom} J^{m-\alpha}_{0+,c}$. Using Propositions 9.23 and 9.14, we find that

$$[D^{\alpha}_{0+,c}(J^{\alpha}_{0+,c}f)]^{\wedge}_{M_c}(c+it) = (-it)^{\alpha}[J^{\alpha}_{0+,c}f]^{\wedge}_{M_c}(c+it) = [f]^{\wedge}_{M_c}(c+it).$$

□

Similarly the conclusions of Theorem 9.26 can be obtained under different hypotheses by methods based on Mellin transform. Details are left to the reader.

9.3 Pointwise Mellin Fractional Derivatives and Interconnections

In this section we introduce a second notion of a pointwise fractional differential operator in the Mellin setting. It is based on fractional differences of order $\alpha > 0$. We begin with the definition of these differences. Later we shall determine a subspace in which the two definitions of pointwise fractional derivatives coincide almost everywhere (see Theorem 9.34).

The studies in Sect. 6.7 show that for $r \in \mathbb{N}$ we have

$$\lim_{h\to 1} \frac{(\Delta^{r,c}_h f)(x)}{(h-1)^r} = \Theta^r_c f(x)$$

under appropriate assumptions on f with $\Delta^{r,c}_h f$ as introduced in Sect. 3.3. It is easily verified that this limit does not change, when in the numerator $(\Delta^{r,c}_h f)(x)$ is replaced by

$$(-1)^r(\Delta^{r,c}_{1/h} f)(x) = \sum_{j=0}^{r}(-1)^j\binom{r}{j}h^{-cj}f(h^{-j}x) =: (\varDelta^{r,c}_h f)(x). \tag{9.14}$$

For an extension to fractional differences, this operator $\varDelta^{r,c}_h$ turns out to be more convenient. Please do not confuse the letters Δ and $\varDelta$.

Definition 9.28 Let $c \in \mathbb{R}$ and $\alpha \in \mathbb{R}^+$. The *Mellin fractional difference* of $f \in X_c$ of *order* α is defined by

$$\begin{aligned}(\varDelta^{\alpha,c}_h f)(x) &:= \sum_{j=0}^{\infty}(-1)^j\binom{\alpha}{j}(\tau^c_{h^{-j}} f)(x)\\ &= \sum_{j=0}^{\infty}(-1)^j\binom{\alpha}{j}h^{-cj}f(h^{-j}x) \quad (x, h \in \mathbb{R}^+),\end{aligned}$$

where

$$\binom{\alpha}{j} := \frac{\alpha(\alpha-1)\cdots(\alpha-j+1)}{j!}.$$

Note that for $\alpha = r \in \mathbb{N}$ the difference $\Delta_h^{\alpha,c} f$ reduces to the finite sum (9.14) since $\binom{r}{j} = 0$ when $j > r$.

First we presents some basic properties of the Mellin fractional difference.

Proposition 9.29 *Let $c \in \mathbb{R}$ and $\alpha, \beta, h \in \mathbb{R}^+$. For $f \in X_c$ the difference $(\Delta_h^{\alpha,c} f)(x)$ exists almost everywhere and has the following properties:*

(i) $\|\Delta_h^{\alpha,c} f\|_{X_c} \le \|f\|_{X_c} \sum_{j=0}^{\infty} \left|\binom{\alpha}{j}\right| < \infty$;

(ii) $[\Delta_h^{\alpha,c} f]^{\wedge}_{M_c}(c+it) = (1-h^{it})^{\alpha}[f]^{\wedge}_{M_c}(c+it) \qquad (t \in \mathbb{R})$
with the power of exponent α being defined by its principal value;

(iii) the differences form a semigroup in the sense that

$$(\Delta_h^{\alpha,c}\Delta_h^{\beta,c} f)(x) = (\Delta_h^{\alpha+\beta,c} f)(x) \qquad (x \in \mathbb{R}^+).$$

Proof Before proving the statements (i)–(iii), we have to care for the existence of the fractional difference. From its definition, we deduce that

$$|(\Delta_h^{\alpha,c} f)(x)| \le \frac{1}{x^c}\sum_{j=0}^{\infty}\left|\binom{\alpha}{j}\right| h^{-cj}x^c|f(h^{-j}x)| \qquad (x, h \in \mathbb{R}^+) \tag{9.15}$$

provided that the series on the right-hand side converges. This is trivial when $\alpha \in \mathbb{N}$. For $\alpha \in \mathbb{R}^+ \setminus \mathbb{N}$ we want to prove convergence by using Beppo Levi's theorem (Proposition 1.3) applied to the sequence of partial sums. Substituting $h^{-j}x = t$, we find that

$$\sum_{j=0}^{\infty}\left|\binom{\alpha}{j}\right| \int_0^{\infty} (h^{-j}x)^c |f(h^{-j}x)| \frac{dx}{x} = \|f\|_{X_c} \sum_{j=0}^{\infty}\left|\binom{\alpha}{j}\right|. \tag{9.16}$$

We have to verify that the right-hand side is finite. For this purpose, we use Euler's formula for the gamma function (1.16), which allows us to conclude that

$$\left|\binom{\alpha}{j}\right| = \frac{1}{|\Gamma(-\alpha)|\, j^{\alpha+1}}\big(1+o(1)\big) \qquad (j \to \infty).$$

With this representation of the generalized binomial coefficients, we find that the series on the right-hand side of (9.16) is finite. Now Beppo Levi's theorem ensures that the series in (9.15) converges almost everywhere on $\mathbb{R}^+$ and

$$\int_0^\infty \sum_{j=0}^{\infty} \left|\binom{\alpha}{j}\right| (h^{-j}x)^c |f(h^{-j}x)| \frac{dx}{x} = \|f\|_{X_c} \sum_{j=0}^{\infty} \left|\binom{\alpha}{j}\right|.$$

Combined with (9.15), we obtain that $(\varDelta_h^{\alpha,c} f)(x)$ exists almost everywhere on $\mathbb{R}^+$ and (i) holds.

Next we turn to (ii). Noting that for $h > 0$ the complex number h^{it} is of modulus 1, we see that the series

$$\sum_{j=0}^{\infty} \binom{\alpha}{j} (-1)^j h^{itj}$$

is absolutely convergent and bounded for $h \in \mathbb{R}^+$. Therefore a theorem of Abel–Stolz on power series (see, e.g., [4]) yields that

$$\sum_{j=0}^{\infty} \binom{\alpha}{j} (-1)^j h^{itj} = (1 - h^{it})^\alpha.$$

Hence

$$\begin{aligned} (1 - h^{it})^\alpha [f]^\wedge_{M_c}(c + it) &= \sum_{j=0}^{\infty} \binom{\alpha}{j} (-1)^j h^{itj} [f]^\wedge_{M_c}(c + it) \\ &= \sum_{j=0}^{\infty} \binom{\alpha}{j} (-1)^j h^{itj} \int_0^\infty f(u) u^{c+it} \frac{du}{u}. \end{aligned} \tag{9.17}$$

Taking the moduli inside the summation and integration, we find that the right-hand side of (9.17) is bounded by

$$\|f\|_{X_c} \sum_{j=0}^{\infty} \left|\binom{\alpha}{j}\right| < \infty.$$

Thus we are allowed to interchange summation and integration in (9.17). Doing so and substituting $u = h^{-j}x$, we obtain

$$\begin{aligned} (1 - h^{it})^\alpha [f]^\wedge_{M_c}(c + it) &= \int_0^\infty \sum_{j=0}^{\infty} \binom{\alpha}{j} (-1)^j h^{-jc} f(h^{-j}x) x^{c+it} \frac{dx}{x} \\ &= [\varDelta_h^{\alpha,c} f]^\wedge_{M_c}(c + it), \end{aligned}$$

and so (ii) is verified.

It remains to prove (iii). Using (ii) repeatedly and noting that the existence of the involved Mellin transforms is guaranteed by (i), we obtain

$$\begin{aligned}[\Delta_h^{\alpha,c}\Delta_h^{\beta,c}f]^\wedge_{M_c}(c+it) &= (1-h^{it})^\alpha[\Delta_h^{\beta,c}f]^\wedge_{M_c}(c+it)\\ &= (1-h^{it})^{\alpha+\beta}[f]^\wedge_{M_c}(c+it)\\ &= [\Delta_h^{\alpha+\beta,c}f]^\wedge_{M_c}(c+it),\end{aligned}$$

and so the assertion follows from the uniqueness theorem for the Mellin transform (see Theorem 6.79). □

For the spaces $X_{[a,b]}$, we state the following results.

Proposition 9.30 *Let $a, b \in \mathbb{R}$ with $a < b$ and let $\alpha \in \mathbb{R}^+$. Suppose that $f \in X_{[a,b]}$ and $c \in]a, b[$.*

(i) If $h > 1$, then $\Delta_h^{\alpha,c} f \in X_{[a,c]}$ and

$$\|\Delta_h^{\alpha,c} f\|_{X_\nu} \le \|f\|_{X_\nu} \sum_{j=0}^{\infty} \left|\binom{\alpha}{j}\right| h^{(\nu-c)j}$$

for all $\nu \in [a, c[$. Moreover,

$$[\Delta_h^{\alpha,c} f]^\wedge_M(\nu+it) = (1-h^{\nu-c+it})^\alpha [f]^\wedge_M(\nu+it) \qquad (t \in \mathbb{R}). \tag{9.18}$$

(ii) If $0 < h \le 1$, then $\Delta_h^{\alpha,c} f \in X_{[c,b]}$ and

$$\|\Delta_h^{\alpha,c} f\|_{X_\mu} \le \|f\|_{X_\mu} \sum_{j=0}^{\infty} \left|\binom{\alpha}{j}\right| h^{(\mu-c)j}$$

for all $\mu \in]c, b]$. Moreover,

$$[\Delta_h^{\alpha,c} f]^\wedge_M(\mu+it) = (1-h^{\mu-c+it})^\alpha [f]^\wedge_M(\mu+it) \qquad (t \in \mathbb{R}). \tag{9.19}$$

Proof We only indicate a proof of (i) since (ii) can be verified in a similar way. Let $\nu \in [a, c[$ be fixed. Using an analogous reasoning as in the proof of Proposition 9.29, we conclude that

$$\|\Delta_h^{\alpha,c} f\|_{X_\nu} \le \int_0^\infty x^{\nu-1} \sum_{j=0}^{\infty} \left|\binom{\alpha}{j}\right| h^{-cj} |f(h^{-j}x)| dx = \|f\|_{X_\nu} \sum_{j=0}^{\infty} \left|\binom{\alpha}{j}\right| h^{j(\nu-c)},$$

the last series being absolutely convergent for $h > 1$. The assertion (9.18) is obtained by interchanging summation and the integration for calculating the Mellin transform just as in the proof of statement (ii) of Proposition 9.29. □

The Mellin fractional difference will now be used for introducing the notion of a pointwise fractional Mellin derivative.

Definition 9.31 Let $c \in \mathbb{R}$ and $\alpha \in \mathbb{R}^+$. The *(left-sided) pointwise Mellin fractional derivative of order* α of a function $f : \mathbb{R}^+ \to \mathbb{C}$ at a point $x \in \mathbb{R}^+$ is defined by

$$\lim_{h\to 1+} \frac{(\Delta_h^{\alpha,c} f)(x)}{(h-1)^\alpha} =: \Theta_c^\alpha f(x)$$

provided that the limit exists.

A notion of right-sided Mellin fractional derivative is briefly described in Sect. 9.5 Notes.

In addition to this pointwise fractional derivative, the following "strong" version is also of interest.

Definition 9.32 Let $c \in \mathbb{R}$, $\alpha \in \mathbb{R}^+$ and let $f \in X_c$. Suppose that there exists a function $g \in X_c$ such that

$$\lim_{h\to 1+} \left\| \frac{\Delta_h^{\alpha,c} f}{(h-1)^\alpha} - g \right\|_{X_c} = 0.$$

Then g is called the *(left-sided) strong fractional Mellin derivative* of f of *order* α and is denoted by $g(x) = \text{s-}\Theta_c^\alpha f(x)$. For $\alpha = 0$ we set $\text{s-}\Theta_c^0 f(x) := f(x)$.

If $f \in X_c$ has a strong fractional derivative of order α and a pointwise Mellin fractional derivative almost everywhere on $\mathbb{R}^+$, then clearly

$$\text{s-}\Theta_c^\alpha f(x) = \Theta_c^\alpha f(x) \qquad (\text{a.e. } x \in \mathbb{R}^+).$$

For integer α the definitions of the pointwise and the strong Mellin derivative reduce to the classical ones where the differences are given by a finite sum provided that we replace $h \to 1+$ with $h \to 1$ (see Sect. 6.7).

For fixed $c \in \mathbb{R}$, we now introduce the following function spaces:

$$W(X_c, \Theta_c^\alpha) := \{f \in X_c \,:\, \Theta_c^\alpha \text{ exists as a function in } X_c\},$$

$$W(X_c, \text{s-}\Theta_c^\alpha) := \{f \in X_c \,:\, \text{s-}\Theta_c^\alpha \text{ exists as a function in } X_c\}$$

and for an interval J,

$$W(X_J, \Theta^\alpha) := \{f \in X_J \ : \ \Theta_c^\alpha \text{ exists for each } c \in J \text{ as a function in } X_J\},$$

$$W(X_J, \text{s-}\Theta^\alpha) := \{f \in X_J \ : \ \text{s-}\Theta_c^\alpha \text{ exists for each } c \in J \text{ as a function in } X_J\}.$$

For further considerations the following subspaces will be useful:

$$\mathscr{W}_c^{\alpha,1} := W(X_c, \Theta_c^\alpha) \cap W(X_c, \text{s-}\Theta_c^\alpha) \qquad (c \in \mathbb{R}^+)$$

and

$$\mathscr{W}_J^{\alpha,1} := W(X_J, \Theta^\alpha) \cap W(X_J, \text{s-}\Theta^\alpha).$$

We state the following result which gives a formula for the Mellin transform of the strong and pointwise Mellin fractional derivatives.

Theorem 9.33 *Let $a, b, c \in \mathbb{R}$ with $a < b$ and let $\alpha \in \mathbb{R}^+$.*

(i) If $f \in W(X_c, s\text{-}\Theta_c^\alpha)$, then

$$[s\text{-}\Theta_c^\alpha f]^\wedge_{M_c}(s) = (-it)^\alpha [f]^\wedge_{M_c}(s) \qquad (s = c + it,\ t \in \mathbb{R}).$$

In particular, if $f \in \mathscr{W}_c^{\alpha,1}$, then

$$[\Theta_c^\alpha f]^\wedge_{M_c}(s) = (-it)^\alpha [f]^\wedge_{M_c}(s) \qquad (s = c + it,\ t \in \mathbb{R}).$$

(ii) If $f \in W(X_{[a,b]}, s\text{-}\Theta^\alpha)$, then

$$[s\text{-}\Theta_c^\alpha f]^\wedge_M(\nu + it) = (c - \nu - it)^\alpha [f]^\wedge_M(\nu + it) \qquad \big(\nu, c \in [a, b],\ t \in \mathbb{R}\big).$$

In particular, if $f \in \mathscr{W}_{[a,b]}^{\alpha,1}$, then

$$[\Theta_c^\alpha f]^\wedge_M(\nu + it) = (c - \nu - it)^\alpha [f]^\wedge_M(\nu + it) \qquad \big(\nu, c \in [a, b],\ t \in \mathbb{R}\big).$$

The powers of exponent α are defined by their principal values.

Proof We begin with (i). Since

$$\lim_{h \to 1+} \left(\frac{1 - h^{it}}{h - 1}\right)^\alpha = (-it)^\alpha,$$

we conclude with the help of Proposition 9.29 (ii) that

$$
\begin{aligned}
&\left|(-it)^\alpha [f]^\wedge_{M_c}(s) - [\text{s-}\Theta^\alpha_c f]^\wedge_{M_c}(s)\right| \\
&= \lim_{h\to 1+} \left| \left(\frac{1-h^{it}}{h-1}\right)^\alpha [f]^\wedge_{M_c}(s) - [\text{s-}\Theta^\alpha_c f]^\wedge_{M_c}(s)\right| \\
&= \lim_{h\to 1+} \left| \left[\frac{\Delta_h^{\alpha,c} f}{(h-1)^\alpha}\right]^\wedge_{M_c}(s) - [\text{s-}\Theta^\alpha_c f]^\wedge_{M_c}(s)\right| \\
&= \lim_{h\to 1+} \left| \left[\frac{\Delta_h^{\alpha,c} f}{(h-1)^\alpha} - \text{s-}\Theta^\alpha_c f\right]^\wedge_{M_c}(s)\right| \\
&\le \lim_{h\to 1+} \left\| \frac{\Delta_h^{\alpha,c} f}{(h-1)^\alpha} - \text{s-}\Theta^\alpha_c f\right\|_{X_c} = 0,
\end{aligned}
$$

and so the assertion follows. Assertion (ii) is verified by a similar reasoning using Proposition 9.30 (i). □

Now we shall show that, under suitable assumptions, the Mellin–Hadamard and the Mellin derivatives coincide almost everywhere. As a consequence, we can obtain an alternative proof of the formula for the Mellin transform of $D^\alpha_{0+,c} f$ under a different set of assumptions.

Theorem 9.34 *Let $\alpha \in \mathbb{R}^+$ and set $m := \lceil \alpha \rceil$. For $c \in \mathbb{R}$ suppose that $f \in \mathscr{W}_c^{\alpha,1} \cap W(X_c, D^\alpha_{0+,c})$ and $\Theta^m_c f \in X_c$. Then*

$$(D^\alpha_{0+,c} f)(x) = \Theta^\alpha_c f(x)$$

almost everywhere on $\mathbb{R}^+$.

Proof By Proposition 9.22 we have

$$(D^\alpha_{0+,c} f)(x) = (J^{m-\alpha}_{0+,c}(\Theta^m_c f))(x),$$

which belongs to X_c. Thus, passing to the Mellin transforms and applying Proposition 9.14, Corollary 6.91 and Theorem 9.33, we obtain

$$
\begin{aligned}
[D^\alpha_{0+,c} f]^\wedge_{M_c}(c+it) &= [(J^{m-\alpha}_{0+,c}(\Theta^m_c f))]^\wedge_{M_c}(c+it) = (-it)^{\alpha-m}[\Theta^m_c f]^\wedge_{M_c}(c+it) \\
&= (-it)^\alpha [f]^\wedge_{M_c}(c+it) = [\Theta^\alpha_c f]^\wedge_{M_c}(c+it)
\end{aligned}
$$

for all $t \in \mathbb{R}$. Hence, $D^\alpha_{0+,c} f$ and $\Theta^\alpha_c f$ have the same Mellin transform along the line $s = c + it$, and so the assertion follows by the uniqueness theorem (Theorem 6.79). □

We conclude this section with some results which constitute a synthesis of the developed theory.

Proposition 9.35 *Let $c \in \mathbb{R}$ and let $\alpha \in \mathbb{R}^+$. Suppose that $f \in X_c$, $g \in \operatorname{dom} J^{\alpha}_{0+,c} \cap X_c$, $J^{\alpha}_{0+,c} g \in X_c$ and $J^{\alpha-\lfloor\alpha\rfloor}_{0+,c} g \in X_c$. Then*

$$(-it)^{\alpha}[f]^{\wedge}_{M_c}(c+it) = [g]^{\wedge}_{M_c}(c+it) \tag{9.20}$$

if and only if

$$f(x) = \left(J^{\alpha}_{0+,c} g\right)(x) \qquad (a.e.\ x \in \mathbb{R}^+). \tag{9.21}$$

Proof If (9.20) holds, then, by Proposition 9.14,

$$\left[J^{\alpha}_{0+,c} g\right]^{\wedge}_{M_c}(c+it) = (-it)^{-\alpha}[g]^{\wedge}_{M_c}(c+it) = [f]^{\wedge}_{M_c}(c+it),$$

and so (9.21) follows from the uniqueness theorem of the Mellin transform.

Conversely, if (9.21) holds, then, again by Proposition 9.14,

$$[f]^{\wedge}_{M_c}(c+it) = \left[J^{\alpha}_{0+,c} g\right]^{\wedge}_{M_c}(c+it) = (-it)^{-\alpha}[g]^{\wedge}_{M_c}(c+it),$$

which implies (9.20). □

Under some side conditions, we can prove that if (9.21) holds, then $g = D^{\alpha}_{0+,c} f$.

Proposition 9.36 *Let $c \in \mathbb{R}$, let $\alpha \in \mathbb{R}^+$ and set $m := \lceil\alpha\rceil$. Suppose that $g \in \operatorname{dom} J^{m}_{0+,c} \cap X_{c,\mathrm{loc}}$ and $J^{\alpha}_{0+,c} g \in W(X_{c,\mathrm{loc}}, D^{\alpha}_{0+,c})$. If (9.21) holds, then*

$$(D^{\alpha}_{0+,c} f)(x) = g(x) \qquad (a.e.\ x \in \mathbb{R}^+).$$

Proof Since $g \in \operatorname{dom} J^{m}_{0+,c}$, it follows from Theorem 9.6 that $g \in \operatorname{dom} J^{\alpha}_{0+,c} \cap \operatorname{dom} J^{m-\alpha}_{0+,c}$ and $J^{m-\alpha}_{0+,c}\left(J^{\alpha}_{0+,c} g\right)(x) = (J^{m}_{0+,c} g)(x)$. Thus, applying $D^{\alpha}_{0+,c}$ on both sides of (9.21) and using Proposition 9.24 (iv), we obtain

$$(D^{\alpha}_{0+,c} f)(x) = D^{\alpha}_{0+,c}\left(J^{\alpha}_{0+,c} g\right)(x) = \Theta^{m}_{c}\left(J^{m}_{0+,c} g\right)(x) = g(x) \quad (a.e.\ x \in \mathbb{R}^+),$$

as was to be shown. □

Proposition 9.37 *Let $c \in \mathbb{R}$ and let $\alpha \in \mathbb{R}^+$. Suppose that $f \in X_c$, $g \in \operatorname{dom} J^{\alpha}_{0+,c} \cap X_c$, $J^{\alpha}_{0+,c} g \in X_c$ and $J^{\alpha-\lfloor\alpha\rfloor}_{0+,c} g \in X_c$. If*

$$\lim_{h\to 1+} \left\| \frac{\Delta^{\alpha,c}_{h} f}{(h-1)^{\alpha}} - g \right\|_{X_c} = 0, \tag{9.22}$$

then

$$f(x) = \left(J^{\alpha}_{0+,c} g\right)(x) \qquad (a.e.\ x \in \mathbb{R}^+).$$

Proof The hypotheses and the validity of (9.22) imply that $f \in W(X_c, \text{s-}\Theta_c^\alpha)$ and $g = \text{s-}\Theta_c^\alpha f$. Hence, by Theorem 9.33, Eq. (9.20) of Proposition 9.35 holds and this proposition yields (9.21). □

Conversely, if (9.21) holds and in addition $f \in W(X_c, \text{s-}\Theta_c^\alpha)$, then (9.22) is satisfied, but this can be considered as trivial.

9.4 Some Examples

In this section we determine Mellin fractional integrals, Mellin–Hadamard fractional derivatives and also pointwise Mellin fractional derivatives for some particular functions. First we take up the former Example 9.3 that had been used in the proofs of Propositions 9.11 and 9.24.

Example 9.38 For $b \in \mathbb{R}$, consider the function $g(x) = x^b$. We have shown in Example 9.3 that for $\alpha > 0$ and $b + c > 0$, there holds $g \in \text{dom } J^\alpha_{0+,c}$ and

$$(J^\alpha_{0+,c}g)(x) = (c+b)^{-\alpha}x^b.$$

In particular, for $b > 0$ and $c = 0$, we obtain $(J^\alpha_{0+}g)(x) = b^{-\alpha}x^b$. Setting $m := \lceil\alpha\rceil$, we know that

$$(D^\alpha_{0+,c}g)(x) = \Theta_c^m(J^{m-\alpha}_{0+,c}g)(x) = \Theta_c^m(c+b)^{\alpha-m}x^b.$$

This easily yields

$$(D^\alpha_{0+,c}g)(x) = (c+b)^\alpha x^b.$$

The formulae for $J^\alpha_{0+,c}g$ and $D^\alpha_{0+,c}g$ well enlighten the fundamental theorem of calculus in the fractional Mellin setting.

It should be noted that we cannot calculate $J^\alpha_{0+}1$ and $D^\alpha_{0+}1$ since the function $g(t) = 1$, corresponding to $b = 0$, is not in the domain of J^α_{0+}. However we can calculate $J^\alpha_{0+,c}1$ and $D^\alpha_{0+,c}1$ for $c > 0$, obtaining $(J^\alpha_{0+,c}1)(x) = c^{-\alpha}$ and $(D^\alpha_{0+,c}1)(x) = c^\alpha$.

Next we calculate the pointwise Mellin fractional derivative. For $g(x) = x^b$ and $h > 1$, we see that

$$(\Delta_h^{\alpha,c}g)(x) = \sum_{j=0}^{\infty}(-1)^j\binom{\alpha}{j}h^{-(c+b)j}x^b = x^b\left(1-h^{-c-b}\right)^\alpha = x^b\left(\frac{h^{c+b}-1}{h^{c+b}}\right)^\alpha.$$

Hence

$$\Theta_c^\alpha g(x) = \lim_{h\to 1+}\frac{(\Delta_h^{\alpha,c}g)(x)}{(h-1)^\alpha} = x^b\lim_{h\to 1+}\left(\frac{h^{c+b}-1}{h-1}\right)^\alpha = (c+b)^\alpha x^b$$

as in the case of the pointwise Mellin–Hadamard derivative. This formula even extends to $c = b = 0$.

Example 9.39 Now we consider the function $g(x) = x^b \chi_{]0,1]}(x)$ for $x \in \mathbb{R}^+$, where again $b \in \mathbb{R}$, $c \in \mathbb{R}^+$ and $c + b > 0$. From the previous example we know already that $(J^{\alpha}_{0+,c}g)(x) = (c+b)^{-\alpha}x^b$ for $0 < x \leq 1$. The definition of fractional Mellin integrals yields

$$(J^{\alpha}_{0+,c}g)(x) = \frac{1}{\Gamma(\alpha)} \int_0^1 \left(\frac{u}{x}\right)^c \left(\log\frac{x}{u}\right)^{\alpha-1} u^b \frac{du}{u}$$

for $x > 1$. This integral can be expressed in terms of the *incomplete gamma function*, defined by

$$\Gamma(\alpha, x) := \int_x^{\infty} e^{-t} t^{\alpha-1} dt;$$

see, e.g., [1, No. 6.5.3]. Indeed, by the transformation $t = (c+b)\log(x/u)$, we obtain

$$\int_0^1 \left(\frac{u}{x}\right)^c \left(\log\frac{x}{u}\right)^{\alpha-1} u^b \frac{du}{u} = \frac{x^b}{(c+b)^{\alpha}} \int_{(c+b)\log x}^{\infty} e^{-t} t^{\alpha-1} dt$$

and so

$$(J^{\alpha}_{0+,c}g)(x) = \begin{cases} (c+b)^{-\alpha}x^b & \text{for } 0 < x \leq 1, \\ \dfrac{x^b\Gamma(\alpha, (c+b)\log x)}{(c+b)^{\alpha}\Gamma(\alpha)} & \text{for } x > 1. \end{cases}$$

Thus, setting $m := \lceil \alpha \rceil$, we see that

$$(D^{\alpha}_{0+,c}g)(x) = \begin{cases} (c+b)^{\alpha}x^b & \text{for } 0 < x \leq 1, \\ \dfrac{\Theta_c^m\big(x^b\Gamma(m-\alpha, (c+b)\log x)\big)}{(c+b)^{m-\alpha}\Gamma(m-\alpha)} & \text{for } x > 1. \end{cases}$$

Let us draw up details for the case $\alpha = 1/2$. Then $m = 1$ and a calculation yields

$$\begin{aligned} \frac{\Theta_c\big(x^b\Gamma(1/2, (c+b)\log x)\big)}{(c+b)^{1/2}\Gamma(1/2)} &= (c+b)^{1/2}x^b \frac{\Gamma(1/2, (c+b)\log x)}{\Gamma(1/2)} \\ &\quad - \frac{1}{\Gamma(1/2)x^c(\log x)^{1/2}} \qquad (9.23) \\ &= (c+b)^{1/2}x^b \frac{\Gamma(-1/2, (c+b)\log x)}{\Gamma(-1/2)}, \end{aligned}$$

where a formula in [160, No. 8.356(5)] has been used in the last step. Hence

$$(D_{0+,c}^{1/2}g)(x) = \begin{cases} (c+b)^{1/2}x^b & \text{for } 0 < x \le 1, \\ (c+b)^{1/2}x^b \, \dfrac{\Gamma(-1/2,(c+b)\log x)}{\Gamma(-1/2)} & \text{for } x > 1. \end{cases}$$

The representation by the function $\Gamma(-1/2,\cdot)$ is more concise than by $\Gamma(1/2,\cdot)$, but we should note that $\Gamma(-1/2,t)$ approaches ∞ as $t \to 0+$ while $\Gamma(1/2,t) \to \Gamma(1/2) = \sqrt{\pi}$.

Next we want to determine the pointwise Mellin fractional derivative of g. From the previous example we know already that $\Theta_c^\alpha g(x) = (c+b)^\alpha x^b$ for $0 < x \le 1$.

Now let $x > 1$ be fixed and let $h > 1$. Then $g(h^{-j}x) \ne 0$ if and only if $h^{-j}x \le 1$, or equivalently, $j \ge \lceil \log x / \log h \rceil =: j_h$ for $j \in \mathbb{N}$. Thus

$$(\Delta_h^{\alpha,c} g)(x) = \sum_{j=j_h}^{\infty} (-1)^j \binom{\alpha}{j} h^{-cj} g(h^{-j}x) = x^b \sum_{j=j_h}^{\infty} (-1)^j \binom{\alpha}{j} h^{-(c+b)j}.$$

From Euler's formula for the gamma function (1.16), we deduce that for fixed $\alpha \in \mathbb{R}^+ \setminus \mathbb{N}$,

$$(-1)^j \Gamma(-\alpha) \binom{\alpha}{j} = \frac{1}{j^{\alpha+1}} \big(1 + o(1)\big) \qquad (j \to \infty).$$

Hence, given $\varepsilon \in]0,1[$, there exists $h_0 > 1$ such that

$$\frac{1-\varepsilon}{j^{\alpha+1}} \le (-1)^j \Gamma(-\alpha) \binom{\alpha}{j} \le \frac{1+\varepsilon}{j^{\alpha+1}}$$

for all $j \ge j_{h_0}$ and so

$$(1-\varepsilon)x^b S(h) \le \Gamma(-\alpha)(\Delta_h^{\alpha,c} g)(x) \le (1+\varepsilon)x^b S(h) \tag{9.24}$$

for all $h \in]1, h_0[$, where

$$S(h) := \sum_{j=j_h}^{\infty} \frac{h^{-(c+b)j}}{j^{\alpha+1}}.$$

Estimating the series by integrals, we obtain

$$\int_{j_h}^{\infty} e^{-(c+b)u \log h} u^{-\alpha-1} du \le S(h) \le \int_{j_h-1}^{\infty} e^{-(c+b)u \log h} u^{-\alpha-1} du. \tag{9.25}$$

Noting that $j_h \leq (\log x / \log h) + 1$ and $j_h - 1 \geq (\log x / \log h) - 1$, we find by substituting $t = (c+b)u \log h$ that the integrals in (9.25) can be replaced by

$$\big((c+b)\log h\big)^{\alpha}\, \Gamma\big(-\alpha, (c+b)(\log x \pm \log h)\big)$$

with the upper sign for the integral on the left-hand side and the lower for the one on the right-hand side. Together with (9.24) this allows us to conclude that

$$\Theta_c^{\alpha} g(x) = \lim_{h\to 1+} \frac{(\Delta_h^{\alpha,c} g)(x)}{(h-1)^{\alpha}} = (c+b)^{\alpha} x^{b}\, \frac{\Gamma(-\alpha, (c+b)\log x)}{\Gamma(-\alpha)} \lim_{h\to 1+} \left(\frac{\log h}{h-1}\right)^{\alpha}.$$

The limit on the right-hand side is easily seen to be 1. Hence

$$\Theta_c^{\alpha} g(x) = \begin{cases} (c+b)^{\alpha} x^{b} & \text{for } 0 < x \leq 1, \\ (c+b)^{\alpha} x^{b}\, \dfrac{\Gamma(-\alpha, (c+b)\log x)}{\Gamma(-\alpha)} & \text{for } x > 1. \end{cases}$$

In particular $(D_{0+,c}^{1/2} g)(x) = \Theta_c^{1/2} g(x)$ for all $x \in \mathbb{R}^+ \setminus \{1\}$. Other than the pointwise Mellin fractional derivative, the pointwise Mellin–Hadamard derivative does not exist at $x = 1$ since $(J_{0+,c}^{1/2} g)(x)$ is not differentiable at that point. Although $g(x) = 0$ for $x > 1$, both pointwise fractional derivatives of order $1/2$ are different from zero. They even approach $-\infty$ when $x \to 1+$ as is seen by Eqs. (9.23). While a classical derivative $g^{(m)}$ at a point $x_0 \in \mathbb{R}^+$ is already determined by the restriction of g to an arbitrarily small neighborhood of x_0, the fractional derivatives of order $\alpha \in \mathbb{R}^+ \setminus \mathbb{N}$ depend on the behavior of g on the whole of $]0, x_0]$.

Example 9.40 Let

$$g(x) = \begin{cases} x^2 & \text{for } 0 < x \leq 1, \\ a_0 + a_1 x & \text{for } x > 1, \end{cases}$$

where $a_0, a_1 \in \mathbb{R}$. Obviously, g is continuous on $\mathbb{R}^+$ if and only if $a_0 + a_1 = 1$ and differentiable if and only $a_0 = -1$ and $a_1 = 2$. Since

$$g(x) = a_0 + a_1 x + (x^2 - a_0 - a_1 x)\chi_{]0,1]}(x) \qquad (x \in \mathbb{R}^+),$$

it is a linear combination of functions studied in Examples 9.38 and 9.39. For $c \in \mathbb{R}^+$ we deduce from the former results that

$$(D_{0+,c}^{1/2} g)(x) = \Theta_c^{1/2} g(x) = (c+2)^{1/2} x^2 \qquad (0 < x < 1)$$

and, employing also (9.23), that

$$(D^{1/2}_{0+,c}g)(x) = \Theta_c^{1/2} g(x) = c^{1/2} a_0 \left(1 - \frac{\Gamma(1/2, c \log x)}{\Gamma(1/2)}\right)$$
$$+ (c+1)^{1/2} a_1 x \left(1 - \frac{\Gamma(1/2, (c+1)\log x)}{\Gamma(1/2)}\right)$$
$$+ (c+2)^{1/2} x^2 \frac{\Gamma(1/2, (c+2)\log x)}{\Gamma(1/2)} - \frac{1 - a_0 - a_1}{x^c (\log x)^{1/2}}$$

for $x > 1$. In addition $\Theta_c^{1/2} g(1) = (c+2)^{1/2}$ exists. However, both pointwise fractional derivatives are bounded on all subintervals of finite length if and only if $a_0 + a_1 = 1$, that is, if and only if g is continuous. In this case $(D^{1/2}_{0+,c}g)(1)$ exists as well and coincides with $\Theta_c^{1/2} g(1)$. Hence the fractional derivatives of order $1/2$ exist even if the graph of g has a breakpoint at $x = 1$.

Example 9.41 Next we consider the function $g_k(x) = \log^k x$ with $k \in \mathbb{N}$. It can be verified that $g_k \in \operatorname{dom} J^\alpha_{0+,c}$ for any $c, \alpha \in \mathbb{R}^+$. By a change of variable of the form $v = \log(x/u)$, an application of the binomial theorem and Euler's formula for the gamma function, we may write

$$(J^\alpha_{0+,c} g_k)(x) = \frac{1}{\Gamma(\alpha)} \int_0^\infty e^{-cv} v^{\alpha-1} (\log x - v)^k dv$$
$$= \frac{1}{\Gamma(\alpha)} \sum_{j=0}^{k} (-1)^{k-j} \binom{k}{j} \frac{\Gamma(\alpha + k - j)}{c^{\alpha+k-j}} \log^j x.$$

Setting

$$B_\alpha(k, j) := \frac{\Gamma(\alpha + k - j)}{\Gamma(\alpha)} = \prod_{\nu=1}^{k-j} (\alpha + k - j - \nu),$$

we obtain

$$(J^\alpha_{0+,c} g_k)(x) = \sum_{j=0}^{k} (-1)^{k-j} \binom{k}{j} \frac{B_\alpha(k, j)}{c^{\alpha+k-j}} \log^j x.$$

With $m := \lceil \alpha \rceil$, we can now calculate the Mellin–Hadamard fractional derivative as

$$(D^\alpha_{0+,c} g_k)(x) = \Theta_c^m (J^{m-\alpha}_{0+,c} g_k)(x) = \sum_{j=0}^{k} (-1)^{k-j} \binom{k}{j} \frac{B_{m-\alpha}(k, j)}{c^{m-\alpha+k-j}} \Theta_c^m \log^j x.$$

In particular, for $k = 1$, we deduce the formulae

$$(J^{\alpha}_{0+,c} g_1)(x) = -\alpha c^{-\alpha-1} + c^{-\alpha} \log x$$

and

$$(D^{\alpha}_{0+,c} g_1)(x) = \alpha c^{\alpha-1} + c^{\alpha} \log x.$$

Their combination easily yields

$$D^{\alpha}_{0+,c}(J^{\alpha}_{0+,c} g_1)(x) = J^{\alpha}_{0+,c}(D^{\alpha}_{0+,c} g_1)(x) = \log x.$$

This is another illustration of the fundamental theorem of calculus in the fractional Mellin setting. Note that the condition $c > 0$ cannot be weakened to $c \geq 0$ since the function $\log x$ does not belong to the domain of J^{α}_{0+}.

Finally we calculate the pointwise Mellin fractional derivative of $g_1(x) = \log x$. For $c \in \mathbb{R}^+$ and $h > 1$, we find that

$$\begin{aligned}
(\Delta_h^{\alpha,c} g_1)(x) &= \sum_{j=0}^{\infty} (-1)^j \binom{\alpha}{j} h^{-cj} \log(h^{-j} x) \\
&= \log x \sum_{j=0}^{\infty} (-1)^j \binom{\alpha}{j} h^{-cj} - \log h \sum_{j=0}^{\infty} (-1)^j \binom{\alpha}{j} j h^{-cj} \\
&=: J_1 - J_2.
\end{aligned}$$

Clearly,

$$J_1 = \log x \left(\frac{h^c - 1}{h^c} \right)^{\alpha}.$$

As regards J_2, we first observe that for $z \in \mathbb{C}$ with $|z| < 1$ we have

$$\sum_{j=0}^{\infty} (-1)^j \binom{\alpha}{j} j z^j = z \frac{d}{dz} \sum_{j=0}^{\infty} (-1)^j \binom{\alpha}{j} z^j = z \frac{d}{dz} (1 - z)^{\alpha} = -\alpha z (1 - z)^{\alpha-1}.$$

Substituting $z = h^{-c}$, we obtain

$$J_2 = -\alpha h^{-c} \log h \left(\frac{h^c - 1}{h^c} \right)^{\alpha-1}.$$

Thus

$$\begin{aligned}\Theta_c^\alpha g_1(x) &= \lim_{h\to 1+} \frac{(\Delta_h^{\alpha,c} g_1)(x)}{(h-1)^\alpha} \\ &= \log x \lim_{h\to 1+} \left(\frac{h^c-1}{h-1}\right)^\alpha + \alpha \lim_{h\to 1+} \frac{\log h}{h-1} \left(\frac{h^c-1}{h-1}\right)^{\alpha-1} \\ &= c^\alpha \log x + \alpha c^{\alpha-1}\end{aligned}$$

as in the case of the pointwise Mellin–Hadamard derivative.

Example 9.42 For $b \in \mathbb{R}$, we consider the function $g(x) = e^{bx}$. It can be shown that $g \in \operatorname{dom} J_{0+,c}^\alpha$ whenever $c, \alpha \in \mathbb{R}^+$. Representing $g(x)$ by its power series and using Lemma 9.10, we obtain

$$(J_{0+,c}^\alpha g)(x) = \sum_{k=0}^\infty (c+k)^{-\alpha} \frac{b^k}{k!} x^k \qquad (x \in \mathbb{R}^+).$$

From the power series in conjunction with Example 9.38, we also deduce that

$$(D_{0+,c}^\alpha g)(x) = \sum_{k=0}^\infty (c+k)^{\alpha} \frac{b^k}{k!} x^k \qquad (x \in \mathbb{R}^+).$$

Again, these formulae do not extend to $c = 0$ since $g(0) \neq 0$ (see Propositions 9.11 and 9.20).

Example 9.43 We consider the sinc function, defined by $\operatorname{sinc} 0 = 1$ and $\operatorname{sinc} x = (\sin \pi x)/(\pi x)$ for $x \neq 0$. It is an entire function. In particular, its Taylor series

$$\operatorname{sinc} x = \sum_{k=0}^\infty (-1)^k \frac{\pi^{2k}}{(2k+1)!} x^{2k}$$

converges on the whole real line. Moreover, it is easily seen that $\operatorname{sinc} \in X_{c,\text{loc}}$ for $c > 0$ while $\operatorname{sinc} \notin X_{0,\text{loc}}$. Using again Lemma 9.10, we obtain

$$(J_{0+,c}^\alpha \operatorname{sinc})(x) = \sum_{k=0}^\infty (-1)^k (c+2k)^{-\alpha} \frac{\pi^{2k}}{(2k+1)!} x^{2k}.$$

For the corresponding Mellin–Hadamard fractional derivative, we find that

$$(D_{0+,c}^\alpha \operatorname{sinc})(x) = \sum_{k=0}^\infty (-1)^k (c+2k)^{\alpha} \frac{\pi^{2k}}{(2k+1)!} x^{2k}.$$

9.5 Notes

The contents of this chapter rests on Bardaro et al. [53] but is presented in a revised form with some modifications and supplements.

The theory of the Mellin fractional integrals is based on the classical Hadamard integrals (see Hadamard [163]) and is treated in several books and articles but without a detailed study of the domain of these operators; see Samko et al. [267, Sects. 18.3, 23.1], Butzer et al. [105–109], Kilbas [191] and Kilbas et al. [192, Chap. 2]. The semigroup property of the Mellin fractional integrals was first studied in the paper by Butzer et al. [105] (see also Kilbas et al. [192, Sect. 2.7]). Several generalizations of these integrals were discussed in the quoted papers. In particular, the following right-sided fractional integral in the Mellin setting was introduced (see e.g. Butzer et al. [107]):

$$(J^{\alpha}_{-,c}f)(x) := \frac{1}{\Gamma(\alpha)}\int_x^{\infty} \left(\frac{x}{u}\right)^c \left(\log\frac{u}{x}\right)^{\alpha-1} f(u)\frac{du}{u} \qquad (x \in \mathbb{R}^+), \tag{9.26}$$

where $c \in \mathbb{R}$ and $\alpha \in \mathbb{R}^+$. Analogously to Theorem 9.14, one can prove that

$$\left[J^{\alpha}_{-,c}f\right]^{\wedge}_{M_c}(c+it) = (it)^{-\alpha}[f]^{\wedge}_{M_c}(c+it) \qquad (t \in \mathbb{R})$$

for any function $f \in X_c$ that belongs to the domain of $J^{\alpha}_{-,c}$.

The notion of pointwise Mellin–Hadamard fractional derivative based on the left-sided Mellin fractional integrals, as introduced in Sect. 9.2, was already developed in the papers by Butzer et al. [105–109, 111]. Using the fractional integral (9.26), it is possible to define the corresponding right-sided fractional derivative by

$$\left(D^{\alpha}_{-,c}f\right)(x) := (-1)^m\Theta^m\left(x^c\left(J^{m-\alpha}_{-,c}f\right)(x)\right) = (-1)^m x^c\Theta^m_c\left(\left(J^{m-\alpha}_{-,c}f\right)(x)\right),$$

where $\alpha \in \mathbb{R}^+$ and $m := \lceil \alpha \rceil$. For functions $f \in X_c$ such that $D^{\alpha}_{-,c}f \in X_c$, one has

$$\left[D^{\alpha}_{-,c}f\right]^{\wedge}_{M_c}(c+it) = (it)^{\alpha}[f]^{\wedge}_{M_c}(c+it) \qquad (t \in \mathbb{R}).$$

This shows that the left-sided and the right-sided derivatives are different concepts of fractional derivatives.

Results related to Theorem 9.25 were given by Kilbas [191] for functions belonging to $L^p(I)$, where I is a bounded interval, and more generally by Kilbas et al. [192, Property 2.28, Lemma 2.35].

Concerning the content of Sect. 9.3, we see that the fractional differences depend crucially on the Mellin translation operator. In the classical theories of Riemann–Liouville and Grünwald–Letnikov fractional calculus, the corresponding differences were based on the classical translation operator, and were first studied in a precise and systematic form by Butzer and Westphal [99, 100]; see also Kilbas et al. [192, Sect. 2.8, p. 121 ff].

It is possible to define a notion of right-sided Mellin fractional derivative using the limit

$$\Theta^{\alpha}_{-,c} f := \lim_{h \to 1-} \frac{\Delta^{\alpha,c}_{1/h} f}{(1-h)^{\alpha}}.$$

Under certain assumptions on the function f, one has

$$[\Theta^{\alpha}_{-,c} f]^{\wedge}_{M_c}(c+it) = (it)^{\alpha}[f]^{\wedge}_{M_c}(c+it) \qquad (t \in \mathbb{R}).$$

This implies that the right-sided Mellin–Hadamard and the right-sided Mellin fractional derivatives coincide on a suitable subspace of functions.

Other generalizations of fractional differences, via the Stirling functions of the first kind, were introduced in Butzer et al. [109, 111].

We mention that for $\alpha > 1$ the Fourier counterparts of the notions of strong fractional derivative (Definition 9.32) and the pointwise one (Definition 9.31) are the same in case of periodic functions in the sense that if the strong derivative exists and belongs to a Lebesgue space, then it coincides almost everywhere with the pointwise derivative. This was proved by Butzer and Westphal [99, Theorem 9.1] as a generalization of a result in the book of Butzer and Nessel [95, Proposition 1.1.10]. It seems to be an unsolved problem if this is also true for $0 < \alpha < 1$. Analogous studies in the Mellin setting may be of interest. So far the spaces $\mathscr{W}^{\alpha}_{c}$ enable us to discuss such interconnections also in the case $\alpha \in]0, 1[$ by using the Mellin transform as a basic tool. Theorem 9.33 together with Theorem 9.34 give conditions under which the fractional derivatives $D^{\alpha}_{0+,c} f$, s-$\Theta^{\alpha}_{c} f$ and $\Theta^{\alpha}_{c} f$ are the same almost everywhere.

In view of the representation of $D^{\alpha}_{0+,c} f$ in terms of generalized Stirling numbers of the second kind, given in Proposition 9.19, we mention an alternative representation in terms of generalized Stirling functions of the second kind which can be defined for $c \in \mathbb{R}$ and $k \in \mathbb{N}_0$ by

$$S_c(\alpha, k) := \frac{1}{k!} \sum_{j=0}^{k} (-1)^{k-j} \binom{k}{j} (c+j)^{\alpha} \qquad (\alpha \in \mathbb{C}).$$

Butzer et al. [108] proved the following statements: *Let $f : \mathbb{R}^+ \to \mathbb{R}$ be infinitely differentiable such that its Taylor series converges and let $c, \alpha \in \mathbb{R}^+$. Then*

$$(D^{\alpha}_{0+,c} f)(x) = \sum_{k=0}^{\infty} S_c(\alpha, k) x^k f^{(k)}(x) \qquad (x \in \mathbb{R}^+).$$

If in addition $f \in \operatorname{dom} J^{\alpha}_{0+,c}$, then

$$(J^{\alpha}_{0+,c} f)(x) = \sum_{k=0}^{\infty} S_c(-\alpha, k) x^k f^{(k)}(x) \qquad (x \in \mathbb{R}^+).$$

Also the limit $c \to 0$ is discussed.

There exist numerous concepts of fractional calculus in the literature. We have already named some of them without giving their definitions. For sake of completeness, we present the one based on Riemann–Liouville fractional integrals since it is considered as a classical approach. As we mentioned at the beginning of Sect. 9.1, the (left-sided) Riemann–Liouville fractional integral of order $\alpha \in \mathbb{R}^+$ is defined by

$$(I_{0+}^{\alpha} f)(x) := \frac{1}{\Gamma(\alpha)} \int_0^x (x-u)^{\alpha-1} f(u)du \qquad (x \in \mathbb{R}^+).$$

With the help of these integrals, the (left-sided) Riemann–Liouville fractional derivative of order $\alpha \in \mathbb{R}^+ \setminus \mathbb{N}$ is introduced as

$$(\mathcal{D}_{0+}^{\alpha} f)(x) := \left(\frac{d}{dx}\right)^m (I_{0+}^{m-\alpha} f)(x) \qquad (x \in \mathbb{R}^+), \tag{9.27}$$

where $m := \lceil \alpha \rceil$. Analogous definitions of right-sided Riemann–Liouville fractional integrals and derivatives are available, obtained by replacing the integration over $(0, x)$ with that over (x, ∞) and $(x - u)$ with $(u - x)$. In Butzer et al. [111, Theorem 6.2] it is shown that under certain hypotheses

$$x^{\alpha} (\mathcal{D}_{0+}^{\alpha} f)(x) = \sum_{k=0}^{\infty} s(\alpha, k)(\Theta^k f)(x) \qquad (x \in \mathbb{R}^+),$$

where $s(\alpha, k)$ are the Stirling functions of the first kind.

In the theory of differential equations a fractional derivative due to Caputo [124] has gained much attention. It can be obtained by interchanging in (9.27) the fractional integration of order $m - \alpha$ and the ordinary differentiation of order m. Indeed, the (left-sided) Caputo derivative $C\mathcal{D}_{0+}^{\alpha} f$ of order $\alpha \in \mathbb{R}^+ \setminus \mathbb{N}$ is defined by

$$(C\mathcal{D}_{0+}^{\alpha} f)(x) := \frac{1}{\Gamma(m-\alpha)} \int_0^x \frac{f^{(m)}(u)}{(x-u)^{\alpha-m+1}} du$$

(see also Kilbas et al. [192, pp. 91–92]). As for the Riemann–Liouville derivatives, a corresponding notion of right-sided derivative can be defined by using a right-sided Riemann–Liouville fractional integral.

An interesting general approach to fractional integrals and derivatives, which includes the classical fractional operators of Riemann–Liouville, Erdélyi–Kober, and the Mellin–Hadamard operators in the case $c = 0$, can be found in Katugampola [187, 188]. In Katugampola [189] the Mellin transforms of these generalized operators are derived with applications to Stirling and Lah numbers and to generalized Laguerre polynomials.

For the role of Mellin transform theory in the broad area of fractional calculus together with a variety of important applications in the realm of special functions, the reader may consult Luchko and Kiryakova [202].

Chapter 10
The Mellin Distance

10.1 A Notion of Distance

We begin by specifying a normed linear space of functions which is tailored for the Mellin setting.

Definition 10.1 For $c \in \mathbb{R}$ and $q \in [1, \infty]$, we denote by G_c^q the normed linear space of all functions $f : \mathbb{R}^+ \to \mathbb{C}$ for which there exists a function $\phi_f \in L^1(\mathbb{R}) \cap L^q(\mathbb{R})$ such that the representation

$$f(x) = \frac{1}{2\pi} \int_{-\infty}^{\infty} \phi_f(v) x^{-c-iv} dv \qquad (x > 0) \tag{10.1}$$

holds and we endow G_c^q with the norm

$$[\![f]\!]_q := \|\phi_f\|_{L^q(\mathbb{R})} = \left(\int_{-\infty}^{\infty} |\phi_f(v)|^q dv \right)^{1/q}.$$

Note that the functional $[\![\cdot]\!]_q$ is really a norm. Indeed, $[\![f]\!]_q = 0$ if and only if $f = 0$ due to the existence and uniqueness of the Mellin inversion. This norm induces the metric

$$\operatorname{dist}_q(f, g) := [\![f - g]\!]_q \qquad (f, g \in G_c^q).$$

For $q = 2$, the norm preserving property of the Mellin transform in X_c^2 (see Proposition 7.1) shows that

$$\operatorname{dist}_2(f, g) = \sqrt{2\pi} \|f - g\|_{X_c^2}.$$

C. Bardaro et al., *Mellin Analysis, Transform Theory, and Applications*,
https://doi.org/10.1007/978-3-031-96672-9_10

Thus this distance reduces to the Euclidean distance in X_c^2, apart from the factor $\sqrt{2\pi}$. As a consequence of the Mellin inversion formula, functions f for which $[f]^\wedge_{M_c}(c+i\cdot) \in L^1(\mathbb{R}) \cap L^q(\mathbb{R})$ belong to G_c^q with $\phi_f = [f]^\wedge_{M_c}(c+i\cdot)$. For $p \in [1,2]$, the space $B^p_{c,\sigma}$ is a subspace of G^q_c since the Mellin transform of $f \in B^p_{c,\sigma}$ has compact support, and so it belongs to any $L^q(\mathbb{R})$.

For $f \in G^q_c$, we define

$$\text{dist}_q(f, B^p_{c,\sigma}) = \inf_{g \in B^p_{c,\sigma}} [\![f - g]\!]_q \,.$$

The following representation theorem holds:

Theorem 10.2 *Let $c \in \mathbb{R}$, $\sigma \in \mathbb{R}^+$, $p \in [1,2]$ and $q \in [1,+\infty]$. Suppose that $f \in G^q_c$. Then*

$$\text{dist}_q(f, B^p_{c,\sigma}) = \left(\int_{|v|\geq\sigma} |\phi_f(v)|^q dv \right)^{1/q} \qquad (1 \leq q < \infty).$$

If ϕ_f is continuous, then

$$\text{dist}_\infty(f, B^p_{c,\sigma}) = \sup_{|v|\geq\sigma} |\phi_f(v)|.$$

Proof First let $q < \infty$. For any $g \in B^p_{c,\sigma}$, we have $|[g]^\wedge_{M^p_c}(c+iv)| = 0$ for $|v| \geq \sigma$, and so

$$\text{dist}_q(f, B^p_{c,\sigma}) = \left\{ \inf_{g \in B^p_{c,\sigma}} \int_{|v|\leq\sigma} |\phi_f(v) - [g]^\wedge_{M^p_c}(c+iv)|^q dv \right.$$
$$\left. + \int_{|v|\geq\sigma} |\phi_f(v)|^q dv \right\}^{1/q}.$$

Therefore we have to prove that

$$I_{p,q} := \inf_{g \in B^p_{c,\sigma}} \int_{|v|\leq\sigma} |\phi_f(v) - [g]^\wedge_{M^p_c}(c+iv)|^q dv = 0.$$

For the given $\sigma > 0$, the space $C^\infty_{\text{comp}}(]-\sigma,\sigma[)$, whose elements are infinitely differentiable functions with compact support in $]-\sigma,\sigma[$, is dense in $L^q(]-\sigma,\sigma[)$ for $1 \leq q < \infty$. Thus, given $\phi_f \in L^q(\mathbb{R})$ and $\varepsilon > 0$, there exists a function $P \in C^\infty_{\text{comp}}(]-\sigma,\sigma[)$ such that

$$\|\phi_f - P\|_{L^q(]-\sigma,\sigma[)} < \varepsilon.$$

Now we define

$$g_\varepsilon(x) := \frac{x^{-c}}{2\pi}\int_{-\sigma}^{\sigma} P(v)x^{-iv}dv \quad (x > 0).$$

Integrating k-times by parts under successive differentiation of P, we easily see that $x^c g_\varepsilon(x) = \mathcal{O}((\log x)^{-k})$ as $x \to \infty$ and $x \to 0+$. This implies that $g_\varepsilon \in X_c^p$ for any $p \geq 1$, and $[g_\varepsilon]^\wedge_{M_c^p}(c+iv) = P(v)$ for $|v| \leq \sigma$ and 0 otherwise. Thus $g_\varepsilon \in B^p_{c,\sigma}$ and therefore

$$I_{p,q}^{1/q} \leq \|\phi_f - [g_\varepsilon]^\wedge_{M_c^p}(c + i\cdot)\|_{L^q(]-\sigma,\sigma[)} = \|\phi_f - P\|_{L^q(]-\sigma,\sigma[)} < \varepsilon.$$

Hence the assertion follows for $1 \leq q < \infty$.

The case $q = \infty$ is treated in a different way. Given $\varepsilon > 0$, there exists a twice continuously differentiable function ψ on $\mathbb{R}$ such that $\sup_{v\in\mathbb{R}} |\phi_f(v) - \psi(v)| < \varepsilon/2$. For example, ψ may be chosen as an appropriate spline function. For $\eta \in]0, \sigma[$, define

$$\psi_1(v) := \begin{cases} 0 & \text{if} \quad |v| \leq \sigma - \eta, \\ \dfrac{\psi(-\sigma)}{-\eta^3}(v + \sigma - \eta)^3 & \text{if} \quad -\sigma \leq v \leq -\sigma + \eta, \\ \dfrac{\psi(\sigma)}{\eta^3}(v - \sigma + \eta)^3 & \text{if} \quad \sigma - \eta \leq v \leq \sigma, \\ \psi(v) & \text{if} \quad |v| \geq \sigma. \end{cases}$$

Note that ψ_1 is continuous on $\mathbb{R}$ and

$$\|\psi_1\|_\infty = \sup_{|v|\geq\sigma-\eta} |\psi_1(v)| = \sup_{|v|\geq\sigma} |\psi(v)| \leq \sup_{|v|\geq\sigma} |\phi_f(v)| + \frac{\varepsilon}{2}.$$

Next define $\psi_0(v) := \psi(v) - \psi_1(v)$ for $v \in \mathbb{R}$. Then ψ_0 is continuous on $\mathbb{R}$, twice continuously differentiable on $]-\sigma, \sigma[$, it vanishes at $\pm\sigma$ and it has support on $[-\sigma, \sigma]$. With these properties, two integrations by parts show that

$$g_\varepsilon(x) := \frac{x^{-c}}{2\pi}\int_{-\sigma}^{\sigma} \psi_0(v)x^{-iv}dv \qquad (x > 0)$$

defines a function $g_\varepsilon \in X_c^p \cap B^p_{c,\sigma}$. Furthermore, the Mellin inversion formula yields that $[g_\varepsilon]^\wedge_{M_c^p}(c + i\cdot) = \psi_0(\cdot)$. Now we conclude that

$$\begin{aligned}\operatorname{dist}_\infty(f, B^p_{c,\sigma}) &= \inf_{g\in B^p_{c,\sigma}} \|\phi_f - [g]^\wedge_{M^p_c}(c+i\cdot)\|_\infty \\ &\le \|\phi_f - [g_\varepsilon]^\wedge_{M^p_c}(c+i\cdot)\|_\infty \\ &= \|\phi_f - \psi_0\|_\infty \\ &\le \|\phi_f - \psi\|_\infty + \|\psi - \psi_0\|_\infty \\ &\le \frac{\varepsilon}{2} + \|\psi_1\|_\infty \\ &\le \varepsilon + \sup_{|v|\ge\sigma} |\phi_f(v)|.\end{aligned}$$

This implies that $\operatorname{dist}_\infty(f, B^p_{c,\sigma}) \le \sup_{|v|\ge\sigma} |\phi_f(v)|$. On the other hand,

$$\begin{aligned}\operatorname{dist}_\infty(f, B^p_{c,\sigma}) &= \max\left\{\inf_{g\in B^p_{c,\sigma}} \sup_{|v|\le\sigma} |\phi_f(v) - [g]^\wedge_{M^p_c}(c+iv)|,\ \sup_{|v|\ge\sigma} |\phi_f(v)|\right\} \\ &\ge \sup_{|v|\ge\sigma} |\phi_f(v)|.\end{aligned}$$

Hence the formula stated in the theorem holds. □

Our next aim is a distance formula for Mellin derivatives. First we establish a representation for Mellin derivatives of functions in G^q_c.

Theorem 10.3 *Under the hypotheses and in the notation of Theorem 10.2 suppose that $f \in G^q_c$ and that $v^r\phi_f(v)$ belongs to $L^1(\mathbb{R})$ as a function of v for some $r \in \mathbb{N}$. Then f has Mellin derivatives up to order r in $C(\mathbb{R}^+)$ and*

$$(\Theta^k_c f)(x) = \frac{(-i)^k}{2\pi}\int_{-\infty}^{\infty} v^k\phi_f(v)x^{-c-iv}dv \qquad (k = 0, 1, \ldots r).$$

Proof Suppose that $r = 1$. For $h \ne 1$, we have

$$\begin{aligned}&\frac{h^c f(hx) - f(x)}{h-1} \\ &= \frac{1}{2\pi(h-1)}\left(\int_{-\infty}^{\infty} \phi_f(v)h^{-iv}x^{-c-iv}dv - \int_{-\infty}^{\infty} \phi_f(v)x^{-c-iv}dv\right) \\ &= \frac{x^{-c}}{2\pi}\int_{-\infty}^{\infty} \phi_f(v)x^{-iv}\frac{h^{-iv}-1}{h-1}dv.\end{aligned}$$

Now, for $h \in [1/2, 3/2] \setminus \{1\}$, the quotient inside the integral may be estimated as

$$\left|\frac{h^{-iv}-1}{h-1}\right| = \frac{2}{|h-1|}\left|\frac{e^{-\frac{iv\log h}{2}} - e^{\frac{iv\log h}{2}}}{2i}\right| = 2\left|\frac{\sin(\frac{v\log h}{2})}{h-1}\right| \le 2|v|. \qquad (10.2)$$

Since

$$\lim_{h\to 1}\frac{h^{-iv}-1}{h-1}=-iv$$

and $v\phi_f(v)$ is absolutely integrable, we conclude with the help of Lebesgue's dominated convergence theorem that

$$(\Theta_c^1 f)(x)=\frac{-i}{2\pi}\int_{-\infty}^{\infty} v\phi_f(v)x^{-c-iv}dv.$$

Finally, Lemma 6.71 shows that $\Theta_c^1 f\in C(\mathbb{R}^+)$.

For general r the proof is accomplished by mathematical induction. □

As a consequence of Theorems 10.2 and 10.3, we immediately obtain the following formula for the distance of $\Theta_c^k f$ from the space $B_{c,\sigma}^p$.

Corollary 10.4 *Let $c\in\mathbb{R}$, $\sigma\in\mathbb{R}^+$, $q\in[1,\infty[$ and $k\in\mathbb{N}$. Suppose that $f\in G_c^q$ with $(\cdot)^k\phi_f(\cdot)\in L^1(\mathbb{R})\cap L^q(\mathbb{R})$. Then, for every $p\in[1,2]$, we have*

$$\text{dist}_q(\Theta_c^k f, B_{c,\sigma}^p)=\left(\int_{|v|\geq\sigma}|v^k\phi_f(v)|^q dv\right)^{1/q},$$

and if ϕ_f is continuous, then

$$\text{dist}_\infty(\Theta_c^k f, B_{c,\sigma}^p)=\sup_{|v|\geq\sigma}|v^k\phi_f(v)|.$$

Now we consider the case of Mellin derivatives of fractional order $\alpha>0$.

Theorem 10.5 *Under the hypotheses of Theorem 10.2 suppose that $f\in G_c^q$ and that $v^\alpha\phi_f(v)$ belongs to $L^1(\mathbb{R})$ as a function of v for some $\alpha\in\mathbb{R}^+$. Then the pointwise Mellin fractional derivative $\Theta_c^\alpha f$ exists and*

$$\Theta_c^\alpha f(x)=\frac{1}{2\pi}\int_{-\infty}^{\infty}(-iv)^\alpha\phi_f(v)x^{-c-iv}dv \qquad (x\in\mathbb{R}^+),$$

where $(-iv)^\alpha=|v|^\alpha e^{-i\,\text{sgn}(v)\pi\alpha/2}$.

Proof For $h>1$ we may write

$$(\varDelta_h^{\alpha,c} f)(x)=\sum_{j=0}^{\infty}(-1)^j\binom{\alpha}{j}h^{-cj}f(h^{-j}x)$$

$$=\frac{1}{2\pi}\sum_{j=0}^{\infty}(-1)^j\binom{\alpha}{j}\int_{-\infty}^{\infty}\phi_f(v)h^{ijv}x^{-c-iv}dv.$$

Since $\left|h^{ijv}\right| = 1$, the integral exists absolutely with a bound that does not depend on j and the series converges absolutely. Interchanging summation and integration, we obtain

$$\frac{(\Delta_h^{\alpha,c} f)(x)}{(h-1)^\alpha} = \frac{1}{2\pi}\int_{-\infty}^{\infty} \phi_f(v)\left(\frac{1-h^{iv}}{h-1}\right)^\alpha x^{-c-iv}dv.$$

Using the estimate (10.2), we find that

$$\left|\left(\frac{1-h^{iv}}{h-1}\right)^\alpha\right| \leq 2^\alpha\, |v|^\alpha$$

for $1 < h \leq 3/2$. Furthermore,

$$\lim_{h\to 1+}\left(\frac{1-h^{iv}}{h-1}\right)^\alpha = (-iv)^\alpha,$$

where the right-hand side has to be specified as in the assertion since the numerator was obtained by a power series which yields the principal value of $(1-h^{iv})^\alpha$. Now Lebegues's dominated convergence theorem allows us to conclude that

$$\lim_{h\to 1+}\frac{(\Delta_h^{\alpha,c} f)(x)}{(h-1)^\alpha} = \frac{1}{2\pi}\int_{-\infty}^{\infty}(-iv)^\alpha\phi_f(v)x^{-c-iv}dv,$$

as was to be shown. □

As an immediate consequence of Theorem 10.5, we obtain the following formula for the distance.

Corollary 10.6 *Let $c \in \mathbb{R}$, $\sigma \in \mathbb{R}^+$, $q \in [1,\infty[$ and $\alpha \in \mathbb{R}^+$. Suppose that $f \in G_c^q$ with $(\cdot)^\alpha\phi_f \in L^1(\mathbb{R}) \cap L^q(\mathbb{R})$. Then for every $p \in [1,2]$, we have*

$$\mathrm{dist}_q(\Theta_c^\alpha f, B_{c,\sigma}^p) = \left(\int_{|v|\geq\sigma} |v^\alpha\phi_f(v)|^q dv\right)^{1/q},$$

and if ϕ_f is continuous, then

$$\mathrm{dist}_\infty(\Theta_c^\alpha f, B_{c,\sigma}^p) = \sup_{|v|\geq\sigma} |v^\alpha\phi_f(v)|.$$

10.2 Estimate of the Mellin Distance and Function Spaces

In this section we will introduce certain basic intermediate function spaces between the Mellin–Paley–Wiener spaces $B_{c,\sigma}^p$ and the space G_c^q. We shall mainly consider the cases $p = 1$ and $p = 2$.

10.2.1 Mellin Inversion Classes

First we specify a subspace of G_c^q which guarantees the existence of a Mellin transform and its inversion.

Definition 10.7 For $c \in \mathbb{R}$ and $p \in [1, 2]$, we define the *Mellin inversion class*, denoted by $\mathcal{M}_c^p$, as the space comprising all functions $f \in X_c^p \cap C(\mathbb{R}^+)$ such that $[f]^\wedge_{M_c^p} \in L^1(c + i\mathbb{R})$.

The following properties of $\mathcal{M}_c^p$ will be useful in the subsequent studies.

Lemma 10.8 *Let $c \in \mathbb{R}$, $p \in [1, 2]$ and $\sigma \in \mathbb{R}^+$. Then*

$$B_{c,\sigma}^p \subset \mathcal{M}_c^p \subset G_c^q$$

for all $q \in [1, p']$, where p' is the conjugate of p, that is, $1/p + 1/p' = 1$. Furthermore, for $f \in \mathcal{M}_c^p$ we have $\phi_f = [f]^\wedge_{M_c^p}(c + i(\cdot))$.

Proof By definition, the functions f of $B_{c,\sigma}^p$ belong to $X_c^p \cap C(\mathbb{R}^+)$ and are Mellin bandlimited to $[-\sigma, \sigma]$. In addition, Proposition 7.3 or 7.14 tells us that $[f]^\wedge_{M_c^p} \in L^{p'}(c + i\mathbb{R})$. Now the bandlimitness together with Hölder's inequality allows us to conclude that $[f]^\wedge_{M_c^p} \in L^1(c + i\mathbb{R})$, and so the first inclusion is verified.

Next let $f \in \mathcal{M}_c^p$. Then, by Mellin inversion, a representation (10.1) holds with $\phi_f := [f]^\wedge_{M_c^p}(c + i(\cdot))$. Thus, for f to belong to G_c^q, it remains to show that $\phi_f \in L^q(\mathbb{R})$ for each $q \in [1, p']$. This is trivial for $q = 1$. For $q = p'$, it follows from Proposition 7.3 or 7.14. Finally, for $q \in]1, p'[$, we may write

$$|\phi_f|^q = |\phi_f|^{(p'-q)/(p'-1)} \, |\phi_f|^{p'(q-1)/(p'-1)}.$$

Now integrating over $\mathbb{R}$ and applying Hölder's inequality to the product on the right-hand side with the admissible pair of exponents

$$(\nu, \mu) := \left(\frac{p' - 1}{p' - q}, \frac{p' - 1}{q - 1} \right),$$

we find that

$$\int_{-\infty}^{\infty} |\phi_f(v)|^q \, dv \le \left(\int_{-\infty}^{\infty} |\phi_f(v)| \, dv \right)^{(p'-q)/(p'-1)} \times \left(\int_{-\infty}^{\infty} |\phi_f(v)|^{p'} \, dv \right)^{(q-1)/(p'-1)} < \infty$$

and so $\phi_f \in L^q(\mathbb{R})$. □

As for the classes $B^p_{c,\sigma}$, we have again the inclusion $\mathcal{M}^p_c \subset \mathcal{M}^q_c$ for $1 \le p < q \le 2$.

For $f \in X_c$, $\alpha = r \in \mathbb{N}$ and any $h > 0$, Proposition 9.29 (ii) and its proof reveal that

$$[\Delta_h^{r,c} f]^\wedge_{M_c}(c+iv) = (h^{-iv} - 1)^r [f]^\wedge_{M_c}(c+iv) \qquad (v \in \mathbb{R}). \tag{10.3}$$

This formula has an analogue for $f \in X_c^2$ noting that $\Delta_h^{r,c} f$ is a (finite) linear combination of translations of f for which the Mellin transform in X_c^2 exists. Formula (10.3) and its variant will be useful in the proof of following two theorems.

Theorem 10.9 *Let $c \in \mathbb{R}$, $\sigma \in \mathbb{R}^+$ and $f \in \mathcal{M}^1_c$. Then, for any $q \in [1, \infty]$ and $r \in \mathbb{N}$, we have*

$$\operatorname{dist}_q(f, B^1_{c,\sigma}) \le D \cdot \begin{cases} \left\{ \displaystyle\int_\sigma^\infty [\omega_r(f, v^{-1}, X_c)]^q dv \right\}^{1/q} & (q < \infty), \\ \omega_r(f, \sigma^{-1}, X_c) & (q = \infty), \end{cases}$$

where D is a constant depending on r and q only.

Proof For $h = e^{\pi/v}$, formula (10.3) gives

$$[\Delta_h^{r,c} f]^\wedge_{M_c}(c+iv) = (-2)^r [f]^\wedge_{M_c}(c+iv).$$

Thus

$$[f]^\wedge_{M_c}(c+iv) = \frac{1}{(-2)^r} \int_0^\infty (\Delta_h^{r,c} f)(u) u^{c+iv-1} du \qquad (h = e^{\pi/v}),$$

and so

$$|[f]^\wedge_{M_c}(c+iv)| \le \frac{1}{2^r} \int_0^\infty |(\Delta_h^{r,c} f)(u)| u^{c-1} du \le \frac{1}{2^r} \omega_r\Big(f, \frac{\pi}{|v|}, X_c\Big). \tag{10.4}$$

Since

$$\omega_r\Big(f, \frac{\pi}{|v|}, X_c\Big) \le (1+\pi)^r \omega_r\Big(f, \frac{1}{|v|}, X_c\Big),$$

as is seen by Proposition 3.17 (iii), we find that

$$|[f]^\wedge_{M_c}(c+iv)| \le \left(\frac{1+\pi}{2}\right)^r \omega_r\Big(f, \frac{1}{|v|}, X_c\Big). \tag{10.5}$$

In view of Theorem 10.2, this implies the assertion for $q < \infty$. The case $q = \infty$ is obtained analogously. □

Theorem 10.10 *Let $c \in \mathbb{R}$, $\sigma \in \mathbb{R}^+$ and $f \in \mathcal{M}_c^2$. Then, for any $q \in [1, 2]$ and $r \in \mathbb{N}$, we have*

$$\mathrm{dist}_q(f, B^2_{c,\sigma}) \le D \left\{ \int_\sigma^\infty v^{-q/2} [\omega_r(f, v^{-1}, X_c^2)]^q dv \right\}^{1/q},$$

where D is a constant depending on r and q only.

Proof First we consider the case $q = 2$. Then (10.3) for the Mellin transform in X_c^2 in conjunction with (10.2) yields

$$|[\Delta_h^{r,c} f]^\wedge_{M_c^2}(c + iv)| = 2^r |\sin((v \log h)/2)|^r |[f]^\wedge_{M_c^2}(c + iv)|.$$

With the help of Lemma 7.5 we find that

$$\|[\Delta_h^{r,c} f]^\wedge_{M_c^2}\|_{L^2(c+i\mathbb{R})} = \sqrt{2\pi}\, \|\Delta_h^{r,c} f\|_{X_c^2} \le \sqrt{2\pi}\, \omega_r(f, |\log h|, X_c^2).$$

Now let $h > 1$. For $v \in [(2\log h)^{-1}, (\log h)^{-1}]$, it is easily verified that

$$\sin((v \log h)/2) \ge \frac{1}{2\pi},$$

and so

$$\int_{1/(2\log h)}^{1/\log h} |[f]^\wedge_{M_c^2}(c + iv)|^2 dv$$
$$\le (2\pi)^{2r} \int_0^\infty |\sin((v \log h)/2)|^{2r} ||[f]^\wedge_{M_c^2}(c + iv)|^2 dv.$$

Analogously we obtain

$$\int_{-1/\log h}^{-1/(2\log h)} |[f]^\wedge_{M_c^2}(c + iv)|^2 dv$$
$$\le (2\pi)^{2r} \int_{-\infty}^0 |\sin((v \log h)/2)|^{2r} ||[f]^\wedge_{M_c^2}(c + iv)|^2 dv.$$

Together with the former relations, this allows us to conclude that

$$\int_{1/(2\log h) \le |v| \le 1/\log h} |[f]^\wedge_{M_c^2}(c + iv)|^2 dv$$
$$\le (2\pi)^{2r} \int_{-\infty}^\infty |\sin((v \log h)/2)|^{2r} ||[f]^\wedge_{M_c^2}(c + iv)|^2 dv$$
$$\le 2\pi^{2r+1} [\omega_r(f, \log h, X_c^2)]^2.$$

Next, set $\sigma_k := \sigma 2^k$ with $k \in \mathbb{N}_0$ and define h by $\log h = 1/\sigma_{k+1}$. Then

$$\int_{\sigma_k \le |v| \le \sigma_{k+1}} |[f]^{\wedge}_{M_c^2}(c+iv)|^2 dv \le 2\pi^{2r+1}\,[\omega_r(f, \sigma_{k+1}^{-1}, X_c^2)]^2, \qquad (10.6)$$

and so summation over k yields

$$\int_{|v|\ge\sigma} |[f]^{\wedge}_{M_c^2}(c+iv)|^2 dv \le 2\pi^{2r+1}\sum_{k=0}^{\infty}[\omega_r(f, \sigma_{k+1}^{-1}, X_c^2)]^2. \qquad (10.7)$$

Noting that $\sigma_{k+1} - \sigma_k = \sigma_k$ and using the monotonicity of ω_r as a function of its second argument, we find that

$$\int_{\sigma_k}^{\sigma_{k+1}} v^{-1}[\omega_r(f, v^{-1}, X_c^2)]^2 dv \ge \frac{\sigma_k}{\sigma_{k+1}}\,[\omega_r(f, \sigma_{k+1}^{-1}, X_c^2)]^2, \qquad (10.8)$$

which gives

$$\sum_{k=0}^{\infty}[\omega_r(f, \sigma_{k+1}^{-1}, X_c^2)]^2 \le 2\int_{\sigma}^{\infty} v^{-1}[\omega_r(f, v^{-1}, X_c^2)]^2 dv.$$

Combining this with (10.7) and recalling Theorem 10.2, we arrive at the assertion for $q = 2$.

Now let $q \in [1, 2[$. Applying first Hölder's inequality with $\mu = 2/(2-q)$ and $\nu = 2/q$ and using (10.6) afterwards, we find for a suitable constant $C_{q,r}$ that

$$\begin{aligned}
&\int_{\sigma_k \le |v| \le \sigma_{k+1}} |[f]^{\wedge}_{M_c^2}(c+iv)|^q dv \\
&\le \left(\int_{\sigma_k \le |v| \le \sigma_{k+1}} dv\right)^{1/\mu}\left(\int_{\sigma_k \le |v| \le \sigma_{k+1}} |[f]^{\wedge}_{M_c^2}(c+iv)|^{q\nu} dv\right)^{1/\nu} \\
&= (2\sigma_k)^{(2-q)/2}\left(\int_{\sigma_k \le |v| \le \sigma_{k+1}} |[f]^{\wedge}_{M_c^2}(c+iv)|^{q\nu} dv\right)^{q/2} \\
&\le C_{q,r}\,\sigma_k^{(2-q)/2}\big[\omega_r(f, \sigma_{k+1}^{-1}, X_c^2)\big]^q.
\end{aligned}$$

Summing over k and using the inequality

$$\int_{\sigma_k}^{\sigma_{k+1}} v^{-q/2}\big[\omega_r(f, v^{-1}, X_c^2)\big]^q dv \ge \frac{\sigma_k}{\sigma_{k+1}^{q/2}}\,[\omega_r(f, \sigma_{k+1}^{-1}, X_c^2)]^q,$$

which is deduced similarly to (10.8), we obtain

$$\int_{|v|\geq\sigma} |[f]^{\wedge}_{M^2_c}(c+iv)|^q dv \leq C_{q,r}\sum_{k=0}^{\infty}\sigma_k^{(2-q)/2}\big[\omega_r(f,\sigma_{k+1}^{-1},X_c^2)\big]^q$$

$$\leq C_{q,r}\,2^{q/2}\sum_{k=0}^{\infty}\int_{\sigma_k}^{\sigma_{k+1}} v^{-q/2}\big[\omega_r(f,v^{-1},X_c^2)\big]^q dv$$

$$= C_{q,r}\,2^{q/2}\int_{\sigma}^{\infty} v^{-q/2}\big[\omega_r(f,v^{-1},X_c^2)\big]^q dv,$$

and so the assertion holds. □

Remark 10.11 The hypotheses of Theorems 10.9 and 10.10 do not guarantee that the integrals involving ω_r exist as a finite number. Thus the conclusion of these theorems may lead to the trivial estimate that for $p \in \{1,2\}$, we have $\mathrm{dist}_q(f, B^p_{c,\sigma}) \leq \infty$. However, the consequences of these theorems stated in the next Corollary 10.12 always give a non-trivial result.

10.2.2 Mellin–Lipschitz Classes

As consequences of Theorems 10.9 and 10.10, we obtain the following corollary which determines the Mellin distance of a function $f \in \mathrm{Lip}_r(\alpha, X_c^p)$ from $B^p_{c,\sigma}$ asymptotically as $\sigma \to \infty$.

Corollary 10.12 *Let $c \in \mathbb{R}$, $\sigma \in \mathbb{R}^+$ and $r \in \mathbb{N}$.*

(i) If $r \geq 2$, $\alpha \in\,]1, r]$ and $f \in \mathrm{Lip}_r(\alpha, X_c \cap C(\mathbb{R}^+))$, then

$$\mathrm{dist}_1(f, B^1_{c,\sigma}) = \mathcal{O}(\sigma^{-\alpha+1}) \qquad (\sigma \to \infty).$$

(ii) If $q \in [1,2]$, $\beta \in\,]\frac{1}{q} - \frac{1}{2}, r]$ and $f \in \mathcal{M}^2_c \cap \mathrm{Lip}_r(\beta, X_c^2)$, then

$$\mathrm{dist}_q(f, B^2_{c,\sigma}) = \mathcal{O}(\sigma^{-\beta-1/2+1/q}) \qquad (\sigma \to \infty).$$

Proof The definition of the Lipschitz classes (see Definition 3.18) provides the asymptotic behavior of the modulus of smoothness ω_r which in turn may be used to estimate the integrals in the conclusions of Theorems 10.9 and 10.10 asymptotically as $\sigma \to \infty$. In the case of statement (i), we have to verify that Theorem 10.9 is applicable. For this we simply note that in view of (10.5) the hypothesis $f \in \mathrm{Lip}_r(\alpha, X_c \cap C(\mathbb{R}^+))$ implies that $[f]^{\wedge}_{M_c} \in L^1(c+i\mathbb{R})$ and so $f \in \mathcal{M}^1_c$. □

For $q = 2$, the estimate of the previous statement (ii) becomes

$$\mathrm{dist}_2(f, B^2_{c,\sigma}) = \mathcal{O}(\sigma^{-\beta}) \qquad (\sigma \to \infty).$$

10.2.3 Mellin–Sobolev Spaces

In this subsection we continue with results for functions belonging to the Mellin–Sobolev spaces $W_c^{r,1}(\mathbb{R}^+)$ or $W_c^{r,2}(\mathbb{R}^+)$, respectively.

Theorem 10.13 *Let $c \in \mathbb{R}$, $\sigma \in \mathbb{R}^+$, $r \in \mathbb{N}$ and $q \in [1, \infty]$. Suppose that $f \in \mathcal{M}_c^1 \cap W_c^{r,1}(\mathbb{R}^+)$. If $r > 1/q$, then*

$$\operatorname{dist}_q(f, B_{c,\sigma}^1) \leq D \, \|\Theta_c^r f\|_{X_c} \cdot \begin{cases} \sigma^{-r+1/q}, & q < \infty, \\ \sigma^{-r}, & q = \infty, \end{cases}$$

where D is a constant depending on r and q only.

If $r > 1 + 1/q$ and, in addition, $v[f]^\wedge_{M_c}(c + iv) \in L^1(\mathbb{R})$ as a function of v, then

$$\operatorname{dist}_q(\Theta_c f, B_{c,\sigma}^1) \leq D' \, \|\Theta_c^r f\|_{X_c} \cdot \begin{cases} \sigma^{-r+1+1/q}, & q < \infty, \\ \sigma^{-r+1}, & q = \infty, \end{cases}$$

where D' is again a constant depending on r and q only.

Proof First we consider the case $q < \infty$. The formula for the Mellin transform of a Mellin derivative (see Theorem 6.14) yields

$$[f]^\wedge_{M_c}(c + iv) = (-iv)^{-r} [\Theta_c^r f]^\wedge_{M_c}(c + iv) \qquad (v \in \mathbb{R} \setminus \{0\}).$$

Thus, from Theorem 10.2 we obtain

$$\operatorname{dist}_q(f, B_{c,\sigma}^1) = \left\{ \int_{|v| \geq \sigma} |v^{-r} [\Theta_c^r f(v)]^\wedge_{M_c}(c + iv)|^q dv \right\}^{1/q}.$$

Since $\Theta_c^r f \in X_c$, its Mellin transform is continuous and bounded on the line $c + i\mathbb{R}$. Therefore

$$\begin{aligned} \operatorname{dist}_q(f, B_{c,\sigma}^1) &\leq \|[\Theta_c^r f]^\wedge_{M_c}\|_{C(c+i\mathbb{R})} \left\{ 2 \int_{v \geq \sigma} v^{-rq} dv \right\}^{1/q} \\ &\leq \|\Theta_c^r f\|_{X_c} \left(\frac{2}{rq - 1} \right)^{1/q} \frac{1}{\sigma^{r-1/q}}, \end{aligned}$$

and so the assertion is proved for $q < \infty$ with $D = (2/(rq - 1))^{1/q}$. For $q = \infty$ we use again Theorem 10.2 and proceed analogously, obtaining $D = 1$.

For the second part, note that under the assumptions on f and $v[f]^\wedge_{M_c}(c + iv) \in L^1(c + i\mathbb{R})$, we have $\Theta_c f \in \mathcal{M}_c^1 \cap W_c^{r-1,1}(\mathbb{R}^+)$. Therefore we can apply the first part of the proof to the function $\Theta_c f$ and obtain the assertion with the constant $D' = (2/(rq - q - 1))^{1/q}$ for $q < \infty$ and $D' = 1$ for $q = \infty$. □

Note that if $f \in \mathcal{M}_c^1 \cap W_c^{r,1}(\mathbb{R}^+)$ satisfies the further condition that $[\Theta_c^r f]^\wedge_{M_c} \in L^q(c+i\mathbb{R})$, then we may write

$$\begin{aligned}\operatorname{dist}_q(f, B^1_{c,\sigma}) &= \left\{\int_{|v|\ge\sigma} |v^{-r}[\Theta_c^r f(v)]^\wedge_{M_c}(c+iv)|^q dv\right\}^{1/q}\\ &\le \frac{1}{\sigma^r}\|[\Theta_c^r f]^\wedge_{M_c}\|_{L^q(c+i\mathbb{R})},\end{aligned}$$

and so

$$\operatorname{dist}_q(f, B^1_{c,\sigma}) = \mathcal{O}(\sigma^{-r}) \qquad (\sigma \to \infty).$$

A corresponding theorem for the Mellin–Sobolev spaces $W_c^{r,2}(\mathbb{R}^+)$ is as follows.

Theorem 10.14 *Let $c \in \mathbb{R}$, $\sigma \in \mathbb{R}^+$, $r \in \mathbb{N}$ and $q \in [1,2]$. Suppose that $f \in \mathcal{M}_c^2 \cap W_c^{r,2}(\mathbb{R}^+)$. Then*

$$\operatorname{dist}_q(f, B^2_{c,\sigma}) \le D\,\|\Theta_c^r f\|_{X_c^2}\,\sigma^{-r-1/2+1/q},$$

where D is a constant depending on r and q only.

If $r > 1 + 1/2 + 1/q$ and, in addition, $v[f]^\wedge_{M_c^2}(c+iv) \in L^1(\mathbb{R})$, then

$$\operatorname{dist}_q(\Theta_c f, B^2_{c,\sigma}) \le D'\,\|\Theta_c^r f\|_{X_c^2}\,\sigma^{-r+1/2+1/q},$$

where D' is again a constant depending on r and q only.

Proof Analogously to the proof of the previous theorem, we may write

$$\operatorname{dist}_q(f, B^2_{c,\sigma}) = \left\{\int_{|v|\ge\sigma} |v^{-r}[\Theta_c^r]^\wedge_{M_c^2}(c+iv)|^q dv\right\}^{1/q}.$$

For $q = 2$, the property that the Mellin transform in X_c^2 is norm-preserving (see Lemma 7.5) yields

$$\begin{aligned}\operatorname{dist}_q(f, B^2_{c,\sigma}) &\le \frac{1}{\sigma^r}\left\{\int_{|v|\ge\sigma} |[\Theta_c^r]^\wedge_{M_c^2}(c+iv)|^2 dv\right\}^{1/2}\\ &\le \frac{1}{\sigma^r}\|[\Theta_c^r f]^\wedge_{M_c}\|_{L^2(c+i\mathbb{R})} = \frac{\sqrt{2\pi}}{\sigma^r}\,\|\Theta_c^r f\|_{X_c^2}.\end{aligned}$$

Thus the assertion follows for $q = 2$ with the constant $D = (2\pi)^{1/2}$. For $q \in [1, 2[$ we use Hölder's inequality with $\mu = 2/(2-q)$, $\nu = 2/q$, obtaining

$$\text{dist}_q(f, B^2_{c,\sigma}) \leq \left\{2\int_\sigma^\infty v^{-rq\mu}dv\right\}^{1/(q\mu)} \left\{\int_{|v|\geq\sigma} |[\Theta^r_c f]^\wedge_{M^2_c}(c+iv)|^{q\nu}dv\right\}^{1/(q\nu)}$$

$$\leq \frac{1}{\sigma^{r+1/2-1/q}}\left\{\frac{4-2q}{(2r+1)q-2}\right\}^{1/q-1/2} \|[\Theta^r_c f]^\wedge_{M^2_c}\|_{L^2(c+i\mathbb{R})}$$

$$= \frac{\sqrt{2\pi}}{\sigma^{r+1/2-1/q}}\left\{\frac{4-2q}{(2r+1)q-2}\right\}^{1/q-1/2} \|\Theta^r_c f\|_{X^2_c}.$$

Hence the first inequality holds with

$$D = \sqrt{2\pi}\left\{\frac{4-2q}{(2r+1)q-2}\right\}^{1/q-1/2}.$$

The second inequality follows by arguments similar to those in the proof of Theorem 10.13. □

Obviously, under the assumptions of the first part of Theorem 10.14, we obtain the asymptotic estimate

$$\text{dist}_q(f, B^2_{c,\sigma}) = \mathcal{O}(\sigma^{-r-1/2+1/q}) \qquad (\sigma \to \infty).$$

10.2.4 Mellin–Hardy Spaces

Here we estimate the distance of a function f belonging to a Mellin–Hardy space $H^p_c(\mathbb{H}_a)$ from the Mellin–Bernstein space for polar-analytic functions $\mathscr{B}^p_{c,\sigma}$ (see Definitions 3.41 and 3.49).

First we note that if $f \in \mathscr{B}^p_{c,\sigma}$ with $p \in [1, 2]$, then $f(\cdot, \theta) \in G^q_c$ for any $q \geq 1$ and $\theta \in \mathbb{R}$ since the Mellin transform has compact support and therefore, by the uniqueness theorem of Mellin transforms, we can take $\phi_f(v) := [f(\cdot, \theta)]^\wedge_{M^p_c}(c + iv)$.

We begin with the following statement in which we restrict ourselves to $p \in \{1, 2\}$.

Theorem 10.15 *Let $c \in \mathbb{R}$, $a \in \mathbb{R}^+$ and $p \in \{1, 2\}$. Suppose that $f \in H^p_c(\mathbb{H}_a)$. Then, for $\alpha \in]0, a[$, we have*

$$|[f(\cdot, 0)]^\wedge_{M^p_c}(c+it)| = e^{-\alpha|t|}|[f(\cdot, \varepsilon\alpha)]^\wedge_{M^p_c}(c+it)| \qquad (a.e.\ for\ t \in \mathbb{R}),$$

where $\varepsilon = 1$ for $t > 0$ and $\varepsilon = -1$ for $t < 0$.

Proof For $f \in H_c^p(\mathbb{H}_a)$, we consider the function g defined by

$$g(r,\theta) := e^{(c-1+it)(\log r + i\theta)} f(r,\theta),$$

which is polar-analytic on $\mathbb{H}_a$. For sake of a simple notation, we do not indicate the dependence of g on c and t explicitly.

First we assume that $t > 0$. For $\rho > 1$, let γ be the positively oriented rectangular curve in $\mathbb{H}$ with vertices at $(1/\rho, 0)$, $(\rho, 0)$, $(1/\rho, \alpha)$ and (ρ, α). By Theorem 2.6, we have

$$\int_\gamma g(r,\theta)e^{i\theta}(dr + ir d\theta) = 0.$$

This equation may be rewritten in terms of ordinary integrals as

$$\int_{1/\rho}^{\rho} g(r,0)dr = \int_{1/\rho}^{\rho} g(r,\alpha)e^{i\alpha}dr + I_+\left(\frac{1}{\rho}, t\right) - I_+(\rho, t),$$

where

$$I_+(r,t) = \int_0^\alpha g(r,\theta) ir e^{i\theta} d\theta = ir^{c+it}\int_0^\alpha e^{-(t-ic)\theta} f(r,\theta)d\theta,$$

and so

$$|I_+(r,t)| \le r^c \int_0^\alpha e^{-t\theta}|f(r,\theta)|d\theta.$$

Now we let $\rho \to \infty$. Then, for $p = 2$ the integrals

$$\int_{1/\rho}^{\rho} g(r,0)dr \quad \text{and} \quad \int_{1/\rho}^{\rho} g(r,\alpha)e^{i\alpha}dr$$

converge in the L^2 sense to the Mellin transforms

$$[f(\cdot,0)]^\wedge_{M_c^p}(c+it) \quad \text{and} \quad e^{-\alpha t}e^{i\alpha c}[f(\cdot,\alpha)]^\wedge_{M_c^p}(c+it),$$

respectively, while for $p = 1$ they converge uniformly.

For $t < 0$, we use the rectangular curve γ with vertices at $(1/\rho, 0)$, $(\rho, 0)$, $(1/\rho, -\alpha)$ and $(\rho, -\alpha)$, and proceed analogously with corresponding integrals over $[-\alpha, 0]$, denoted by $I_-(r,t)$ for $r = 1/\rho$ and $r = \rho$. In any case we obtain an equation of the form

$$\int_{1/\rho}^{\rho} r^{c-1+it} f(r,0)dr = e^{-\alpha|t|} e^{i\varepsilon\alpha c} \int_{1/\rho}^{\rho} r^{c-1+it} f(r,\varepsilon\alpha)dr + I\left(\frac{1}{\rho}, t\right) - I(\rho,t), \tag{10.9}$$

where $I(r,t)$ are integrals which may be estimated as

$$|I(r,t)| \le r^c \int_{-\alpha}^{\alpha} e^{-|t\theta|} |f(r,\theta)| d\theta.$$

Using Proposition 3.52, we can show that $\|I(r,\cdot)\|_{L^{p'}(\mathbb{R})}$ exists for $p' \in \{2,\infty\}$ and approaches 0 as $r \to 0+$ or $r \to \infty$. Hence (10.9) implies the assertion. □

Using Proposition 3.50 and Theorem 10.15, we see that the restriction to the positive real line of a function in the Hardy space $H_c^p(\mathbb{H}_a)$ is an element of $\mathcal{M}_c^p$.

Now we define the distance $\operatorname{dist}_q(f, \mathscr{B}_{c,\sigma}^p)$ of a function $f \in H_c^p(\mathbb{H}_a)$ from the space $\mathscr{B}_{c,\sigma}^p$ by considering the distance between the restrictions to the positive real line of the function f and the Mellin Bernstein spaces $\mathscr{B}_{c,\sigma}^p$.

The following result gives estimates of the Mellin distance $\operatorname{dist}_q(f, \mathscr{B}_{c,\sigma}^p)$ for a function $f \in H_c^p(\mathbb{H}_a)$.

Theorem 10.16 *Let $c \in \mathbb{R}$, $a \in \mathbb{R}^+$, $\sigma \in \mathbb{R}^+$ and $p \in \{1,2\}$. Suppose that $f \in H_c^p(\mathbb{H}_a)$. Then, for $p = 1$, $q \in [1,\infty]$,*

$$\operatorname{dist}_q(f, \mathscr{B}_{c,\sigma}^1) \le 2\,\|f\|_{H_c^1(\mathbb{H}_a)} \left(\frac{2}{aq}\right)^{1/q} e^{-a\sigma}$$

and for $p = 2$,

$$\operatorname{dist}_2(f, \mathscr{B}_{c,\sigma}^2) \le 2\sqrt{\pi}\,\|f\|_{H_c^2(\mathbb{H}_a)} e^{-a\sigma}.$$

Furthermore, for $p = 2$, $q \in [1,2[$,

$$\operatorname{dist}_q(f, \mathscr{B}_{c,\sigma}^2) \le 2\sqrt{\pi}\left(\frac{2-q}{aq}\right)^{(1/q)-(1/2)} \|f\|_{H_c^2(\mathbb{H}_a)} e^{-a\sigma}.$$

Proof For $p = 1$, we consider the functions $f(\cdot,\varepsilon\alpha)$ for $\varepsilon \in \{-1,1\}$. It is easy to see that

$$|[f(\cdot,\varepsilon\alpha)]^{\wedge}_{M_c}(c+it)| \le \|f(\cdot,\varepsilon\alpha)\|_{X_c} \le 2\|f\|_{H_c^1(\mathbb{H}_a)}.$$

Then, by Theorem 10.15,

$$\begin{aligned}
\operatorname{dist}_q(f, \mathscr{B}^1_{c,\sigma}) &= \left(\int_{|t|\geq\sigma} |[f(\cdot,0)]^\wedge_{M_c}(c+it)|^q dt\right)^{1/q} \\
&= \left(\int_{-\infty}^{-\sigma} e^{-q\alpha|t|} \left|[f(\cdot,-\alpha)]^\wedge_{M_c}(c+it)\right|^q dt\right. \\
&\quad \left. + \int_{\sigma}^{\infty} e^{-q\alpha|t|} \left|[f(\cdot,\alpha)]^\wedge_{M_c}(c+it)\right|^q dt\right)^{1/q} \\
&\leq 2\,\|f\|_{H^1_c(\mathbb{H}_a)} \left(2\int_{\sigma}^{\infty} e^{-q\alpha t} dt\right)^{1/q} \\
&= 2\,\|f\|_{H^1_c(\mathbb{H}_a)} \left(\frac{2}{\alpha q}\right)^{1/q} e^{-\alpha\sigma}
\end{aligned}$$

for every $\alpha \in]0, a[$. Now the assertion follows for $\alpha \to a-$.

Next, let $p = 2$. For $q = 2$ we may employ Theorem 10.15 in conjunction with Lemma 7.8 to conclude that

$$\begin{aligned}
\operatorname{dist}_2(f, \mathscr{B}^2_{c,\sigma}) &= \left(\int_{|t|\geq\sigma} \left|[f(\cdot,0)]^\wedge_{M^2_c}(c+it)\right|^2 dt\right)^{1/2} \\
&= \left(\int_{|t|\geq\sigma} e^{-2\alpha|t|} \left|[f(\cdot,\varepsilon\alpha)]^\wedge_{M^2_c}(c+it)\right|^2 dt\right)^{1/2} \\
&\leq e^{-\alpha\sigma}\left(\int_{-\infty}^{-\sigma} \left|[f(\cdot,-\alpha)]^\wedge_{M^2_c}(c+it)\right|^2 dt\right. \\
&\quad \left. + \int_{\sigma}^{\infty} \left|[f(\cdot,\alpha)]^\wedge_{M^2_c}(c+it)\right|^2 dt\right)^{1/2} \\
&\leq e^{-\alpha\sigma}\sqrt{2\pi}\left(\|f(\cdot,-\alpha)\|^2_{X^2_c} + \|f(\cdot,\alpha)\|^2_{X^2_c}\right)^{1/2} \\
&\leq e^{-\alpha\sigma} 2\sqrt{\pi}\, \|f\|_{H^2_c(\mathbb{H}_a)},
\end{aligned}$$

and the assertion follows for $\alpha \to a-$.

Finally, for $q \in [1, 2[$, we modify the previous proof by using Hölder's inequality with the conjugate exponents $\mu = 2/(2-q)$ and $\nu = 2/q$. This leads us to

$$\begin{aligned}
&\operatorname{dist}_q(f, \mathscr{B}^2_{c,\sigma}) \\
&= \left(\int_{|t|\geq\sigma} e^{-q\alpha|t|} \left|[f(\cdot,\varepsilon\alpha)]^\wedge_{M^2_c}(c+it)\right|^q dt\right)^{1/q} \\
&\leq \left(\int_{|t|\geq\sigma} e^{-q\mu\alpha|t|} dt\right)^{1/(q\mu)} \left(\int_{|t|\geq\sigma} \left|[f(\cdot,\varepsilon\alpha)]^\wedge_{M^2_c}(c+it)\right|^{q\nu} dt\right)^{1/(q\nu)}.
\end{aligned}$$

The first integral on the right-hand side can be easily calculated. The integrand of the second integral has the exponent 2. Therefore, we can use Lemma 7.8 as in the previous paragraph. Thus we obtain

$$\operatorname{dist}_q(f, \mathscr{B}^2_{c,\sigma}) \le 2\sqrt{\pi}\, \|f\|_{H^2_c(\mathbb{H}_a)} \left(\frac{2-q}{\alpha q}\right)^{(1/q)-(1/2)} e^{-\alpha\sigma},$$

and the assertion follows again by letting $\alpha \to a-$. □

10.3 Characterizations of Distance by Function Spaces

We are going to establish a characterization of the distances from Mellin–Paley–Wiener spaces by functions spaces guaranteeing a certain regularity.

For doing this, we have to characterize prescribed distances by function spaces. As a general hypothesis, we suppose that $f \in \mathcal{M}^1_c$.

We begin with the following characterization which is an immediate consequence of Theorem 10.2.

Theorem 10.17 (Exactness) *Let $c \in \mathbb{R}$ and $\sigma \in \mathbb{R}^+$. Suppose that $f \in \mathcal{M}^1_c$. Then*

$$f \in B^1_{c,\sigma} \iff \operatorname{dist}_\infty(f, B^1_{c,\tau}) = 0$$

for all $\tau > \sigma$.

The following theorem characterizes the exponential speed of convergence of the distance from the space $B^1_{c,\sigma}$. As before, we agree that $\varphi(\cdot) := f(\cdot, 0)$ for $f : \mathbb{H}_a \to \mathbb{C}$.

Theorem 10.18 (Exponential Speed) *Let $c \in \mathbb{R}$ and $\sigma, a \in \mathbb{R}^+$. Suppose that $\varphi \in \mathcal{M}^1_c$. Then φ is the restriction to $\mathbb{R}^+$ of a function $f \in H^*_c(\mathbb{H}_a)$ if and only if $\operatorname{dist}_\infty(\varphi, B^1_{c,\sigma}) = \mathcal{O}(e^{-a\sigma})$ as $\sigma \to \infty$.*

Proof This is essentially the generalized Paley–Wiener theorem (Theorem 8.20), taking into account that, under the present assumptions, we have, by Lemmas 8.3 and 8.14, that $\varphi \in X_c \cap X^2_c$ and so, by the consistency property for the Mellin transform in X^2_c, we obtain $M^*_c[\varphi] = M_c[\varphi] = M^2_c[\varphi]$. □

In conjunction with Proposition 8.18, Theorem 10.18 yields as a corollary that

$$f \in H^1_c(\mathbb{H}_a) \Longrightarrow \operatorname{dist}_\infty(f(\cdot, 0), B^1_{c,\sigma}) = \mathcal{O}(e^{-a\sigma}) \quad (\sigma \to \infty),$$

a result contained in Theorem 10.16.

Theorem 10.13 says that functions from Mellin–Sobolev spaces have distances from $B^1_{c,\sigma}$ that behave like $\mathcal{O}(\sigma^{-r})$. However, the familiar Mellin–Sobolev spaces are not appropriate for a true characterization. The following modified space will provide the desired equivalence. Subsequently, the fractional derivatives $\Theta_c^\alpha f$ are given by Definition 9.31.

Definition 10.19 For $c, \gamma \in \mathbb{R}$ with $\gamma > 1$, we introduce

$$W_c^{\gamma,*}(\mathbb{R}^+) := \left\{ f \in X_c \ : \ \sup_{v\in\mathbb{R}} |v^\gamma [f]^\wedge_{M_c}(c+iv)| < \infty \right\}.$$

Obviously, the functional

$$\|\cdot\|_{W_c^{\gamma,*}} \ : \ f \longmapsto \sup_{v\in\mathbb{R}} |v^\gamma [f]^\wedge_{M_c}(c+iv)|$$

is defined on $W_c^{\gamma,*}(\mathbb{R}^+)$ and endows this space with a norm.

Remark 10.20 In order to establish a connection between the space $W_c^{\gamma,*}(\mathbb{R}^+)$ and other Sobolev type spaces, we note that for $\gamma > 1$ the inclusion

$$\mathscr{W}_c^{\gamma,1} \subset W_c^{\gamma,*}(\mathbb{R}^+) \tag{10.10}$$

holds. Indeed, for $f \in \mathscr{W}_c^{\gamma,1}$ we find by Theorem 9.33 (i) that

$$[\Theta_c^\gamma f]^\wedge_{M_c}(c+iv) = (-iv)^\gamma [f]^\wedge_{M_c}(c+iv) \qquad (v \in \mathbb{R}).$$

Since the left-hand side is a bounded function of $v \in \mathbb{R}$, we easily conclude that $f \in W_c^{\gamma,*}(\mathbb{R}^+)$.

The desired equivalence can now be stated as follows.

Theorem 10.21 (Polynomial Speed) *Let $c, r, \alpha \in \mathbb{R}$ with $r > 1$ and $\alpha \geq 0$. Suppose that $f \in \mathcal{M}_c^1$ and $(\cdot)^\alpha [f]^\wedge_{M_c}(c+i\cdot) \in L^1(\mathbb{R})$. Then*

$$f \in W_c^{r+\alpha,*}(\mathbb{R}^+) \iff \mathrm{dist}_\infty(\Theta_c^\alpha f, B^1_{c,\sigma}) = \mathcal{O}(\sigma^{-r}) \qquad (\sigma \to \infty).$$

Proof Let us write $[f]^\wedge_{M_c}(c+i\cdot) =: \varphi(\cdot)$ for short. The hypotheses on f imply that $\varphi \in C(\mathbb{R})$ and $(\cdot)^\alpha \varphi(\cdot) \in L^1(\mathbb{R} \cap C(\mathbb{R})$. Theorem 10.5 and Corollary 10.6 yield that $\Theta_c^\alpha f$ exists and

$$\mathrm{dist}_\infty(\Theta_c^\alpha f, B^1_{c,\sigma}) = \sup_{|v|\geq\sigma} \left|v^\alpha \varphi(v)\right|.$$

Now assume that $f \in W_c^{r+\alpha,*}(\mathbb{R}^+)$. Then

$$|v^\alpha \varphi(v)| \leq \frac{\|f\|_{W_c^{r+\alpha,*}}}{|v|^r} \qquad (v \neq 0)$$

and so

$$\operatorname{dist}_\infty(\Theta_c^\alpha f, B_{c,\sigma}^1) \le \frac{\|f\|_{W_c^{r+\alpha,*}}}{\sigma^r} \qquad (\sigma > 0)$$

which shows the claimed asymptotic behavior.

Conversely, if $\operatorname{dist}_\infty(\Theta_c^\alpha f, B_{c,\sigma}^1) = \mathcal{O}(\sigma^{-r})$ as $\sigma \to \infty$, then there exist $\sigma_0 > 0$ and $C > 0$ such that

$$\sup_{|v|\ge\sigma} \left|v^\alpha \varphi(v)\right| \le C\sigma^{-r} \qquad (\sigma \ge \sigma_0).$$

In particular, $|(\pm\sigma)^\alpha \varphi(\pm\sigma)| \le C\sigma^{-r}$ for all $\sigma \ge \sigma_0$, or equivalently, $|v^\alpha \varphi(v)| \le C|v|^{-r}$ for $|v| \ge \sigma_0$. This implies that

$$\sup_{|v|\ge\sigma_0} |v^{r+\alpha}\varphi(v)| \le C.$$

By continuity we also have

$$\sup_{|v|\le\sigma_0} |v^{r+\alpha}\varphi(v)| < \infty.$$

Combining these inequalities, we see that $f \in W_c^{r+\alpha,*}(\mathbb{R}^+)$. □

10.4 Notes

For a motivation of the notion of distance introduced in this chapter, we begin with a look at Fourier analysis. There exist various basic relations (equations and inequalities) that are valid for bandlimited functions f. Some examples are the classical sampling formula, Valiron's formula, the reproducing kernel formula, Boas' differentiation formula, Parseval's formula, the compound trapezoidal rule on $\mathbb{R}$ and Bernstein's inequality. When the function f is not bandlimited, they are no longer valid, but they may be used in an approximate sense by introducing a remainder. One would expect that the remainder is of small modulus if and only if f is close to a bandlimited function and therefore try to estimate the remainder in terms of the distance of f from the subspace B_σ^p of $L^p(\mathbb{R})$ which comprises all functions that are bandlimited to $[-\sigma, \sigma]$. Since $L^p(\mathbb{R})$ is a normed space, it looks natural to use a notion of distance defined by

$$\mathrm{d}_p(f, g) := \|f - g\|_{L^p(\mathbb{R})}.$$

Unfortunately this metric does not lead to satisfactory error estimates. Except for $p = 2$, one can construct functions that produce a small remainder although their distance from B_σ^p is large. Since bandlimitedness is defined via the Fourier transform

$\widehat{f}$ of f, there is some hope that a notion of distance based on the Fourier transform may lead to more realistic error estimates. Such reflections inspired Butzer et al. [116] (also see [115, 118, 120] for continuations) to introduce and work with a notion of distance defined by

$$\mathrm{dist}_p(f, g) := \|\widehat{f} - \widehat{g}\|_{L^p(\mathbb{R})}.$$

In particular, for f belonging to certain intermediate spaces between B^p_σ and $L^p(\mathbb{R})$, such as Lipschitz classes, Sobolev spaces and Hardy spaces for strips, these authors determined or estimated

$$\mathrm{dist}_q(f, B^p_\sigma) := \inf_{g \in B^p_\sigma} \mathrm{dist}_q(f, g),$$

where $p \in [1, 2]$ and $q \in [1, \infty]$. Furthermore, for the approximate forms of the aforementioned relations, they estimated the remainders in terms of $\mathrm{dist}_q(f, B^p_\sigma)$. These results turned out to be very realistic. In some cases these authors succeeded in characterizing immediate spaces by the asymptotic behavior of $\mathrm{dist}_q(f, B^p_\sigma)$ as $\sigma \to \infty$. That the shortcomings of the "natural" distance $\mathrm{d}_p(f, g)$ do not occur for $p = 2$ can be explained by the isometry of the Fourier transform in $L^2(\mathbb{R})$ which implies that $\mathrm{d}_2(f, g) = \mathrm{dist}_2(f, g)$.

In view of these results it was clear that in Mellin analysis one should use a notion of distance that is based on the Mellin transform in an analogous way. Corresponding studies were performed by Bardaro et al. in [56, 58–60].

Section 10.1 is based on [56, Sect. 3], except for Theorem 10.5 and Corollary 10.6 which were obtained in [59, Sect. 2]. Sections 10.2.1–10.2.3 originate from [56, Sect. 4] and Sect. 10.2.4 from [58, Sects. 6–7]. Finally, the characterizations of distances by function spaces in Sect. 10.3 was first obtained in [59, Sect. 4]. A special case of the Mellin inversion classes of Sect. 10.2.1 was also considered by Yakubovich [305] for the construction of a class of certain non-convolution transforms.

Chapter 11
Mellin Transform Methods for Partial Differential Equations

In this chapter, we first use Mellin transforms for partial differential equations of a classical type, namely variants of wave and heat equations on the wedge-shaped domain $\mathbb{R}^+ \times \mathbb{R}^+$. This approach seems to be very suitable for describing problems of this type. Then we apply the theory developed in Chap. 9 for studying some fractional partial differential equations.

We shall mainly consider initial-value problems but the Mellin transform method can also be used for boundary-value problems. For further information, see Sect. 11.4 Notes.

11.1 An Analogue of the One-Dimensional Wave Equation

We consider a version of the one-dimensional wave equation in which the second partial derivative with respect to the time is replaced by a Mellin derivative of order two, namely

$$(\Theta^2 w(\cdot, y))(x) = \frac{\partial^2}{\partial y^2} w(x, y) \qquad (x, y \in \mathbb{R}^+), \tag{11.1}$$

or equivalently,

$$x^2 \frac{\partial^2}{\partial x^2} w(x, y) + x \frac{\partial}{\partial x} w(x, y) = \frac{\partial^2}{\partial y^2} w(x, y) \qquad (x, y \in \mathbb{R}^+), \tag{11.2}$$

under the initial conditions $w(x, 0) = f(x)$, $\frac{\partial}{\partial y} w(x, 0) = g(x)$. Here we denote the time variable by y in order to avoid confusion with the parameter t of the Mellin transform. We have to interpret this initial-value problem in the following more precise sense: For given functions $f, g \in X_0$, we look for a function

C. Bardaro et al., *Mellin Analysis, Transform Theory, and Applications*,
https://doi.org/10.1007/978-3-031-96672-9_11

$$w : \mathbb{R}^+ \times \mathbb{R}^+ \longrightarrow \mathbb{C}$$

with the following properties:

(i) $\frac{\partial}{\partial x} w(x, y)$, $\frac{\partial^2}{\partial x^2} w(x, y)$ and $\frac{\partial^2}{\partial y^2} w(x, y)$ exist for all $x, y \in \mathbb{R}^+$;
(ii) $w(x, y)$ satisfies Eq. (11.2) for all $x, y \in \mathbb{R}^+$;
(iii) for each $y \in \mathbb{R}^+$, $w(\cdot, y)$ and $\frac{\partial}{\partial y} w(\cdot, y)$ belong to X_0 and

$$\lim_{y\to 0+} \|w(\cdot, y) - f(\cdot)\|_{X_0} = 0, \qquad \lim_{y\to 0+} \left\| \frac{\partial}{\partial y} w(\cdot, y) - g(\cdot)\right\|_{X_0} = 0;$$

(iv) for each $y \in \mathbb{R}^+$, $w(\cdot, y)$ belongs to $W_0^{2,1}(\mathbb{R}^+)$;
(v) for each $y \in \mathbb{R}^+$, $\frac{\partial^2}{\partial y^2} w(\cdot, y)$ belongs to X_0 and

$$\lim_{h\to 0} \left\| \frac{w(\cdot, y+h) - w(\cdot, y)}{h} - \frac{\partial}{\partial y} w(\cdot, y)\right\|_{X_0} = 0, \tag{11.3}$$

$$\lim_{h\to 0} \left\| \frac{\frac{\partial}{\partial y} w(\cdot, y+h) - \frac{\partial}{\partial y} w(\cdot, y)}{h} - \frac{\partial^2}{\partial y^2} w(\cdot, y)\right\|_{X_0} = 0. \tag{11.4}$$

To solve this problem, we first assume that there exists a function $w(x, y)$ with the above properties and apply the Mellin transform with respect to the first variable, namely

$$[w(\cdot, y)]^\wedge_{M_0}(s) \equiv [w]^\wedge_{M_0}(s, y) := \int_0^\infty w(u, y) u^{s-1} du \qquad (s = it, t \in \mathbb{R}),$$

to both sides of (11.1). By Theorem 6.14, we obtain for the left-hand side

$$\begin{aligned} \left[(\cdot)^2 \frac{\partial^2}{\partial x^2} w(\cdot, y) + (\cdot)\frac{\partial}{\partial x} w(\cdot, y)\right]^\wedge_{M_0}(s) &= \left[\Theta^2 w(\cdot, y)\right]^\wedge_{M_0}(s) \\ &= s^2 [w]^\wedge_{M_0}(s, y) \qquad (y \in \mathbb{R}^+). \end{aligned}$$

For the right-hand side, we conclude that

$$\left[\frac{\partial}{\partial y} w(\cdot, y)\right]^\wedge_{M_0}(s) = \frac{\partial}{\partial y} [w]^\wedge_{M_0}(s, y), \qquad \left[\frac{\partial^2}{\partial y^2} w(\cdot, y)\right]^\wedge_{M_0}(s) = \frac{\partial^2}{\partial y^2} [w]^\wedge_{M_0}(s, y)$$

since by (6.2) and (11.3),

$$\left|\frac{[w]^\wedge_{M_0}(s, y+h) - [w]^\wedge_{M_0}(s, y)}{h} - \left[\frac{\partial}{\partial y} w(\cdot, y)\right]^\wedge_{M_0}(s)\right|$$

$$\leq \left\|\frac{w(\cdot, y+h) - w(\cdot, y)}{h} - \frac{\partial}{\partial y} w(\cdot, y)\right\|_{X_0} = o(1)$$

as $h \to 0$ and likewise for $\frac{\partial^2 w}{\partial y^2}$. Hence $\frac{\partial}{\partial y}[w]^\wedge_{M_0}(s, y)$ exists for each $s = it$ and $y \in \mathbb{R}^+$ and is equal to the Mellin transform of $\frac{\partial w}{\partial y}(x, y)$, that is, we may interchange partial differentiation with the Mellin transform. In particular, $[w]^\wedge_{M_0}(s, y)$ is a continuous function of y on $\mathbb{R}^+$ for each $s = it$. Hence the transformed function $[w]^\wedge_{M_0}(s, y)$ satisfies for each $s = it$ the ordinary differential equation

$$\frac{\partial^2}{\partial y^2}[w]^\wedge_{M_0}(s, y) = s^2 [w]^\wedge_{M_0}(s, y) \qquad (s = it,\ t \in \mathbb{R},\ y \in \mathbb{R}^+).$$

This equation has the general solution

$$[w]^\wedge_{M_0}(s, y) = \begin{cases} A(s)e^{sy} + B(s)e^{-sy}, & s \neq 0, \\ A(0) + B(0)y, & s = 0, \end{cases} \tag{11.5}$$

where $A(s)$ and $B(s)$ are arbitrary complex constants not depending on y. Next we want to determine these constants by taking the initial conditions into account. By (6.2) and (iii), we obtain

$$|[w]^\wedge_{M_0}(s, y) - [f]^\wedge_{M_0}(s)| \leq \|w(\cdot, y) - f(\cdot)\|_{X_0} = o(1),$$

$$\left|\frac{\partial}{\partial y}[w]^\wedge_{M_0}(s, y) - [g]^\wedge_{M_0}(s)\right| \leq \left\|\frac{\partial}{\partial y} w(\cdot, y) - g(\cdot)\right\|_{X_0} = o(1)$$

as $y \to 0+$. Hence

$$\lim_{y \to 0+} [w]^\wedge_{M_0}(s, y) = [f]^\wedge_{M_0}(s), \qquad \lim_{y \to 0+} \frac{\partial}{\partial y}[w]^\wedge_{M_0}(s, y) = [g]^\wedge_{M_0}(s),$$

uniformly with respect to $s = it$. In view of the general solution (11.5), this implies

$$[f]^\wedge_{M_0}(s) = \begin{cases} A(s) + B(s), & s \neq 0, \\ A(0), & s = 0 \end{cases}$$

and

$$[g]^\wedge_{M_0}(s) = \begin{cases} sA(s) - sB(s), & s \neq 0, \\ B(0), & s = 0. \end{cases}$$

These equations allow us to express $A(s)$ and $B(s)$ in terms of $[f]^{\wedge}_{M_0}(s)$ and $[g]^{\wedge}_{M_0}(s)$. Substituted in (11.5), we find that the Mellin transform of any solution $w(x, y)$ of the initial-value problem for (11.2), as specified by (i)–(v), is given by

$$[w]^{\wedge}_{M_0}(s, y) = \begin{cases} [f]^{\wedge}_{M_0}(s)\dfrac{e^{sy}+e^{-sy}}{2} + [g]^{\wedge}_{M_0}(s)\dfrac{e^{sy}-e^{-sy}}{2s}, & s \neq 0, \\ [f]^{\wedge}_{M_0}(0) + [g]^{\wedge}_{M_0}(0)y, & s = 0. \end{cases}$$

Now we recall that, by Proposition 6.8 (b), we may write $e^{\pm sy}[f]^{\wedge}_{M_0}(s) = [f((\cdot)e^{\mp y})]^{\wedge}_{M_0}(s)$. Concerning $e^{\pm sy}[g]^{\wedge}_{M_0}(s)/s$, we first observe that $[g]^{\wedge}_{M_0}(s)/s = -[Jg]^{\wedge}_{M_0}(s)$, where Jg is the Mellin integral of g (see Theorem 6.95), and use Proposition 6.8 (b) once again. Altogether, we obtain

$$[w]^{\wedge}_{M_0}(s, y) = \left[\frac{f(\cdot\, e^{y}) + f(\cdot\, e^{-y})}{2} + \frac{1}{2}\int_{(\cdot)e^{-y}}^{(\cdot)e^{y}} g(u)\frac{du}{u}\right]^{\wedge}_{M_0}(s, y).$$

In order to represent the solution in the original (non-transformed) function space, the uniqueness theorem (Theorem 6.79) now plays the role of the inversion theorem. It gives for each $y \in \mathbb{R}^+$ the solution

$$w(x, y) = \frac{1}{2}\big(f(xe^{y}) + f(xe^{-y})\big) + \frac{1}{2}\int_{xe^{-y}}^{xe^{y}} g(u)\frac{du}{u}. \tag{11.6}$$

Thus, if the initial-value problem for (11.2) has a solution $w(x, y)$, then it is of the form (11.6).

In order to show that (11.6) is actually a solution, we have to impose certain conditions on the functions f and g.

Theorem 11.1 *Let $f \in W_0^{2,1}(\mathbb{R}^+)$ and $g \in W_0^{1,1}(\mathbb{R}^+)$ such that $f''(x)$ and $g'(x)$ exist for all $x \in \mathbb{R}^+$. Then the initial-value problem for (11.2), as specified by (i)–(v), has a unique solution which is given by (11.6) and holds for all $x, y \in \mathbb{R}^+$.*

The proof, which is based on straightforward calculations, will be left to the reader.

11.2 An Equation Related to the Heat Problems

We consider modifications of the one and two-dimensional heat equations in which the second-order spatial derivative with respect to x is replaced by a Mellin derivative. We begin with the heat flow in a semi-infinite rod,

$$(\Theta^2 w(\cdot, y))(x) = \frac{\partial}{\partial y} w(x, y) \qquad (x, y \in \mathbb{R}^+), \tag{11.7}$$

or equivalently,

$$x^2 \frac{\partial^2}{\partial x^2} w(x, y) + x \frac{\partial}{\partial x} w(x, y) = \frac{\partial}{\partial y} w(x, y) \qquad (x, y \in \mathbb{R}^+) \tag{11.8}$$

with a given initial temperature distribution $w(x, 0) = f(x)$. We solve this problem in the following more precise form: Given a function $f \in X_0$, we look for a function

$$w : \mathbb{R}^+ \times \mathbb{R}^+ \longrightarrow \mathbb{C}$$

with the following properties:

(i) $\frac{\partial}{\partial x} w(x, y)$, $\frac{\partial^2}{\partial x^2} w(x, y)$, $\frac{\partial}{\partial y} w(x, y)$ exist for all $x, y \in \mathbb{R}^+$;
(ii) $w(x, y)$ satisfies Eq. (11.8) for all $x, y \in \mathbb{R}^+$;
(iii) for each $y \in \mathbb{R}^+$, $w(\cdot, y)$ belongs to X_0 and

$$\lim_{y \to 0+} \|w(\cdot, y) - f(\cdot)\|_{X_0} = 0;$$

(iv) for each $y \in \mathbb{R}^+$, $w(\cdot, y)$ belongs to $W_0^{2,1}(\mathbb{R}^+)$;
(v) for each $y \in \mathbb{R}^+$, $\frac{\partial}{\partial y} w(\cdot, y)$ belongs to X_0 and

$$\lim_{h \to 0} \left\| \frac{w(\cdot, y + h) - w(\cdot, y)}{h} - \frac{\partial}{\partial y} w(\cdot, y) \right\|_{X_0} = 0.$$

Assuming that there exists a function $w(x, y)$ with these properties, we again apply the Mellin transform for $s = it$, $t \in \mathbb{R}^+$ with respect to x to both sides of (11.7). On using Theorem 6.14, this yields, as in the foregoing application,

$$\frac{\partial}{\partial y} [w]^{\wedge}_{M_0}(s, y) = s^2 [w]^{\wedge}_{M_0}(s, y) \qquad (s = it,\ t \in \mathbb{R},\ y \in \mathbb{R}^+),$$

where we made use of conditions (iv) and (v). This ordinary differential equation has for each $s \in i\mathbb{R}$ the solution

$$[w]^{\wedge}_{M_0}(s, y) = A(s) e^{s^2 y}$$

with a constant $A(s)$ being independent of y. The determination of $A(s)$ follows from condition (iii), giving the solution

$$[w]^{\wedge}_{M_0}(s, y) = [f]^{\wedge}_{M_0}(s) e^{s^2 y} \qquad (s = it,\ y \in \mathbb{R}^+).$$

In order to represent w in terms of the original functions, we use the inversion theorem (Theorem 6.75). It is applicable since the boundedness of $[f]^{\wedge}_{M_0}$ implies that $[f]^{\wedge}_{M_0}(it)e^{-t^2 y} \in L^1(\mathbb{R})$ as a function of t. Thus

$$w(x, y) = \frac{1}{2\pi} \int_{-\infty}^{\infty} [f]^{\wedge}_{M_0}(it)e^{-t^2 y} x^{-it} dt \tag{11.9}$$

for all $x, y \in \mathbb{R}^+$. Now, using the Mellin–Parseval formula (Theorem 6.81) together with (6.29) for $\varrho = \frac{1}{\sqrt{y}}$, we obtain

$$w(x, y) = \frac{1}{\sqrt{4\pi y}} \int_0^{\infty} f\left(\frac{x}{u}\right) u^{-\frac{1}{4y}\log u} \frac{du}{u}. \tag{11.10}$$

Hence, if a solution exists, then it is unique by Theorem 6.79 and can be represented in this way.

These considerations lead us to the following theorem.

Theorem 11.2 *For $f \in X_0$ as initial distribution, the problem (11.8) on the flow of heat in a semi-infinite rod, as specified by (i)–(v), has a unique solution. It is given by the Mellin-convolution integral (11.10) of Gauss-Weierstrass type associated with f.*

Proof By the commutativity of the Mellin convolution, we may rewrite (11.10) as

$$w(x, y) = \frac{1}{\sqrt{4\pi y}} \int_0^{\infty} f(u) \left(\frac{x}{u}\right)^{-\frac{1}{4y}\log(x/u)} \frac{du}{u}.$$

Now the proof that $w(x, y)$ is actually a solution, can be performed by differentiation inside the integral, which is legitimate by the hypotheses. Also the validity of the initial condition can be verified. Details are left to the reader. □

A similar example of a partial differential equation with a solution involving the Mellin–Abel–Poisson kernel P_ϱ is given by

$$x^2 \frac{\partial^2}{\partial x^2} w(x, y) + x \frac{\partial}{\partial x} w(x, y) = -\frac{\partial^2}{\partial y^2} w(x, y) \qquad (x, y \in \mathbb{R}^+). \tag{11.11}$$

It is related to the problem of determining the temperature $w(x, y)$ of a semi-infinite plate $\mathbb{R}^+ \times \mathbb{R}^+$ for a given temperature distribution $w(x, 0) = f(x)$ on the horizontal boundary $\mathbb{R}^+$. Since x and y are interpreted as spatial variables, we have an incomplete boundary-value problem. In terms of Mellin derivatives, Eq. (11.11) takes the form

$$(\Theta^2 w(\cdot, y))(x) = -\frac{\partial^2}{\partial y^2} w(x, y) \qquad (x, y \in \mathbb{R}^+). \tag{11.12}$$

As in the preceding example, we apply the Mellin transform to both sides of (11.12). Under suitable conditions, this leads us to

$$(-it)^2[w]^{\wedge}_{M_0}(it, y) = -\frac{\partial^2}{\partial y^2}[w]^{\wedge}_{M_0}(it, y) \qquad (t \in \mathbb{R},\, y \in \mathbb{R}^+).$$

Admitting only solutions which are bounded with respect to t, we obtain

$$[w]^{\wedge}_{M_0}(it, y) = [f]^{\wedge}_{M_0}(it)e^{-y|t|} = [f]^{\wedge}_{M_0}(it)\,\frac{2}{y}\,P_y(e^t).$$

An application of Proposition 6.30 and Lemma 6.76 together with the inversion formula finally provide the solution of (11.11), namely

$$w(x, y) = \frac{y}{\pi}\int_0^{\infty} \frac{f(u)}{y^2 + (\log x - \log u)^2}\,\frac{du}{u} \qquad (x, y \in \mathbb{R}^+)$$

for a given temperature $w(x, 0) = f(x)$ on $\mathbb{R}^+$.

11.3 Mellin Transform Methods for Fractional Differential Equations

In this section we consider an initial-value problem for two fractional differential equations in two variables and construct the exact solution by using the Mellin transform method. We begin with an integro-differential equation which generates a differential equation with a Mellin–Hadamard fractional partial derivative and a first order partial derivative. The second example contains a Mellin–Hadamard fractional partial derivative and a second order partial derivative. In both examples the (unique) solution is obtained in terms of a Mellin convolution operator.

11.3.1 An Integro-Differential Equation

For fixed $\alpha \in\,]0, 1[$, let

$$K_\alpha(x, u) := \frac{1}{\Gamma(1-\alpha)}\left(\log\frac{x}{u}\right)^{-\alpha}\chi_{]0,x[}(u) \qquad (x, u \in \mathbb{R}^+).$$

We consider the following problem: Find a function $w : \mathbb{R}^+ \times \mathbb{R}^+ \to \mathbb{C}$ such that $w(x, 0) = f(x)$ for given initial data $f : \mathbb{R}^+ \to \mathbb{C}$ and

$$Ax\frac{\partial}{\partial x}\int_0^{\infty} K_\alpha(x, u)w(u, y)\frac{du}{u} + B\frac{\partial}{\partial y}w(x, y) = 0, \tag{11.13}$$

where A and B are positive constants.

Writing (11.13) as a fractional partial differential equation in the Mellin–Hadamard sense (see Definition 9.17) and assuming for simplicity that $A = B = 1$, we obtain the equation

$$(D^{\alpha}_{0+} w(\cdot, y))(x) = -\frac{\partial}{\partial y} w(x, y) \qquad (x, y \in \mathbb{R}^+) \tag{11.14}$$

with the initial data $w(x, 0) = f(x)$ for $x > 0$.

Interpreting the initial condition in the sense of norm convergence, we require that the solution $w : \mathbb{R}^+ \times \mathbb{R}^+ \to \mathbb{C}$ of (11.14) has the following properties:

(1) for a fixed $a < 0$ and $c, \nu \in [a, 0]$, we have $w(\cdot, y) \in W(X_{[a,0]}, D^{\alpha}_{0+})$ and $J^{1-\alpha}_{0+,c} w(\cdot, y) \in W^{1,1}_{\nu}$ for every $y \in \mathbb{R}^+$;
(2) there is a function $K \in X_\nu$, where $\nu \in [a, 0[$, such that

$$\left| \frac{\partial}{\partial y} w(x, y) \right| \le K(x) \qquad (x, y \in \mathbb{R}^+);$$

(3) for given $f \in X_\nu$, we have $\lim_{y \to 0+} \| w(\cdot, y) - f(\cdot) \|_{X_\nu} = 0$.

Assuming that such a function exists, we apply the Mellin transform with respect to the variable x on the line $\nu + it$ to both sides of (11.14) and obtain

$$[D^{\alpha}_{0+} w(\cdot, y)]^{\wedge}_{M_\nu}(\nu + it) = -\left[\frac{\partial}{\partial y} w(\cdot, y) \right]^{\wedge}_{M_\nu} (\nu + it).$$

Next Proposition 9.23(ii) with $c = 0$ and $\nu \in]a, 0[$ yields

$$[D^{\alpha}_{0+} w(\cdot, y)]^{\wedge}_{M_\nu}(\nu + it) = (-\nu - it)^{\alpha} [w(\cdot, y)]^{\wedge}_{M_\nu}(\nu + it),$$

where the principal value of the power with exponent α has to be taken. Furthermore, property (2) allows us to write

$$\left[\frac{\partial}{\partial y} w(\cdot, y) \right]^{\wedge}_{M_\nu} (\nu + it) = \int_0^\infty x^{\nu + it - 1} \frac{\partial}{\partial y} w(x, y) dx = \frac{\partial}{\partial y} [w(\cdot, y)]^{\wedge}_{M_\nu}(\nu + it).$$

Thus Eq. (11.14) can transformed into the first order ordinary differential equation

$$(-\nu - it)^{\alpha} [w(\cdot, y)]^{\wedge}_{M_\nu}(\nu + it) = -\frac{\partial}{\partial y} [w(\cdot, y)]^{\wedge}_{M_\nu}(\nu + it)$$

which has the solution

$$[w(\cdot, y)]^{\wedge}_{M_\nu}(\nu + it) = A(\nu + it) e^{-(-\nu - it)^{\alpha} y},$$

where $A(\nu + it)$ is independent of y. From (3) we see that $[w(\cdot, y)]^{\wedge}_{M_\nu}(\nu + it) \to [f]^{\wedge}_{M_\nu}(\nu + it)$ uniformly for $y \to 0+$ and $t \in \mathbb{R}$. Thus $A(\nu + it) = [f]^{\wedge}_{M_\nu}(\nu + it)$, and so

$$[w(\cdot, y)]^{\wedge}_{M_\nu}(\nu + it) = [f]^{\wedge}_{M_\nu}(\nu + it)e^{-(-\nu-it)^\alpha y}.$$

Since $-\nu > 0$, $y > 0$ and $\alpha \in]0, 1[$, we find by a short calculation that

$$0 < \Re(-\nu - it)^\alpha y = \left(\nu^2 + t^2\right)^{\alpha/2} y \cos(\pi\alpha/2)\left(1 + \mathcal{O}(t^{-1})\right) \qquad (t \to \pm\infty).$$

Therefore the inverse Mellin transform of the function $\nu + it \mapsto e^{-(-\nu-it)^\alpha y}$ exists and is given by

$$G(x, y) := \frac{1}{2\pi}\int_{-\infty}^{\infty} e^{-(-\nu-it)^\alpha y} x^{-\nu-it} dt. \tag{11.15}$$

Thus, if (11.14) has a solution with the properties (1)–(3), then, by the Mellin–Parseval equation (Theorem 6.81), it can be represented as

$$w(x, y) = \int_0^\infty f(v)G\left(\frac{x}{v}, y\right)\frac{dv}{v} \qquad (x, y \in \mathbb{R}^+).$$

In order to verify that the function $w(x, y)$ is actually a solution, we substitute (11.15) and differentiate inside the integral, which leads us to

$$\begin{aligned}
-\frac{\partial w}{\partial y}(x, y) &= \int_0^\infty f(v)\left[\frac{1}{2\pi}\int_{-\infty}^{\infty}\left(\frac{x}{v}\right)^{-\nu-it}(-\nu - it)^\alpha e^{-(-\nu-it)^\alpha y} dt\right]\frac{dv}{v} \\
&= \frac{1}{2\pi}\int_{-\infty}^{\infty}(-\nu - it)^\alpha x^{-\nu-it} e^{-(-\nu-it)^\alpha y}[f]^{\wedge}_{M_\nu}(\nu + it)dt.
\end{aligned} \tag{11.16}$$

Next we consider

$$(D^\alpha_{0+} w(\cdot, y))(x) = \Theta(J^{1-\alpha}_{0+} w(\cdot, y))(x)$$

and conclude that

$$\begin{aligned}
&(D^\alpha_{0+} w(\cdot, y))(x) \\
&= \left(x\frac{\partial}{\partial x}\right)\left[\frac{1}{\Gamma(1-\alpha)}\int_0^x\left(\log\frac{x}{u}\right)^{-\alpha}\left(\int_0^\infty f(v)G\left(\frac{u}{v}, y\right)\frac{dv}{v}\right)\frac{du}{u}\right] \\
&= \left(x\frac{\partial}{\partial x}\right)\left[\frac{1}{\Gamma(1-\alpha)}\int_1^\infty(\log z)^{-\alpha}\left(\int_0^\infty f(v)G\left(\frac{x}{zv}, y\right)\frac{dv}{v}\right)\frac{dz}{z}\right] \\
&= \frac{x}{\Gamma(1-\alpha)}\int_1^\infty(\log z)^{-\alpha}\left(\int_0^\infty f(v)\frac{\partial}{\partial x}G\left(\frac{x}{zv}, y\right)\frac{dv}{v}\right)\frac{dz}{z}.
\end{aligned}$$

Noting that

$$\frac{\partial}{\partial x}G(x,y) = \frac{1}{2\pi}\int_{-\infty}^{\infty} e^{-(-\nu-it)^{\alpha}y}(-\nu-it)x^{-\nu-it-1}dt$$

and setting $s = -\nu - it$, we obtain

$$\begin{aligned}
&(D_{0+}^{\alpha}w(\cdot,y))(x)\\
&= \frac{x}{\Gamma(1-\alpha)}\int_1^{\infty}(\log z)^{-\alpha}\\
&\quad\times\left(\int_0^{\infty} f(v)\frac{1}{zv}\left(\frac{1}{2\pi}\int_{-\infty}^{\infty} e^{-s^{\alpha}y}s\left(\frac{x}{zv}\right)^{s-1}dt\right)\frac{dv}{v}\right)\frac{dz}{z}\\
&= \frac{1}{2\pi}\frac{1}{\Gamma(1-\alpha)}\int_1^{\infty}(\log z)^{-\alpha}\left(\int_{-\infty}^{\infty} e^{-s^{\alpha}y}s\left(\frac{x}{z}\right)^{s}[f]^{\wedge}_{M}(-s)dt\right)\frac{dz}{z}\\
&= \frac{1}{2\pi}\frac{1}{\Gamma(1-\alpha)}\int_{-\infty}^{\infty}[f]^{\wedge}_{M}(-s)e^{-s^{\alpha}y}s\left(\int_0^{x}\left(\log\frac{x}{u}\right)^{-\alpha}u^{s}\frac{du}{u}\right)dt.
\end{aligned}$$

As in Example 9.3 for $c = 0$ but with complex b of positive real part, we find that

$$\frac{1}{\Gamma(1-\alpha)}\int_0^{x}\left(\log\frac{x}{u}\right)^{-\alpha}u^{s}\frac{du}{u} = s^{\alpha-1}x^{s},$$

and so

$$(D_{0+}^{\alpha}w(\cdot,y))(x) = \frac{1}{2\pi}\int_{-\infty}^{\infty} s^{\alpha}x^{s}e^{-s^{\alpha}y}[f]^{\wedge}_{M_{\nu}}(\nu+it)dt.$$

Comparison with (11.16) shows that (11.14) holds. So we have proved the following statement.

Theorem 11.3 *Under the hypotheses (1)–(3), Eq. (11.14) with the initial data f has a unique solution given by*

$$w(x,y) = \int_0^{\infty} f(v)G\left(\frac{x}{v},y\right)\frac{dv}{v} \qquad (x,y\in\mathbb{R}^{+}),$$

where $G(x,y)$ is defined in (11.15).

We mention that for $\alpha = 1/2$ there exists a closed form for $G(x,y)$. Indeed, employing [230, formula 3.7, p. 174], we obtain

$$G(x,y) = \frac{y}{2\sqrt{\pi}}(-\log x)^{-3/2}\exp\left(\frac{y^2}{4\log x}\right)\chi_{]0,1[}(x)$$

and so

$$w(x,y) = \frac{y}{2\sqrt{\pi}} \int_0^1 f\left(\frac{x}{v}\right) (-\log v)^{-3/2} \exp\left(\frac{y^2}{4\log v}\right) \frac{dv}{v}.$$

11.3.2 A Fractional Equation with a Second Order Derivative

For $\alpha > 0$ we consider the fractional diffusion equation

$$(D^{\alpha}_{0+} w(\cdot, y))(x) = \frac{\partial^2}{\partial y^2} w(x,y) \qquad (x, y \in \mathbb{R}^+) \tag{11.17}$$

with an initial condition in the sense of the norm of X_0 as specified subsequently.

We look for a solution $w : \mathbb{R}^+ \times \mathbb{R}^+ \to \mathbb{C}$ of (11.17) satisfying the following conditions:

(1) for $m = \lceil \alpha \rceil$, we have $w(\cdot, y) \in W(X_0, D^{\alpha}_{0+,0})$, $J^{m-\alpha}_{0+} w(\cdot, y) \in W^{m,1}_0$, and there exists $N > 0$ such that $\|w(\cdot, y)\|_{X_0} \leq N$ for every $y \in \mathbb{R}^+$;
(2) there exist functions $K_1, K_2 \in X_0$ such that

$$\left|\frac{\partial}{\partial y} w(x,y)\right| \leq K_1(x), \qquad \left|\frac{\partial^2}{\partial y^2} w(x,y)\right| \leq K_2(x) \qquad (x, y \in \mathbb{R}^+);$$

(3) for a given $f \in X_0$, we have $\lim_{y \to 0+} \|w(\cdot, y) - f(\cdot)\|_{X_0} = 0$.

Proceeding as in the previous subsection, we take the Mellin transforms on both sides of Eq. (11.17) and obtain

$$[D^{\alpha}_{0+} w(\cdot, y)]^{\wedge}_{M_0}(it) = -\left[\frac{\partial^2}{\partial y^2} w(\cdot, y)\right]^{\wedge}_{M_0}(it).$$

Using Proposition 9.23(i) with $c = 0$, we may write

$$[D^{\alpha}_{0+} w(\cdot, y)]^{\wedge}_{M_0}(it) = (-it)^{\alpha} [w(\cdot, y)]^{\wedge}_{M_0}(it).$$

Moreover, property (2) allows us to conclude that

$$\left[\frac{\partial^2}{\partial y^2} w(\cdot, y)\right]^{\wedge}_{M_0}(it) = \frac{\partial^2}{\partial y^2} [w(\cdot, y)]^{\wedge}_{M_0}(it).$$

Thus Eq. (11.17) has been transformed into the second order linear ordinary differential equation

$$z_t''(y) = (-it)^{\alpha} z_t(y) \qquad (y > 0) \tag{11.18}$$

with respect to the function

$$z_t(y) := [w(\cdot, y)]^{\wedge}_{M_0}(it) \qquad (t \in \mathbb{R}).$$

For $t = 0$, the solutions are affine functions of the form $z_0(y) = A(0) + B(0)y$. For $t \neq 0$, the characteristic equation associated with (11.18), namely $\lambda^2 = \exp(\alpha \log(-it))$, has two complex solutions

$$\lambda_1 := |t|^{\alpha/2} \left(\cos \frac{\alpha\pi}{4} + i \sin \frac{\alpha\pi}{4}(-\operatorname{sgn} t)\right)$$

and

$$\lambda_2 := -|t|^{\alpha/2} \left(\cos \frac{\alpha\pi}{4} + i \sin \frac{\alpha\pi}{4}(-\operatorname{sgn} t)\right).$$

Thus, for $t \neq 0$, we obtain the general solution as

$$\begin{aligned} z_t(y) = A(t)e^{-|t|^{\alpha/2}(\cos(\alpha\pi/4)+i(-\operatorname{sgn} t)\sin(\alpha\pi/4))y} \\ + B(t)e^{|t|^{\alpha/2}(\cos(\alpha\pi/4)+i(-\operatorname{sgn} t)\sin(\alpha\pi/4))y}. \end{aligned}$$

Now, suppose that $\cos(\alpha\pi/4) > 0$. From condition (3), we deduce that $z_t(y)$ converges uniformly to $[f]^{\wedge}_{M_0}(it)$ as $y \to 0+$. Moreover, by condition (1), there exists a constant $N > 0$ such that $|z_t(y)| \leq N$ for every $t \in \mathbb{R}$. These observations imply that $A(t) = [f]^{\wedge}_{M_0}(it)$ and $B(t) = 0$ for every $t \in \mathbb{R}$. Thus

$$z_t(y) = [w(\cdot, y)]^{\wedge}_{M_0}(it) = [f]^{\wedge}_{M_0}(it)e^{-|t|^{\alpha/2}(\cos(\alpha\pi/4)+i(-\operatorname{sgn} t)\sin(\alpha\pi/4))y}.$$

Clearly, the function

$$t \longmapsto e^{-|t|^{\alpha/2}(\cos(\alpha\pi/4)+i(-\operatorname{sgn} t)\sin(\alpha\pi/4))y}$$

belongs to $L^1(\mathbb{R})$, and its inverse Mellin transform is given by

$$G(x, y) := \frac{1}{2\pi} \int_{-\infty}^{\infty} e^{-|t|^{\alpha/2}(\cos(\alpha\pi/4)+i(-\operatorname{sgn} t)\sin(\alpha\pi/4))y} x^{-it} dt.$$

Employing again the Mellin–Parseval equation (Theorem 6.81), we conclude that if a solution exists, then it has the form

$$w(x, y) = \int_0^{\infty} f(u) G\left(\frac{x}{u}, y\right) \frac{du}{u} \qquad (x, y \in \mathbb{R}^+).$$

Analogously, if $\cos(\alpha\pi/4) < 0$, then $A(t) = 0$ and $B(t) = [f]^{\wedge}_{M_0}(it)$ for every $t \in \mathbb{R}$ while the corresponding function $G(x, y)$ takes the form

$$G(x, y) = \frac{1}{2\pi} \int_{-\infty}^{\infty} e^{-|t|^{\alpha/2}(|\cos(\alpha\pi/4)|+i(-\operatorname{sgn} t)\sin(\alpha\pi/4))y} x^{-it} dt.$$

That the above function is really a solution can again be proved by a direct substitution into the differential equation.

The function $G(x, y)$ can be written in a simpler form valid for both signs of $\cos(\alpha\pi/4)$. Indeed, setting $a := |\cos(\alpha\pi/4)|$, $b := \sin(\alpha\pi/4)$ and using Euler's formula, we see that

$$\begin{aligned} G(x, y) &= \frac{1}{2\pi} \int_0^{\infty} e^{-t^{\alpha/2}(a-ib)y} \big(\cos(t \log x) - i \sin(t \log x)\big) dt \\ &\quad + \frac{1}{2\pi} \int_0^{\infty} e^{-t^{\alpha/2}(a+ib)y} \big(\cos(t \log x) + i \sin(t \log x)\big) dt \\ &= \frac{1}{\pi} \int_0^{\infty} e^{-t^{\alpha/2}ay} \big[\cos(t \log x)\cos(t^{\alpha/2}by) + \sin(t \log x)\sin(t^{\alpha/2}by)\big] dt \\ &= \frac{1}{\pi} \int_0^{\infty} e^{-t^{\alpha/2}ay} \cos(t \log x - t^{\alpha/2}by) dt. \end{aligned} \tag{11.19}$$

Now we want to consider this formula for some special values of α. For $\alpha = 1$ we obtain the (non-fractional) equation

$$x \frac{\partial w}{\partial x}(x, y) = \frac{\partial^2 w}{\partial y^2}(x, y) \qquad (x, y \in \mathbb{R}^+).$$

Its initial-value problem has a unique solution of the form

$$w(x, y) = \int_0^{\infty} f(u) G\left(\frac{x}{u}, y\right) \frac{du}{u} \qquad (x, y \in \mathbb{R}^+),$$

where

$$G(x, y) = \frac{1}{\pi} \int_0^{\infty} \exp\left(-\frac{\sqrt{2t}y}{2}\right) \cos\left(\frac{\sqrt{2t}y}{2} - t \log x\right) dt.$$

This integral can be calculated in closed form. By an elementary substitution, we obtain

$$\begin{aligned} I &:= \int_0^{\infty} \exp\left(-\frac{\sqrt{2t}y}{2}\right) \cos\left(\frac{\sqrt{2t}y}{2} - t \log x\right) dt \\ &= 2 \int_0^{\infty} \exp\left(-\frac{\sqrt{2}yu}{2}\right) \cos\left(\frac{\sqrt{2}yu}{2} - u^2 \log x\right) u du. \end{aligned}$$

The latter integral can be evaluated by using one of the formulae (see [160, p. 533, formulae 3.966])

$$\int_0^\infty ve^{-pv}\cos(2v^2 - pv)dv = \frac{p\sqrt{\pi}}{8}\exp\left(-\frac{p^2}{4}\right),$$

$$\int_0^\infty ve^{-pv}\cos(2v^2 + pv)dv = 0,$$

both holding for $p > 0$. Indeed, substituting $u = \sqrt{2/|\log x|}v$, we obtain with the help of the first formula that

$$I = \frac{\sqrt{\pi}y}{2(\log x)^{3/2}}\exp\left(-\frac{y^2}{4\log x}\right) \qquad \text{for } x > 1,$$

while the second formula yields $I = 0$ for $0 < x < 1$. Therefore,

$$G(x, y) = \begin{cases} \dfrac{1}{2\sqrt{\pi}}\dfrac{y}{(\log x)^{3/2}}\exp\left(-\dfrac{y^2}{4\log x}\right), & x > 1,\ y > 0, \\ 0, & 0 < x \leq 1,\ y > 0. \end{cases}$$

For $\alpha = 1/2$, Eq. (11.17) becomes

$$(D_{0+}^{1/2}w(\cdot, y))(x) = \frac{\partial^2}{\partial y^2}w(x, y) \qquad (x, y \in \mathbb{R}^+).$$

By (11.19), its associated function $G(x, y)$ can be represented as

$$G(x, y) = \frac{1}{\pi}\int_0^\infty \exp\left(-\sqrt[4]{t}ay\right)\cos\left(t\log x - \sqrt[4]{t}by\right)dt,$$

where $a = \cos(\pi/8) = \sqrt{2+\sqrt{2}}/2$ and $b = \sin(\pi/8) = \sqrt{2-\sqrt{2}}/2$.

For $\alpha = 4$, Eq. (11.17) can be written as

$$\sum_{k=0}^{4} S(4, k)x^k\left(\frac{\partial}{\partial x}\right)^{(k)} w(x, y) = \frac{\partial^2 w}{\partial y^2}(x, y) \qquad (x, y \in \mathbb{R}^+),$$

where $S(4, k)$ are Stirling numbers of the second kind (see Proposition 3.21). Their substitution yields

$$x^4\frac{\partial^4 w}{\partial x^4}(x, y) + 6x^3\frac{\partial^3 w}{\partial x^3}(x, y) + 7x^2\frac{\partial^2 w}{\partial x}(x, y) + x\frac{\partial w}{\partial x}(x, y)$$
$$= \frac{\partial^2 w}{\partial y^2}(x, y) \qquad (x, y \in \mathbb{R}^+).$$

The initial-value problem of this equation has a unique solution of the form

$$w(x, y) = \int_0^\infty f(u) G\left(\frac{x}{u}, y\right) \frac{du}{u} \qquad (x, y \in \mathbb{R}^+).$$

Since (11.19) applies with $a = 1$ and $b = 0$, we find by using a known formula for the integral (see, e.g., [160, p. 515, formula 3.896(4)]) that

$$G(x, y) = \frac{1}{\pi} \int_0^\infty e^{-t^2 y} \cos(t \log x) dt = \frac{1}{2\sqrt{\pi y}} \exp\left(\frac{-\log^2 x}{4y}\right).$$

The case $\alpha = 5/2$ is related to $\alpha = 1/2$ since $\cos(5\pi/8) = -\sin(\pi/8)$ and $\sin(5\pi/8) = \cos(\pi/8)$. Therefore (11.19) yields

$$G(x, y) = \frac{1}{\pi} \int_0^\infty \exp\left(-t^{5/4} a y\right) \cos\left(t \log x - t^{5/4} b y\right) dt,$$

where $a = \sqrt{2 - \sqrt{2}}/2$ and $b = \sqrt{2 + \sqrt{2}}/2$.

The approach based on $G(x, y)$ only fails when $\cos(\alpha\pi/4) = 0$ as it happens for $\alpha = 2$. Note that in the latter case the resulting equation (11.17), namely

$$x^2 \frac{\partial^2}{\partial x^2} w(x, y) + x \frac{\partial}{\partial x} w(x, y) = \frac{\partial^2}{\partial y^2} w(x, y) \qquad (x, y \in \mathbb{R}^+),$$

has been studied in Sect. 11.1 with different side conditions.

11.4 Notes

The material in Sects. 11.1 and 11.2 is borrowed from Butzer and Jansche [89, Sect. 9]. It can serve for illustrating what is called the *Mellin transform method* for solving certain types of differential equations. This approach has been described in various books; see, e.g., Sneddon [275] and [276, Chap. 4], Colombo [128, Chap. II], Colombo and Lavoine [129], Uflyand [294], and Zemanian [310, Chap. 4]. For an early research paper studying the boundary behavior of the solution of Dirichlet's problem on a wedge-shaped region by using the Mellin transform method, see Kolbe and Nessel [194].

A remarkable monograph about the Mellin transform method for ordinary and partial differential equations, published in 1992, is due to Szmydt and Ziemian [284]. In [284, Chap. III, Sect. 13] a detailed description of the application of the Mellin transform for solving differential equations is given in the general setting of Mellin distributions. For a polynomial P, the authors consider the operator

$P(x\frac{d}{dx}) = P(\Theta)$ and call it an *ordinary differential operator of Fuchsian type with constant coefficients*. Equations of the form

$$P(\Theta)u = f$$

are called *ordinary differential equations of Fuchsian type with constant coefficients*. Solutions of this kind of equations are obtained under suitable assumptions on the Mellin distributions by using the Mellin transform method. Also generalizations with the coefficients of P being replaced by smooth functions are considered.

For extensions to partial differential equations, the authors consider a polynomial R in n variables written as an n-tuple $\boldsymbol{x} = (x_1, \ldots, x_n)$ and the differential operator

$$\boldsymbol{\Theta}_{\boldsymbol{x}} := \left(x_1 \frac{\partial}{\partial x_1}, \ldots, x_n \frac{\partial}{\partial x_n} \right).$$

Then $R(\boldsymbol{\Theta}_{\boldsymbol{x}})$ is called a *partial differential operator of Fuchsian type with constant coefficients*. The corresponding equation

$$R(\boldsymbol{\Theta}_{\boldsymbol{x}})u = f$$

is called a *partial differential equation of Fuchsian type with constant coefficients*. Such equations are again treated by using the Mellin transform of Mellin distributions. As a further generalization, the authors consider the case where the coefficients of R are replaced by functions of $\boldsymbol{x}$ which are analytic in a neighborhood of the origin; see [284, Chap. III, Sects. 14–15].

Independently, Annaby and Butzer [16] investigated systems of first order ordinary differential equations of Fuchsian type being of the form

$$\Theta_c y_j = f_j(x, y_0, \cdots, y_n) \qquad (c \in \mathbb{R},\ j = 0, \cdots, n)$$

and gained results on existence and uniqueness of solutions but without using the Mellin transform method.

An abstract approach to partial differential equations with Mellin derivatives in case $c = 0$, is developed in Dománski and Langenbruch [140, 141], where the Mellin derivative $\Theta = \Theta_0$ is called *Euler derivative*. In [140] the authors describe all Euler partial differential operators with analytic coefficients acting on the space of real analytic functions and identify these operators among the Taylor multipliers on these spaces while in [141] they give criteria for global analytic solvability for some Euler type partial differential equations, employing a very general Mellin type theory for analytic functionals as distributions with compact support in place of the classical Mellin transform.

An interesting field of applications of the theory of Mellin transforms is financial mathematics. In Panini and Srivastav [235], Choi et al. [126], Frontczak [153], Guardasoni et al. [161], and Jeon and Kim [180], the authors developed mathematical models for the study of options pricing in finance science, mainly based on certain inhomogeneous final-value problems (initial-value problems with respect to $\tau := T - t$) of the form

$$\begin{cases} \mathscr{L} v(x,t) = f(x,t), & x \in \mathbb{R}^+,\ t \in [0,T[\\ v(x,T) = g(x), & x \in \mathbb{R}^+. \end{cases}$$

Here $\mathscr{L}$ is the so-called Black–Scholes second order linear differential operator, v is the pay-off function and f, g are suitable prescribed functions. The crux is that $\mathscr{L}$, when written in terms of the Mellin differential operator $\Theta = \Theta_0$, becomes a linear second order differential operator with *constant* coefficients. Therefore this kind of problems are natural candidates for the Mellin setting. Indeed, by employing the Mellin transform, the pay-off function v can be represented as a Mellin convolution integral; see, e.g., Guardasoni et al. [161] and the references therein.

Garcia [154] presents a methodology, named *group quantization formalism*, for constructing a financial theory based on an appropriate symmetry group. Group quantization generates representations of a financial theory in different functional spaces, allowing for alternative pricing frameworks such as the use of Laplace and Mellin transforms.

The results on the Mellin transform method for fractional equations presented in Sect. 11.3 are based on Bardaro et al. [53, Sect. 7]. For earlier contributions, mainly for fractional versions of diffusion equations, see, e.g., Wyss [304], Schneider and Wyss [274], Hilfer [177], and the books of Kilbas et al. [192] and Baleanu et al. [27]. However, these authors use the Mellin transform method for different fractional concepts such as those of Riemann–Liouville and Caputo. An analogue of Eq. (11.14) of Sect. 11.3.1 with a fractional Caputo derivative is discussed in [192, p. 355]. Equations similar to that of Sect. 11.3.1 but involving Riemann–Liouville or Caputo fractional derivatives and employing Laplace or Fourier transforms are studied in [192, Sects. 6.2–6.3].

Some contributions to fractional differential equations (ordinary and partial) dealing with Mellin–Hadamard derivatives, but without using Mellin transforms are given in Kilbas et al. [192, Sect. 3.6], Kou et al. [196], Qassim et al. [248], and Glushak and Manaenkova [158].

Boyadjiev and Luchko [75] use the Mellin transform for studying an analogue of the multidimensional wave and diffusion equations with a Caputo time-fractional derivative and a fractional Laplacian as spatial derivative. The solution is given in terms of Mellin–Barnes integrals. For the latter see the notes of Chap. 6.

The book of Baleanu et al. [27] presents numerical methods for an approximate solution of fractional differential equation, mainly in case of Caputo and Riemann–Liouville fractional derivatives.

Chapter 12
Exponential Sampling Theory

One of the basic results of signal and Fourier analysis is the Whittaker–Shannon–Kotel'nikov sampling theorem. It states that if f is (Fourier) bandlimited to an interval $[-\pi T, \pi T]$ for some $T > 0$, i.e., the (normalized) Fourier transform $\widehat{f}(v) := (1/\sqrt{2\pi}) \int_{\mathbb{R}} f(u)e^{-ivu}du$ vanishes outside $[-\pi T, \pi T]$, then f can be completely reconstructed for all $u \in \mathbb{R}$ from its sampled values $f(k/T)$ taken at the equally spaced nodes k/T with $k \in \mathbb{Z}$. The reconstruction formula reads as

$$f(u) = \sum_{k=-\infty}^{\infty} f\left(\frac{k}{T}\right) \text{sinc}(Tu - k) \qquad (u \in \mathbb{R}), \tag{12.1}$$

where $\text{sinc}(u) := \sin(\pi u)/\pi u$ for $u \neq 0$ and $\text{sinc}(0) = 1$.

In optical physics and engineering the so-called exponential sampling method turned out to be of great importance, especially in problems of particle sizing in quasi-elastic light scattering (photon correlation spectroscopy) and small angle diffraction scattering (Fraunhofer diffraction). Here the samples are not equally spaced as in the Whittaker–Shannon–Kotel'nikov case but exponentially spaced. Such spacing is useful in those applications where independent pieces of information accumulate near the time $x = 0$.

These kind of problems were solved in a satisfactory way by using the Mellin transform (in particular that on the space X_c^2) and the notion of Mellin bandlimited functions. If f is Mellin bandlimited to the interval $[-\pi T, \pi T]$ with $T > 0$, then the exponential sampling theorem states that

$$f(x) = \sum_{k=-\infty}^{\infty} f(e^{k/T}) \text{lin}_{c/T}(e^{-k}x^T) \qquad (x \in \mathbb{R}^+), \tag{12.2}$$

C. Bardaro et al., *Mellin Analysis, Transform Theory, and Applications*,
https://doi.org/10.1007/978-3-031-96672-9_12

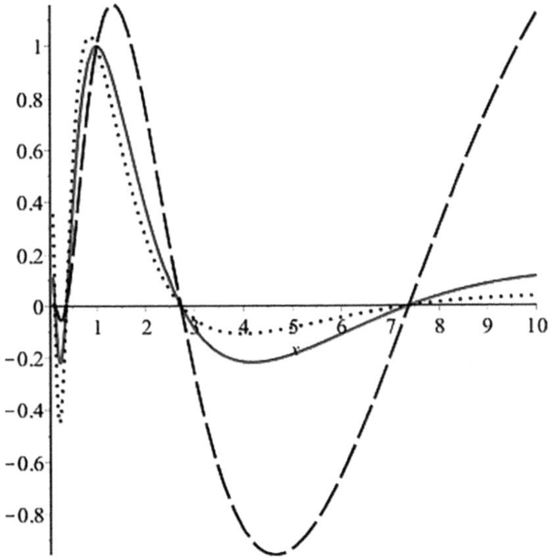

Fig. 12.1 The lin_c function for $c = 0$ (solid line), $c = -1$ (dashed line) and $c = 1/2$ (dotted line)

the role of the sinc function now being taken by the so-called lin_c function, given by

$$\mathrm{lin}_c(x) := x^{-c}\mathrm{sinc}(\log x) = \frac{x^{-c}}{2\pi i}\frac{x^{\pi i} - x^{-\pi i}}{\log x}$$

with $\mathrm{lin}_c 1 = 1$ (Fig. 12.1). Note that the sample points are now given by $1,\ e^{1/T},\ e^{2/T},\cdots$ lying to the right of 1, and $e^{-1/T},\ e^{-2/T},\ e^{-3/T},\ldots$ lying between 0 and 1 with 0 being an accumulation point.

The exponential sampling theorem can be formally deduced from that of Whittaker–Shannon–Kotel'nikov and vice versa. Indeed, writing (12.1) in terms of $F(u) := e^{cu} f(e^u)$, we obtain

$$F(u) = e^{cu} f(e^u) = \sum_{k=-\infty}^{\infty} e^{ck/T} f(e^{k/T})\,\mathrm{sinc}(Tu - k).$$

Since

$$\mathrm{sinc}(Tu - k) = e^{(Tu-k)c/T}\mathrm{lin}_{c/T}(e^{Tu-k}),$$

we find that

$$f(e^u) = \sum_{k=-\infty}^{\infty} f(e^{k/T})\mathrm{lin}_{c/T}(e^{Tu-k}).$$

Finally, substituting $x = e^u$, we arrive at (12.2). However, for a thorough study of the exponential sampling theorem and its surroundings, a quick approach by substitutions sometimes leads to cumbersome results that disguise the true nature of exponential sampling. We therefore prefer the setting of Mellin analysis.

Remark 12.1 In sampling theory, many mathematicians and engineers express the band-width of a bandlimited function as a multiple of π. The benefit is that in the corresponding reconstruction formula the sample points contain no π, as we have seen in the previous considerations. We follow this custom in this chapter and Chap. 13 consistently in order to simplify a comparison of the various types of formulae. For this reason, we shall have assumptions like $f \in B^2_{c,\pi T}$ instead of $f \in B^2_{c,T}$. This is just a matter of scaling and should not unsettle a reader who prefers a different convention.

12.1 The Exponential Sampling Formula

The aim of this section is to study the exponential sampling theorem for both, functions in X_c and in X_c^2 spaces.

First we consider again the lin_c function. We note that for every $c \in \mathbb{R}$ it is real-valued on $\mathbb{R}^+$. Furthermore, we present the following properties, which can be easily verified.

Lemma 12.2 *For $c \in \mathbb{R}$, the lin_c function has the following properties:*

(a) For any $\tilde{c} \in \mathbb{R}$, one has $\text{lin}_c \notin X_{\tilde{c}}$.
(b) There hold the representations

$$\text{lin}_c(x) = \frac{x^{-c}}{2\pi} \int_{-\pi}^{\pi} x^{-it} dt \qquad (x \in \mathbb{R}^+), \tag{12.3}$$

$$\text{lin}_c(x) = x^{-c} \text{sinc}\,(\log x) \qquad (x \in \mathbb{R}^+). \tag{12.4}$$

(c) For any $c \in \mathbb{R}$, one has $\text{lin}_c \in C(\mathbb{R}^+) \cap X_c^2$ and

$$[\text{lin}_c]^{\wedge}_{M_c^2}(c + it) = \chi_{[-\pi,\pi]}(t) \qquad (t \in \mathbb{R}). \tag{12.5}$$

(d) The estimate $|\text{lin}_c(x)| \le x^{-c}$ is valid for all $x \in \mathbb{R}^+$.
(e) There holds the orthogonality relation

$$\text{lin}_c(e^k) = \delta_{k,0} \qquad (k \in \mathbb{Z}), \tag{12.6}$$

where $\delta_{k,0}$ denotes Kronecker's delta, i.e., $\delta_{0,0} = 1$ and $\delta_{k,0} = 0$ for $k \in \mathbb{Z}\setminus\{0\}$.

Note that in view of the inversion theory of the Mellin transform in X_c^2, the representation (12.3) implies (12.5).

12.1.1 X_c-theory

As we have mentioned, the function lin_c does not belong to X_c. However, its product with a Mellin bandlimited function of X_c belongs to X_0. More precisely, the following statement holds.

Lemma 12.3 *Let $c \in \mathbb{R}$ and $f \in B^1_{c,\pi}$. Then, for every fixed $y \in \mathbb{R}^+$, the function $g_y(\cdot) := f(\cdot)\mathrm{lin}_c(y/\cdot)$ belongs to $B^1_{0,2\pi}$ and its Mellin transform can be represented as*

$$\left[f(\cdot)\mathrm{lin}_c\left(\frac{y}{\cdot}\right)\right]^{\wedge}_{M_0}(it) = \frac{1}{2\pi}\int_{-\pi}^{\pi}[f]^{\wedge}_{M_c}(c+i(t+v))y^{-c-iv}\,dv.$$

Proof Since by the hypotheses f is continuous, so is $f(\cdot)\mathrm{lin}_c(y/\cdot)$, and by Lemma 12.2 (d) we obtain

$$\left\|f(\cdot)\mathrm{lin}_c\left(\frac{y}{\cdot}\right)\right\|_{X_0} \le \frac{1}{y^c}\|f\|_{X_c}.$$

Hence $f(\cdot)\mathrm{lin}_c(y/\cdot) \in X_0 \cap C(\mathbb{R}^+)$. Since $f \in X_c$, we can apply Fubini's theorem to conclude that

$$\begin{aligned}\left[f(\cdot)\mathrm{lin}_c\left(\frac{y}{\cdot}\right)\right]^{\wedge}_{M_0}(it) &= \int_0^{\infty} f(x)\left\{\frac{1}{2\pi}\int_{-\pi}^{\pi}\left(\frac{y}{x}\right)^{-c-iv}dv\right\}x^{it-1}\,dx\\ &= \frac{1}{2\pi}\int_{-\pi}^{\pi} y^{-c-iv}\left\{\int_0^{\infty} f(x)x^{c+i(t+v)-1}\,dx\right\}dv\\ &= \frac{1}{2\pi}\int_{-\pi}^{\pi}[f]^{\wedge}_{M_c}(c+i(t+v))y^{-c-iv}\,dv.\end{aligned}$$

Now, for $|t| > 2\pi$ and $|v| \le \pi$, we have $|t+v| > \pi$. Therefore the hypotheses on f imply that $[f]^{\wedge}_{M_c}(c+i(t+v)) = 0$ for all $|t| > 2\pi$ and $|v| \le \pi$. Consequently, we obtain $[f(\cdot)\mathrm{lin}_c(y/\cdot)]^{\wedge}_{M_0}(it) = 0$ for all $|t| > 2\pi$, as remained to be shown. □

Now we are ready for a proof of the exponential sampling theorem in X_c based on Mellin transform methods

Theorem 12.4 (Exponential Sampling Formula) *Let $c \in \mathbb{R}$, $T \in \mathbb{R}^+$ and $f \in B^1_{c,\pi T}$. Then the series*

$$x^c \sum_{k=-\infty}^{\infty} f(e^{k/T})\mathrm{lin}_{c/T}(e^{-k}x^T) \tag{12.7}$$

converges absolutely and uniformly on $\mathbb{R}^+$ *and the representation*

$$f(x) = \sum_{k=-\infty}^{\infty} f(e^{k/T})\mathrm{lin}_{c/T}(e^{-k}x^T) \qquad (x \in \mathbb{R}^+) \tag{12.8}$$

holds.

Proof By Proposition 6.8 (d), the hypothesis $f \in B^1_{c,\pi T}$ implies that $f((\cdot)^{1/T}) \in B^1_{c/T,\pi}$. Now, according to Lemma 12.3, the function

$$g_y(\cdot) := f((\cdot)^{1/T})\mathrm{lin}_{c/T}\Big(\frac{y}{\cdot}\Big)$$

belongs to $B^1_{0,2\pi}$ for all $y \in \mathbb{R}^+$. In particular, $g_y \in X_0$ and, recalling that the Mellin transform of a function in X_0 is continuous, there holds $[g_y]^\wedge_{M_0}(2\pi ik) = 0$ for all $k \in \mathbb{Z} \setminus \{0\}$. Taking also Lemma 8.3 and Theorem 8.4 into account, we see that g_y satisfies the hypotheses of Corollary 6.100 for $c = 0$ and $T = 1$. Thus we obtain

$$\sum_{k=-\infty}^{\infty} g_y(e^k) = [g_y]^\wedge_{M_0}(0). \tag{12.9}$$

Calculating the Mellin transform of g_y and substituting $u = y/x$, we find that

$$\begin{aligned}
[g_y]^\wedge_{M_0}(it) &= \int_0^\infty g_y(x)x^{it}\,\frac{dx}{x} = \int_0^\infty f(x^{1/T})\mathrm{lin}_{c/T}\left(\frac{y}{x}\right)x^{it}\,\frac{dx}{x} \\
&= y^{it}\int_0^\infty f\left(\left(\frac{y}{u}\right)^{1/T}\right)\mathrm{lin}_{c/T}(u)u^{-it}\,\frac{du}{u} \\
&= y^{it}\int_0^\infty g_y\left(\frac{y}{u}\right)u^{-it}\,\frac{du}{u}.
\end{aligned}$$

Since along with g_y also $g_y(y/\cdot) \in X_0$, we deduce that

$$[g_y]^\wedge_{M_0}(it) = y^{it}[g_y(y/\cdot)]^\wedge_{M_0}(-it).$$

In particular, $g_y(y/\cdot)$ belongs to $B^1_{0,2\pi}$ and satisfies the hypotheses of Corollary 6.100 for $c = 0$ and $T = 1$. Therefore,

$$[g_y]^\wedge_{M_0}(0) = \left[g_y\Big(\frac{y}{\cdot}\Big)\right]^\wedge_{M_0}(0) = \sum_{k=-\infty}^{\infty} g_y(e^{-k}y).$$

Combining this equation with (12.9) and using Lemma 12.2 (e), we obtain

$$\sum_{k=-\infty}^{\infty} f(e^{k/T})\mathrm{lin}_{c/T}(e^{-k}y) = \sum_{k=-\infty}^{\infty} f(e^{-k/T}y^{1/T})\mathrm{lin}_{c/T}(e^k) = f(y^{1/T}).$$

Substituting $x = y^{1/T}$, we arrive at Eq. (12.8).

Finally, Lemma 12.2 (d) shows that

$$x^c \sum_{k=-\infty}^{\infty} \left| f(e^{k/T}) \mathrm{lin}_{c/T}(e^{-k}x^T) \right| \le x^c \sum_{k=-\infty}^{\infty} \left| f(e^{k/T}) \right| e^{kc/T} x^{-c}$$
$$= \sum_{k=-\infty}^{\infty} \left| f(e^{k/T}) \right| e^{kc/T}.$$

Since by Corollary 6.100 the latter series converges, we have proved that the series (12.7) converges absolutely and uniformly on $\mathbb{R}^+$. □

In practice, one cannot use formula (12.8) directly, but has to reduce the infinite series to a finite sum. Using the partial sum

$$(S_N f)(x) := \sum_{|k| \le N} f(e^{k/T}) \,\mathrm{lin}_{c/T}(e^{-k}x^T).$$

for some $N \in \mathbb{N}$, we produce a pointwise *truncation error*, defined by

$$(T_N f)(x) := f(x) - (S_N f)(x) = \sum_{|k| \ge N+1} f(e^{k/T}) \,\mathrm{lin}_{c/T}(e^{-k}x^T) \qquad (x \in \mathbb{R}^+).$$

As an example, we estimate this truncation error for a particular Mellin bandlimited function, namely the Mellin–Fejér kernel

$$F_\rho^c(x) := \frac{x^{-c}}{2\pi\rho}\left(\frac{x^{\rho i/2} - x^{-\rho i/2}}{\log x}\right)^2 \qquad (x \in \mathbb{R}^+ \setminus \{1\}), \tag{12.10}$$

with the continuous extension $F_\rho^c(1) := \rho/(2\pi)$ (see Definition 6.83). By Lemma 6.84, it belongs to $B^1_{c,\rho}$ for $c \in \mathbb{R}$ and $\rho \in \mathbb{R}^+$. Therefore the representation (12.8) holds with $T = \rho/\pi$ and takes the form

$$F_\rho^c(x) = \sum_{k=-\infty}^{\infty} F_\rho^c(e^{k\pi/\rho}) \,\mathrm{lin}_{c\pi/\rho}(e^{-k}x^{\rho/\pi})$$
$$= \frac{2\rho}{\pi^3 x^c} \sum_{k=-\infty}^{\infty} \frac{\sin^2(k\pi/2)}{k^2} \,\mathrm{sinc}\left(\frac{\rho}{\pi}\log x - k\right). \tag{12.11}$$

The truncation error is given by

$$(T_N F_\rho^c)(x) = \frac{2\rho}{\pi^3 x^c} \sum_{|k| \ge N+1} \frac{\sin^2(k\pi/2)}{k^2} \,\mathrm{sinc}\left(\frac{\rho}{\pi}\log x - k\right).$$

Choosing N such that

$$N \geq 2\max\left\{x^{-c}, \frac{\rho}{\pi}|\log x|\right\},$$

we have

$$\left|\operatorname{sinc}\left(\frac{\rho}{\pi}\log x - k\right)\right| \leq \frac{1}{\pi(|k| - |(\rho/\pi)\log x|)} \leq \frac{2}{\pi|k|} \qquad (|k| \geq N+1).$$

This implies

$$\left|(T_N F_\rho^c)(x)\right| \leq \frac{4\rho}{\pi^4 x^c} \sum_{|k|\geq N+1} \frac{1}{|k|^3} \leq \frac{8\rho}{\pi^4 x^c}\int_N^\infty \frac{du}{u^3} \leq \frac{4\rho}{\pi^4 x^c N^2} \leq \frac{2\rho}{\pi^4 N},$$

or equivalently,

$$\left|F_\rho^c(x) - (S_N F_\rho^c)(x)\right| \leq \frac{2\rho}{\pi^4 N} \qquad \left(N \geq 2\max\left\{x^{-c}, \frac{\rho}{\pi}|\log x|\right\}\right).$$

12.1.2 X_c^2-theory

In order to study the exponential sampling theorem for functions in X_c^2, we follow a different approach, which gives an interesting extension of the classical sampling theorem in the Kramer form (see Sect. 12.9 Notes).

For $c \in \mathbb{R}$, we introduce the function

$$K_c(x,u) := x^{-c}u^{-c-i\log x} \qquad (x \in \mathbb{R}^+,\ u \in [e^{-\pi}, e^{\pi}]). \tag{12.12}$$

Substituting $x_k := e^k$ with $k \in \mathbb{Z}$, we obtain

$$K_c(x_k,u) = e^{-ck}\varphi_{c,k}(u) \qquad (u \in [e^{-\pi}, e^{\pi}]), \tag{12.13}$$

where $\varphi_{c,k}(u) = u^{-c-ik}$ (see also Proposition 5.5). The functions $K_c(x_k,\cdot)$, defined on $[e^{-\pi}, e^{\pi}]$, have a c-recurrent continuation on $\mathbb{R}^+$. Moreover, the orthogonality of the system $\{\varphi_{c,k}\}$ entails that $\{K_c(x_k,\cdot)\}_{k\in\mathbb{Z}}$ also forms an orthogonal system in Y_c^2 with respect to the scalar product $\langle f, g\rangle_{Y_c^2}$ as defined in (3.6). The Mellin–Fourier coefficients of the function $K_c(x,\cdot)$ are given by

$$\left[K_c(x,\cdot)\right]^\wedge_{\mathscr{M}^c}(k) = \langle K_c(x,\cdot), \varphi_{c,k}\rangle_{Y_c^2} = e^{ck}\langle K_c(x,\cdot), K_c(x_k,\cdot)\rangle_{Y_c^2}. \tag{12.14}$$

Recalling Theorem 5.23, we may use (12.14) in order to express the kernel K_c by its Mellin–Fourier series, obtaining

$$K_c(x, u) = \frac{1}{2\pi} \sum_{k \in \mathbb{Z}} \left[K_c(x, \cdot)\right]^{\wedge}_{\mathscr{M}^c}(k) u^{-c-ik} = \sum_{k \in \mathbb{Z}} e^{-ck} S_{c,k}(x) \varphi_{c,k}(u), \quad (12.15)$$

where

$$S_{c,k}(x) := \frac{e^{2ck}}{2\pi} \langle K_c(x, \cdot), K_c(x_k, \cdot)\rangle_{Y_c^2} \qquad (k \in \mathbb{Z},\ x \in \mathbb{R}^+). \quad (12.16)$$

In (12.15) convergence is understood in the sense of the Y_c^2 norm.

Now we are ready for a Kramer type sampling theorem in the Mellin setting. We begin with the following preliminary version stated as a proposition.

Proposition 12.5 *Let $f : \mathbb{R}^+ \to \mathbb{C}$ be representable as*

$$f(x) = \int_{e^{-\pi}}^{e^{\pi}} g(u) \overline{K_c(x, u)}\, u^{2c} \frac{du}{u} \qquad (x \in \mathbb{R}^+) \quad (12.17)$$

for some $c \in \mathbb{R}$ and $g \in Y_c^2$. Then f can be reconstructed from its sampled values $f(x_k) = f(e^k)$ by the formula

$$f(x) = \sum_{k \in \mathbb{Z}} f(e^k) S_{c,k}(x) \qquad (x \in \mathbb{R}^+), \quad (12.18)$$

where the functions $S_{c,k}$ are defined in (12.16).

Proof Employing (12.17), (12.13), (12.15) and Hölder's inequality, we conclude that for any $x \in \mathbb{R}^+$ and $N \in \mathbb{N}$,

$$\begin{aligned} &\left| f(x) - \sum_{|k| \le N} f(e^k) S_{c,k}(x) \right| \\ &= \left| \langle g, K_c(x, \cdot)\rangle_{Y_c^2} - \sum_{|k| \le N} \langle g, K_c(e^k, \cdot)\rangle_{Y_c^2} S_{c,k}(x) \right| \\ &= \left| \left\langle g, K_c(x, \cdot) - \sum_{|k| \le N} S_{c,k}(x) e^{-ck} \varphi_{c,k} \right\rangle_{Y_c^2} \right| \\ &\le \|g\|_{Y_c^2} \left\| K_c(x, \cdot) - \frac{1}{2\pi} \sum_{|k| \le N} \left[K_c(x, \cdot)\right]^{\wedge}_{\mathscr{M}^c}(k) \varphi_{c,k} \right\|_{Y_c^2}. \end{aligned} \quad (12.19)$$

Since the second norm approaches zero as $N \to \infty$, we obtain the desired result. □

Remark 12.6 The functions $S_{c,k}$ have a simple representation in terms of the lin_c function, namely

$$S_{c,k}(x) = \mathrm{lin}_c(e^{-k}x) \qquad (x \in \mathbb{R}^+). \tag{12.20}$$

Indeed, using the substitution $u = e^t$ and recalling (12.3), we find that

$$\begin{aligned} S_{c,k}(x) &= \frac{e^{2ck}}{2\pi}\int_{e^{-\pi}}^{e^{\pi}} K_c(x,u)\overline{K_c(e^k,u)}\,u^{2c}\,\frac{du}{u} \\ &= \frac{e^{ck}}{2\pi}x^{-c}\int_{e^{-\pi}}^{e^{\pi}} u^{-i\log x+ik}\,\frac{du}{u} = \frac{1}{2\pi}\int_{-\pi}^{\pi}\left(e^{-k}x\right)^{-c-it}dt = \mathrm{lin}_c\left(e^{-k}x\right). \end{aligned}$$

Thus the expansion (12.18) can be rewritten as

$$f(x) = \sum_{k\in\mathbb{Z}} f\left(e^k\right)\mathrm{lin}_c\left(e^{-k}x\right) \qquad (x \in \mathbb{R}^+). \tag{12.21}$$

By appropriate scalings of the variables u and x, we deduce from Proposition 12.5 the following more general version of the Mellin–Kramer sampling theorem, in which $\widetilde{K}_c(x,\cdot)$ denotes the c-recurrent continuation of $K_c(x,\cdot)$.

Theorem 12.7 *Let $c \in \mathbb{R}$, $T \in \mathbb{R}^+$ and let f be representable as*

$$f(x) = \int_{e^{-\pi T}}^{e^{\pi T}} g(u)\,\overline{\widetilde{K}_c(x,u)}\,u^{2c}\,\frac{du}{u} \qquad (x \in \mathbb{R}^+) \tag{12.22}$$

for some function g such that $g\left((\cdot)^T\right) \in Y^2_{cT}$. Then

$$f(x) = \sum_{k\in\mathbb{Z}} f(e^{k/T})\mathrm{lin}_{c/T}\left(e^{-k}x^T\right) \qquad (x \in \mathbb{R}^+). \tag{12.23}$$

Proof Let us define the functions

$$F(x) := f\left(x^{1/T}\right) \quad \text{and} \quad G(u) := Tg\left(u^T\right)u^{cT-c/T}.$$

As is easily verified, the hypotheses on g imply that $G \in Y^2_{c/T}$. Next, using the identity

$$K_c\left(x^{1/T},u^T\right)u^{2cT} = u^{cT-c/T}K_{c/T}(x,u)u^{2c/T},$$

we deduce from (12.22) that

$$F(x) = \int_{e^{-\pi}}^{e^{\pi}} G(u)\overline{K_{c/T}(x,u)}u^{2c/T}\,\frac{du}{u}. \tag{12.24}$$

Hence the function F satisfies the hypotheses of Proposition 12.5 with the role of c being taken by c/T. Now the representation (12.23) follows from (12.18) and (12.20). □

Remark 12.8 Note that in Theorem 12.7 the hypotheses on g are equivalent to

$$g(u) = e^{2\pi cT} g(e^{2\pi T} u) \qquad (u \in \mathbb{R}^+) \tag{12.25}$$

and

$$\int_{e^{-\pi T}}^{e^{\pi T}} |g(u)|^2 u^{2c} \frac{du}{u} < \infty. \tag{12.26}$$

This means that g belongs to a generalization of the space Y_c^2 in which the role of the interval $[e^{-\pi}, e^{\pi}]$ is taken by $[e^{-\pi T}, e^{\pi T}]$.

Now, using Theorem 12.7, we establish the exponential sampling formula for Mellin bandlimited functions in X_c^2.

Theorem 12.9 *Let $f \in B^2_{c,\pi T}$, where $c \in \mathbb{R}$ and $T \in \mathbb{R}^+$. Then*

$$f(x) = \sum_{k\in\mathbb{Z}} f(e^{k/T}) \mathrm{lin}_{c/T}\left(e^{-k} x^T\right) \qquad (x \in \mathbb{R}^+). \tag{12.27}$$

Proof Since f belongs to $B^2_{c,\pi T}$, the Mellin transform of f is integrable and so the inversion theorem yields

$$f(x) = \frac{1}{2\pi} \int_{-\pi T}^{\pi T} [f]^{\wedge}_{M_c^2}(c+it) x^{-c-it} dt. \tag{12.28}$$

Now our aim is to rewrite (12.28) in the form (12.22). Defining

$$g(u) := \frac{1}{2\pi} u^{-c} [f]^{\wedge}_{M_c^2}(c - i \log u) \qquad (u \in [e^{-\pi T}, e^{\pi T}]) \tag{12.29}$$

and substituting $t = -\log u$ on the right-hand side of (12.28), we obtain

$$f(x) = \frac{x^{-c}}{2\pi} \int_{e^{-\pi T}}^{e^{\pi T}} [f]^{\wedge}_{M_c^2}(c - i \log u)\, u^{i \log x} \frac{du}{u} = \int_{e^{-\pi T}}^{e^{\pi T}} g(u) \overline{\widetilde{K}_c(x,u)}\, u^{2c} \frac{du}{u}.$$

Clearly g, defined in (12.29), can be continued from the interval $[e^{-\pi T}, e^{\pi T}]$ to the whole of $\mathbb{R}^+$ such that (12.25) is satisfied. Furthermore, by the change of variable $\log u = -y$, we obtain

$$\int_{e^{-\pi T}}^{e^{\pi T}} |g(u)|^2 u^{2c} \frac{du}{u} = \frac{1}{2\pi} \int_{-\pi T}^{\pi T} |[f]^{\wedge}_{M_c^2}(c+iy)|^2 dy < \infty,$$

which shows that (12.26) holds. Hence, by Remark 12.8, Theorem 12.7 is applicable and yields the assertion. □

Remark 12.10 Later, as a consequence of the Mellin–Parseval formula, we shall obtain that the series in (12.27) converges absolutely and uniformly on compact subsets of $\mathbb{R}^+$. Moreover, when multiplied by x^c, it converges uniformly on the whole $\mathbb{R}^+$ (see Corollary 12.54).

12.2 A Mellin Analogue of Valiron's Sampling Formula

In Fourier analysis, there exists a sampling formula of Valiron—sometimes also attributed to Tschakaloff [289] (see, e.g., Higgins [172], [173, p. 60], Butzer et al. [115])—that improves upon the classical sampling formula. It applies to a wider class of functions and its series converges faster. The only price one has to pay for this improvement is that, apart from the samples of the classical sampling formula, also the value of the derivative at zero is needed.

Here we want to establish an analogue of Valiron's formula in the Mellin setting. Our approach is very similar to the derivation of an analogue of Boas' differentiation formula (see Theorem 3.46).

Theorem 12.11 *Let $f \in \mathscr{B}^{\infty}_{c,\pi T}$, where $c \in \mathbb{R}$ and $T \in \mathbb{R}^+$. Then*

$$r^c f(r,0) = \sin(\pi T \log r)\Bigg[\frac{(\widetilde{\Theta}_c f)(1,0)}{\pi T} + \frac{f(1,0)}{\pi T \log r} + T \log r \sum_{k\in\mathbb{Z}\setminus\{0\}} \frac{(-1)^{k+1} e^{kc/T} f(e^{k/T},0)}{k\pi(k - T\log r)}\Bigg].$$

Under continuous continuation, this formula holds for all $r \in \mathbb{R}^+$ and the convergence is absolute and uniform on compact subsets of $\mathbb{R}^+$.

Proof First we suppose that $f \in \mathscr{B}^{\infty}_{c,1}$, which is the case of $T = 1/\pi$. Let $t \in \mathbb{R}^+ \setminus \{e^{k\pi} : k \in \mathbb{Z}\}$ be arbitrary but fixed and consider the function

$$F : (r,\theta) \longmapsto \frac{f(r,\theta)}{(\log(r/t) + i\theta)(\log r + i\theta)\sin(\log r + i\theta)}.$$

It is polar-analytic on $\mathbb{H}$ except for isolated singularities at the points where the denominator vanishes. These points are $(t, 0)$ and $(e^{k\pi}, 0)$ for all $k \in \mathbb{Z}$. The point $(1, 0)$, obtained for $k = 0$, is a pole of order at most two while all other isolated singularities are poles of order at most one.

Let $n \in \mathbb{N}$ be such that $e^{-n\pi} < t < e^{n\pi}$, and denote by $\partial\mathcal{R}_n$ the positively oriented rectangular curve with vertices at the points $(e^{\pm(n+1/2)\pi}, \pm(n + 1/2)\pi)$. By Theorem 3.39, we have

$$\frac{1}{2\pi i}\int_{\partial\mathcal{R}_n}\left(re^{i\theta}\right)^{c-1}F(r,\theta)e^{i\theta}(dr+ird\theta)=(\mathrm{res}_cF)(t,0)+\sum_{k=-n}^{n}(\mathrm{res}_cF)(e^{k\pi},0).$$

First we want to show that the integral on the left-hand side approaches zero as $n\to\infty$. Since f was assumed to belong to $\mathscr{B}^{\infty}_{c,1}$, we have

$$|f(r,\theta)|\ \le\ C_f r^{-c}e^{|\theta|},$$

for a suitable constant $C_f>0$. Furthermore it can be verified that on $\partial\mathcal{R}_n$

$$|\sin(\log r+i\theta)|\ \ge\ \frac{e^{|\theta|}}{3}. \tag{12.30}$$

Thus

$$\left|r^{c-1}F(r,\theta)\right|\ \le\ \frac{3C_f}{r\left(\log^2(r/t)+\theta^2\right)^{1/2}\left(\log^2 r+\theta^2\right)^{1/2}}. \tag{12.31}$$

For an estimate of the contributions coming from the horizontal parts of $\partial\mathcal{R}_n$, we have to substitute $\theta=\pm(n+1/2)\pi$ in (12.31). Replacing in addition the squared logarithms by zero, we obtain for the integrals under consideration the upper bound

$$\frac{3}{\pi}C_f\int_{e^{-(n+1/2)\pi}}^{e^{(n+1/2)\pi}}\frac{1}{(n+1/2)^2\pi^2}\frac{dr}{r},$$

which approaches zero as $n\to\infty$. For the contributions coming from the vertical parts we have to substitute $r=e^{\pm(n+1/2)\pi}$ in (12.31). Replacing in addition θ^2 by zero, we obtain an upper bound for the corresponding integrals which again approaches zero as $n\to\infty$. Hence the above residue formula yields

$$(\mathrm{res}_cF)(t,0)\ =\ -\sum_{k\in\mathbb{Z}}(\mathrm{res}_cF)(e^{k\pi},0). \tag{12.32}$$

Next we calculate the residues (see Definition 3.38) . It is easily seen that

$$(\mathrm{res}_cF)(t,0)\ =\ \frac{t^c f(t,0)}{\log t\,\sin(\log t)}.$$

For the residues on the right-hand side of (12.32), we employ the sinc function for factoring the sine as

$$\sin z\ =\ (-1)^k(z-k\pi)\,\mathrm{sinc}\left(\frac{z}{\pi}-k\right)\qquad (z\in\mathbb{C},\ k\in\mathbb{Z}). \tag{12.33}$$

Then

$$(\mathrm{res}_c F)(e^{k\pi}, 0) = (-1)^k \frac{e^{ck\pi} f(e^{k\pi}, 0)}{k\pi(k\pi - \log t)} \qquad (k \neq 0).$$

Noting that in a neighborhood of $(1, 0)$ we may write

$$F(r, \theta) = \frac{1}{(\log r + i\theta)^2} \frac{f(r, \theta)}{(\log(r/t) + i\theta)\, \mathrm{sinc}(\log r^{1/\pi} + i\theta/\pi)} =: \frac{g(r, \theta)}{(\log r + i\theta)^2},$$

we obtain

$$(\mathrm{res}_c F)(1, 0) = (\widetilde{\Theta}_c g)(1, 0) = (D_{\mathrm{pol}} g)(1, 0) + c g(1, 0).$$

Since

$$g(r, \theta) = \frac{f(r, \theta)}{\big(\log(re^{i\theta}) - \log t\big)\mathrm{sinc}\big(\log(re^{i\theta})/\pi\big)},$$

we find by polar differentiation that

$$(\mathrm{res}_c F)(1, 0) = -\left[\frac{(\widetilde{\Theta}_c f)(1, 0)}{\log t} + \frac{f(1, 0)}{\log^2 t}\right].$$

With these values for the residues, formula (12.32) may be rewritten as

$$t^c f(t, 0) = \sin(\log t) \times \left[(\widetilde{\Theta}_c f)(1, 0) + \frac{f(1, 0)}{\log t} + \log t \sum_{k \in \mathbb{Z}\setminus\{0\}} \frac{(-1)^{k+1} e^{ck\pi} f(e^{k\pi}, 0)}{k\pi(k\pi - \log t)}\right].$$

It holds for $f \in \mathscr{B}^{\infty}_{c,1}$ and all $t \in \mathbb{R}^+$, even at the points $t = e^{k\pi}$, since by (12.33)

$$\frac{\sin(\log t)}{k\pi - \log t} = (-1)^{k+1} \mathrm{sinc}\left(\frac{\log t}{\pi} - k\right),$$

which shows that there exists a continuous continuation to the whole of $\mathbb{R}^+$. Thus we have proved the desired formula for $T = 1/\pi$.

When $f \in \mathscr{B}^{\infty}_{c,\pi T}$ with an arbitrary $T \in \mathbb{R}^+$, then, by Proposition 3.44, the function

$$h : (r, \theta) \longmapsto f\big(r^{1/(\pi T)}, \theta/(\pi T)\big)$$

belongs to $\mathscr{B}^{\infty}_{c/(\pi T),1}$. Therefore we can apply the achieved formula to h with c replaced by $c/(\pi T)$. Then, expressing h by f and substituting $t = r^{\pi T}$, we obtain the formula stated in Theorem 12.11.

Finally, noting that $e^{kc/T} \left| f(e^{k/T}, 0) \right| \le C_f$ for all $k \in \mathbb{Z}$, we easily verify the assertion on the convergence. □

Remark 12.12 We can write the formula of Theorem 12.11 in terms of the $\mathrm{lin}_{c/T}$ function (see Sect. 12.1) as

$$f(r,0) = \mathrm{lin}_{c/T}(r^T)\left[\log r \left(\widetilde{\Theta}_c f\right)(1,0) + f(1,0)\right]$$
$$+ \log(r^T) \sum_{k\in\mathbb{Z}\setminus\{0\}} \frac{f(e^{k/T},0)}{k} \mathrm{lin}_{c/T}(e^{-k}r^T).$$

This formula easily compares with the exponential sampling formula (Theorem 12.4), and its right-hand side is defined for all $r \in \mathbb{R}^+$. On the other hand, it is very inconvenient when we want to estimate f using property (iii) of the Mellin–Bernstein space since all factors depending on c are hidden in the $\mathrm{lin}_{c/T}$ function. An alternative way is to use lin_0 only. It gives

$$r^c f(r,0) = \mathrm{lin}_0(r^T)\left[\log r \left(\widetilde{\Theta}_c f\right)(1,0) + f(1,0)\right]$$
$$+ \log(r^T) \sum_{k\in\mathbb{Z}\setminus\{0\}} \frac{e^{kc/T} f(e^{k/T},0)}{k} \mathrm{lin}_0(e^{-k}r^T).$$

Now the factors depending on c can go with f as needed in (iii).

Valiron's formula and its Mellin analogue will serve primarily for the reconstruction of a function f from samples. Among the latter, there is also a sample of the derivative of f. One may therefore think of interpreting this formula as a representation of the derivative of f by samples of f. In the following corollary, we establish such a differentiation formula by skillful use of Theorem 12.11. It has some advantages over the corresponding analogue of Boas' differentiation formula. It seems that even a Fourier version of this result has not yet been mentioned in books.

Corollary 12.13 *Let $f \in \mathscr{B}^{\infty}_{c,\pi T}$, where $c \in \mathbb{R}$ and $T \in \mathbb{R}^+$. Then*

$$(\widetilde{\Theta}_c f)(r,\theta) = \frac{\pi T}{2}\left[e^{c/(2T)} f(re^{1/(2T)},\theta) - e^{-c/(2T)} f(re^{-1/(2T)},\theta)\right]$$
$$+ T \sum_{k\in\mathbb{Z}\setminus\{0\}} \frac{(-1)^k e^{kc/T} f(re^{k/T},\theta)}{k(4k^2-1)}$$

for $(r, \theta) \in \mathbb{H}$*. Multiplied by* r^c*, the series converges absolutely and uniformly on strips of bounded width parallel to the* r*-axis of* $\mathbb{H}$*.*

Proof Since by Proposition 3.44 the Mellin–Bernstein space is invariant under Mellin translations, the formula of Theorem 12.11 applies to

$$g : (r, \theta) \longmapsto t^c f(tr, \theta) \qquad (t > 0).$$

Rewritten in terms of f, it takes the form

$$r^c f(tr, 0) = \sin(\pi T \log r)\left[\frac{(\widetilde{\Theta}_c f)(t, 0)}{\pi T} + \frac{f(t, 0)}{\pi T \log r} + T \log r \sum_{k \in \mathbb{Z} \setminus \{0\}} \frac{(-1)^{k+1} e^{kc/T} f(te^{k/T}, 0)}{k\pi(k - T \log r)}\right].$$

Now we substitute $r = e^{\pm 1/(2T)}$ and obtain the two equations:

$$e^{c/(2T)} f(te^{1/(2T)}, 0) = \frac{(\widetilde{\Theta}_c f)(t, 0)}{\pi T} + \frac{2}{\pi} f(t, 0) + \frac{1}{2\pi} \sum_{k \in \mathbb{Z} \setminus \{0\}} \frac{(-1)^{k+1} e^{kc/T} f(te^{k/T}, 0)}{k(k - 1/2)}$$

and

$$e^{-c/(2T)} f(te^{-1/(2T)}, 0) = -\frac{(\widetilde{\Theta}_c f)(t, 0)}{\pi T} + \frac{2}{\pi} f(t, 0) + \frac{1}{2\pi} \sum_{k \in \mathbb{Z} \setminus \{0\}} \frac{(-1)^{k+1} e^{kc/T} f(te^{k/T}, 0)}{k(k + 1/2)}.$$

Next, by subtraction of these equations, we find that

$$(\widetilde{\Theta}_c f)(t, 0) = \frac{\pi T}{2}\left[e^{c/(2T)} f(te^{1/(2T)}, 0) - e^{-c/(2T)} f(te^{-1/(2T)}, 0)\right] + T \sum_{k \in \mathbb{Z} \setminus \{0\}} \frac{(-1)^k e^{kc/T} f(te^{k/T}, 0)}{k(4k^2 - 1)}.$$

Finally, employing Proposition 3.45, we can transform the argument of the polar Mellin derivative from $(t, 0)$ to an arbitrary point $(t, \alpha) \in \mathbb{H}$. This leads us to

$$(\widetilde{\Theta}_c f)(t,\alpha) = \frac{\pi T}{2}\Big[e^{c/(2T)} f(te^{1/(2T)},\alpha) - e^{-c/(2T)} f(te^{-1/(2T)},\alpha)\Big]$$

$$+ T \sum_{k\in\mathbb{Z}\setminus\{0\}} \frac{(-1)^k e^{kc/T} f(te^{k/T},\alpha)}{k(4k^2-1)},$$

which is the formula of the corollary with (r,θ) replaced by (t,α). Recalling property (iii) in the definition of $\mathscr{B}^{\infty}_{c,\pi T}$, we easily verify the statement on convergence. □

Remark 12.14 In order to write the differentiation formula of Corollary 12.13 in a simpler and more suggestive way, we introduce the *central Mellin difference* $\delta_{c,h}$ *with increment* $h > 0$ by

$$(\delta_{c,h} f)(r,\theta) := h^c f(hr,\theta) - h^{-c} f(h^{-1}r,\theta).$$

Then, setting $h := e^{1/T}$, we may write the formula of Corollary 12.13 as

$$(\widetilde{\Theta}_c f)(r,\theta) = \pi T\left[\frac{1}{2}\big(\delta_{c,h^{1/2}} f\big)(r,\theta) + \frac{1}{\pi}\sum_{k=1}^{\infty} \frac{(-1)^k}{k(4k^2-1)}\big(\delta_{c,h^k} f\big)(r,\theta)\right]. \tag{12.34}$$

Likewise the Mellin analogue of Boas' differentiation formula may be expressed as

$$(\widetilde{\Theta}_c f)(r,\theta) = \frac{4T}{\pi}\sum_{k=0}^{\infty} \frac{(-1)^k}{(2k+1)^2}\big(\delta_{c,h^{k+1/2}} f\big)(r,\theta), \tag{12.35}$$

where again $f \in \mathscr{B}^{\infty}_{c,\pi T}$ and $h = e^{1/T}$. We observe that the summation in (12.34) has coefficients decaying like $\mathcal{O}(k^{-3})$ as $k \to \infty$ while those of (12.35) decay like $\mathcal{O}(k^{-2})$ only.

Remark 12.15 For sake of completeness, let us have a quick look at the classical Fourier case. For f belonging to the (Fourier) Bernstein space $B^{\infty}_{\pi w}$, Valiron's sampling formula states that

$$f(t) = \sin(\pi wt)\left[\frac{f'(0)}{\pi w} + \frac{f(0)}{\pi wt} + wt \sum_{k\in\mathbb{Z}\setminus\{0\}} \frac{(-1)^k f(k/w)}{k\pi(wt-k)}\right]$$

for $t \in \mathbb{R}$ under continuous continuation of the right-hand side. Now we may imitate the proof of Corollary 12.13 in a Fourier version, that is, we apply this formula to $f(\cdot + x)$, substitute in the resulting formula $t = \pm 1/(2w)$ and subtract the two resulting equations. Thus we arrive at the (new) differentiation formula

$$f'(x) = \frac{\pi w}{2}\left[f(x + 1/(2w)) - f(x - 1/(2w))\right] + w \sum_{k\in\mathbb{Z}\setminus\{0\}} \frac{(-1)^k f(x + k/w)}{k(4k^2 - 1)}$$

$$= \frac{\pi w}{2}\left[f(x + 1/(2w)) - f(x - 1/(2w))\right]$$

$$+ w \sum_{k=1}^{\infty} \frac{(-1)^k}{k(4k^2 - 1)}\left[f(x + k/w) - f(x - k/w)\right].$$

Here the derivative is represented by an alternating series of central differences with coefficients decaying like $\mathcal{O}(k^{-3})$ as $k \to \infty$. Of course, in this formula we can replace the real variable x with the complex variable z.

We conclude the section with an interesting consequence of Theorem 12.11, which states an identity theorem in Mellin–Bernstein spaces.

Corollary 12.16 *Let $f \in \mathscr{B}^{\infty}_{c,\pi T}$, where $c \in \mathbb{R}$ and $T \in \mathbb{R}^+$. If f vanishes on the set $\{(e^{k/T}, 0) : k \in \mathbb{Z}\}$, then*

$$f(r, \theta) \equiv a\left(re^{i\theta}\right)^{-c} \sin\left(\pi T(\log r + i\theta)\right)$$

for some $a \in \mathbb{C}$.

Proof It follows from the Mellin analogue of Valiron's sampling formula (Theorem 12.11) that

$$f(r, 0) = ar^{-c} \sin(\pi T \log r)$$

for all $r > 0$, where $a = (\widetilde{\Theta}_c f)(1, 0)/(\pi T)$.

Now consider the function $g : \mathbb{H} \to \mathbb{C}$ defined by

$$g(r, \theta) = a\left(re^{i\theta}\right)^{-c} \sin\left(\pi T(\log r + i\theta)\right).$$

Obviously it belongs to $\mathscr{B}^{\infty}_{c,\pi T}$ and $g(r, 0) = f(r, 0)$ for all $r > 0$. Thus $\varphi := f - g$ also belongs to $\mathscr{B}^{\infty}_{c,\pi T}$ and $\varphi(r, 0) = 0$ for all $r > 0$. By Theorem 3.33, it follows that $\varphi(r, \theta) \equiv 0$ and so $f(r, \theta) \equiv g(r, \theta)$ on $\mathbb{H}$. □

As a consequence of Corollary 12.16, we obtain the following result.

Corollary 12.17 *Let $f \in \mathscr{B}^{p}_{c,\pi T}$, where $c \in \mathbb{R}$, $T \in \mathbb{R}^+$ and $p \in [1, \infty[$. If f vanishes on the set $\{(e^{k/T}, 0) : k \in \mathbb{Z}\}$, then f is identically zero.*

Proof Since $\mathscr{B}^{p}_{c,\pi T} \subset \mathscr{B}^{\infty}_{c,\pi T}$, Corollary 12.16 applies to f and yields that

$$f(r, \theta) \equiv a\left(re^{i\theta}\right)^{-c} \sin\left(\pi T(\log r + i\theta)\right)$$

for some $a \in \mathbb{C}$. The membership of f in $\mathscr{B}^p_{c,\pi T}$ requires that $f(\cdot, 0) \in X^p_c$. This is possible if and only if $a = 0$. □

12.3 The Derivative Sampling Formula

As a further application of the residue theorem (Theorem 3.39) for polar-analytic functions, we establish an exponential sampling formula for Mellin bandlimited functions that involves samples of Mellin derivatives up to order m.

Theorem 12.18 *Let $f \in B^2_{c,\pi T}$ with $c \in \mathbb{R}$ and $T \in \mathbb{R}^+$. Then, for $m \in \mathbb{N}_0$ and $t \in \mathbb{R}^+$, there holds the representation*

$$f(t) = t^{-c} \sum_{k\in\mathbb{Z}} e^{c(m+1)k/T}$$
$$\times \sin^{m+1}(\log t^{\pi T/(m+1)}) \frac{(\Theta^m_{c(m+1)/(\pi T)} L_{k,t})(e^{k\pi})}{m!} \qquad (t \in \mathbb{R}^+),$$

where

$$L_{k,t}(r) := \frac{(-1)^{(m+1)k} f(r^{(m+1)/(\pi T)})}{\log(t^{\pi T/(m+1)} r^{-1}) \operatorname{sinc}^{m+1}(\log(e^{-k} r^{1/\pi}))}.$$

The series converges uniformly on compact subsets of $\mathbb{R}^+$.

Proof For simplicity, we first suppose that $f \in B^2_{c,m+1}$ or equivalently $T = (m+1)/\pi$. By the Mellin version of the Paley–Wiener theorem for polar-analytic functions (Theorem 8.10), f is the restriction to $\mathbb{R}^+$ of a function $\widetilde{f} \in \mathscr{B}^2_{c,m+1}$.

Now let $\mathcal{L}$ be a compact subset of $\mathbb{R}^+$. For $t \in \mathcal{L}$, consider the function

$$F_t : (r, \theta) \longmapsto \frac{\widetilde{f}(r, \theta)}{\big(\log(r/t) + i\theta\big) \sin^{m+1}(\log r + i\theta)}.$$

It is polar-analytic on $\mathbb{H}$ except for poles at the point $(t, 0)$ and the points $(r_k, 0)$, where $r_k := e^{k\pi}$ with $k \in \mathbb{Z}$. Assuming for the moment that $t \in \mathcal{L} \setminus \{r_k : k \in \mathbb{Z}\}$, we see that the pole at $(t, 0)$ is of order at most 1 and those at $(r_k, 0)$ are of order at most $m + 1$.

Next let $\lambda > 1$ be a real number and let K and N be integers at least as large as $\lambda^{3/2}$. Denote by $\partial\mathcal{R}_\lambda$ the positively oriented rectangular curve with vertices at $(e^{-(K+1/2)\pi}, \pm\lambda\pi)$ and $(e^{(N+1/2)\pi}, \pm\lambda\pi)$. We choose λ so large that $\mathcal{L}$ is located inside $\partial\mathcal{R}_\lambda$. Then, by the residue theorem for polar-analytic functions (see Theorem 3.39), we have

$$I_\lambda(t) := \frac{1}{2\pi i}\int_{\partial\mathcal{R}_\lambda}\left(re^{i\theta}\right)^{c-1}F_t(r,\theta)e^{i\theta}(dr+ird\theta)$$

$$= (\mathrm{res}_c F_t)(t,0) + \sum_{k=-K}^{N}(\mathrm{res}_c F_t)(r_k,0). \tag{12.36}$$

Under our restrictions on t, we see that

$$(\mathrm{res}_c F_t)(t,0) = \frac{t^c f(t)}{\sin^{m+1}(\log t)}. \tag{12.37}$$

For the calculation of the c-residues at $(r_k, 0)$, we use again the factorization (12.33). It allows us the representation

$$F_t(r,0) = \frac{H_{k,t}(r)}{(\log r - k\pi)^{m+1}}$$

with

$$H_{k,t}(r) := \frac{(-1)^{(m+1)k} f(r)}{\log(r/t)\,\mathrm{sinc}^{m+1}(\log(e^{-k}r^{1/\pi}))}.$$

In view of Remarks 3.30 and 3.40, we find that

$$(\mathrm{res}_c F_t)(r_k,0) = r_k^c\,\frac{(\Theta_c^m H_{k,t})(r_k)}{m!}. \tag{12.38}$$

Substituting (12.37) and (12.38) in (12.36) and introducing

$$A_{m,k}(t) := -\left(\frac{r_k}{t}\right)^c \sin^{m+1}(\log t)\,\frac{(\Theta_c^m H_{k,t})(r_k)}{m!}, \tag{12.39}$$

we may rewrite (12.36) as

$$f(t) - \sum_{k=-K}^{N} A_{m,k}(t) = t^{-c}\sin^{m+1}(\log t)\,I_\lambda(t). \tag{12.40}$$

So far, in this formula we have excluded those $t \in \mathcal{L}$ with $t = r_j$ for some integer $j \in [-K, N]$. Now consider the limit $t \to r_j$ in (12.40). Obviously, for $j \neq k$ the pole of F_t at $(r_k, 0)$ does not change its order as t approaches r_j. Therefore (12.38) and (12.39) show that $\lim_{t\to r_j} A_{m,k}(t) = 0$ for $j \neq k$. Clearly,

$$\lim_{t\to r_j} t^{-c}\sin^{m+1}(\log t)\,I_\lambda(t) = 0$$

for all $r_j \in \mathcal{L}$. Hence (12.40) implies that $\lim_{t\to r_j} A_{m,j}(t) = f(r_j)$. Thus, defining at the exceptional points

$$A_{m,k}(r_j) = f(r_j)\,\delta_{k,j}$$

with Kronecker's delta, we obtain a continuous continuation of the functions $A_{m,k}$ for $k \in \mathbb{Z}$ such that (12.40) is valid for all $t \in \mathcal{L}$.

Next we want to study the integral $I_\lambda(t)$ as $\lambda \to \infty$. For this, we split it into four parts

$$I_\lambda(t) = I^-_{\text{hor}} + I^{\text{r}}_{\text{vert}} + I^+_{\text{hor}} + I^{\ell}_{\text{vert}} \tag{12.41}$$

corresponding to the lower horizontal, right vertical, upper horizontal and left vertical line segments of $\partial\mathcal{R}_\lambda$, respectively.

It can be verified that on $\partial\mathcal{R}_\lambda$ the estimate (12.30) is valid. Therefore

$$\left|I^\pm_{\text{hor}}\right| \le \frac{3^{m+1}}{2\pi\, e^{(m+1)\lambda\pi}} \int_{e^{-(N+1/2)\pi}}^{e^{(N+1/2)\pi}} \frac{r^c\left|\widetilde{f}(r, \pm\lambda\pi)\right|}{\sqrt{\log^2(r/t) + \lambda^2\pi^2}}\,\frac{dr}{r}\,.$$

Employing the Cauchy–Schwarz inequality on the right-hand side, we deduce that

$$\left|I^\pm_{\text{hor}}\right| \le \frac{3^{m+1}}{2\pi\, e^{(m+1)\lambda\pi}}\,\left\|\widetilde{f}(\cdot, \pm\lambda\pi)\right\|_{X^2_c}\left(\int_0^\infty \frac{1}{\log^2(r/t) + \lambda^2\pi^2}\,\frac{dr}{r}\right)^{1/2}.$$

By Theorem 3.42, we have

$$\left\|\widetilde{f}(\cdot, \pm\lambda\pi)\right\|_{X^2_c} \le e^{(m+1)\lambda\pi}\,\|f\|_{X^2_c}\,.$$

Furthermore, by an appropriate substitution, we see that

$$\int_0^\infty \frac{1}{\log^2(r/t) + \lambda^2\pi^2}\,\frac{dr}{r} = \frac{1}{\lambda\pi}\int_{-\infty}^\infty \frac{dx}{x^2+1} = \frac{1}{\lambda}\,.$$

Thus

$$\left|I^\pm_{\text{hor}}\right| \le \frac{3^{m+1}}{2\pi\sqrt{\lambda}}\,\|f\|_{X^2_c}\,,$$

which shows that $I^\pm_{\text{hor}} \to 0$ as $\lambda \to \infty$ uniformly with respect to t.

It remains to estimate the contributions from the vertical line segments. Using property (iii) of Mellin–Bernstein spaces (see Definition 3.41) in conjunction with (12.30), we find that for some constant $C_{\widetilde{f}} > 0$, we have

$$\left|I_{\text{vert}}^{\text{r}}\right| \le \frac{3^{m+1}C_{\widetilde{f}}}{2\pi}\int_{-\lambda\pi}^{\lambda\pi}\frac{d\theta}{\sqrt{((N+1/2)\pi-\log t)^2+\theta^2}}$$

$$\le \frac{3^{m+1}\lambda\, C_{\widetilde{f}}}{|(N+1/2)\pi-\log t|} \le \frac{3^{m+1}\lambda\, C_{\widetilde{f}}}{\left|(\lambda^{3/2}+1/2)\pi-\log t\right|} \longrightarrow 0 \qquad (\lambda\to\infty)$$

uniformly with respect to $t\in\mathcal{L}$. By an analogous consideration, the same result is obtained for I_{vert}^{ℓ}.

Altogether we have proved that

$$f(t)=\sum_{k\in\mathbb{Z}}A_{m,k}(t) \tag{12.42}$$

for $f\in B^2_{c,m+1}$ and $t\in\mathbb{R}^+$ with $A_{m.k}$ given by (12.39). The series converges uniformly on compact subsets of $\mathbb{R}^+$.

When $f\in B^2_{c,\pi T}$, we consider $g : r\mapsto f(r^{(m+1)/(\pi T)})$ and note that $g\in B^2_{c(m+1)/(\pi T),m+1}$ (see Proposition 3.44). Therefore formula (12.42) applies to g if the role of c is taken by $c(m+1)/(\pi T)$. Finally, expressing g in terms of f and replacing t with $t^{\pi T/(m+1)}$, we arrive at the desired result. □

Let us now consider the cases $m=0$ and $m=1$. For $m=0$ and $f\in B^2_{c,\pi T}$ the formula of Theorem 12.18 takes the form

$$f(t)=t^{-c}\sum_{k\in\mathbb{Z}}f(e^{k/T})e^{ck/T}\operatorname{sinc}\bigl(\log(e^{-k}t^T)\bigr)$$

$$=\sum_{k\in\mathbb{Z}}f(e^{k/T})\operatorname{lin}_{c/T}(e^{-k}t^T) \qquad (t\in\mathbb{R}^+).$$

This shows that Theorem 12.9 is contained in Theorem 12.18 even with uniform convergence on compact subsets of $\mathbb{R}^+$.

For $m=1$ the formula of Theorem 12.18 reads as

$$f(t)=t^{-c}\sum_{k\in\mathbb{Z}}e^{c2k/T}\sin^2(\log t^{\pi T/2})\bigl(\Theta_{2c/(\pi T)}L_{k,t}\bigr)(e^{k\pi}). \tag{12.43}$$

By the definition of the Mellin derivative, we have

$$\bigl(\Theta_{2c/(\pi T)}L_{k,t}\bigr)(e^{k\pi})=e^{k\pi}L'_{k,t}(e^{k\pi})+\frac{2c}{\pi T}L_{k,t}(e^{k\pi}).$$

A straightforward calculation yields

$$e^{k\pi}L'_{k,t}(e^{k\pi})=\frac{f(e^{2k/T})}{\log^2(e^{-k\pi}t^{\pi T/2})}+\frac{2}{\pi T}\frac{e^{2k/T}f'(e^{2k/T})}{\log(e^{-k\pi}t^{\pi T/2})}.$$

Adding $(2c/(\pi T))L_{k,t}(e^{k\pi})$ on both sides, we obtain

$$\big(\Theta_{2c/(\pi T)}L_{k,t}\big)(e^{k\pi}) = \frac{f(e^{2k/T})}{\log^2(e^{-k\pi}t^{\pi T/2})} + \frac{2}{\pi T}\frac{(\Theta_c f)(e^{2k/T})}{\log(e^{-k\pi}t^{\pi T/2})}.$$

When we substitute this in (12.43) and note that

$$\sin(\log t^{\pi T/2}) = (-1)^k \sin\big(\log(e^{-k\pi}t^{\pi T/2})\big),$$

we can build lin functions and arrive at the following result.

Corollary 12.19 *Let $f \in B^2_{c,\pi T}$ with $c \in \mathbb{R}$ and $T \in \mathbb{R}^+$. Then*

$$f(t) = \sum_{k\in\mathbb{Z}}\Big[f(e^{2k/T})\mathrm{lin}^2_{c/T}(e^{-k}t^{T/2}) + (-1)^k \frac{2}{\pi T}(\Theta_c f)(e^{2k/T})\sin(\log t^{\pi T/2})\mathrm{lin}_{2c/T}(e^{-k}t^{T/2})\Big]$$

for $t \in \mathbb{R}^+$. The series converges uniformly on compact subsets of $\mathbb{R}^+$.

Remark 12.20

(i) We note that in Theorem 12.18 uniform convergence cannot be guaranteed on an interval $]0, a[\subset \mathbb{R}^+$ although it is of finite length with respect to the Lebesgue measure on $\mathbb{R}^+$. This is a peculiarity of Mellin analysis, taking into account that the invariant measure of a Borel subset A of the multiplicative group $(\mathbb{R}^+, \cdot)$ is given by $\mu(A) = \int_A dt/t$, and we have $\mu(]0, a[) = \infty$.

(ii) It can be shown that

$$\big(\Theta^m_{c(m+1)/(\pi T)}L_{k,t}\big)(e^{k\pi}) = \sum_{j=0}^{m}\big(\Theta^j_c f\big)(e^{(m+1)k/T})\,\varphi_{m,k,j,T}(t),$$

where $\varphi_{m,k,j,T}$ are functions on $\mathbb{R}^+$ depending on m, k, j and T but not on f. Thus the sampling formula of Theorem 12.18 involves samples of the functions $f, \Theta_c f, \ldots, \Theta^m_c f$. The benefit of choosing $m > 0$ is that the sample points $e^{(m+1)k/T}$ $(k \in \mathbb{Z})$ are less dense as m increases.

12.4 Approximate Exponential Sampling Formula

By the Mellin–Paley–Wiener theorem, the assumption that f is Mellin bandlimited is quite restrictive. In this section we will extend the exponential sampling theorem to the case when the function f is not necessarily Mellin bandlimited, but belongs to the inversion class $\mathcal{M}^2_c$.

12.4.1 Pointwise Estimates

We begin with two auxiliary results.

Lemma 12.21 *Let $f \in \mathcal{M}_c^2$ with $c \in \mathbb{R}$. Then*

$$f(x) = \frac{1}{2\pi} \int_{-\infty}^{\infty} [f]^{\wedge}_{M_c^2}(c + it) x^{-c-it} dt$$
$$= \lim_{T \to \infty} \frac{1}{2\pi} \int_{-T}^{T} [f]^{\wedge}_{M_c^2}(c + it) x^{-c-it} dt \tag{12.44}$$

for all $x \in \mathbb{R}^+$.

Proof In view of Definition 7.7, Theorem 7.10 and Remark 7.11, Eqs. (12.44) hold almost everywhere on $\mathbb{R}^+$. By the hypotheses, f is continuous on $\mathbb{R}^+$ and $[f]^{\wedge}_{M_c^2} \in L^1(c + i\mathbb{R})$. According to Lemma 6.71 (a), the latter property entails that the limit on the right hand side of (12.44) exists as a continuous function of x. Hence (12.44) holds for all $x \in \mathbb{R}^+$. □

Now we recall the function $K_c(x, u)$ defined in (12.12) and denote by $\widetilde{K}_c(x, u)$ its c-recurrent continuation to $\mathbb{R}^+$ with respect to u. Thus, $\widetilde{K}_c(x, \cdot)$ belongs to Y_c^2. The next lemma shows the boundedness of the Mellin–Fourier series of $\widetilde{K}_c(x, \cdot)$ on $[e^{-\pi}, e^{\pi}]$.

Lemma 12.22 *The Mellin–Fourier series of $\widetilde{K}_c(x, \cdot)$ is given by*

$$\widetilde{K}_c(x, u) = \sum_{k \in \mathbb{Z}} e^{-ck} \operatorname{lin}_c\big(e^{-k} x\big) u^{-c-ik} \qquad (x,\ u \in \mathbb{R}^+). \tag{12.45}$$

Moreover, there exists a universal constant $M > 0$ such that

$$\left| \sum_{k=-n}^{n} e^{-ck} \operatorname{lin}_c\big(e^{-k} x\big) u^{-c-ik} \right| \le (xu)^{-c}\big(1 + M|\log x|\big) \tag{12.46}$$

for all $n \in \mathbb{N}$, $u \in [e^{-\pi}, e^{\pi}]$ and $x \in \mathbb{R}^+$.

Proof The expansion (12.45) follows from (12.15) and (12.20). Concerning (12.46), we first note that $(\Theta_c K_c(x, \cdot))(u) = -i \log x\, K_c(x, u)$. Then, recalling inequality (5.31) of Corollary 5.22, we find that the left-hand side of (12.46) is bounded from above by

$$u^{-c}\left(|K_c(x, 1)| + M \operatorname*{ess\,sup}_{y \in [e^{-\pi}, e^{\pi}]} |y^c (\Theta_c K_c(x, \cdot))(y)| \right) = (xu)^{-c}\big(1 + M|\log x|\big),$$

as was to be shown. □

Now we are ready for an estimate of the difference between f and its exponential sampling series in terms of the Mellin transform in X_c^2.

Theorem 12.23 *Let $f \in \mathcal{M}_c^2$ with $c \in \mathbb{R}$. Then*

$$\lim_{N\to\infty}\left| f(x) - \sum_{k=-N}^{N} f(e^{k/T})\,\mathrm{lin}_{c/T}\big(e^{-k}x^{T}\big)\right| \leq \frac{x^{-c}}{\pi}\int_{|t|\geq \pi T} \big|[f]^{\wedge}_{M_c^2}(c+it)\big|dt \tag{12.47}$$

for all $x, T \in \mathbb{R}^+$.

Proof Substituting $t = \log u$ in (12.44), we obtain

$$f(x) = \frac{1}{2\pi}\int_0^{\infty} [f]^{\wedge}_{M_c^2}(c + i\log u)\, x^{-c-i\log u}\,\frac{du}{u}. \tag{12.48}$$

With this representation of f and any $N \in \mathbb{N}$, we may write

$$\sum_{k=-N}^{N} f(e^k)\,\mathrm{lin}_c\big(e^{-k}x\big) = \frac{1}{2\pi}\int_0^{\infty} [f]^{\wedge}_{M_c^2}(c+i\log u)\left\{\sum_{k=-N}^{N} e^{-ck}\mathrm{lin}_c\big(e^{-k}x\big)u^{-ik}\right\}\frac{du}{u}. \tag{12.49}$$

Note that the expression in braces does not change when u is replaced by $e^{2\pi j}u$ for some $j \in \mathbb{Z}$. In view of (12.46), it is therefore bounded by $x^{-c}(1 + M\,|\log x|)$ on the whole of $\mathbb{R}^+$. Since by the hypotheses $[f]^{\wedge}_{M_c^2} \in L^1(c + i\mathbb{R})$, we conclude that

$$x^{-c}\,\big|[f]^{\wedge}_{M_c^2}(c+i\log u)\big|\,(1 + M|\log x|)\,u^{-1}$$

as a function of u is an $L^1(\mathbb{R}^+)$ majorant of the integrand in (12.49). Hence for the limit $N \to \infty$, Lebesgue's dominated convergence theorem is applicable. By (12.45) it yields

$$\sum_{k\in\mathbb{Z}} f(e^k)\mathrm{lin}_c\big(e^{-k}x\big) = \frac{1}{2\pi}\int_0^{\infty} [f]^{\wedge}_{M_c^2}(c+i\log u)\,\widetilde{K}_c(x,u)u^c\,\frac{du}{u}. \tag{12.50}$$

Now, for $u \in [e^{-\pi}, e^{\pi}]$, we have $\widetilde{K}_c(x,u)u^c = K_c(x,u)u^c = x^{-c-i\log u}$. Moreover, for any $u \in \mathbb{R}^+$, we find that

$$\big|x^{-c-i\log u} - \widetilde{K}_c(x,u)u^c\big| \leq \big|x^{-c-i\log u}\big| + \big|\widetilde{K}_c(x,u)u^c\big| \leq 2x^{-c}.$$

Thus subtracting (12.50) from (12.48), and writing $J := [e^{-\pi}, e^{\pi}]$, we obtain

$$\begin{aligned}
&\Big|f(x) - \sum_{k\in\mathbb{Z}} f(e^k)\mathrm{lin}_c\big(e^{-k}x\big)\Big| \\
&\quad \le \frac{1}{2\pi}\int_{\mathbb{R}^+\setminus J} \Big|[f]^{\wedge}_{M_c^2}(c + i\log u)\big(x^{-c-i\log u} - \widetilde{K}_c(x,u)u^c\big)\Big|\,\frac{du}{u} \\
&\quad \le \frac{x^{-c}}{\pi}\int_{|t|\ge\pi} \big|[f]^{\wedge}_{M_c^2}(c+it)\big|dt. \qquad (12.51)
\end{aligned}$$

This is (12.47) for $T = 1$. For arbitrary $T > 0$, we apply (12.51) to the function $F(x) := f(x^{1/T})$, which belongs to $\mathcal{M}^2_{c/T}$, and we readily arrive at (12.47). □

The following corollary is an easy consequence.

Corollary 12.24 *Let $f \in \mathcal{M}^2_c$ with $c \in \mathbb{R}$. Then*

$$f(x) = \lim_{T\to\infty}\sum_{k\in\mathbb{Z}} f\big(e^{k/T}\big)\mathrm{lin}_{c/T}\big(e^{-k}x^T\big) \qquad (x \in \mathbb{R}^+).$$

The convergence for $T \to \infty$ is uniform on compact subsets of $\mathbb{R}^+$. After multiplying the equation by x^c, the convergence becomes uniform on the whole of $\mathbb{R}^+$.

Remark 12.25 The above results still holds for any function $f \in X_c \cap \mathcal{M}^p_c$ with $1 \le p \le 2$ since $\mathcal{M}^p_c \subset \mathcal{M}^2_2$ for $p \in [1, 2[$. In this instance the right-hand side of (12.47) can be written in terms of the Mellin transform of $f \in X_c$, due to the consistency property (7.31).

In practice, the pointwise approximation of a functions $f \in \mathcal{M}^2_c$ by the exponential sampling series involves two errors, namely a *truncation error* caused by using a partial sum $S_N f$ of the series and an *aliasing error*, which is estimated by the right-hand side of (12.47). As an example, we want to study these errors for f being the Mellin–Abel–Poisson kernel (see Example 6.29) defined by

$$P_\rho(x) = \begin{cases} \frac{\rho}{2}x^{\rho}, & 0 < x \le 1, \\ \frac{\rho}{2}x^{-\rho}, & x > 1. \end{cases}$$

with $\rho > 0$. We recall that the Mellin transform of P_ρ, evaluated on the line $c + it$ with $t \in \mathbb{R}$ is given by

$$[P_\rho]^{\wedge}_{M_c}(s) = \frac{\rho^2}{\rho^2 - s^2} \qquad \big(s = c + it,\ c \in\,]-\rho, \rho[,\ t \in \mathbb{R}\big),$$

which belongs to $L^1(c+i\mathbb{R})$ as a function of $t \in \mathbb{R}$. Thus P_ρ satisfies the hypotheses of Theorem 12.23 for each $c \in\,]-\rho, \rho[$, as taken subsequently.

For $N \in \mathbb{N}$, the (symmetric) partial sum of the exponential sampling series is given by

$$
\begin{aligned}
(S_N P_\rho)(x) &= \frac{\rho}{2} \sum_{|k|\le N} e^{-\rho|k|/T} \operatorname{lin}_{c/T}(e^{-k}x^T) \\
&= \frac{\rho x^{-c}}{2} \sum_{|k|\le N} e^{(-\rho|k|+ck)/T} \operatorname{sinc}(T\log x - k).
\end{aligned}
$$

where $T > 0$. For estimating the truncation error, we choose $N \ge 2T|\log x|$ and note that

$$
|\operatorname{sinc}(T\log x - k)| \le \frac{2}{\pi|k|} \qquad (|k| \ge N+1).
$$

Now, by Theorem 12.23, the entire error is bounded by

$$
\begin{aligned}
|P_\rho(x) - (S_N P_\rho)(x)| \le{}& \frac{\rho x^{-c}}{\pi} \sum_{|k|\ge N+1} \frac{e^{(-\rho|k|+ck)/T}}{|k|} \\
&+ \frac{x^{-c}}{\pi} \int_{|t|>\pi T} |[P_\rho]^\wedge_{M_c}(c+it)|dt =: I_1 + I_2,
\end{aligned}
$$

where I_1 is a bound for the truncation error and I_2 a bound for the aliasing error. Setting $r := \min\{\rho+c, \rho-c\}$, which implies $0 < r \le \rho$ for the considered numbers c, we can write

$$
\begin{aligned}
I_1 &\le \frac{\rho x^{-c}}{\pi} \left[\sum_{k=-\infty}^{-N-1} \frac{e^{-(\rho+c)|k|/T}}{|k|} + \sum_{k=N+1}^{\infty} \frac{e^{-(\rho-c)k/T}}{k} \right] \\
&\le \frac{2\rho x^{-c}}{\pi} \sum_{k=N+1}^{\infty} \frac{e^{-rk/T}}{k} \le \frac{2\rho x^{-c}}{\pi} \int_{Nr/T}^{\infty} \frac{e^{-v}}{v}\, dv.
\end{aligned}
$$

The last integral is the well-known exponential integral (see [1, p. 229])

$$
E_1(\xi) := \int_\xi^\infty \frac{e^{-u}}{u}\, du \qquad (\xi > 0),
$$

which can obviously be estimated as $E_1(\xi) \le e^{-\xi}/\xi$. Thus,

$$
I_1 \le \frac{2\rho x^{-c} T}{\pi r N} e^{-rN/T} \qquad (N \ge 2T|\log x|).
$$

For I_2 we obtain

$$I_2 = \frac{x^{-c}}{\pi} \int_{|t|>\pi T} \frac{\rho^2}{|\rho^2 - (c+it)^2|}\,dt \le \frac{x^{-c}}{\pi} \int_{|t|>\pi T} \frac{\rho^2}{t^2}\,dt = \frac{2x^{-c}\rho^2}{\pi^2 T}.$$

Thus for every fixed $x > 0$,

$$|P_\rho(x) - (S_N P_\rho)(x)| \le \frac{2\rho x^{-c}}{\pi} \left[\frac{T}{rN} e^{-rN/T} + \frac{\rho}{\pi T} \right] \qquad (N \ge 2T|\log x|).$$

If we choose $T = \lambda^r \log \lambda$ and $N = \lceil \lambda^r \log^2 \lambda \rceil$ with $\lambda \in \mathbb{R}^+$, then for sufficiently large λ the condition $N \ge 2T|\log x|$ is satisfied. Furthermore, the two error bounds are of the same asymptotic order and yield

$$|P_\rho(x) - (S_N P_\rho)(x)| = \mathcal{O}\left(\frac{1}{\lambda^r \log \lambda} \right) \qquad (\lambda \to \infty)$$

uniformly on compact subsets of $\mathbb{R}^+$. Note that for $c = 0$ the exponent r attains its largest value ρ.

Figure 12.2 shows the graphs of $P_{1/2}$ and of the partial sums $S_N P_{1/2}$ for $c = 0$, $T = 4$ and $N \in \{6, 8\}$. Note that on the interval $]0, \frac{2}{5}[$ the condition $N > 2T|\log x|$ is violated.

Finally, Fig. 12.3 shows the graphs of P_1 and the corresponding partial sums for the same values of c, T and N.

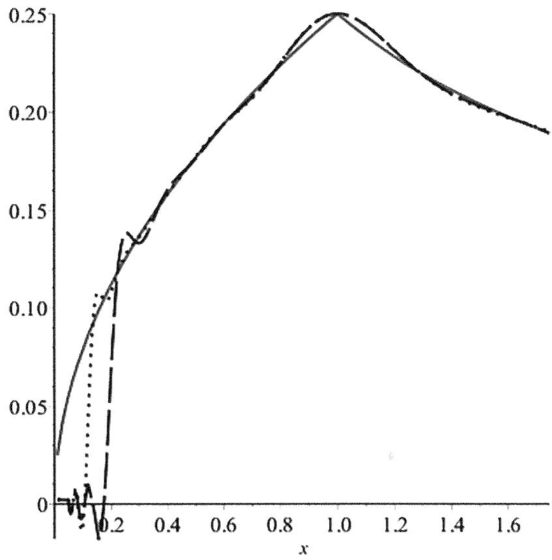

Fig. 12.2 The function $P_{1/2}$ (solid line) and $(S_6 P_{1/2})$ (dashed line), $(S_8 P_{1/2})$ (dotted line)

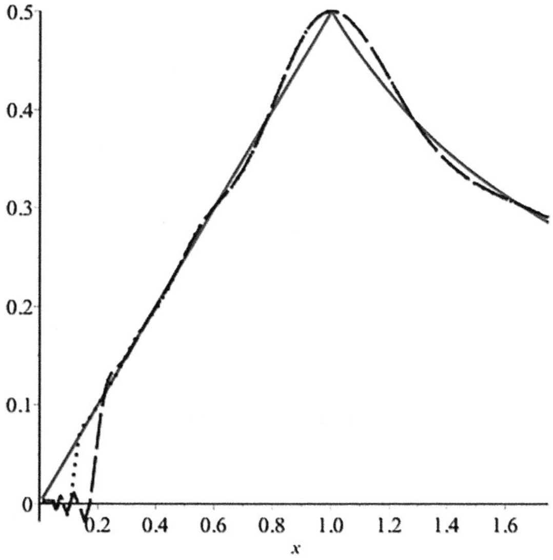

Fig. 12.3 The function P_1 (solid line) and $(S_6 P_1)$ (dashed line), $(S_8 P_1)$ (dotted line)

12.4.2 Uniform Estimates

In this subsection we shall provide uniform estimates for the right-hand side of (12.47) when the function f belongs to a suitable function space. They will be in terms of the Mellin distance of f from the Mellin–Paley–Wiener space $B^2_{c,\pi T}$.

Indeed, the estimate of Theorem 12.23 permits a metric interpretation by using the notion of Mellin distance developed in Chap. 10.

Writing

$$f(x) = \lim_{n\to\infty} \sum_{k=-n}^{n} f\left(e^{k/T}\right) \mathrm{lin}_{c/T}\left(e^{-k}x^{T}\right) + (R_{\pi T} f)(x), \qquad (12.52)$$

with a remainder $(R_{\pi T} f)(x)$, we may express (12.47) as

$$\left|(R_{\pi T} f)(x)\right| \le \frac{x^{-c}}{\pi} \mathrm{dist}_1(f, B^2_{c,\pi T}) \qquad (x \in \mathbb{R}^+),$$

or equivalently,

$$\|R_{\pi T} f\|^*_{X_c^\infty} \le \frac{1}{\pi} \mathrm{dist}_1(f, B^2_{c,\pi T}). \qquad (12.53)$$

This relation is a trivial equality when $f \in B^2_{c,\pi T}$. But equality can also occur when $f \notin B^2_{c,\pi T}$. For example, consider the function

$$f(x) := x^{-c}\mathrm{sinc}(2T \log x - 1).$$

By a straight forward calculation, we find that

$$[f]^{\wedge}_{M^2_c}(c+iv) = \frac{e^{iv/(2T)}}{2T}\,\mathrm{rect}\left(\frac{v}{2T}\right),$$

where rect denotes the rectangle function defined by

$$\mathrm{rect}(x) := \begin{cases} 1 & \text{if } |x| < \pi, \\ \frac{1}{2} & \text{if } |x| = \pi, \\ 0 & \text{if } |x| > \pi. \end{cases}$$

Thus, $f \notin B^2_{c,\pi T}$ and

$$\mathrm{dist}_1(f, B^2_{c,\pi T}) = \int_{|v|\geq \pi T} |[f]^{\wedge}_{M^2_c}(c+iv)|dv = \frac{1}{2T}\int_{|v|\geq \pi T} \mathrm{rect}\left(\frac{v}{2T}\right) dv = \pi.$$

Furthermore,

$$f(e^{k/T}) = e^{-kc/T}\,\mathrm{sinc}(2k-1) = 0$$

for all $k \in \mathbb{Z}$. Therefore $(R_{\pi T} f)(x) = f(x)$, which shows that

$$\|R_{\pi T} f\|^*_{X^\infty_c} = \sup_{x>0} |\mathrm{sinc}(2T\log x - 1)| = 1,$$

and so equality occurs in (12.53).

Now, employing Theorem 10.2, Corollary 10.12, Theorem 10.13 and Theorem 10.14, we obtain the following theorem.

Theorem 12.26 *Let $c \in \mathbb{R}$, $T \in \mathbb{R}^+$ and $r \in \mathbb{N}$. For the remainder of the approximate exponential sampling formula (12.52), the following asymptotic estimates hold:*

(1) If $f \in \mathrm{Lip}_r(\alpha, X_c \cap C(\mathbb{R}^+))$ with $r \geq 2$ and $1 < \alpha \leq r$, then

$$\|(R_{\pi T} f)\|^*_{X^\infty_c} = \mathcal{O}(T^{-\alpha+1}) \qquad (T \to \infty).$$

(2) If $f \in \mathcal{M}^2_c \cap \mathrm{Lip}_r(\beta, X^2_c)$ with $1/2 < \beta \leq r$, then

$$\|(R_{\pi T} f)\|^*_{X^\infty_c} = \mathcal{O}(T^{-\beta+1/2}) \qquad (T \to \infty).$$

(3) If $f \in \mathcal{M}^1_c \cap W^{r,1}_c(\mathbb{R}^+)$ with $r > 1$, then

$$\|(R_{\pi T} f)\|^*_{X^\infty_c} = \mathcal{O}(T^{-r+1}) \qquad (T \to \infty).$$

(4) If $f \in \mathcal{M}_c^2 \cap W_c^{r,2}(\mathbb{R}^+)$*, then*

$$\|(R_{\pi T} f)\|_{X_c^\infty}^* = \mathcal{O}(T^{-r+1/2}) \qquad (T \to \infty).$$

Corresponding results hold in Mellin–Sobolev spaces of fractional order as can be shown with the help of Corollary 10.6 and Theorem 10.21.

Theorem 10.16 provides estimates for polar-analytic functions belonging to Mellin–Hardy spaces. We will consider the Mellin–Hardy space $H_c^2(\mathbb{H}_a)$ for $c \in \mathbb{R}$ and $a > 0$.

We note that if $f \in H_c^2(\mathbb{H}_a)$, then its restriction to the positive real axis $g(\cdot) := f(\cdot, 0)$ belongs to the space $\mathcal{M}_c^2$, and so Theorem 12.23 applies.

In general the exponential sampling series in Theorem 12.23 converges only in the ordinary sense. However, if $f \in H_c^2(\mathbb{H}_a)$, then by Proposition 3.51 one can deduce that the series for $g(\cdot) := f(\cdot, 0)$ converges absolutely and uniformly on compact subsets of $\mathbb{R}^+$, and

$$|(R_{\pi T} g)(x)| \le \frac{x^{-c}}{\pi} \operatorname{dist}_1(g, B_{c,\pi T}^2) \qquad (x > 0),$$

or equivalently,

$$\|R_{\pi T} g\|_{X_c^\infty}^* \le \frac{1}{\pi} \operatorname{dist}_1(g, B_{c,\pi T}^2).$$

Thus, Theorem 10.16 yields the following statement.

Corollary 12.27 *Let* $f \in H_c^2(\mathbb{H}_a)$ *with* $c \in \mathbb{R}$ *and* $a \in \mathbb{R}^+$*. Then*

$$\|R_{\pi T} f(\cdot, 0)\|_{X_c^\infty}^* = \mathcal{O}(e^{-a\pi T}) \qquad (T \to \infty).$$

12.5 Mellin Reproducing Kernel Formula

In this section, we establish an integral version of the exponential sampling formula, called *Mellin reproducing kernel formula*, for both, functions in X_c and in X_c^2. We begin with the following theorem.

Theorem 12.28 (Mellin Reproducing Kernel Formula in X_c**)** *Let* $f \in B_{c,\pi T}^1$ *with* $c \in \mathbb{R}$ *and* $T \in \mathbb{R}^+$*. Then*

$$f(x) = T \int_0^\infty f(y) \operatorname{lin}_{c/T}\left(\left(\frac{x}{y}\right)^T\right) \frac{dy}{y},$$

the integral being absolutely convergent on $\mathbb{R}^+$*.*

Proof Since $f \in X_c$ and

$$|\mathrm{lin}_{c/T}(x)| = x^{-c/T}|\mathrm{sinc}(\log x)| \le x^{-c/T}$$

for every $x \in \mathbb{R}^+$, the absolute convergence of the integral follows immediately.

Next, introducing $h(y) := f(y^{1/T})$, we have $h \in B^1_{c/T,\pi}$. Moreover, the function $h(\cdot)\mathrm{lin}_{c/T}(x/\cdot)$ is continuous and belongs to X_0 since

$$\left\| h(\cdot)\mathrm{lin}_{c/T}\left(\frac{x}{\cdot}\right)\right\|_{X_0} \le \frac{T}{x^{c/T}}\,\|f\|_{X_c}.$$

Now Lemma 12.3 with h and c/T in place of f and c, respectively, and with interchanged roles of x and y shows that

$$\left[h(\cdot)\mathrm{lin}_{c/T}\left(\frac{x}{\cdot}\right)\right]^{\wedge}_{M_0}(it) = \frac{x^{-c/T}}{2\pi}\int_{-\pi}^{\pi}[h]^{\wedge}_{M_{c/T}}\left(\frac{c}{T}+i(t+v)\right)x^{-iv}dv \quad (t \in \mathbb{R}).$$

For $t = 0$ the inversion theorem (see Theorem 6.75) yields

$$\int_0^\infty h(y)\mathrm{lin}_{c/T}\left(\frac{x}{y}\right)\frac{dy}{y} = \frac{x^{-c/T}}{2\pi}\int_{-\pi}^{\pi}[h]^{\wedge}_{M_{c/T}}\left(\frac{c}{T}+iv\right)x^{-iv}dv = h(x).$$

Thus

$$f(x^{1/T}) = \int_0^\infty f(y^{1/T})\mathrm{lin}_{c/T}\left(\frac{x}{y}\right)\frac{dy}{y} = T\int_0^\infty f(y)\mathrm{lin}_{c/T}\left(\frac{x}{y^T}\right)\frac{dy}{y},$$

and the assertion is obtained by replacing x with x^T. □

Next, using the Mellin transform of X_c^2 functions, we prove the Mellin reproducing kernel formula for functions in $B^2_{c,\pi T}$.

Theorem 12.29 (Mellin Reproducing Kernel Formula in X_c^2) *Let $f \in B^2_{c,\pi T}$ with $c \in \mathbb{R}$ and $T \in \mathbb{R}^+$. Then*

$$f(x) = T\int_0^\infty f(y)\mathrm{lin}_{c/T}\left(\left(\frac{x}{y}\right)^T\right)\frac{dy}{y}, \tag{12.54}$$

the integral being absolutely convergent on $\mathbb{R}^+$.

Proof Roughly speaking, we proceed by an L^2 modification of the previous proof. Since $f \in X_c^2$, we find with the help of Hölder's inequality that

$$\int_0^\infty |f(y)|\left|\mathrm{lin}_{c/T}\left(\left(\frac{x}{y}\right)^T\right)\right|\frac{dy}{y} = \int_0^\infty |f(y)|\left(\frac{x}{y}\right)^{-c}\left|\mathrm{sinc}\left(T\log\left(\frac{x}{y}\right)\right)\right|\frac{dy}{y}$$

$$\leq x^{-c}\left(\int_0^\infty |f(y)|^2 y^{2c}\frac{dy}{y}\right)^{1/2}\left(\int_0^\infty \left|\mathrm{sinc}\left(T\log\left(\frac{x}{y}\right)\right)\right|^2\frac{dy}{y}\right)^{1/2}$$

$$= \frac{x^{-c}}{T^{1/2}}\|f\|_{X_c^2}\,\|\mathrm{sinc}\|_{L^2(\mathbb{R})},$$

which shows the absolute convergenve of the integral.

Next, introducing again $h(y) := f(y^{1/T})$, we have $h \in B^2_{c/T,\pi}$. Moreover, the function $h(\cdot)\mathrm{lin}_{c/T}(x/\cdot)$ is continuous and belongs to $X_0 \cap X_0^2$ since

$$\left\| h(\cdot)\mathrm{lin}_{c/T}\left(\frac{x}{\cdot}\right)\right\|_{X_0} \leq \frac{T^{1/2}}{x^{c/T}}\|f\|_{X_c^2}\,\|\mathrm{sinc}\|_{L^2(\mathbb{R})}$$

and

$$\left\| h(\cdot)\mathrm{lin}_{c/T}\left(\frac{x}{\cdot}\right)\right\|_{X_0^2} \leq \frac{T^{1/2}}{x^{c/T}}\|f\|_{X_c^2}.$$

In the following steps, we use the notation

$$[g_\rho]^\wedge_{M_c^2}(it) := \int_{1/\rho}^{\rho} g(y)y^{c+it}\frac{dy}{y},$$

where $\rho > 1$ and g can be any function for which the right-hand side exists. Hence the subscript ρ reduces the integration of a Mellin transform to an interval of finite length. Therefore, Lemma 12.3 and its proof show us that

$$\left[\left(h(\cdot)\mathrm{lin}_{c/T}\left(\frac{x}{\cdot}\right)\right)_\rho\right]^\wedge_{M_0^2}(it)$$
$$= \frac{x^{-c/T}}{2\pi}\int_{-\pi}^{\pi}[h_\rho]^\wedge_{M^2_{c/T}}\left(\frac{c}{T} + i(t+v)\right)x^{-iv}dv \qquad (t \in \mathbb{R}).$$

Thus, using the generalized Minkowski inequality (Proposition 1.9), we obtain

$$\left\|\frac{x^{-c/T}}{2\pi}\int_{-\pi}^{\pi}[h]^\wedge_{M^2_{c/T}}\left(\frac{c}{T} + i(\cdot+v)\right)x^{-iv}dv - \left[\left(h(\cdot)\mathrm{lin}_{c/T}\left(\frac{x}{\cdot}\right)\right)_\rho\right]^\wedge_{M_0^2}(i\,\cdot)\right\|_{L^2(\mathbb{R})}$$

$$= \left\|\frac{x^{-c/T}}{2\pi}\int_{-\pi}^{\pi}\left\{[h]^\wedge_{M^2_{c/T}}\left(\frac{c}{T} + i(\cdot+v)\right)\right.\right.$$

$$-[h_\rho]^\wedge_{M^2_{c/T}}\left(\frac{c}{T}+i(\cdot+v)\right)\Big\}x^{-iv}dv\Big\|_{L^2(\mathbb{R})}$$

$$\leq \frac{x^{-c/T}}{2\pi}\int_{-\pi}^{\pi}\left\|[h]^\wedge_{M^2_{c/T}}\left(\frac{c}{T}+i(\cdot+v)\right)-[h_\rho]^\wedge_{M^2_{c/T}}\left(\frac{c}{T}+i(\cdot+v)\right)\right\|_{L^2(\mathbb{R})}dv.$$

This allows us to conclude that

$$\lim_{\rho\to\infty}\left[\left(h(\cdot)\mathrm{lin}_{c/T}\left(\frac{x}{\cdot}\right)\right)_\rho\right]^\wedge_{M^2_0}(it)=\frac{x^{-c/T}}{2\pi}\int_{-\pi}^{\pi}[h]^\wedge_{M^2_{c/T}}\left(\frac{c}{T}+i(t+v)\right)x^{-iv}dv$$

almost everywhere for $t\in\mathbb{R}$. Since $h(\cdot)\mathrm{lin}_{c/T}(x/\cdot)$ belongs to X_0, we can write this limit relation as

$$\int_0^\infty h(y)\mathrm{lin}_{c/T}\left(\frac{x}{y}\right)y^{it}\frac{dy}{y}$$
$$=\frac{x^{-c/T}}{2\pi}\int_{-\pi}^{\pi}[h]^\wedge_{M^2_{c/T}}\left(\frac{c}{T}+i(t+v)\right)x^{-iv}dv \qquad (\text{a.e. } t\in\mathbb{R})$$

and know that the left-hand side is a continuous function of t. Since $[h]^\wedge_{M^2_{c/T}}\in L^1(\mathbb{R})$ as a function of t, we can show that the right-hand side is also continuous. Therefore the last equation holds for all $t\in\mathbb{R}$. In particular, for $t=0$ the inversion theorem of the Mellin transform on X^2_c shows that

$$\int_0^\infty h(y)\mathrm{lin}_{c/T}\left(\frac{x}{y}\right)\frac{dy}{y}=\frac{x^{-c/T}}{2\pi}\int_{-\pi}^{\pi}[h]^\wedge_{M^2_{c/T}}\left(\frac{c}{T}+iv\right)x^{-iv}dv=h(x).$$

Thus

$$f(x^{1/T})=T\int_0^\infty f(y)\mathrm{lin}_{c/T}\left(\frac{x}{y^T}\right)\frac{dy}{y},$$

and the assertion is obtained by replacing x with x^T. □

As an important consequence of Theorem 12.29, we state the following orthogonality property of the lin_c functions:

Corollary 12.30 *Let $c\in\mathbb{R}$ and $T\in\mathbb{R}^+$. Then, for $j,k\in\mathbb{Z}$,*

$$Te^{-2ck/T}\left\langle \mathrm{lin}_{c/T}((\cdot)^Te^{-j}),\mathrm{lin}_{c/T}((\cdot)^Te^{-k})\right\rangle_{X^2_c}=\mathrm{lin}_{c/T}(e^{k-j})=\delta_{j,k} \qquad (12.55)$$

with Kronecker's delta.

Proof By the definition of the lin_c function, it is easily verified that

$$\text{lin}_{c/T}\left(\left(\frac{x}{y}\right)^T\right) = \left(\frac{y}{x}\right)^{2c} \text{lin}_{c/T}\left(\left(\frac{y}{x}\right)^T\right)$$

for every $x, y \in \mathbb{R}^+$. Thus, by Theorem 12.29,

$$\begin{aligned} f(x) &= T\int_0^\infty f(y)\text{lin}_{c/T}\left(\left(\frac{x}{y}\right)^T\right)\frac{dy}{y} = T\int_0^\infty f(y)\left(\frac{y}{x}\right)^{2c} \text{lin}_{c/T}\left(\left(\frac{y}{x}\right)^T\right)\frac{dy}{y} \\ &= Tx^{-2c}\left\langle f, \text{lin}_{c/T}\left(\left(\frac{\cdot}{x}\right)^T\right)\right\rangle_{X_c^2} \end{aligned}$$

for any $f \in B^2_{c,\pi T}$. For $x = e^{k/T}$ with $k \in \mathbb{Z}$, we obtain

$$f(e^{k/T}) = Te^{-2ck/T}\langle f, \text{lin}_{c/T}((\cdot)^T e^{-k})\rangle_{X_c^2}.$$

Now for any $j \in \mathbb{Z}$, the function $f(x) := \text{lin}_{c/T}(x^T e^{-j})$ is admissible as can be seen with the help of the Mellin–Paley–Wiener theorem (Theorem 8.10) under a polar-analytic continuation by replacing x with $re^{i\theta}$. Finally, noting that $f(e^{k/T}) = \text{lin}_{c/T}(e^{k-j})$, we arrive at the assertion. □

12.6 Approximate Mellin Reproducing Kernel Formula

Just as the exponential sampling formula, the Mellin reproducing kernel formula holds approximately with a so-called *aliasing error* in case of not necessarily Mellin bandlimited functions. This may be stated as follows.

Theorem 12.31 (Approximate Mellin Reproducing Kernel Formula) *Let $c \in \mathbb{R}$ and $f \in \mathcal{M}_c^2$. Then, for any $T > 0$,*

$$f(x) = T\int_0^\infty f(y)\text{lin}_{c/T}\left(\left(\frac{x}{y}\right)^T\right)\frac{dy}{y} + (\widetilde{R}_{\pi T} f)(x) \qquad (x \in \mathbb{R}^+),$$

where the aliasing error $(\widetilde{R}_{\pi T} f)(x)$ is given by

$$(\widetilde{R}_{\pi T} f)(x) = \frac{x^{-c}}{2\pi}\int_{|t|\geq \pi T} [f]^\wedge_{M_c^2}(c+it)x^{-it}dt \qquad (x \in \mathbb{R}^+).$$

Furthermore, there holds the error estimate

$$|(\widetilde{R}_{\pi T} f)(x)| \leq \frac{x^{-c}}{2\pi}\int_{|t|\geq \pi T} |[f]^\wedge_{M_c^2}(c+it)|dt \qquad (x \in \mathbb{R}^+).$$

Proof Setting $G(x) := \mathrm{lin}_{c/T}(x^T)$, we know that $G \in X_c^2$ and can write the first integral as the Mellin convolution $(f * G)(x)$. Under our hypotheses, it is easily verified that all integrals in the assertion exist absolutely and are continuous functions of x. Using the integral representation of the lin_c function (see Lemma 12.2), we may write

$$G(x) = \frac{x^{-c}}{2\pi} \int_{-\pi}^{\pi} x^{-itT} dt = \frac{x^{-c}}{2\pi T} \int_{-\pi T}^{\pi T} x^{-iu} du.$$

From this we deduce that the Mellin transform of G in X_c^2 is given by

$$[G]^{\wedge}_{M_c^2}(c + it) = \frac{1}{T} \chi_{[-\pi T, \pi T]}(t) \qquad (t \in \mathbb{R}).$$

Next we introduce $G_\rho(x) := G(x)\chi_{[1/\rho, \rho]}(x)$, where $\rho > 1$ is a real parameter. Then $G_\rho \in X_c \cap X_c^2$, and by Proposition 7.6, we obtain

$$[f * G_\rho]^{\wedge}_{M_c^2}(c + it) = [f]^{\wedge}_{M_c^2}(c + it)\, [G_\rho]^{\wedge}_{M_c^2}(c + it) \qquad (t \in \mathbb{R}).$$

For the right-hand side the Mellin inversion formula applies and yields

$$\int_0^\infty f(y) G_\rho\left(\frac{x}{y}\right) \frac{dy}{y} = \frac{x^{-c}}{2\pi} \int_{-\infty}^{\infty} [f]^{\wedge}_{M_c^2}(c+it)[G_\rho]^{\wedge}_{M_c^2}(c+it) x^{-it} dt \qquad (12.56)$$

for all $x \in \mathbb{R}^+$. We want to show that this formula also holds when G_ρ is replaced by G. Indeed, by Theorem 4.3,

$$\left| \int_0^\infty f(y) G_\rho\left(\frac{x}{y}\right) \frac{dy}{y} - \int_0^\infty f(y) G\left(\frac{x}{y}\right) \frac{dy}{y} \right| \le \|f\|_{X_c^2} \|G_\rho - G\|_{X_c^2} \longrightarrow 0$$

as $\rho \to \infty$ and, by Hölder's inequality,

$$\begin{aligned}
&\left| \int_{-\infty}^{\infty} [f]^{\wedge}_{M_c^2}(c + it)[G_\rho]^{\wedge}_{M_c^2}(c + it) x^{-it} dt \right. \\
&\qquad \left. - \int_{-\infty}^{\infty} [f]^{\wedge}_{M_c^2}(c + it)[G]^{\wedge}_{M_c^2}(c + it) x^{-it} dt \right| \\
&\qquad \le \int_{-\infty}^{\infty} \left| [f]^{\wedge}_{M_c^2}(c + it) \left([G_\rho]^{\wedge}_{M_c^2}(c + it) - [G]^{\wedge}_{M_c^2}(c + it) \right) \right| dt \\
&\qquad \le \left\| [f]^{\wedge}_{M_c^2} \right\|_{L^2(c+i\mathbb{R})} \left\| [G_\rho]^{\wedge}_{M_c^2} - [G]^{\wedge}_{M_c^2} \right\|_{L^2(c+i\mathbb{R})} \\
&\qquad = 2\pi \|f\|_{X_c^2} \|G_\rho - G\|_{X_c^2} \longrightarrow 0 \qquad (\rho \to \infty).
\end{aligned}$$

Thus, taking the limit $\rho \to \infty$ on both sides of (12.56), we obtain

$$\int_0^\infty f(y) G\left(\frac{x}{y}\right) \frac{dy}{y} = \frac{x^{-c}}{2\pi} \int_{-\infty}^{\infty} [f]^\wedge_{M_c^2}(c+it)[G]^\wedge_{M_c^2}(c+it) x^{-it} dt$$
$$= \frac{x^{-c}}{2\pi T} \int_{-\pi T}^{\pi T} [f]^\wedge_{M_c^2}(c+it) x^{-it} dt.$$

On the other hand, by Mellin inversion, f may be represented as

$$f(x) = \frac{x^{-c}}{2\pi} \int_{-\infty}^{\infty} [f]^\wedge_{M_c^2}(c+it) x^{-it} dt \qquad (x \in \mathbb{R}^+).$$

Combining these equations, we arrive at

$$f(x) - T \int_0^\infty f(y) G\left(\frac{x}{y}\right) \frac{dy}{y} = \frac{x^{-c}}{2\pi} \int_{|t| \geq \pi T} [f]^\wedge_{M_c^2}(c+it) x^{-it} dt \qquad (x \in \mathbb{R}^+),$$

as was to be shown. Finally, the error estimate is obvious. □

As an immediate consequence of Theorem 12.31, we obtain the following result on pointwise convergence.

Corollary 12.32 *If $f \in \mathcal{M}_c^2$, then*

$$\lim_{T \to \infty} x^{-c} |(\widetilde{R}_{\pi T} f)(x)| = 0 \qquad (x \in \mathbb{R}^+).$$

Furthermore, if $f \in B^1_{c,\pi T_0}$ for some $T_0 > 0$, then

$$|(\widetilde{R}_{\pi T} f)(x)| = 0 \qquad (x \in \mathbb{R}^+)$$

for all $T > T_0$.

Another interesting consequence of Theorem 12.31 is the following observation, which we state in a geometric setting.

Corollary 12.33 *Let $c \in \mathbb{R}$ and $T \in \mathbb{R}^+$. Then the mapping*

$$\mathcal{P}_{c,\pi T} : f \longmapsto T \int_0^\infty f(y) \mathrm{lin}_{c/T}\left(\left(\frac{\cdot}{y}\right)^T\right) \frac{dy}{y}$$

is an orthogonal projection of $\mathcal{M}_c^2$ onto its subspace $B^2_{c,\pi T}$, that is, $\mathcal{P}_{c,\pi T} f \in B^2_{c,\pi T}$ and $\langle f - \mathcal{P}_{c,\pi T} f, \mathcal{P}_{c,\pi T} f \rangle_{X_c^2} = 0$ for all $f \in \mathcal{M}_c^2$.

Proof We may split $f \in \mathcal{M}_c^2$ as $f = f_1 + f_2$, where

$$f_1(x) := \frac{x^{-c}}{2\pi} \int_{-\pi T}^{\pi T} [f]^{\wedge}_{M_c^2}(c+it)x^{-it}dt$$

and

$$f_2(x) := \frac{x^{-c}}{2\pi} \int_{|t|\geq \pi T} [f]^{\wedge}_{M_c^2}(c+it)x^{-it}dt.$$

Then the formula of Theorem 12.31 may be rewritten as

$$f(x) = (\mathcal{P}_{c,\pi T} f)(x) + f_2(x) \qquad (x \in \mathbb{R}^+)$$

and so $\mathcal{P}_{c,\pi T} f = f_1$, which is easily seen to belong to $B^2_{c,\pi T}$. Since $f - \mathcal{P}_{c,\pi T} f = f_2$, the orthogonal relation follows immediately from Theorem 7.10. □

Further interpretations of a geometric nature can be obtained by using the notion of distance introduced in Chap. 10. Obviously, the error estimate of Theorem 12.31 can be rewritten as

$$|(\widetilde{R}_{\pi T} f)(x)| \leq \frac{x^{-c}}{2\pi} \text{dist}_1(f, B^2_{c,\pi T}) \qquad (x > 0),$$

or equivalently,

$$\|\widetilde{R}_{\pi T} f\|^*_{X_c^\infty} \leq \frac{1}{2\pi} \text{dist}_1(f, B^2_{c,\pi T}). \tag{12.57}$$

This is a sharp inequality. Trivially, equality is attained for $f \in B^2_{c,\pi T}$. But there exist also non-trivial cases of equality. Indeed, consider $f(x) := x^{-c}\text{sinc}(2T \log x)$. Then $f \in \mathcal{M}_c^2 \setminus B^2_{c,\pi T}$. By a calculation, we find that $\text{dist}_1(f, B^2_{c,\pi T}) = \pi$ and $\|\widetilde{R}_{\pi T} f\|^*_{X_c^\infty} = 1/2$. Hence equality occurs in (12.57).

Using the estimates of the distance obtained in Chap. 10 for Mellin–Lipschitz and Mellin–Sobolev spaces, we can deduce from (12.57) the following counterpart of Theorem 12.26.

Theorem 12.34 *Let $c \in \mathbb{R}$, $T \in \mathbb{R}^+$ and $r \in \mathbb{N}$. For the remainder $\widetilde{R}_{\pi T} f$ of the approximate Mellin reproducing kernel formula, the following asymptotic estimates hold:*

(1) If $f \in \text{Lip}_r(\alpha, X_c \cap C(\mathbb{R}^+))$ with $r \geq 2$ and $1 < \alpha \leq r$, then

$$\|(\widetilde{R}_{\pi T} f)\|^*_{X_c^\infty} = \mathcal{O}(T^{-\alpha+1}) \qquad (T \to \infty).$$

(2) If $f \in \mathcal{M}_c^2 \cap \mathrm{Lip}_r(\beta, X_c^2)$ *with* $1/2 < \beta \le r$, *then*

$$\|(\widetilde{R}_{\pi T} f)\|_{X_c^\infty}^* = \mathcal{O}(T^{-\beta+1/2}) \qquad (T \to \infty).$$

(3) If $f \in \mathcal{M}_c^1 \cap W_c^{r,1}(\mathbb{R}^+)$ *with* $r > 1$, *then*

$$\|(\widetilde{R}_{\pi T} f)\|_{X_c^\infty}^* = \mathcal{O}(T^{-r+1}) \qquad (T \to \infty).$$

(4) If $f \in \mathcal{M}_c^2 \cap W_c^{r,2}(\mathbb{R}^+)$, *then*

$$\|(\widetilde{R}_{\pi T} f)\|_{X_c^\infty}^* = \mathcal{O}(T^{-r+1/2}) \qquad (T \to \infty).$$

12.7 Interconnections Between the Classical and the Approximate Exponential Sampling Formulae

In this section we will show that the exponential sampling theorem (Theorem 12.9) and its approximate version (Theorem 12.23) are equivalent in the sense that each can be obtained as a consequence of the other. We shall demonstrate this for both, convergence in the uniform norm and in the X_c^p norm. For the reader's convenience, we first recall the two results under consideration.

Theorem A (Exponential Sampling Formula) *Let* $f \in B_{c,\pi T}^2$, *where* $c \in \mathbb{R}$ *and* $T \in \mathbb{R}^+$. *Then*

$$f(x) = (S_T f)(x) := \sum_{k\in\mathbb{Z}} f(e^{k/T}) \mathrm{lin}_{c/T}(e^{-k} x^T) \qquad (x \in \mathbb{R}^+).$$

Theorem B (Approximate Exponential Sampling Formula) *Let* $f \in \mathcal{M}_c^2$ *with* $c \in \mathbb{R}$. *Then*

$$\left| f(x) - \sum_{k\in\mathbb{Z}} f(e^{k/T}) \mathrm{lin}_{c/T}(e^{-k} x^T) \right| \le \frac{x^{-c}}{\pi} \int_{|t|\ge\pi T} \left|[f]^{\wedge}_{M_c^2}(c+it)\right| dt$$

for all $x, T \in \mathbb{R}^+$. *In particular,*

$$f(x) = \lim_{T\to\infty} \sum_{k\in\mathbb{Z}} f(e^{k/T}) \mathrm{lin}_{c/T}(e^{-k} x^T).$$

Remark 12.35 Let $1 < p \le 2$. Then $B_{c,\pi T}^p \subset B_{c,\pi T}^2$ and $\mathcal{M}_c^p \subset \mathcal{M}_c^2$ (see Sect. 10.2.1). Therefore Theorems A and B also hold with $B_{c,\pi T}^2$ and $\mathcal{M}_c^2$ replaced by $B_{c,\pi T}^p$ and $\mathcal{M}_c^p$, respectively.

Obviously Theorem A is an immediate consequence of Theorem B since $B^2_{c,\pi T}$ is the subspace of $\mathcal{M}^2_c$ which comprises all functions that are Mellin bandlimited to the interval $[-\pi T, \pi T]$. The reverse implication is not so trivial.

12.7.1 Studies in Uniform Norm

In Sect. 12.4 we gave a direct proof of Theorem B. Now we want to show that Theorem A can serve as an essential tool for its own extension to Theorem B.

Theorem B Obtained via Theorem A

If $f \in \mathcal{M}^2_c$, then $[f]^\wedge_{M^2_c}$ exists and belongs to $L^1(c+i\mathbb{R}) \cap L^2(c+i\mathbb{R})$. Therefore we can apply the inversion formula of Theorem 7.17 and split f for any $T > 0$ as

$$f(x) = f_1(x) + f_2(x),$$

where

$$f_1(x) := \frac{x^{-c}}{2\pi} \int_{-\pi T}^{\pi T} [f]^\wedge_{M^2_c}(c+it)x^{-it}dt$$

and

$$f_2(x) := \frac{x^{-c}}{2\pi} \int_{|t|\geq \pi T} [f]^\wedge_{M^2_c}(c+it)x^{-it}dt.$$

Since the Mellin transform is a bijective operator from $\mathcal{M}^2_c$ onto $L^2(c+i\mathbb{R})$, it follows that $f_1 \in B^2_{c,\pi T}$ and $f_2 \in X^2_c$. Moreover, by Theorem A we have

$$f_1(x) = \sum_{k\in\mathbb{Z}} f_1(e^{k/T}) \operatorname{lin}_{c/T}(e^{-k}x^T). \tag{12.58}$$

As to the exponential sampling series $(S_T f_2)(x)$, we show that it is convergent for all $x \in \mathbb{R}^+$ and deduce an integral representation. To this end, fix $x \in \mathbb{R}^+$ and define $g_x(t)$ as the $2\pi T$-periodic extension of the function $t \mapsto x^{-it} = e^{-it\log x}$ from $]-\pi T, \pi T]$ to the whole real axis, that is,

$$g_x(t) = e^{i(t-2\pi nT)\log x} \qquad \big(t \in](2n-1)\pi T, (2n+1)\pi T];\ n \in \mathbb{Z}\big).$$

Since the Fourier coefficients of the function g_x are given by

$$(g_x)_k = \frac{1}{2\pi T}\int_{-\pi T}^{\pi T} e^{-it\log x} e^{-ikt/T} dt = \mathrm{sinc}(T\log x + k) \qquad (k \in \mathbb{Z})$$

and the function g_x is of bounded variation on every bounded interval, we obtain

$$g_x(t) = \sum_{k\in\mathbb{Z}} \mathrm{sinc}(T\log x + k)e^{ikt/T} = \sum_{k\in\mathbb{Z}} \mathrm{sinc}(T\log x - k)e^{-ikt/T} \tag{12.59}$$

at each continuity point of g_x, i. e., for all $t \neq (2n+1)\pi T$, $n \in \mathbb{Z}$. Moreover, the symmetric partial sums are uniformly bounded (see Proposition 1.53).

Now it follows by the definition of f_2 that

$$\sum_{k=-N}^{N} f_2(e^{k/T})\mathrm{lin}_{c/T}(e^{-k}x^T)$$

$$= \sum_{k=-N}^{N} \left\{ \frac{e^{-kc/T}}{2\pi} \int_{|t|\geq \pi T} [f]^{\wedge}_{M_c^2}(c+it)\, e^{-ikt/T} dt \right\} e^{kc/T} x^{-c} \mathrm{sinc}\big(\log(e^{-k}x^T)\big)$$

$$= \frac{x^{-c}}{2\pi} \int_{|t|\geq \pi T} [f]^{\wedge}_{M_c^2}(c+it) \sum_{k=-N}^{N} e^{-ikt/T} \mathrm{sinc}(T\log x - k) dt.$$

Letting $N \to \infty$ and using (12.59), we conclude that

$$\lim_{N\to\infty} \sum_{k=-N}^{N} f_2(e^{k/T})\mathrm{lin}_{c/T}(e^{-k}x^T)$$

$$= \frac{x^{-c}}{2\pi} \int_{|t|\geq \pi T} [f]^{\wedge}_{M_c^2}(c+it) \sum_{k=-\infty}^{\infty} e^{-ikt/T} \mathrm{sinc}(T\log x - k) dt$$

$$= \frac{x^{-c}}{2\pi} \int_{|t|\geq \pi T} [f]^{\wedge}_{M_c^2}(c+it) g_x(t) dt,$$

the interchange of the limit and the integral being justified by the uniform boundedness of the partial sums of (12.59) and Lebesgue's dominated convergence theorem. This shows the convergence of the sampling series $(S_T f_2)(x)$ and gives the desired integral representation. The last equation may also expressed as

$$(S_T f_2)(x) = \frac{x^{-c}}{2\pi} \int_{|t|\geq \pi T} [f]^{\wedge}_{M^2_c}(c+it) g_x(t) dt$$

$$= \frac{x^{-c}}{2\pi} \sum_{n\in\mathbb{Z}\setminus\{0\}} \int_{(2n-1)\pi T}^{(2n+1)\pi T} [f]^{\wedge}_{M^2_c}(c+it) x^{-i(t-2n\pi T)} dt.$$

Thus, we can write

$$f(x) = f_1(x) + f_2(x) = (S_T f_1)(x) + f_2(x) = (S_T f)(x) + (R_T f)(x),$$

where

$$(R_T f)(x) = f(x) - (S_T f)(x) = f_2(x) - (S_T f_2)(x)$$

$$= \frac{x^{-c}}{2\pi} \sum_{n\in\mathbb{Z}} \left(1 - x^{2in\pi T}\right) \int_{(2n-1)\pi T}^{(2n+1\pi T} [f]^{\wedge}_{M^2_c}(c+it) x^{-it} dt.$$

From this representation, we obtain the error estimate of Theorem B. □

12.7.2 Studies in X^p_c Norm

In this subsection we will investigate the convergence of the exponential sampling series and its approximate version in the space X^p_c.

In the following, we will denote by ℓ^p_c, where $c \in \mathbb{R}$ and $p \in [1, \infty[$, the *Mellin–Lebesgue sequence space* comprising all sequences $x = (x_n)_{n\in\mathbb{Z}}$ such that

$$\|x\|_{\ell^p_c} := \left\{ \sum_{n\in\mathbb{Z}} |x_n|^p e^{ncp} \right\}^{1/p} < \infty.$$

Since the exponential sampling series is not a bounded linear operator on the whole of X^p_c, we have to introduce suitable subspaces which guarantee the boundedness. For further details see the subsequent Sect. 13.3.

Subspaces of X^p_c

A partition $\Sigma = \{x_j\}_{j\in\mathbb{Z}}$ of $\mathbb{R}^+$ is called *admissible*, if

$$0 < \underline{\Delta}_\Sigma := \inf_{j\in\mathbb{Z}} \left(\log x_j - \log x_{j-1}\right) \leq \sup_{j\in\mathbb{Z}} \left(\log x_j - \log x_{j-1}\right) =: \overline{\Delta}_\Sigma < \infty. \tag{12.60}$$

It follows that $\lim_{j\to-\infty}\log x_j = -\infty$ and so $\lim_{j\to-\infty} x_j = 0$; similarly, $\lim_{j\to\infty} x_j = \infty$.

For $c \in \mathbb{R}$, $p \geq 1$ and any admissible partition $\Sigma = \{x_j\}_{j\in\mathbb{Z}}$ of $\mathbb{R}^+$, we denote by $\ell_c^p(\Sigma)$ the space of all functions $f : \mathbb{R}^+ \to \mathbb{C}$ such that

$$|f|_{\ell_c^p(\Sigma)} := \left\{ \sum_{k\in\mathbb{Z}} \left| f(x_k) x_k^c \right|^p (\log x_k - \log x_{k-1}) \right\}^{1/p} < \infty.$$

In view of (12.60), we see that for an admissible partition $\Sigma = \{x_j\}_{j\in\mathbb{Z}}$, we have $f \in \ell_c^p(\Sigma)$ if and only if $\sum_{k\in\mathbb{Z}} \left| f(x_k) x_k^c \right|^p < \infty$.

For $c \in \mathbb{R}$ and $p \geq 1$, we denote by Λ_c^p the space of all measurable and bounded functions $f : \mathbb{R}^+ \to \mathbb{C}$ such that $|f|_{\ell_c^p(\Sigma)} < \infty$ for every admissible partition Σ of $\mathbb{R}^+$.

For $c \in \mathbb{R}$, we associate with $f : \mathbb{R}^+ \to \mathbb{C}$ the function $f_c : \mathbb{R}^+ \to \mathbb{C}$ defined by $f_c(t) := t^c f(t)$ for $t \in \mathbb{R}^+$. Obviously, $f_c \in \Lambda_0^p$ if and only if $f \in \Lambda_c^p$.

For an admissible partition $\Sigma = \{x_j\}_{j\in\mathbb{Z}}$ of $\mathbb{R}^+$, we associate with a bounded function $g : \mathbb{R}^+ \to \mathbb{C}$ the function $g_\Sigma^* : \mathbb{R}^+ \to \mathbb{C}$ piecewise defined by

$$g_\Sigma^*(t) := \sup_{u\in[x_{j-1},x_j]} |g(u)| \qquad (t \in]x_{j-1}, x_j],\ j \in \mathbb{Z}).$$

Lemma 12.36 *Let $c \in \mathbb{R}$ and $p \geq 1$. The following statements are equivalent:*

(i) $f \in \Lambda_c^p$;
(ii) $(f_c)_\Sigma^* \in \Lambda_0^p$ *for each admissible partition Σ of $\mathbb{R}^+$;*
(iii) *for each admissible partition $\Sigma = \{x_j\}_{j\in\mathbb{Z}}$ of $\mathbb{R}^+$ and any choice of $\xi_j \in [x_{j-1}, x_j]$, $j \in \mathbb{Z}$, we have*

$$\sum_{j\in\mathbb{Z}} |f_c(\xi_j)|^p \left(\log x_j - \log x_{j-1}\right) < \infty.$$

Proof (i) $\Rightarrow$ (ii): Let $\Sigma = \{y_j\}_{j\in\mathbb{Z}}$ be an admissible partition of $\mathbb{R}^+$. We first show that $(f_c)_\Sigma^* \in \ell_0^p(\Sigma)$ by proceeding as follows. For each $k \in \mathbb{Z}$, there exists a point $t_k \in [y_{k-1}, y_k]$ such that

$$(f_c)_\Sigma^*(y_k) \leq |f_c(t_k)| + \frac{1}{1+k^2}\,.$$

By Minkowski's inequality, this implies that

$$\left\{ \sum_{k\in\mathbb{Z}} (f_c)_\Sigma^*(y_k)^p \right\}^{1/p} \leq \left\{ \sum_{k\in\mathbb{Z}} |f_c(t_k)|^p \right\}^{1/p} + \left\{ \sum_{k\in\mathbb{Z}} \left(\frac{1}{1+k^2} \right)^p \right\}^{1/p}. \tag{12.61}$$

Clearly, the second term on the right-hand side is bounded. By statement (i), the first term would also be bounded if $\{t_j\}_{j\in\mathbb{Z}}$ were an admissible partition of $\mathbb{R}^+$. Unfortunately, this cannot be guaranteed since $\inf_{j\in\mathbb{Z}}(\log t_j - \log t_{j-1})$ may be zero. However, it is easily verified that $\Sigma_0 := \{t_{2j}\}_{j\in\mathbb{Z}}$ and $\Sigma_1 := \{t_{2j+1}\}_{j\in\mathbb{Z}}$ are admissible partitions of $\mathbb{R}^+$ with $\underline{\Delta}_{\Sigma_m} \geq \underline{\Delta}_\Sigma$ and $\overline{\Delta}_{\Sigma_m} \leq 3\overline{\Delta}_\Sigma$ for $m \in \{0, 1\}$. Thus, as a consequence of (i),

$$\left\{\sum_{k\in\mathbb{Z}} |f_c(t_k)|^p\right\}^{1/p} \leq \left\{\sum_{k\in\mathbb{Z}} |f_c(t_{2k})|^p\right\}^{1/p} + \left\{\sum_{k\in\mathbb{Z}} |f_c(t_{2k+1})|^p\right\}^{1/p} < \infty.$$

Combined with (12.61), we find that $(f_c)^*_\Sigma \in \ell^p_0(\Sigma)$.

Now let $\Sigma' = \{x_j\}_{j\in\mathbb{Z}}$ be an arbitrary admissible partition of $\mathbb{R}^+$. Then

$$\begin{aligned}\sum_{j\in\mathbb{Z}} (f_c)^*_\Sigma (x_j)^p &= \sum_{k\in\mathbb{Z}} \sum_{x_j\in]y_{k-1},y_k]} (f_c)^*_\Sigma (x_j)^p \\ &= \sum_{k\in\mathbb{Z}} (f_c)^*_\Sigma (y_k)^p \sum_{x_j\in]y_{k-1},y_k]} 1 \\ &\leq \left(1 + \frac{\overline{\Delta}_\Sigma}{\underline{\Delta}_{\Sigma'}}\right) \sum_{k\in\mathbb{Z}} (f_c)^*_\Sigma (y_k)^p.\end{aligned}$$

Thus $(f_c)^*_\Sigma \in \ell^p_0(\Sigma)$ implies that $(f_c)^*_\Sigma \in \ell^p_0(\Sigma')$ for any admissible partition Σ' of $\mathbb{R}^+$, and so we have proved that $(f_c)^*_\Sigma \in \Lambda^p_0$.

(ii) $\Rightarrow$ (iii): Let $\Sigma = \{x_j\}_{j\in\mathbb{Z}}$ be an admissible partition of $\mathbb{R}^+$ and $\xi_j \in [x_{j-1}, x_j]$. Then

$$\sum_{j\in\mathbb{Z}} \left|f_c(\xi_j)\right|^p (\log x_j - \log x_{j-1}) \leq \sum_{j\in\mathbb{Z}} (f_c)^*_\Sigma (x_j)^p (\log x_j - \log x_{j-1}),$$

which shows that (iii) is a consequence of (ii).

(iii) $\Rightarrow$ (i): Under the validity of (iii), we may choose $\xi_j = x_j$, which gives (i) immediately. □

Next we establish a relation between the spaces Λ^p_c and X^p_c. Furthermore we describe how $\|f\|_{X^p_c}$ can be approximated arbitrarily close by $|f|_{\ell^p_c(\Sigma)}$ for appropriate admissible partitions Σ. For this, we denote by R_{comp} the set of all functions $g : \mathbb{R}^+ \to \mathbb{C}$ which are Riemann integrable on compact intervals of $\mathbb{R}^+$.

Lemma 12.37 *Let $c \in \mathbb{R}$ and $p \geq 1$. The following statements hold:*

(a) *Λ^p_c is a proper linear subspace of X^p_c.*

(b) *Let* $f \in \Lambda_c^p \cap R_{\text{comp}}$ *and for each* $\nu > 0$ *let* $\Sigma_\nu := \{x_j^{(\nu)}\}_{j\in\mathbb{Z}}$ *be an admissible partition of* $\mathbb{R}^+$ *such that* $\overline{\Delta}_{\Sigma_\nu}$ *tends to zero as* $\nu \to \infty$. *Suppose that* $\xi_j^{(\nu)}$ *is an arbitrary point in* $[x_{j-1}^{(\nu)}, x_j^{(\nu)}]$ *for each* $j \in \mathbb{Z}$. *Then*

$$\lim_{\nu\to\infty} |f|_{\ell_c^p(\Sigma_\nu)} = \lim_{\nu\to\infty} \left\{ \sum_{j\in\mathbb{Z}} \left| f_c(\xi_j^{(\nu)}) \right|^p \Delta_j^{(\nu)} \right\}^{1/p} = \|f\|_{X_c^p}, \tag{12.62}$$

where $\Delta_j^{(\nu)} = \log x_j^{(\nu)} - \log x_{j-1}^{(\nu)}$.

Proof (a) Obviously, Λ_c^p is a linear manifold and $\Sigma_0 := \{e^j\}_{j\in\mathbb{Z}}$ is an admissible partition of $\mathbb{R}^+$. Therefore, if $f \in \Lambda_c^p$, then by Lemma 12.36,

$$\int_0^\infty |f(x)x^c|^p \frac{dx}{x} = \sum_{j\in\mathbb{Z}} \int_{e^{j-1}}^{e^j} |f_c(x)|^p \frac{dx}{x} \le \sum_{j\in\mathbb{Z}} (f_c)^*_{\Sigma_0} (e^j)^p < \infty.$$

This shows that $f \in \Lambda_c^p$ implies $f \in X_c^p$.

Since the function $g : \mathbb{R}^+ \to \mathbb{C}$, defined by

$$g(x) = \begin{cases} e^{-cj} & \text{if } x = e^j,\ j \in \mathbb{Z}, \\ 0 & \text{otherwise,} \end{cases}$$

belongs to X_c^p but not to $\ell_c^p(\Sigma_0)$, we see that Λ_c^p is a proper subspace of X_c^p.

(b) Let $\varepsilon > 0$ be given. Since $f \in X_c^p$, there exists an $N_0 \in \mathbb{N}$ such that

$$\int_{\mathbb{R}^+\setminus[e^{-N},e^N]} |f_c(x)|^p \frac{dx}{x} < \frac{\varepsilon}{3} \tag{12.63}$$

for all integers $N \ge N_0$. We note that

$$\int_{[e^{-N},e^N]} |f_c(x)|^p \frac{dx}{x} = \int_{e^{-N}}^{e^N} |f_c(x)|^p \, d\log x. \tag{12.64}$$

Since $f \in R_{\text{comp}}$, the integral on the left-hand side exists as a Riemann integral. This allows us to conclude that the integral on the right-hand side exists as a Riemann–Stieltjes integral. We want to show that this integral can approximated arbitrarily close by a truncation of the series $\sum_{j\in\mathbb{Z}} \left| f_c\left(\xi_j^{(\nu)}\right)\right|^p \Delta_j^{(\nu)}$ that comprises those terms for which $\xi_j^{(\nu)} \in\,]e^{-N}, e^N]$.

First we consider the complementary part of the truncation called the *remainder*. By the assumption on the partitions Σ_ν, there exists a $\nu_0 > 0$ such that $\overline{\Delta}_{\Sigma_\nu} \le 1$ for all $\nu \ge \nu_0$. With Σ_0 as in part (a) of the proof, we can estimate the remainder for $\nu \ge \nu_0$ as follows:

$$\begin{aligned}
\sum_{\xi_j^{(\nu)} \notin]e^{-N}, e^N]} \left| f_c(\xi_j^{(\nu)}) \right|^p \Delta_j^{(\nu)} &= \sum_{k \in \mathbb{Z} \setminus]-N, N]} \sum_{\xi_j^{(\nu)} \in]e^{k-1}, e^k]} \left| f_c(\xi_j^{(\nu)}) \right|^p \Delta_j^{(\nu)} \\
&\leq \sum_{k \in \mathbb{Z} \setminus]-N, N]} \left| (f_c)^*_{\Sigma_0} (e^k) \right|^p \sum_{\xi_j^{(\nu)} \in]e^{k-1}, e^k]} \Delta_j^{(\nu)} \\
&\leq (1 + 2\overline{\Delta}_{\Sigma_\nu}) \sum_{k \in \mathbb{Z} \setminus]-N, N]} \left| (f_c)^*_{\Sigma_0} (e^k) \right|^p \\
&\leq 3 \sum_{k \in \mathbb{Z} \setminus]-N, N]} \left| (f_c)^*_{\Sigma_0} (e^k) \right|^p .
\end{aligned}$$

By Lemma 12.36, the series in the last line converges. Hence there exists an integer $N_1 \geq N_0$ such that

$$\sum_{\xi_j^{(\nu)} \notin]e^{-N}, e^N]} \left| f_c(\xi_j^{(\nu)}) \right|^p \Delta_j^{(\nu)} < \frac{\varepsilon}{3} \tag{12.65}$$

for $\nu \geq \nu_0$ and all integers $N \geq N_1$.

Now, for a fixed integer $N \geq N_1$, we denote by m_ν the smallest integer j such that $\xi_j^{(\nu)} > e^{-N}$ and by n_ν the largest integer j such that $\xi_j^{(\nu)} \leq e^N$. Then we can write

$$\sum_{\xi_j^{(\nu)} \in]e^{-N}, e^N]} \left| f_c(\xi_j^{(\nu)}) \right|^p \Delta_j^{(\nu)} = S_\nu + T_\nu,$$

where

$$\begin{aligned}
S_\nu := \left| f_c(\xi_{m_\nu}^{(\nu)}) \right|^p \left(\log x_{m_\nu}^{(\nu)} + N \right) + \sum_{j=m_\nu+1}^{n_\nu-1} \left| f_c(\xi_j^{(\nu)}) \right|^p \Delta_j^{(\nu)} \\
+ \left| f_c(\xi_{n_\nu}^{(\nu)}) \right|^p \left(N - \log x_{n_\nu-1}^{(\nu)} \right)
\end{aligned}$$

and

$$T_\nu := - \left| f_c(\xi_{m_\nu}^{(\nu)}) \right|^p \left(N + \log x_{m_\nu-1}^{(\nu)} \right) + \left| f_c(\xi_{n_\nu}^{(\nu)}) \right|^p \left(\log x_{n_\nu}^{(\nu)} - N \right).$$

It is easily verified that

$$0 \leq \log x_{m_\nu}^{(\nu)} + N < 2\overline{\Delta}_{\Sigma_\nu} \quad \text{and} \quad 0 \leq N - \log x_{n_\nu-1}^{(\nu)} \leq 2\overline{\Delta}_{\Sigma_\nu}.$$

Since $\overline{\Delta}_{\Sigma_\nu} \to 0$ as $\nu \to \infty$, we see that S_ν is a Riemann-Stieltjes sum such that

$$\lim_{\nu\to\infty} S_\nu = \int_{e^{-N}}^{e^N} |f_c(x)|^p \, d\,\log x.$$

As regards T_ν, we note that

$$\left|N + \log x_{m_\nu-1}^{(\nu)}\right| \le \overline{\Delta}_{\Sigma_\nu} \quad \text{and} \quad \left|\log x_{n_\nu}^{(\nu)} - N\right| \le \overline{\Delta}_{\Sigma_\nu},$$

which allows us to conclude that $\lim_{\nu\to\infty} T_\nu = 0$. Thus there exists a $\nu_1 \ge \nu_0$ such that

$$\left| \int_{e^{-N}}^{e^N} |f_c(x)|^p \, d\,\log x - \sum_{\xi_j^{(\nu)} \in]e^{-N}, e^N]} \left| f_c(\xi_j^{(\nu)}) \right|^p \Delta_j^{(\nu)} \right| < \frac{\varepsilon}{3} \tag{12.66}$$

for all $\nu \ge \nu_1$.

Combining (12.63)–(12.66), we find that

$$\begin{aligned}
&\left| \int_0^\infty |f_c(x)|^p \frac{dx}{x} - \sum_{j\in\mathbb{Z}} \left| f_c(\xi_j^{(\nu)}) \right|^p \Delta_j^{(\nu)} \right| \\
&\quad \le \int_{\mathbb{R}^+ \setminus [e^{-N}, e^N]} |f_c(x)|^p \frac{dx}{x} + \sum_{\xi_j^{(\nu)} \notin]e^{-N}, e^N]} \left| f_c(\xi_j^{(\nu)}) \right|^p \Delta_j^{(\nu)} \\
&\quad + \left| \int_{e^{-N}}^{e^N} |f_c(x)|^p \, d\,\log x - \sum_{\xi_j^{(\nu)} \in]e^{-N}, e^N]} \left| f_c(\xi_j^{(\nu)}) \right|^p \Delta_j^{(\nu)} \right| \\
&\quad < \varepsilon
\end{aligned}$$

for $\nu \ge \nu_1$. This proves the second equality in (12.62). The special choice $\xi_j^{(\nu)} = x_j^{(\nu)}$ for $j \in \mathbb{Z}$ yields the first equality immediately. □

In the previous lemma, we identified Λ_c^p as a proper subspace of X_c^p. Now we show that Λ_c^p in turn contains two familiar subspaces. Thus it is an intermediate member in a hierarchy of spaces of functions defined on $\mathbb{R}^+$.

Proposition 12.38 *For $c \in \mathbb{R}$, $\sigma \in \mathbb{R}^+$, $p \in]1, 2]$ and $r \in \mathbb{N}$, we have*

$$B_{c,\sigma}^p \subset W_c^{r,p} \subset \Lambda_c^p, \tag{12.67}$$

where the inclusions are strict.

Proof The first inclusion follows from the definition of the Mellin–Sobolev spaces and the results of Chap. 10. As to the second inclusion, it suffices to prove it for $r = 1$. Let $f \in W_c^{1,p}$ and let $\Sigma = \{x_j\}_{j\in\mathbb{Z}}$ be any admissible partition of $\mathbb{R}^+$. Putting $\Delta_j := \log x_j - \log x_{j-1}$, we have to show that

$$|f|_{\ell_c^p(\Sigma)} = \left\{\sum_{j\in\mathbb{Z}} |f_c(x_j)|^p \Delta_j\right\}^{1/p} < \infty.$$

Since f is continuous and $1/x$ is positive on $\mathbb{R}^+$, we see with the help of the mean-value theorem for integrals that for each $j \in \mathbb{Z}$, there exists a $\xi_j \in [x_{j-1}, x_j]$ giving

$$\int_{x_{j-1}}^{x_j} |f_c(x)|^p \frac{dx}{x} = |f_c(\xi_j)|^p \Delta_j$$

and so

$$\|f\|_{X_c^p} = \|f_c\|_{X_0^p} = \left\{\sum_{j\in\mathbb{Z}} |f_c(\xi_j)|^p \Delta_j\right\}^{1/p}.$$

Thus, by Minkowski's inequality,

$$|f|_{\ell_c^p(\Sigma)} \leq \|f\|_{X_c^p} + \left\{\sum_{j\in\mathbb{Z}} |f_c(x_j) - f_c(\xi_j)|^p \Delta_j\right\}^{1/p}. \tag{12.68}$$

Next we estimate the second term on the right-hand side. Since $f \in W_c^{1,p}$, we conclude that the Mellin derivative $\Theta_c f$ exists almost everywhere and

$$f_c(x_j) - f_c(\xi_j) = \int_{\xi_j}^{x_j} x^c(\Theta_c f)(x)\frac{dx}{x}.$$

Now by Hölder's inequality, we find that

$$\begin{aligned}
|f_c(x_j) - f_c(\xi_j)| &\leq \int_{x_{j-1}}^{x_j} |x^c(\Theta_c f)(x)|\frac{dx}{x} \\
&\leq \left\{\int_{x_{j-1}}^{x_j} |x^c(\Theta_c f)(x)|^p \frac{dx}{x}\right\}^{1/p} \left\{\int_{x_{j-1}}^{x_j} \frac{dx}{x}\right\}^{1/p'} \\
&= \left\{\int_{x_{j-1}}^{x_j} |x^c(\Theta_c f)(x)|^p \frac{dx}{x}\right\}^{1/p} \Delta_j^{1/p'}
\end{aligned}$$

with p' being the conjugate exponent of p. This shows that

$$|f_c(x_j) - f_c(\xi_j)|^p \Delta_j \le \int_{x_{j-1}}^{x_j} |x^c(\Theta_c f)(x)|^p \frac{dx}{x} \,\overline{\Delta}_\Sigma^{1+p/p'}.$$

Combining this estimate with (12.68) and noting that $\Theta_c f \in X_c^p$, we obtain

$$|f|_{\ell_c^p(\Sigma)} \le \|f\|_{X_c^p} + \overline{\Delta}_\Sigma \, \|\Theta_c f\|_{X_c^p} < \infty.$$

This completes the proof of $W_c^{1,p} \subset \Lambda_c^p$.

Concerning the proof that the inclusions (12.67) are strict, one may take $f(x) = 1$ for $x \in I$, and $f(x) = 0$ elsewhere, I being a compact subinterval of $\mathbb{R}^+$. Then f obviously belongs to Λ_c^p but not to $W_c^{1,p}$, which shows that the second inclusion is strict. As to the first inclusion, we simply observe that a function belonging to $B_{c,\sigma}^p$ is infinitely differentiable, whereas functions in $W_c^{r,p}$ need only be differentiable r-times. □

Hilbert Transforms

In the following we shall need a notion of "mixed" Mellin–Hilbert transform which maps sequences to functions. First we consider two types of discrete Hilbert transforms. We begin with the following definition.

Definition 12.39 The *discrete Hilbert transform* of a sequence $z = (z_k)_{k\in\mathbb{Z}}$ is a sequence $Hz = ((Hz)_j)_{j\in\mathbb{Z}}$ with elements defined by

$$(Hz)_j := \sum_{k\in\mathbb{Z}\setminus\{j\}} \frac{z_k}{j-k} \qquad (j \in \mathbb{Z}).$$

It is well known that this transform is a bounded linear operator in the sequence space ℓ^p for $1 < p < \infty$, that is,

$$\|Hz\|_{\ell^p} \le C\|z\|_{\ell^p} \tag{12.69}$$

with a constant $C > 0$. For more details, see Sect. 12.9 Notes.

We also need the related transformation $\mathcal{H} : z \mapsto ((\mathcal{H}z)_j)_{j\in\mathbb{Z}}$ with elements defined by

$$(\mathcal{H}z)_j := \sum_{k\in\mathbb{Z}\setminus\{j\}} \frac{z_k}{(j-k)^2} \qquad (j \in \mathbb{Z}).$$

It is a bounded linear operator in ℓ^p for $1 \le p \le \infty$. Indeed, by a shift of the index k we may write

$$(\mathcal{H}z)_j = \sum_{n\in\mathbb{Z}\setminus\{0\}} \frac{z_{j+n}}{n^2} \qquad (j \in \mathbb{Z}).$$

Thus $\mathcal{H}z = \sum_{n\in\mathbb{Z}\setminus\{0\}} n^{-2} z^{(n)}$, where $z^{(n)} := (z_{j+n})_{j\in\mathbb{Z}}$ is a shifted form of the sequence z. Clearly $\|z^{(n)}\|_{\ell^p} = \|z\|_{\ell^p}$ for all $n \in \mathbb{Z}$. Hence by Minkowski's inequality for infinite summations (see Proposition 1.10), we have

$$\|\mathcal{H}z\|_{\ell^p} \leq \sum_{n\in\mathbb{Z}\setminus\{0\}} \frac{1}{n^2} \|z^{(n)}\|_{\ell^p} = 2\sum_{n=1}^{\infty} \frac{1}{n^2} \|z\|_{\ell^p},$$

and so

$$\|\mathcal{H}z\|_{\ell^p} \leq \frac{\pi^3}{3} \|z\|_{\ell^p}. \tag{12.70}$$

Now we turn to the main objective of this subsection.

Definition 12.40 For $c \in \mathbb{R}$ and a given sequence $y = (y_k)_{k\in\mathbb{Z}}$, we define

$$(H_M y)(x) := x^{-c} \sum_{k\neq k(x)} \frac{e^{kc} y_k}{\log(xe^{-k})} = x^{-c} \sum_{k\neq k(x)} \frac{e^{kc} y_k}{\log x - k} \qquad (x \in \mathbb{R}^+),$$

where $k(x)$ is the unique integer such that $x \in [e^{k(x)-1/2}, e^{k(x)+1/2}[$ or equivalently, $\log x \in [k(x)-1/2, k(x)+1/2[$. We call $H_M y$ the *mixed Mellin–Hilbert transform of the sequence* y.

The following lemma shows an important property of the mixed Mellin–Hilbert transform.

Lemma 12.41 *For $c \in \mathbb{R}$ and $1 < p < \infty$, the mixed Mellin–Hilbert transform is a bounded linear operator from ℓ^p_c to X^p_c.*

Proof Let $y = (y_k)_{k\in\mathbb{Z}}$ be a sequence in ℓ^p_c. We can write

$$\frac{1}{\log(xe^{-k})} = \frac{1}{k(x)-k} + \frac{1}{(k(x)-k)^2}\left[\frac{\log(e^{k(x)}x^{-1})(k(x)-k)}{\log(xe^{-k})}\right].$$

Now, taking the definition of $k(x)$ into account, we note that the term in square brackets is bounded for $k \in \mathbb{Z} \setminus \{k(x)\}$. Hence there exists a constant $C > 0$ such that

$$\left|\sum_{k\neq k(x)} x^{-c} \frac{e^{kc} y_k}{\log(xe^{-k})}\right| \leq \left|\sum_{k\neq k(x)} x^{-c} \frac{e^{kc} y_k}{k(x)-k}\right| + C\left|\sum_{k\neq k(x)} x^{-c} \frac{e^{kc} y_k}{(k(x)-k)^2}\right|$$

$$\leq x^{-c}\left|\sum_{k\neq k(x)} \frac{e^{kc} y_k}{k(x)-k}\right| + Cx^{-c} \sum_{k\neq k(x)} \frac{e^{kc} |y_k|}{(k(x)-k)^2}.$$

Taking the norm in X_c^p and noting that $k(x) = j$ when $x \in [e^{j-1/2}, e^{j+1/2}[$, we have

$$\left\{\int_0^\infty \left|\sum_{k\neq k(x)} \frac{e^{kc}\, y_k}{\log(xe^{-k})}\right|^p \frac{dx}{x}\right\}^{1/p}$$

$$\leq \left\{\int_0^\infty \left(\left|\sum_{k\neq k(x)} \frac{e^{kc}\, y_k}{k(x)-k}\right| + C\sum_{k\neq k(x)} \frac{e^{kc}|y_k|}{(k(x)-k)^2}\right)^p \frac{dx}{x}\right\}^{1/p}$$

$$= \left\{\sum_{j=-\infty}^{\infty} \int_{e^{j-1/2}}^{e^{j+1/2}} \left(\left|\sum_{k\neq k(x)} \frac{e^{kc}\, y_k}{k(x)-k}\right| + C\sum_{k\neq k(x)} \frac{e^{kc}|y_k|}{(k(x)-k)^2}\right)^p \frac{dx}{x}\right\}^{1/p}$$

$$= \left\{\sum_{j=-\infty}^{\infty} \left(\left|\sum_{k\neq j} \frac{e^{kc}\, y_k}{j-k}\right| + C\sum_{k\neq j} \frac{e^{kc}|y_k|}{(j-k)^2}\right)^p \right\}^{1/p}.$$

Now, by Minkowski's inequality,

$$\left\{\int_0^\infty \left|\sum_{k\neq k(x)} \frac{e^{kc}\, y_k}{\log(xe^{-k})}\right|^p \frac{dx}{x}\right\}^{1/p}$$

$$\leq \left\|\left(\sum_{k\neq j} \frac{e^{kc} y_k}{j-k}\right)_{j\in\mathbb{Z}}\right\|_{\ell^p} + C\left\|\left(\sum_{k\neq j} \frac{e^{kc}|y_k|}{(j-k)^2}\right)_{j\in\mathbb{Z}}\right\|_{\ell^p}.$$

Setting $z = (z_k)_{k\in\mathbb{Z}}$ with $z_k := e^{kc} y_k$, $k \in \mathbb{Z}$, we have $\|z\|_{\ell^p} = \|y\|_{\ell^p_c}$, and so $z \in \ell^p$. Using the boundedness of the discrete Hilbert transforms (12.69) and (12.70), we finally obtain

$$\|H_M y\|_{X_c^p} \leq C_1 \|y\|_{\ell_c^p}$$

with a constant $C_1 > 0$, as was to be shown. □

Using Lemma 12.41, we can prove the boundedness of certain exponential sampling operators acting on Mellin–Lebesgue sequence spaces.

For any sequence $y = (y_k)_{k\in\mathbb{Z}}$ and $c \in \mathbb{R}$, $T \in \mathbb{R}^+$, we formally define the exponential series

$$(S_T y)(x) := \sum_{k\in\mathbb{Z}} y_k \operatorname{lin}_{c/T}(e^{-k} x^T) \qquad (x \in \mathbb{R}^+).$$

Subsequently, we shall need the following result on the operator S_T.

Theorem 12.42 *Let $c \in \mathbb{R}$, $T \in \mathbb{R}^+$ and $1 < p < \infty$. Then S_T is a bounded linear operator from $\ell^p_{c/T}$ to X^p_c, the bound being of the form $C/T^{1/p}$, where C is a positive constant not depending on T.*

Proof We first prove the theorem for $T = 1$. Let $y = (y_k)_{k\in\mathbb{Z}}$ be a sequence of complex numbers belonging to the space ℓ^p_c. Then

$$(S_1 y)(x) := \sum_{k\in\mathbb{Z}} y_k \mathrm{lin}_c(e^{-k}x) = x^{-c} \sum_{k\in\mathbb{Z}} e^{kc} y_k \mathrm{sinc}(\log x - k).$$

Using the notation $k(x)$ as introduced in Definition 12.40, we find that

$$\begin{aligned} &|(S_1 y)(x)| \\ &= x^{-c} \left| \sum_{k\neq k(x)} (-1)^k y_k e^{kc} \frac{\sin(\pi \log x)}{\pi(\log x - k)} + e^{ck(x)} y_{k(x)} \frac{\sin(\pi(\log x - k(x)))}{\pi(\log x - k(x))} \right| \\ &\leq x^{-c} \left| \sum_{k\neq k(x)} (-1)^k e^{kc} \frac{y_k}{\pi(\log x - k)} \right| + x^{-c} e^{ck(x)} |y_{k(x)}| =: I_1 + I_2. \end{aligned}$$

By Lemma 12.41, the X^p_c norm of the first term I_1 may be estimated as

$$\|I_1\|_{X^p_c} \leq C_1 \|y\|_{\ell^p_c}$$

with a constant $C_1 > 0$. For the second term, we have

$$\|I_2\|^p_{X^p_c} = \sum_{j\in\mathbb{Z}} \int_{e^{j-1/2}}^{e^{j+1/2}} |y_{k(x)}|^p e^{ck(x)p} \frac{dx}{x} = \sum_{j\in\mathbb{Z}} \int_{e^{j-1/2}}^{e^{j+1/2}} |y_j|^p e^{cjp} \frac{dx}{x} = \|y\|^p_{\ell^p_c}.$$

Thus we obtain

$$\|S_1 y\|_{X^p_c} \leq C \|y\|_{\ell^p_c}, \tag{12.71}$$

where $C = C_1 + 1$. This is the assertion for $T = 1$.

When $T > 0$ but $T \neq 1$, a suitable substitution allows us to deduce from (12.71) that

$$\|S_T y\|_{X^p_c} \leq \frac{C}{T^{1/p}} \|y\|_{\ell^p_{c/T}}.$$

This completes the proof. □

As a consequence of Theorem 12.42, we obtain the following statement.

Corollary 12.43 *Let $c \in \mathbb{R}$, $T \in \mathbb{R}^+$ and $1 < p < \infty$. Suppose that $f \in \Lambda_c^p$. Then*

$$\|S_T f\|_{X_c^p} \le C|f|_{\ell_c^p(\Sigma_T)}, \tag{12.72}$$

where Σ_T is the partition $\{e^{k/T}\}_{k\in\mathbb{Z}}$ and C is a positive constant independent of T.

Proof The result follows by applying Theorem 12.42 to the sequence $y = (y_k)_{k\in\mathbb{Z}} = \big(f(e^{k/T})\big)_{k\in\mathbb{Z}}$. □

Now we are ready for the main result of this subsection. In view of Remark 12.35, the proof of the following theorem shows that Theorem A can also serve as a tool for its extension to a version of Theorem B with convergence in the norm of X_c^p.

Theorem 12.44 *Let $c \in \mathbb{R}$ and $p \in]1, 2]$. Suppose that $f \in \Lambda_c^p \cap R_{\mathrm{comp}}$. Then*

$$\lim_{T\to\infty} \|S_T f - f\|_{X_c^p} = 0. \tag{12.73}$$

Proof For $c \in \mathbb{R}$ and $\rho \in \mathbb{R}^+$, let F_ρ^c be the Mellin–Fejér kernel, defined by

$$F_\rho^c(x) := -\frac{x^{-c}}{2\pi\rho}\left(\frac{x^{i\rho/2} - x^{-i\rho/2}}{\log x}\right)^2 \qquad (x \in \mathbb{R}^+ \setminus \{1\})$$

with the continuous extension $F_\rho^c(1) = \rho/(2\pi)$. The function F_ρ^c belongs to the Mellin–Paley–Wiener space $B_{c,\rho}^1$ (see Lemma 6.84). Furthermore, for $f \in X_c^p$, the Mellin convolution $f * F_\rho^c$ belongs to $B_{c,\rho}^p$ (see Theorem 7.15) and by Theorem 6.19 we have

$$\lim_{\rho\to\infty} \|f * F_\rho^c - f\|_{X_c^p} = 0.$$

Hence for a fixed $f \in \Lambda_c^p \cap R_{\mathrm{comp}}$ and $\varepsilon > 0$, there exists a $T_0 > 0$ and a function $g_T \in B_{c,\pi T}^p$, in fact $g_T := f * F_{\pi T}^c$, such that $\|f - g_T\|_{X_c^p} < \varepsilon$ for all $T \ge T_0$. Moreover, since $f - g_T \in \Lambda_c^p \cap R_{\mathrm{comp}}$ as well, there exists, by Lemma 12.37 (b), a $T_1 \ge T_0$ such that

$$|f - g_T|_{\ell_c^p(\Sigma_T)} < \|f - g_T\|_{X_c^p} + \varepsilon$$

for all $T \ge T_1$, where here Σ_T is the partition specified in Corollary 12.43. Since $g_T \in B_{c,\pi T}^p$, Proposition 12.38 implies that $g_T \in \Lambda_c^p$. Thus, by Corollary 12.43 and Theorem A, we have

$$\begin{aligned}\|S_T f - f\|_{X_c^p} &\le \|S_T(f - g_T)\|_{X_c^p} + \|S_T g_T - g_T\|_{X_c^p} + \|g_T - f\|_{X_c^p}\\ &\le C|f - g_T|_{\ell_c^p(\Sigma_T)} + \|g_T - f\|_{X_c^p} \le (2C + 1)\varepsilon\end{aligned}$$

for all $T \ge T_1$, and so the assertion follows. □

12.8 The Mellin–Parseval Formula

This section deals with another interesting formula, namely the Mellin–Parseval formula, which establishes a discrete representation of the scalar product in X_c^2 of two Mellin bandlimited functions $f, g \in X_c^2$. We shall prove it as a consequence of the exponential sampling formula. The proof requires a number of lemmas. We begin with the following basic lemma.

Lemma 12.45 *For $c \in \mathbb{R}$, let $f, g \in X_c^2$ and let $(f_n)_{n\in\mathbb{N}}, (g_n)_{n\in\mathbb{N}} \subset X_c^2$ be two sequences which converge in the norm of X_c^2 to f, g, respectively. Then*

$$\lim_{n\to+\infty} \int_0^\infty f_n(u)\overline{g_n(u)}u^{2c}\frac{du}{u} = \int_0^\infty f(u)\overline{g(u)}u^{2c}\frac{du}{u}.$$

Proof By Hölder's inequality (Proposition 1.7), we have

$$\begin{aligned}
&\left|\int_0^\infty f_n(u)\overline{g_n(u)}u^{2c}\frac{du}{u} - \int_0^\infty f(u)\overline{g(u)}u^{2c}\frac{du}{u}\right| \\
&\leq \int_0^\infty |f_n(u) - f(u)|\,|g_n(u)|u^{2c}\frac{du}{u} + \int_0^\infty |g_n(u) - g(u)|\,|f(u)|u^{2c}\frac{du}{u} \\
&\leq \|f_n - f\|_{X_c^2}\|g_n\|_{X_c^2} + \|g_n - g\|_{X_c^2}\|f\|_{X_c^2}.
\end{aligned}$$

From this the assertion follows immediately. □

A discrete version of Lemma 12.45 for the space of sequences ℓ_c^2 (see Sect. 12.7.2) is the following Lemma, the proof of which is similar.

Lemma 12.46 *For $c \in \mathbb{R}$, let $\alpha := (a_k)_{k\in\mathbb{Z}}$, $\beta := (b_k)_{k\in\mathbb{Z}}$ be two sequences in ℓ_c^2, and for each $n \in \mathbb{N}$, let $\alpha_n := (a_{n,k})_{k\in\mathbb{Z}}, \beta_n := (a_{n,k})_{k\in\mathbb{Z}} \in \ell_c^2$ with $\lim_{n\to+\infty} \|\alpha_n - \alpha\|_{\ell_c^2} = 0$ and $\lim_{n\to+\infty} \|\beta_n - \beta\|_{\ell_c^2} = 0$. Then*

$$\lim_{n\to+\infty} \sum_{k=-\infty}^{\infty} a_{n,k}\overline{b_{n,k}}e^{2ck} = \sum_{k=-\infty}^{\infty} a_k\overline{b_k}e^{2ck}.$$

The next lemma is an extension to the Mellin setting of a classical inequality due to Nikol'skii (see Proposition 1.39).

Lemma 12.47 *Let $c \in \mathbb{R}$ and $T \in \mathbb{R}^+$. Suppose that $f \in B^2_{c,\pi T}$. Then*

$$\left\{\frac{1}{T}\sum_{k=-\infty}^{\infty} |f(e^{k/T})|^2 e^{2ck/T}\right\}^{1/2} \leq (1+\pi)\|f\|_{X_c^2}.$$

Proof Since $f \in B^2_{c,\pi T}$, it is easily seen the function $F : u \mapsto f(e^u)e^{cu}$ belongs to the classical Bernstein space $B^2_{\pi T}$ (see Definition 1.34).Therefore Proposition 1.39 applies to F and yields that, for any $h > 0$,

$$\left\{ h \sum_{k=-\infty}^{\infty} |F(hk)|^2 \right\}^{1/2} \leq (1 + \pi T h)\|F\|_2.$$

Since $\|F\|_2 = \|f\|_{X^2_c}$, the assertion is obtained by putting $h = 1/T$. □

In the following, we shall denote by $S_T f$ the exponential sampling series as in Theorem A of Sect. 12.7.

Lemma 12.48 *For $c \in \mathbb{R}$, let $f \in X^2_c$ be a continuous function such that $u^c f(u) = \mathcal{O}((\log u)^{-2})$ as $u \to 0+$ and $u \to +\infty$. Then, for any $T > 0$, the exponential sampling series $S_T f$ converges in X^2_c. Moreover, if $f \in B^2_{c,\pi T}$, then the series $S_T f$ converges in the norm of X^2_c to the function f.*

Proof Let $T > 0$ and let $n, m \in \mathbb{N}$ with $n < m$. Then

$$\begin{aligned} I &:= \left\| \sum_{n \leq |k| \leq m} f(e^{k/T}) \mathrm{lin}_{c/T}(e^{-k}(\cdot)^T) \right\|_{X^2_c} \\ &\leq \sum_{n \leq |k| \leq m} |f(e^{k/T})| \, \|\mathrm{lin}_{c/T}(e^{-k}(\cdot)^T)\|_{X^2_c}. \end{aligned}$$

The norms on the right-hand side may be expressed as

$$\|\mathrm{lin}_{c/T}(e^{-k}(\cdot)^T)|\|^2_{X^2_c} = \int_0^\infty u^{2c} |\mathrm{lin}_{c/T}(e^{-k}u^T)|^2 \frac{du}{u} = \frac{e^{2kc/T}}{T} \|\mathrm{sinc}\|^2_2.$$

Hence, by the assumption on f, there exists an absolute constant $C > 0$ such that

$$I \leq \frac{1}{\sqrt{T}} \|\mathrm{sinc}\|_2 \sum_{n \leq |k| \leq m} |f(e^{k/T})| e^{kc/T} \leq C T^{3/2} \sum_{n \leq |k| \leq m} \frac{1}{k^2}.$$

This shows the convergence in X^2_c. The supplement for $f \in B^2_{c,\pi T}$ follows in conjunction with the exponential sampling formula of Theorem 12.9. □

Next, we present a basic density lemma for the spaces $B^2_{c,\pi T}$.

Lemma 12.49 *Let $c \in \mathbb{R}$ and $T \in \mathbb{R}^+$. Suppose that $f \in B^2_{c,\pi T}$. Then there exists a sequence of functions $f_n \in B^2_{c,\pi T}$ with $u^c f_n(u) = \mathcal{O}((\log u)^{-2})$ as $u \to 0+$ and $u \to +\infty$, such that $\lim_{n \to +\infty} \|f_n - f\|_{X^2_c} = 0$.*

Proof For $n \in \mathbb{N}$, let $a_n := 1 - 1/n$, $b_n := 1/(2n)$ and consider the function

$$f_n : u \longmapsto f(u^{a_n}) \operatorname{lin}^2_{c/T}(u^{b_n T}) \qquad (u \in \mathbb{R}^+).$$

Since $f \in B^2_{c,\pi T}$, we can show that $f((\cdot)^{a_n}) \in B^2_{ca_n,\pi a_n T}$ (see Proposition 3.44). Furthermore,

$$\operatorname{lin}^2_{c/T}\big((\cdot)^{b_n T}\big) = (\cdot)^{-2cb_n} \operatorname{sinc}^2\big(b_n T \log(\cdot)\big) \in B^2_{2cb_n, 2\pi b_n T}.$$

As $a_n + 2b_n = 1$, we conclude via the Mellin–Paley–Wiener theorem (Theorem 8.10) that $f_n \in B^2_{c,\pi T}$.

Recalling Definition 3.41 (iii), we find for $\theta = 0$ that

$$\begin{aligned} \left|u^c f_n(u)\right| &= \left|u^{ca_n} f(u^{a_n}) \operatorname{sinc}^2(b_n T \log u)\right| \\ &\le C_f \left|\operatorname{sinc}^2(b_n T \log u)\right| \le \frac{C_f}{(\pi b_n T \log u)^2} \qquad (u \in \mathbb{R}^+ \setminus \{1\}), \end{aligned} \tag{12.74}$$

and so f_n has the desired asymptotic behavior.

It remains to show convergence in the norm of X^2_c. For any $r > 1$ put $U_r := [1/r, r]$. It is easily seen that $(f_n)_{n\in\mathbb{N}}$ converges pointwise to f and uniformly on U_r. Using (12.74) and taking into account that $\operatorname{sinc}^2(b_n T \log u) \le 1$ for $u \in \mathbb{R}^+$, we deduce for the complement of U_r that

$$\begin{aligned} \int_{\mathbb{R}^+\setminus U_r} u^{2c} \left|f_n(u)\right|^2 \frac{du}{u} &\le \int_{\mathbb{R}^+\setminus U_r} u^{2ca_n} \left|f(u^{a_n})\right|^2 \frac{du}{u} \\ &= \frac{1}{a_n} \int_{\mathbb{R}^+\setminus U_{r^{a_n}}} w^{2c} \left|f(w)\right|^2 \frac{dw}{w}. \\ &\le \frac{1}{a_2} \int_{\mathbb{R}^+\setminus U_{r^{a_2}}} w^{2c} \left|f(w)\right|^2 \frac{dw}{w} =: I(r) \end{aligned}$$

for $n \ge 2$. Since $a_2 = 1/2$ and $U_{r^{1/2}} \subset U_r$, there also holds

$$\int_{\mathbb{R}^+\setminus U_r} u^{2c} \left|f(u)\right|^2 \frac{du}{u} \le I(r).$$

Now, given any $\varepsilon > 0$, there exist an $r_0 > 1$ such that $I(r) \le \varepsilon^2/4$ for all $r \ge r_0$. Combining these estimates, we conclude that

$$\begin{aligned}\|f_n - f\|_{X_c^2} &\le \left(\int_{U_r} u^{2c}|f_n(u) - f(u)|^2 \frac{du}{u}\right)^{1/2} \\ &\quad + \left(\int_{\mathbb{R}^+\setminus U_r} u^{2c}|f_n(u)|^2 \frac{du}{u}\right)^{1/2} + \left(\int_{\mathbb{R}^+\setminus U_r} u^{2c}|f(u)|^2 \frac{du}{u}\right)^{1/2} \\ &< \left(\int_{U_r} u^{2c}|f_n(u) - f(u)|^2 \frac{du}{u}\right)^{1/2} + \varepsilon\end{aligned}$$

for $n \ge 2$ and all $r \ge r_0$. In view of the uniform convergence on U_r, his implies that $\lim_{n\to+\infty} \|f_n - f\|_{X_c^2} = 0$. □

We are now ready to prove the Mellin–Parseval formula for bandlimited functions.

Theorem 12.50 *Let $c \in \mathbb{R}$ and $T \in \mathbb{R}^+$. Suppose that $f, g \in B^2_{c,\pi T}$. Then*

$$\langle f, g\rangle_{X_c^2} = \int_0^\infty f(u)\overline{g(u)}u^{2c}\frac{du}{u} = \frac{1}{T}\sum_{k=-\infty}^{\infty} f(e^{k/T})\overline{g(e^{k/T})}e^{2ck/T}. \qquad (12.75)$$

Proof Formally, the proof follows from the exponential sampling formula in X_c^2, by interchanging integration and summation. In fact, we have

$$f(x) = \sum_{k=-\infty}^{\infty} f(e^{k/T})\,\mathrm{lin}_{c/T}(e^{-k}x^T), \qquad g(x) = \sum_{j=-\infty}^{\infty} g(e^{j/T})\,\mathrm{lin}_{c/T}(e^{-j}x^T),$$

and so

$$\begin{aligned}&\int_0^\infty f(u)\overline{g(u)}u^{2c}\frac{du}{u} \\ &\quad = \int_0^\infty \sum_{k=-\infty}^{\infty} f(e^{k/T})\,\mathrm{lin}_{c/T}(e^{-k}u^T) \sum_{j=-\infty}^{\infty} \overline{g(e^{j/T})}\,\mathrm{lin}_{c/T}(e^{-j}u^T)u^{2c}\frac{du}{u} \\ &\quad = \sum_{j,k\in\mathbb{Z}} f(e^{k/T})\overline{g(e^{j/T})}\left\langle \mathrm{lin}_{c/T}(e^{-k}(\cdot)^T), \mathrm{lin}_{c/T}(e^{-j}(\cdot)^T)\right\rangle_{X_c^2}.\end{aligned}$$

Now the assertion follows in view of the orthogonality of the lin_c functions (see Corollary 12.30).

However, this approach is not acceptable unless we justify the interchange of integration and summation. For a rigorous proof we proceed in two steps.

Step 1 We assume in addition that f and g satisfy the hypotheses of Lemma 12.48. For $n \in \mathbb{N}$ we define

$$F_n(x) = \sum_{|k|\le n} f(e^{k/T})\,\mathrm{lin}_{c/T}(e^{-k}x^T), \qquad G_n(x) = \sum_{|j|\le n} g(e^{j/T})\,\mathrm{lin}_{c/T}(e^{-j}x^T).$$

Then $F_n, G_n \in B^2_{c,\pi T}$ and the previous procedure is admissible since we have finite sums. Thus we obtain

$$\int_0^\infty F_n(u)\overline{G_n(u)}u^{2c}\frac{du}{u} = \frac{1}{T}\sum_{|k|\le n} f(e^{k/T})\overline{g(e^{k/T})}e^{2ck/T}. \tag{12.76}$$

By Lemma 12.48, the function F_n, G_n converge in the norm of X^2_c to f, g, respectively, as $n \to \infty$. Hence Lemma 12.45 can be employed to show that the limit $n \to \infty$ in (12.76) leads to (12.75).

Step 2 Now we turn to the general case. By what we have proved in Step 1, Eq. (12.75) is valid for the functions f_n and g_n of Lemma 12.49, that is,

$$\int_0^\infty f_n(u)\overline{g_n(u)}u^{2c}\frac{du}{u} = \frac{1}{T}\sum_{k=-\infty}^{\infty} f_n(e^{k/T})\overline{g_n(e^{k/T})}e^{2ck/T} \qquad (n \in \mathbb{N}). \tag{12.77}$$

Again the hypotheses of Lemma 12.45 are satisfied and so the left-hand side of (12.77) converges to that of (12.75) as $n \to \infty$. As regards the right-hand side, we note that Lemma 12.47 applies to the function $f_n - f$ and yields

$$\left\{\frac{1}{T}\sum_{k=-\infty}^{\infty} |f_n(e^{k/T}) - f(e^{k/T})|^2 e^{2ck/T}\right\}^{1/2} \le (1+\pi)\|f_n - f\|_{X^2_c}.$$

An analogous inequality holds for $g_n - g$. Since the right-hand sides converge to 0 as $n \to \infty$, Lemma 12.46 allows us to conclude that the right-hand side of (12.77) converges to that of (12.75). This completes the proof.

□

Theorem 12.50 has several interesting consequences. For $f = g$ it provides the following improvement of Lemma 12.47.

Corollary 12.51 *Let $c \in \mathbb{R}$ and $T \in \mathbb{R}^+$. Suppose that $f \in B^2_{c,\pi T}$. Then*

$$\left\{\frac{1}{T}\sum_{k=-\infty}^{\infty} |f(e^{k/T})|^2 e^{2ck/T}\right\}^{1/2} = \|f\|_{X^2_c}.$$

Partial sums on the left-hand side can serve as a quadrature formula for an approximate computation of $\|f\|_{X_c^2}$.

Another interesting consequence is the convergence of a class of series which contains the exponential sampling series as a special case.

Corollary 12.52 *Let $c \in \mathbb{R}$ and $T \in \mathbb{R}^+$. Suppose that $\alpha = (a_k)_{k\in\mathbb{Z}}$ is a sequence in $\ell^2_{c/T}$. Then the series*

$$\sum_{k=-\infty}^{\infty} a_k x^c \operatorname{lin}_{c/T}(e^{-k}x^T) \qquad (x \in \mathbb{R}^+)$$

converges absolutely and uniformly on $\mathbb{R}^+$.

Proof Let $\varepsilon > 0$ be given. By the hypothesis on the sequence α, there exists an $n_0 \in \mathbb{N}$ such that for all integers $m > n \geq n_0$, we have

$$\left\{ \sum_{n\leq|k|\leq m} |a_k|^2 e^{2ck/T} \right\}^{1/2} < \varepsilon.$$

Therefore, by the Cauchy–Schwarz inequality,

$$\begin{aligned}
&\sum_{n\leq|k|\leq m} \left| a_k x^c \operatorname{lin}_{c/T}(e^{-k}x^T) \right| \\
&\quad = \sum_{n\leq|k|\leq m} \left| a_k e^{ck/T} \operatorname{lin}_0(e^{-k}x^T) \right| \\
&\quad \leq \left\{ \sum_{n\leq|k|\leq m} |a_k|^2 e^{2ck/T} \right\}^{1/2} \left\{ \sum_{n\leq|k|\leq m} \left| \operatorname{lin}_0(e^{-k}x^T) \right|^2 \right\}^{1/2} \\
&\quad < \varepsilon \left\{ \sum_{k=-\infty}^{\infty} \left| \operatorname{lin}_0(e^{-k}x^T) \right|^2 \right\}^{1/2}.
\end{aligned}$$

Now for arbitrary but fixed $x \in \mathbb{R}^+$, consider the function $g : u \mapsto \operatorname{lin}_0((x/u)^T)$. Then $g \in B^2_{0,\pi T}$ and $g(e^{k/T}) = \operatorname{lin}_0(e^{-k}x^T)$. Hence by Corollary 12.51 for $c = 0$,

$$\begin{aligned}
\left\{ \sum_{k=-\infty}^{\infty} \left| \operatorname{lin}_0(e^{-k}x^T) \right|^2 \right\}^{1/2} &= T^{1/2} \left\| \operatorname{lin}_0\left(\left(\frac{x}{\cdot} \right)^T \right) \right\|_{X_0^2} \\
&= T^{1/2} \left\{ \int_0^\infty \left| \operatorname{sinc}\left(T \log\left(\frac{x}{u} \right) \right) \right|^2 \frac{du}{u} \right\}^{1/2} \\
&= \|\operatorname{sinc}\|_2,
\end{aligned}$$

where the last equation is verified by substituting $w = T\log(x/u)$. Combining these relations, we obtain

$$\sum_{n\le|k|\le m} \left|a_k x^c \operatorname{lin}_{c/T}(e^{-k}x^T)\right| < \varepsilon\|\operatorname{sinc}\|_2.$$

Since the right-hand side does not depend on x, we have shown the absolute and uniform convergence on the whole of $\mathbb{R}^+$. □

Example 12.53 Let $f : \mathbb{R}^+ \to \mathbb{C}$ be a continuous function such that $x^c f(x) = \mathcal{O}(|\log x|^{-\gamma})$ as $x \to 0+$ and $x \to +\infty$, where $\gamma > 1/2$. Then it can be easily seen that $f \in X_c^2$ and the sequence $(f(e^{k/T}))_{k\in\mathbb{Z}}$ belongs to $\ell^2_{c/T}$. Hence by Corollary 12.52 the exponential sampling series

$$\sum_{k=-\infty}^{\infty} f(e^{k/T}) \operatorname{lin}_{c/T}(e^{-k}x^T) \qquad (x \in \mathbb{R}^+)$$

converges absolutely and uniformly on compact subsets of $\mathbb{R}^+$.

For Mellin bandlimited functions, Corollary 12.51 in conjunction with Corollary 12.52 gives the following supplement to Theorem 12.9.

Corollary 12.54 *Let $c \in \mathbb{R}$ and $T \in \mathbb{R}^+$. Suppose that $f \in B^2_{c,\pi T}$. Then the exponential sampling series*

$$\sum_{k=-\infty}^{\infty} f(e^{k/T}) \operatorname{lin}_{c/T}(e^{-k}x^T) \qquad (x \in \mathbb{R}^+)$$

converges absolutely and uniformly on compact subsets of $\mathbb{R}^+$. Multiplied by x^c, it converges uniformly on the whole of $\mathbb{R}^+$.

12.9 Notes

The classical sampling theorem, usually named after Whittaker, Kotel'nikov and Shannon, is a basic result of Fourier analysis due to its diverse applications in signal processing and its interconnections with some fundamental theorems of other branches of mathematics. To enveil its origins is a matter of interpretation. Already in 1827, Cauchy published a theorem on partial fraction decomposition of meromorphic functions, which contains the sampling theorem and some of its variants as special cases. However, Cauchy did not explicitly disclose his formula as a tool for the reconstruction of functions from samples, which may prevent us from seeing him as a father of sampling theory. The aspect of interpolation and representation of functions clearly shows in the paper of Whittaker [299] in 1915 and his further studies in [300, 301]. For historical discussions, see Butzer et al. [104, Sect. 2.10], Higgins [174] and Butzer et al. [112]. For a survey article on the sampling theorem and its extensions, see Butzer et al. [102]. An extensive study of sampling theory can be found in the books by Zayed [308] and Higgins [173]. For an abstract approach in the setting of harmonic analysis, see Beaty and Dodson [69]. A survey on sampling theory in the sense of Kramer for functions in $L^2(\mathbb{R})$ was given by Butzer and Nasri-Roudsari [94]. Bardaro et al. [47] used the sampling series for studying the approximation of discontinuous signals in the L^p norm.

Connections of the sampling theorem with fundamental theorems of other mathematical disciplines such as complex analysis, numerical analysis and number theory can be found in the papers by Butzer and Stens [96], Butzer et al. [103], Rahman and Schmeisser [257], Higgins et al. [176], Butzer [85], and Butzer et al. [113, 117, 119]. Altogether these papers show that the sampling theorem, the reproducing kernel formula,, Poisson's summation formula, the Euler–Maclaurin formula, the Abel–Plana summation formula, Valiron's interpolation formula, a differentiation formula of Boas, the general Parseval formula, the Paley–Wiener theorem expressed by a formula, Cauchy's integral formula and a functional equation of Riemann's zeta function are interconnected in the sense that each of these formulae can be derived from the others by using only basic standard tools as side results. Some of these formulae hold for bandlimited functions f only. When f is not bandlimited they hold in an approximate sense and can be made exact by introducing a remainder. It is remarkable that the interconnections extend to some approximate versions (see Butzer et al. [110, 119]). For this, one splits the function as $f = f_1 + f_2$, where f_1 is a function of prescribed bandwidth closest to f while f_2 generates a remainder. Closeness is measured by an appropriately defined metric (see Butzer et al. [116, 118]).

A collection of applications of sampling theory in various areas of science can be found in the book of Higgins and Stens [175].

During the eighties of the last century, a group of optical physicists and engineers introduced a mathematical model for the study of certain phenomena related to light scattering, diffraction, radio-astronomy and more; see Casasent [125], Bertero and Pike [70], Gori [159], and Ostrowsky et al. [233]. Their approach was based on the solution of integral equations of form

$$g(t) = \int_0^\infty K(ts)f(s)ds,$$

where g is the data function, K is a kernel depending on the particular problem under consideration and f is the unknown function. The most convenient mathematical tool for these problems was found in the theory of Mellin transform which provided a solution in terms of a sampling series with sample points that are exponentially spaced on the positive real line $\mathbb{R}^+$.

This motivated Butzer and Jansche to carry out a rigorous mathematical investigation of the exponential sampling series for Mellin bandlimited functions, along with its approximate versions for functions which are not necessarily Mellin bandlimited.

Section 12.1.1 is based on Butzer and Jansche [90]. The estimate of the truncation error in case of the Mellin–Fejér kernel is borrowed from Bardaro et al. [52, Sect. 6 (a)]. Section 12.1.2, where the sampling formula in X_c^2 is estabished by a Kramer type method, originates from Butzer and Jansche [92, Sect. 4, Subsect. 5.1]. For an approach in the terminology of electrical engineers, see Bertrand et al. [71, Chap. 12].

The results of Sects. 12.2 and 12.3 were obtained in Bardaro et al. [65]. Valiron's interpolation formula in its classical form was first described by Valiron [295] in 1925. That it can be transformed into a differentiation formula which improves upon the one of Boas (see Corollary 12.13 for the Mellin case and Remark 12.15 for the classical case) seems to have not been mentioned by other authors. The Mellin version of Valiron's formula was independently obtained by Pesenson [242, Theorem 6] in the setting of groups of operators in Banach spaces (see Butzer and Berens [86]). For this, the Mellin derivative Θ_0 is seen as the infinitesimal generator of the one-parameter group of operators $U(t)$ on the Banach space X_0^p, where

$$U(t)f(x) = f(e^t x) \qquad (t \in \mathbb{R},\ f \in X_0^p).$$

The results of Sect. 12.4.1 originate from Butzer and Jansche [92, Subsect. 5.2]. The considerations for the Mellin–Abel–Poisson kernel are an extended presentation of an example in Bardaro et al. [52, Sect. 6 (b)]. As sources for the results in Sect. 12.4.2, we cite Bardaro et al. [56, Subsect. 5.1] and Bardaro et al. [58, Corollary 8.2].

The results of Sects. 12.5 and 12.6 up to Corollary 12.32 are based on Bardaro et al. [52, Subsect. 4.2]. Corollary 12.33 is new. The remaining material of Sect. 12.6 originates from Bardaro et al. [56, Subsect. 5.2]. In [52] it is also shown how the exponential sampling formula and its integral form, namely the Mellin reproducing kernel formula, can be deduced from each other.

The results of Sect. 12.7 were obtained in Bardaro et al. [67]. Key tools are subspaces on which the sampling operator is bounded and the notions of discrete and mixed Hilbert transforms. The boundedness of the discrete Hilbert transform (see Definition 12.39) on the sequence spaces ℓ^p can be found, e.g., in Riesz [262], Butzer and Nessel [95, p. 332] and Marsden et al. [214]. The notion introduced in Definition 12.40 is the Mellin counterpart of the notion studied in [214] (also see Bardaro et al. [47]). The subspaces Λ_c^p were already introduced in Bardaro et al. [63]. They are Mellin versions of those studied in Bardaro et al. [47].

The contents of Sect. 12.8 is an extended presentation of results in Bardaro et al. [54] with new corollaries. In [54] it is also shown how the exponential sampling formula and the Mellin–Parseval formula can be deduced from each other.

Chapter 13
Generalized Exponential Sampling Theory

In this chapter we present a different approach to approximate sampling formulae by using general kernels in place of the lin_c function. This allows us to establish approximate reconstruction formulae under weak assumptions on the function f.

13.1 Definition of the Generalized Exponential Sampling Series

The function lin_c figuring in the exponential sampling formula of the previous chapter has the disadvantage that it does not belong to X_d for any $d \in \mathbb{R}$. Now we introduce a class of kernels with desirable properties. We shall obtain generalized exponential sampling formulae for the reconstruction of not necessarily Mellin bandlimited, bounded, continuous functions in an approximate sense.

Definition 13.1 We denote by Φ the class of all the continuous functions $\varphi : \mathbb{R}^+ \to \mathbb{R}$ with the following properties:

(i) for every $u \in \mathbb{R}^+$, we have $\sum_{k=-\infty}^{\infty} \varphi(e^{-k}u) = 1$;

(ii) $M_0(\varphi) := \sup_{u \in \mathbb{R}^+} \sum_{k=-\infty}^{\infty} |\varphi(e^{-k}u)| < \infty$;

(iii) $\lim_{r \to \infty} \sum_{|k - \log u| > r} |\varphi(e^{-k}u)| = 0$, uniformly with respect $u \in \mathbb{R}^+$.

For $\varphi \in \Phi$ we define the *generalized exponential sampling operator* as

$$(S_w^{\varphi} f)(x) = \sum_{k=-\infty}^{\infty} f(e^{k/w})\varphi(e^{-k}x^w) \qquad (x \in \mathbb{R}^+, w > 0), \tag{13.1}$$

where we admit all Lebesgue measurable functions $f : \mathbb{R}^+ \to \mathbb{C}$ for which the series converges absolutely almost everywhere on $\mathbb{R}^+$. The set of admissible

C. Bardaro et al., *Mellin Analysis, Transform Theory, and Applications*,
https://doi.org/10.1007/978-3-031-96672-9_13

functions will be called the *domain* of S_w^φ and denoted by dom S_w^φ. It is easily seen that every bounded function f belongs to dom S_w^φ since by the above property (ii) we have

$$\sum_{k\in\mathbb{Z}} |f(e^{k/w})||\varphi(e^{-k}x^w)| \leq M_0(\varphi)\, \|f\|^*_{X_0^\infty}.$$

However, in Sect. 13.3.1 we shall describe larger subsets of dom S_w^φ.

The above property (i) allows us to represent of deviation of $S_w^\varphi f$ from f as

$$(S_w^\varphi f)(x) - f(x) = \sum_{k=-\infty}^{\infty} \varphi(e^{-k}x^w)\big(f(e^{k/w}) - f(x)\big) \qquad (x \in \mathbb{R}^+). \tag{13.2}$$

This is a useful identity for the study of convergence and error estimates.

13.2 Pointwise and Uniform Convergence

We begin with a result on pointwise convergence valid under weak assumptions.

Theorem 13.2 *Let $\varphi \in \Phi$ and let $f : \mathbb{R}^+ \to \mathbb{C}$ be a bounded function. Suppose that f is continuous at some point $x \in \mathbb{R}^+$. Then, at this point, $\lim_{w\to\infty}(S_w^\varphi f)(x) = f(x)$.*

Proof The hypotheses imply that the composition $f \circ \exp$ is continuous at $\log x$. Hence, given any $\varepsilon > 0$, there exists $\delta > 0$ such that

$$\left|f(x) - f(e^{k/w})\right| = \left|f(e^{\log x}) - f(e^{k/w})\right| < \varepsilon$$

whenever $|\log x - k/w| < \delta$ or, equivalently, $|k - w\log x| < \delta w$. Now, using (13.2) and splitting the summation, we conclude that

$$\begin{aligned}
&|(S_w^\varphi f)(x) - f(x)| \\
&\quad = \left|\sum_{k=-\infty}^{\infty} \varphi(e^{-k}x^w)\big(f(e^{k/w}) - f(x)\big)\right| \\
&\quad \leq \Bigg(\sum_{|k-w\log x|\leq\delta w} + \sum_{|k-w\log x|>\delta w}\Bigg)|\varphi(e^{-k}x^w)||f(e^{k/w}) - f(x)| \\
&\quad \leq M_0(\varphi)\varepsilon + 2\|f\|^*_{X_0^\infty} \sum_{|k-w\log x|>\delta w} |\varphi(e^{-k}x^w)|.
\end{aligned}$$

The proof is completed by letting $w \to \infty$ and employing property (iii) of the class Φ. $\square$

Under a stronger hypothesis on f, the same method of proof provides uniform convergence. For this result we first recall that $\mathcal{C}(\mathbb{R}^+)$ denotes the space of all the log-uniformly continuous functions on $\mathbb{R}^+$ (see Definition 3.15).

Theorem 13.3 *Let $\varphi \in \Phi$ and let $f \in \mathcal{C}(\mathbb{R}^+)$. Then* $\lim_{w\to\infty} \|S_w^\varphi f - f\|_{X_0^\infty} = 0$.

Proof Given $\varepsilon > 0$, the hypothesis on f guarantees the existence of $\delta > 0$ such that for all $x \in \mathbb{R}^+$ we have $\left|f(x) - f(e^{k/w})\right| < \varepsilon$ whenever $|k - w\log x| < \delta w$. With these inequalities the previous proof leads to uniform convergence. $\square$

Our next aim is to estimate the approximation error for $f \in \mathcal{C}(\mathbb{R}^+)$ in terms of the logarithmic modulus of continuity given by

$$\omega(f,\delta) := \sup\{|f(s) - f(t)| : |\log s - \log t| \le \delta\} \qquad (\delta > 0)$$

(see Definition 6.52). For this, we introduce the *(algebraic) discrete moment of order* $j \in \mathbb{N}$ of a kernel $\varphi \in \Phi$ by

$$m_j(\varphi, x) := \sum_{k=-\infty}^{\infty} \varphi(e^{-k}x)\log^j(e^k x^{-1}) = \sum_{k=-\infty}^{\infty} \varphi(e^{-k}x)(k - \log x)^j,$$

and the *absolute discrete moments of order* $\alpha \in \mathbb{R}^+$ by

$$M_\alpha(\varphi, x) := \sum_{k=-\infty}^{\infty} |\varphi(e^{-k}x)||k - \log x|^\alpha,$$

whenever these series converge. Finally, we introduce

$$M_\alpha(\varphi) := \sup_{x\in\mathbb{R}^+} M_\alpha(\varphi, x),$$

which may not necessarily be a finite number.

In this notation, an estimate of the approximation error can be stated as follows.

Theorem 13.4 *Let $\varphi \in \Phi$ be such that $M_1(\varphi) < \infty$ and let $f \in \mathcal{C}(\mathbb{R}^+)$. Then, for every $\delta > 0$, there holds*

$$|(S_w^\varphi f)(x) - f(x)| \le M_0(\varphi)\omega(f,\delta) + M_1(\varphi)\frac{\omega(f,\delta)}{w\delta} \qquad (w, x \in \mathbb{R}^+).$$

Proof For any $\delta > 0$, the logarithmic modulus of continuity and its property (6.50) allow us to conclude that

$$\left|f(e^{k/w}) - f(x)\right| \leq \omega\left(f, \left|\frac{k}{w} - \log x\right|\right) = \omega\left(f, \frac{\delta\,|k - \log x^w|}{w\delta}\right)$$
$$\leq \left(1 + \frac{|k - \log x^w|}{w\delta}\right)\omega(f,\delta) \qquad (k \in \mathbb{Z},\, w \in \mathbb{R}^+).$$

Using this estimate in conjunction with (13.2), we obtain

$$\left|(S_w^{\varphi} f)(x) - f(x)\right| \leq \omega(f,\delta) \sum_{k=-\infty}^{\infty} \left|\varphi(e^{-k}x^w)\right| \left(1 + \frac{|k - \log x^w|}{w\delta}\right)$$
$$\leq M_0(\varphi)\omega(f,\delta) + M_1(\varphi)\frac{\omega(f,\delta)}{w\delta},$$

as was to be shown. □

The following result is an immediate consequence.

Corollary 13.5 *Under the hypotheses of Theorem 13.4, there holds*

$$|(S_w^{\varphi} f)(x) - f(x)| \leq C\omega\left(f, \frac{1}{w}\right) \qquad (w, x \in \mathbb{R}^+),$$

where C is a constant depending only on φ.

Proof Choosing in Theorem 13.4 $\delta = 1/w$, we obtain the assertion with $C = M_0(\varphi) + M_1(\varphi)$. □

Remark 13.6 When $M_1(\varphi) = \infty$ but $M_\alpha(\varphi) < \infty$ for some $\alpha \in]0, 1[$, we can similarly find an error bound for functions f satisfying a *logarithmic Hölder condition of order* α, that is

$$|f(s) - f(t)| \leq C|\log s - \log t|^\alpha \qquad (s, t \in \mathbb{R}^+)$$

with a constant C depending only on f. In fact,

$$|(S_w^{\varphi} f)(x) - f(x)| \leq \sum_{k=-\infty}^{\infty} |\varphi(e^{-k}x^w)||f(e^{k/w}) - f(x)|$$
$$\leq C \sum_{k=-\infty}^{\infty} |\varphi(e^{-k}x^w)| \left|\frac{k - \log x^w}{w}\right|^\alpha$$
$$\leq M_\alpha(\varphi)\frac{C}{w^\alpha} = \mathcal{O}\left(\frac{1}{w^\alpha}\right) \qquad (w \to \infty).$$

Next we look for sufficient conditions on the kernel function φ that enable a higher speed of convergence than in the previous results. For this, we introduce a subclass of Φ.

Definition 13.7 For $n \in \mathbb{N}$, we specify the subclass Φ_n of Φ as the set comprising all $\varphi \in \Phi$ with the following additional properties:

(iv) for $j = 0, 1, \ldots, n$ the moments $m_j(\varphi, x) =: m_j(\varphi)$ are independent of x;
(v) $M_n(\varphi) < \infty$;
(vi) $\lim_{r\to\infty} \sum_{|k-\log u|>r} |\varphi(e^{-k}u)||k - \log u|^n = 0$ uniformly with respect $u \in \mathbb{R}^+$.

We mention that property (v) can replace (ii) and (iii) of Definition 13.1 since it implies that

$$\sum_{|k-\log u|>r} |\varphi(e^{-k}u)| \leq \frac{1}{r^n} M_n(\varphi).$$

Now we are ready for an asymptotic formula that gives the precise order of pointwise approximation when we have a continuous, bounded function f which belongs locally to the class C^n; see Definition 6.42 for the notation used in the following statement.

Theorem 13.8 *Let $\varphi \in \Phi_n$, where $n \in \mathbb{N}$. Suppose that $f \in C(\mathbb{R}^+)$ is a bounded function that belongs to $C^n(x)$ for some $x \in \mathbb{R}^+$. Then, at this point x,*

$$(S_w^\varphi f)(x) - f(x) = \sum_{j=1}^{n} \frac{(\Theta^j f)(x)}{j!} \frac{m_j(\varphi)}{w^j} + o(w^{-n}) \qquad (w \to \infty).$$

Proof By the Taylor formula in the Mellin setting as presented in Proposition 6.43, we have

$$f(e^{k/w}) - f(x) = \sum_{j=1}^{n} \frac{(\Theta^j f)(x)}{j!} \log^j\left(\frac{e^{k/w}}{x}\right) + h\left(\frac{e^{k/w}}{x}\right) \log^n\left(\frac{e^{k/w}}{x}\right),$$

where $k \in \mathbb{Z}$, $w \in \mathbb{R}^+$ and $h : \mathbb{R}^+ \to \mathbb{C}$ is a bounded function such that $h(t) \to 0$ as $t \to 1$. In conjunction with (13.2), we obtain

$$\begin{aligned}(S_w^\varphi f)(x) - f(x) \\ = \sum_{k=-\infty}^{\infty} \varphi(e^{-k}x^w)\left[\sum_{j=1}^{n} \frac{(\Theta^j f)(x)}{j!} \log^j\left(\frac{e^{k/w}}{x}\right) + h\left(\frac{e^{k/w}}{x}\right) \log^n\left(\frac{e^{k/w}}{x}\right)\right].\end{aligned}$$

Noting that

$$\sum_{k=-\infty}^{\infty} \varphi(e^{-k}x^w) \log^j \left(\frac{e^{k/w}}{x}\right) = \frac{m_j(\varphi)}{w^j},$$

we may rewrite the approximation error as

$$(S_w^\varphi f)(x) - f(x) = \sum_{j=1}^{n} \frac{(\Theta^j f)(x)}{j!} \frac{m_j(\varphi)}{w^j} + R$$

with a remainder R given by

$$R = \sum_{k=-\infty}^{\infty} \varphi(e^{-k}x^w) h\left(\frac{e^{k/w}}{x}\right) \log^n \left(\frac{e^{k/w}}{x}\right).$$

It remains to estimate $|R|$. For any $\varepsilon > 0$, there exists a $\delta > 0$ such that $|h(e^{k/w}x^{-1})| < \varepsilon$ whenever $|k - w \log x| \leq \delta w$. Hence

$$\begin{aligned} w^n|R| < \varepsilon & \sum_{|k-w\log x|\leq \delta w} |\varphi(e^{-k}x^w)||k - w\log x|^n \\ & + \|h\|_\infty^* \sum_{|k-w\log x|>\delta w} |\varphi(e^{-k}x^w)||k - w\log x|^n \\ & \leq M_n(\varphi)\varepsilon + \|h\|_\infty^* \sum_{|k-w\log x|>\delta w} |\varphi(e^{-k}x^w)||k - w\log x|^n. \end{aligned}$$

Using property (vi) of the class Φ_n, we conclude that $w^n R \to 0$ as $w \to \infty$. This completes the proof. □

As an immediate consequence of Theorem 13.8, we obtain the following result.

Corollary 13.9 *Under the hypotheses of Theorem 13.8, we have*

$$\lim_{w\to\infty} w[(S_w^\varphi f)(x) - f(x)] = \Theta f(x) m_1(\varphi).$$

If in addition $m_j(\varphi) = 0$ *for* $j = 1, \ldots, n-1$, *then*

$$\lim_{w\to\infty} w^n[(S_w^\varphi f)(x) - f(x)] = \frac{\Theta^n f(x)}{n!} m_n(\varphi).$$

13.3 Generalized Exponential Sampling Series in X_c^p

In this section we shall study the convergence properties in the space X_c^p. We begin with a discussion on the domain of the generalized exponential sampling series S_w^φ as defined in (13.1). In the following we assume that $\varphi \in \Phi$.

13.3.1 On the Domain of S_w^φ

As we have seen already, the bounded functions build a subset of the domain of S_w^φ. However, there exist larger subsets that are contained in dom S_w^φ. We can admit functions which grow like a power of the logarithm as their argument approaches infinity or 0+ provided that certain absolute moments are finite. Indeed, the following result holds.

Proposition 13.10 *Let $\varphi \in \Phi$ and suppose that $M_r(\varphi) < \infty$ for some $r \in \mathbb{N}$. If $f : \mathbb{R}^+ \to \mathbb{C}$ and $|f(e^x)| \leq a + b|x|^r$ for some $a, b \in \mathbb{R}^+$ and all $x \in \mathbb{R}$, then $f \in \mathrm{dom}\, S_w^\varphi$.*

Proof Given $x, w \in \mathbb{R}^+$, let $k_0 \in \mathbb{N}$ be such that $|w \log x| / k_0 \leq 1/2$. By the growth condition on f, we have

$$\sum_{|k| \geq k_0} \left| f(e^{k/w})\varphi(e^{-k}x^w) \right| \leq a M_0(\varphi) + b \sum_{|k| \geq k_0} \left| \varphi(e^{-k}x^w) \right| \left| \frac{k}{w} \right|^r .$$

For estimating the series on the right-hand side, we use the absolute moment of order r and proceed as follows:

$$\begin{aligned} M_r(\varphi) &\geq \sum_{|k| \geq k_0} \left| \varphi(e^{-k}x^w) \right| |k - w \log x|^r \\ &= \sum_{|k| \geq k_0} \left| \varphi(e^{-k}x^w) \right| |k|^r \left| 1 - \frac{w \log x}{k} \right|^r \geq 2^{-r} \sum_{|k| \geq k_0} \left| \varphi(e^{-k}x^w) \right| |k|^r . \end{aligned}$$

Thus,

$$\sum_{k \in \mathbb{Z}} \left| f(e^{k/w})\varphi(e^{-k}x^w) \right| \leq \sum_{|k| < k_0} \left| f(e^{k/w})\varphi(e^{-k}x^w) \right| + a M_0(\varphi) + b \left(\frac{2}{w} \right)^r M_r(\varphi).$$

This shows absolute convergence of the series $(S_w^\varphi f)(x)$. □

We mention that if $f \in X_c^p$ for some $c \in \mathbb{R}$ and $p \geq 1$, then $f \notin \mathrm{dom}\, S_w^\varphi$ in general.

Example 13.11 Let $c = 1$, $p = 1$ and consider the function $f : \mathbb{R}^+ \to \mathbb{R}$ defined by

$$f(x) = \begin{cases} k^r & \text{if } x \in \left[e^k - 2^{-k}, e^k + 2^{-k}\right] \text{ for } k \in \mathbb{N}, \\ 0 & \text{otherwise,} \end{cases}$$

where $r \in \mathbb{N}$ with $r \geq 2$. Then $f \in X_1$ since

$$\|f\|_{X_1} = \sum_{k=1}^{\infty} \frac{k^r}{2^{k-1}} < \infty,$$

but for $w = 1$,

$$(S_1^{\varphi} f)(x) = \sum_{k=1}^{\infty} \varphi(e^{-k}x)k^r$$

does not converge in general.

Next we want to show that, using the notations of Sect. 12.7.2, for a fixed $w > 0$ and $\Sigma_w := \{e^{k/w}\}_{k \in \mathbb{Z}}$ the space $\ell_c^p(\Sigma_w)$ is a subset of dom S_w^{φ}. More precisely, we have the following statement.

Theorem 13.12 *Let $\varphi \in \Phi \cap X_{cp/w}$, where $c \in \mathbb{R}$, $p \geq 1$ and $w > 0$. If $f \in \ell_c^p(\Sigma_w)$, where $\Sigma_w = \{e^{k/w}\}_{k \in \mathbb{Z}}$, then $f \in$ dom S_w^{φ}, $S_w^{\varphi} f \in X_c^p$ and*

$$\|S_w^{\varphi} f\|_{X_c^p} \leq D(w) |f|_{\ell_c^p(\Sigma_w)}, \tag{13.3}$$

with

$$D(w) := M_0(\varphi)^{1-1/p} \|\varphi\|_{X_{cp/w}}^{1/p}. \tag{13.4}$$

Proof For $n \in \mathbb{N}$, define

$$A_n(x) := \sum_{|k| \leq n} |\varphi(e^{-k}x^w)|$$

and

$$T_n(x) := \sum_{|k| \leq n} \left| f(e^{k/w})\varphi(e^{-k}x^w) \right|.$$

Obviously $A_n(x) > 0$ for sufficiently large n and then we may write

$$\big(T_n(x)\big)^p = \left(A_n(x) \sum_{|k|\le n} |f(e^{k/w})| \frac{|\varphi(e^{-k}x^w)|}{A_n(x)} \right)^p.$$

Since for $p \ge 1$ the function $\psi : u \to u^p$ is convex, we conclude with the help of Jensen's inequality for sums (see Proposition 1.12) that

$$\begin{aligned}\big(T_n(x)\big)^p &\le A_n(x)^{p-1} \sum_{|k|\le n} |f(e^{k/w})|^p |\varphi(e^{-k}x^w)| \\ &\le M_0(\varphi)^{p-1} \sum_{|k|\le n} |f(e^{k/w})|^p |\varphi(e^{-k}x^w)|,\end{aligned}$$

where we have used that $A_n(x) \le M_0(\varphi)$ in the last step. From this inequality we deduce by integration that

$$\begin{aligned}\|T_n\|_{X_c^p}^p &= \int_0^\infty \big(T_n(x)\big)^p x^{cp} \frac{dx}{x} \\ &\le M_0(\varphi)^{p-1} \sum_{|k|\le n} \left|f(e^{k/w})\right|^p \int_0^\infty |\varphi(e^{-k}x^w)|\, x^{cp} \frac{dx}{x} \\ &= \frac{M_0(\varphi)^{p-1}}{w} \sum_{|k|\le n} \left|f(e^{k/w})\right|^p e^{kcp/w} \int_0^\infty |\varphi(t)|\, t^{cp/w} \frac{dt}{t} \\ &\le M_0(\varphi)^{p-1} \,\|\varphi\|_{X_{cp/w}} \,|f|_{\ell_c^p(\Sigma_w)}^p.\end{aligned}$$

The last upper bound on the right-hand side does not depend on n. Therefore, by Proposition 1.3,

$$\lim_{n\to\infty} T_n(x) = \sum_{k\in\mathbb{Z}} \left| f(e^{k/w})\varphi(e^{-k}x^w)\right| =: T(x)$$

exists almost everywhere on $\mathbb{R}^+$ and $\lim_{n\to\infty} \|T_n\|_{X_c^p} = \|T\|_{X_c^p}$. Hence $f \in \operatorname{dom} S_w^\varphi$. Furthermore, since $\left|(S_w^\varphi f)(x)\right| \le T(x)$, we conclude that $S_w^\varphi f \in X_c^p$ and

$$\|S_w^\varphi f\|_{X_c^p} \le \|T\|_{X_c^p} \le M_0(\varphi)^{1-1/p} \,\|\varphi\|_{X_{cp/w}}^{1/p} \,|f|_{\ell_c^p(\Sigma_w)}.$$

This completes the proof. □

Theorem 13.12 implies that S_w^φ is a continuous linear operator from $\ell_c^p(\Sigma_w)$ to X_c^p under the assumptions on the kernel function φ. This consequence may be stated as follows.

Corollary 13.13 *Let $\varphi \in \Phi \cap X_{cp/w}$, where $c \in \mathbb{R}$, $p \geq 1$ and $w > 0$. Then, for $f, g \in \ell_c^p(\Sigma_w)$, where $\Sigma_w = \{e^{k/w}\}_{k \in \mathbb{Z}}$, we have*

$$\|S_w^\varphi f - S_w^\varphi g\|_{X_c^p} \leq D(w)|f - g|_{\ell_c^p(\Sigma_w)} \tag{13.5}$$

with $D(w)$ given by (13.4).

13.3.2 Norm Convergence

As a simple consequence of Theorem 13.3, we state the following local convergence result.

Corollary 13.14 *Let $\varphi \in \Phi$ and $f \in C_{\text{comp}}(\mathbb{R}^+)$. Then, for every compact interval $A \subset \mathbb{R}^+$ and any $c \in \mathbb{R}$, $p \geq 1$, there holds*

$$\lim_{w \to \infty} \|(S_w^\varphi f - f)\chi_A\|_{X_c^p} = 0.$$

Proof Let $A = [\alpha, \beta]$ be fixed. Then

$$\|(S_w^\varphi f - f)\chi_A\|_{X_c^p}^p \leq \|(S_w^\varphi f - f)\chi_A\|_{X_0^\infty}^p \int_\alpha^\beta x^{cp-1}dx = C\,\|(S_w^\varphi f - f)\chi_A\|_{X_0^\infty}^p$$

with a constant C not depending on f and w. Now the assertion follows by Theorem 13.3. □

Under an additional assumption on φ, the result of Corollary 13.14 can be extended to a global convergence theorem.

Theorem 13.15 *Let $\varphi \in \Phi \cap X_{cp/w_0}$, where $c \in \mathbb{R}$, $p \geq 1$ and $w_0 > 0$. Then, for any $f \in C_{\text{comp}}(\mathbb{R}^+)$, we have*

$$\lim_{w \to \infty} \|S_w^\varphi f - f\|_{X_c^p} = 0. \tag{13.6}$$

Proof Let $w_1 := \max\{1, w_0\}$. First we consider the case $c > 0$. Since Theorem 13.3 guarantees uniform convergence, we will obtain the desired convergence in the norm with the help of Vitali's theorem (see Proposition 1.5) if we succeed in showing that the family $\{S_w^\varphi f \ : \ w \geq w_1\}$ is *uniformly integrable* with respect to the measure $\mu(\cdot) := \int_{(\cdot)} x^{cp-1}dx$ on $\mathcal{B}(\mathbb{R}^+)$. In the situation of $c > 0$, this means that we have to verify the following two conditions:

(i) For every $\varepsilon > 0$ there exists $\delta > 0$ such that, if $A \in \mathcal{B}(\mathbb{R}^+)$ and $\int_A x^{cp-1}dx < \delta$, then

$$\int_A x^{cp-1} \left|(S_w^{\varphi} f)(x)\right|^p dx < \varepsilon$$

for all $w \geq w_1$.

(ii) For every $\varepsilon > 0$ there exists $M > 1$ such that

$$\int_M^{\infty} x^{cp-1} \left|(S_w^{\varphi} f)(x)\right|^p dx < \varepsilon$$

for all $w \geq w_1$.

It is easily seen that (i) holds. Indeed, since $f \in C_{\mathrm{comp}}(\mathbb{R}^+)$, we have $\left|(S_w^{\varphi} f)(x)\right| \leq M_0(\varphi)\|f\|_{X_0^\infty}$, and so $\delta := (M_0(\varphi)\|f\|_{X_0^\infty})^{-p}\,\varepsilon$ yields the desired estimate.

To show that (ii) is also fullfilled needs more efforts. Let $r > 1$ be such that the support of the function f is contained in the interval $[1/r, r]$ and define

$$C := 2M_0(\varphi)^{p-1}(\log r + 1)r^{cp}\|f\|_{X_0^\infty}^p, \tag{13.7}$$

which is positive if we exclude the trivial case that $\|f\|_{X_0^\infty} = 0$. Since $\varphi \in X_{cp/w_0}$, there exits for every $\varepsilon > 0$ a real number $\widetilde{M} > 1$ such that

$$\int_{\widetilde{M}}^{\infty} x^{cp/w_0-1} |\varphi(x)|\, dx < \frac{\varepsilon}{C}. \tag{13.8}$$

Now let $M > r\widetilde{M}$. By using Jensen's inequality and proceeding as in the proof of Theorem 13.12, we conclude that

$$\begin{aligned}
&\int_M^{\infty} x^{cp-1}|(S_w^{\varphi} f)(x)|^p dx \\
&\leq M_0(\varphi)^{p-1}\sum_{k\in\mathbb{Z}}|f(e^{k/w})|^p \int_M^{\infty} x^{cp}|\varphi(e^{-k}x^w)|\frac{dx}{x} \\
&= \frac{M_0(\varphi)^{p-1}}{w}\sum_{k\in\mathbb{Z}}|f(e^{k/w})|^p e^{kcp/w}\int_{e^{-k}M^w}^{\infty} t^{cp/w}|\varphi(t)|\frac{dt}{t}.
\end{aligned}$$

Counting the integers k such that $e^{k/w} \in [1/r, r]$, we find that the series has at most $2\lfloor w \log r\rfloor + 1$ non-zero terms. Noting that for $w \geq w_1$ and $e^{k/w} \in [1/r, r]$, there holds

$$e^{-k}M^w \geq \left(Mr^{-1}\right)^w > \widetilde{M}^w \geq \widetilde{M} > 1,$$

we see that $t > e^{-k}M^w$ implies $t^{cp/w} \le t^{cp/w_0}$. Hence for $w \ge w_1$ the preceding estimate of the integral can be continued to

$$\int_M^\infty x^{cp-1}|(S_w^\varphi f)(x)|^p dx$$

$$\le M_0(\varphi)^{p-1}\,\frac{2\lfloor w\log r\rfloor + 1}{w}\,\|f\|^p_{X_0^\infty}\, r^{cp}\int_{\widetilde{M}}^\infty t^{cp/w_0-1}|\varphi(t)|dt.$$

Finally, observing that $(2\lfloor w\log r\rfloor + 1)/w \le 2(\log r + 1)$ and recalling (13.7) and (13.8), we arrive at

$$\int_M^\infty x^{cp-1}|(S_w^\varphi f)(x)|^p dx \le 2M_0(\varphi)^{p-1}(\log r + 1)r^{cp}\,\|f\|^p_{X_0^\infty}\,\frac{\varepsilon}{C} = \varepsilon.$$

Thus (ii) is true as well. This completes the proof for $c > 0$.

In the case $c < 0$, the substitution $x = 1/t$ transforms the relevant integrals into a form that corresponds to the case $c > 0$ and allows us to use the previous method of proof.

When $c = 0$, condition (ii) of uniform integrability has to be replaced by

(ii*) For every $\varepsilon > 0$ there exists $M > 1$ such that

$$\int_{\mathbb{R}^+\setminus[1/M,M]} \left|(S_w^\varphi f)(x)\right|^p \frac{dx}{x} < \varepsilon$$

for all $w \ge w_1$.

Splitting the integral as $\int_0^{1/M} + \int_M^\infty$, we find that the second integral can be handled by a simplified form of the previous proof while the first one takes the form of the second by the substitution $x = 1/t$. □

13.3.3 Norm Convergence and Bounded Coarse Variation

First we introduce a space that is based on the notion of bounded coarse variation (see Sect. 1.2.2) but is tailored for the study of convergence in the norm of X_c^p.

Definition 13.16 For $c \in \mathbb{R}$ and $p \ge 1$, we denote by $\mathrm{BCV}_c^p(\mathbb{R}^+)$ the space of all bounded functions $f : \mathbb{R}^+ \to \mathbb{C}$ such that the function $g : t \mapsto |f(e^t)|^p e^{pct}$ is of bounded coarse variation on $\mathbb{R}$, that is, $g \in \mathrm{BCV}(\mathbb{R})$.

Now we want to show an asymptotic connection between $|\cdot|_{\ell_c^p(\Sigma_w)}$ and $\|\cdot\|_{X_c^p}$. For the subsequent notation R_{comp}, see p. 361.

Proposition 13.17 *Let $f \in X_c^p \cap \mathrm{BCV}_c^p(\mathbb{R}^+) \cap R_{\mathrm{comp}}$, where $c \in \mathbb{R}$ and $p \geq 1$, and let $\Sigma_w = \{e^{k/w}\}_{k\in\mathbb{Z}}$. Then $f \in \ell_c^p(\Sigma_w)$ and*

$$\lim_{w\to\infty} |f|^p_{\ell_c^p(\Sigma_w)} = \|f\|^p_{X_c^p}. \tag{13.9}$$

Proof Indeed, we have formally

$$|f|^p_{\ell_c^p(\Sigma_w)} = \frac{1}{w}\sum_{k\in\mathbb{Z}} |f(e^{k/w})|^p e^{kcp/w}$$

$$= \sum_{k\in\mathbb{Z}} \left| f(e^{\log(e^{k/w})})\right|^p \left(e^{\log e^{k/w}}\right)^{cp} \left(\log e^{k/w} - \log e^{(k-1)/w}\right).$$

Putting $y_k := \log e^{k/w}$, we obtain a partition $\{y_k\}_{k\in\mathbb{Z}}$ of $\mathbb{R}$ which shows that $|f|^p_{\ell_c^p(\Sigma_w)}$ is an infinite, convergent Riemann sum of the integral

$$\int_{-\infty}^{\infty} e^{tcp}|f(e^t)|^p dt = \int_0^{\infty} x^{cp}|f(x)|^p \frac{dx}{x} = \|f\|^p_{X_c^p} < \infty;$$

convergence being guaranteed by Proposition 1.16. Now Proposition 1.17 is applicable and yields (13.9). □

Definition 13.18 For $c \in \mathbb{R}$, $p \geq 1$, $Q \geq 1$ and $\Sigma_w = \{e^{k/w}\}_{k\in\mathbb{Z}}$, we denote by $\mathcal{L}_c^p(Q)$ the set of all functions $f \in \ell_c^p(\Sigma_w)\cap X_c^p$ such that $\limsup_{w\to\infty} |f|^p_{\ell_c^p(\Sigma_w)} \leq Q\|f\|^p_{X_c^p}$.

Remark 13.19 Note that the class $\mathcal{L}_c^p(Q)$ is not a vector space. If $f, g \in \mathcal{L}_c^p(Q)$, it may happen that $f + g \notin \mathcal{L}_c^p(Q)$. However, if $f \in \mathcal{L}_c^p(Q)$ and $g \in \mathcal{L}_c^p(Q')$ with $Q, Q' \geq 1$, and there exists a constant $M > 1$ such that $\|f\|^p_{X_c^p} + \|g\|^p_{X_c^p} \leq M\|f + g\|^p_{X_c^p}$, then $f + g \in \mathcal{L}_c^p(Q^*)$ for some constant $Q^* > 1$ depending on f and g.

If $f \in X_c^p \cap \mathrm{BCV}_c^p(\mathbb{R}^+) \cap R_{\mathrm{comp}}$, then, by Proposition 13.17, $f \in \mathcal{L}_c^p(Q)$ with $Q = 1$.

The next lemma will be useful for the main convergence result for functions in the class $\mathcal{L}_c^p(Q)$.

Lemma 13.20 *Let $\varphi \in \Phi \cap X_0 \cap X_{cp/w_0}$, where $c \in \mathbb{R}$, $p \geq 1$ and $w_0 > 0$. Then there exists a constant $C > 0$ such that*

$$\|\varphi\|_{X_{cp/w}} \leq C \tag{13.10}$$

for all $w \geq w_0$.

Proof First suppose that $c \geq 0$. Then, for $w \geq w_0$,

$$\|\varphi\|_{X_{cp/w}} = \int_0^\infty t^{cp/w}|\varphi(t)|\frac{dt}{t} = \int_0^1 t^{cp/w}|\varphi(t)|\frac{dt}{t} + \int_1^\infty t^{cp/w}|\varphi(t)|\frac{dt}{t}$$

$$\leq \int_0^1 |\varphi(t)|\frac{dt}{t} + \int_1^\infty t^{cp/w_0}|\varphi(t)|\frac{dt}{t} \leq \|\varphi\|_{X_0} + \|\varphi\|_{X_{cp/w_0}} =: C.$$

For $c < 0$ the proof is similar. □

Remark 13.21 Under the hypotheses of Lemma 13.20 the constant $D(w)$ appearing in Theorem 13.12 and Corollary 13.13 can be replaced by

$$D := M_0(\varphi)^{1-1/p}C^{1/p}.$$

The latter does not depend on w.

We are ready to prove the main convergence result.

Theorem 13.22 *Let $\varphi \in \Phi \cap X_0 \cap X_{cp/w_0}$, where $c \in \mathbb{R}$, $p \geq 1$ and $w_0 > 0$. Suppose that $f \in \mathcal{L}_c^p(Q)$ with $Q \geq 1$ and $f - g \in \mathcal{L}_c^p(Q)$ for all $g \in C_{\mathrm{comp}}^\infty(\mathbb{R}^+)$. Then*

$$\lim_{w\to\infty} \|S_w^\varphi f - f\|_{X_c^p} = 0. \tag{13.11}$$

Proof By Theorem 13.15, the assertion (13.11) holds for every function $g \in C_{\mathrm{comp}}^\infty(\mathbb{R}^+)$. Note that every such g belongs to $\mathcal{L}_c^p(Q)$ for any $Q \geq 1$ since g is of bounded variation and so (13.9) holds. Now, by Proposition 3.4, there exists a sequence $(g_n)_{n\in\mathbb{N}} \subset C_{\mathrm{comp}}^\infty(\mathbb{R}^+)$ such that

$$\lim_{n\to\infty} \|g_n - f\|_{X_c^p} = 0.$$

Thus, given $\varepsilon > 0$, there exists $n_0 \in \mathbb{N}$ such that

$$\|g_n - f\|_{X_c^p} \leq \varepsilon$$

for all $n \geq n_0$. By Theorem 13.15, we have

$$\lim_{w\to\infty} \|S_w^\varphi g_{n_0} - g_{n_0}\|_{X_c^p} = 0.$$

Now, employing Corollary 13.13 and Lemma 13.20, thereby taking Remark 13.21 into account, we conclude that

$$\|S_w^\varphi f - f\|_{X_c^p} \leq \|S_w^\varphi f - S_w^\varphi g_{n_0}\|_{X_c^p} + \|S_w^\varphi g_{n_0} - g_{n_0}\|_{X_c^p} + \|g_{n_0} - f\|_{X_c^p}$$

$$\leq D|f - g_{n_0}|_{\ell_c^p(\Sigma_w)} + \|S_w^\varphi g_{n_0} - g_{n_0}\|_{X_c^p} + \varepsilon.$$

Since $f - g_{n_0} \in \mathcal{L}_c^p(Q)$ with a constant $Q \geq 1$, we see that

$$\limsup_{w\to\infty} \|S_w^{\varphi} f - f\|_{X_c^p} \leq DQ\|f - g_{n_0}\|_{X_c^p} + \varepsilon \leq (DQ+1)\varepsilon,$$

and so the assertion follows. □

In conjunction with Proposition 13.17, Theorem 13.22 implies the following statement.

Corollary 13.23 *Suppose that the hypotheses of Theorem 13.22 on φ are satisfied and that $f \in X_c^p \cap \mathrm{BCV}_c^p(\mathbb{R}^+) \cap R_{\mathrm{comp}}$. Then*

$$\lim_{w\to\infty} \|S_w^{\varphi} f - f\|_{X_c^p} = 0. \tag{13.12}$$

In particular (13.12) holds if the function $f \in X_c^p$ is such that $t \mapsto |f(e^t)|^p e^{pct}$ is of bounded variation on $\mathbb{R}$.

13.4 Some Examples

In this section, we give some examples of kernels $\varphi \in \Phi$ satisfying the assumptions employed in the previous sections. In a final section we shall study a multiplier technique which leads to further cases of generalized exponential sampling series.

(a) Mellin Splines

An important class of functions in various branches of analysis are B-splines. For $n+1$ equally spaced knots located symmetrically to the origin, the B-spline of order n is given by

$$\mathcal{B}_n(x) = \frac{1}{(n-1)!}\sum_{j=0}^{n}(-1)^j\binom{n}{j}\left(\frac{n}{2} - j + x\right)_+^{n-1} \qquad (x \in \mathbb{R}),$$

where we use the notation $r_+ := (r + |r|)/2$ for $r \in \mathbb{R}$. The support of $\mathcal{B}_n$ is the interval $[-n/2, n/2]$. A natural analogue for the needs of Mellin analysis will be

$$B_n(t) := \mathcal{B}_n(\log t) = \frac{1}{(n-1)!}\sum_{j=0}^{n}(-1)^j\binom{n}{j}\left(\frac{n}{2} - j + \log t\right)_+^{n-1} \qquad (t \in \mathbb{R}^+).$$

The support of B_n is the interval $[e^{-n/2}, e^{n/2}]$. Therefore $B_n \in X_c^p$ for any $c \in \mathbb{R}$ and $p \geq 1$. Either by a direct calculation or via known results on the Fourier transform of $\mathcal{B}_n$, we find that

$$[B_n]^{\wedge}_{M_0}(iv) = \operatorname{sinc}^n\left(\frac{v}{2\pi}\right) \qquad (v \in \mathbb{R}).$$

We want to show that B_n is well suited for serving as a kernel function φ of a generalized exponential sampling series. For an arbitrary but fixed $x \in \mathbb{R}^+$, we consider the scaled function $B_{n,x} : t \mapsto B_n(tx)$. Then, by Proposition 6.8 (b), we have

$$[B_{n,x}]^\wedge_{M_0}(iv) = [B_n]^\wedge_{M_0}(iv)x^{-iv} = \mathrm{sinc}^n\left(\frac{v}{2\pi}\right)x^{-iv}.$$

Now we can apply the Mellin–Poisson summation formula of Corollary 6.100 to $B_{n,x}$. For $c = 0$ and $T = 1$ it yields

$$\sum_{k=-\infty}^{\infty} B_n(e^k x) = \sum_{k=-\infty}^{\infty} \mathrm{sinc}^n(k)x^{-2\pi ki} = 1.$$

We also note that the absolute moments $M_k(B_n)$ are finite numbers for $k \in \mathbb{N}_0$ since they are defined by a series which has only a finite number of non-zero terms due to the compact support of B_n. This allows us to conclude that $\varphi := B_n$ belongs to the class Φ (see Definition 13.1) and satisfies the conditions (v) and (vi) of Definition 13.7.

Next we turn to conditon (iv). Consider the function $\psi_j(u) := B_n(u)(-\log u)^j$. Then

$$[\psi_j]^\wedge_{M_0}(iv) = \int_0^\infty B_n(u)(-\log u)^j u^{iv}\frac{du}{u}$$
$$= i^j \frac{d^j}{dv^j}\int_0^\infty B_n(u)u^{iv}\frac{du}{u} = i^j\left[\mathrm{sinc}^n\left(\frac{\cdot}{2\pi}\right)\right]^{(j)}(v).$$

Now, applying the Mellin–Poisson summation formula to the function $t \mapsto \psi_j(tx)$, we obtain

$$\sum_{k=-\infty}^{\infty} B_n(e^{-k}x)\log^j(e^k x^{-1}) = i^j \sum_{k=-\infty}^{\infty}\left[\mathrm{sinc}^n\left(\frac{\cdot}{2\pi}\right)\right]^{(j)}(2\pi k)x^{-2\pi ki}$$
$$= i^j\left[\mathrm{sinc}^n\left(\frac{\cdot}{2\pi}\right)\right]^{(j)}(0) \qquad (j = 0, 1, \ldots, n-1).$$

This shows that the moments $m_j(B_n, x)$ do not depend on x for $j = 0, 1, \ldots, n-1$ and so $B_n \in \Phi_{n-1}$. Moreover, among these moments those with an odd index j are zero since sinc^n is an even function.

For $n = 2$ we find that

$$B_2(t) = \begin{cases} 1 + \log t, & e^{-1} < t < 1, \\ 1 - \log t, & 1 < x < e, \\ 0, & \text{otherwise on } \mathbb{R}^+. \end{cases}$$

Hence we may simply write $B_2(t) = (1 - |\log t|)_+$, valid for all $t \in \mathbb{R}^+$. Since $m_1(B_2) = 0$, the asymptotic formula of Corollary 13.9 yields

$$(S_w^{B_2} f)(x) - f(x) = o\left(w^{-1}\right) \qquad (w \to \infty).$$

(b) The Mellin–Fejér Kernel

In the previous example we profited from a spline function of compact support with a Mellin transform in terms of the sinc function. One may also think of a converse construction, where one has a power of a sinc function and profits from a Mellin transform of compact support. Such a situation is achieved by the Mellin–Fejér kernels given by

$$F_\rho^c(x) := \frac{\rho}{2\pi} x^{-c} \operatorname{sinc}^2\left(\frac{\rho}{2\pi}\log x\right) \qquad (c \in \mathbb{R},\ \rho, x \in \mathbb{R}^+)$$

(see Definition 6.83). Then $F_\rho^c \in X_c^p$ for any $c \in \mathbb{R}$, $p \geq 1$ and (see Lemma 6.84)

$$[F_\rho^c]^\wedge_{M_c}(c + iv) = \begin{cases} 1 - \dfrac{|v|}{\rho}, & |v| \leq \rho, \\ 0, & |v| > \rho. \end{cases}$$

In the following we restrict ourselves to $c = 0$ and $\rho = 1$. Then the Mellin–Poisson summation formula applied to the function $t \mapsto F_1^0(tx)$ yields

$$\sum_{k=-\infty}^{\infty} F_1^0(e^k x) = \sum_{k=-\infty}^{\infty} [F_1^0]^\wedge_{M_0}(2k\pi i) x^{-2k\pi i} = [F_1^0]^\wedge_{M_0}(0) = 1 \qquad (x \in \mathbb{R}^+).$$

Since $F_1^0(e^k x) \geq 0$ for $k \in \mathbb{Z}$ and $x \in \mathbb{R}^+$, we have shown that condition (i) and (ii) of Definition 13.1 are satisfied for $\varphi = F_1^0$. As regards (iii), we choose an arbitrary $r > 1$ and observe that

$$\begin{aligned}
\sum_{|k-\log u|>r} \left|F_1^0(e^{-k}u)\right| &= \frac{1}{2\pi} \sum_{|k-\log u|>r} \operatorname{sinc}^2\left(\frac{1}{2\pi}\log(e^{-k}u)\right) \\
&\leq \frac{2}{\pi} \sum_{|k-\log u|>r} \frac{1}{|k - \log u|^2} \leq \frac{4}{\pi}\sum_{j=0}^{\infty} \frac{1}{|r+j|^2} \\
&\leq \frac{4}{\pi}\int_{r-1}^{\infty} \frac{dt}{t^2} = \frac{4}{\pi(r-1)}.
\end{aligned}$$

This shows that (iii) is also satisfied and so $F_1^0 \in \Phi$. Thus F_1^0 is admissible as a kernel function φ in Theorems 13.2 and 13.3. However, $M_1(F_1^0) = \infty$ and so F_1^0 is not admissible in Theorems 13.4 and 13.8. But Remark 13.6 provides an alternative since $M_\alpha(F_1^0) < \infty$ for $\alpha \in]0, 1[$.

(c) Generalized Mellin–Fejér Kernel
This kernel is a generalization of the Mellin–Fejér kernel with an additional parameter which modifies the power of the sinc function. In the literature it has been introduced as

$$J^c_{\gamma,\beta}(x) := d_{\gamma,\beta} x^{-c} \operatorname{sinc}^{2\beta}\left(\frac{\log x}{2\pi\gamma\beta}\right) \qquad (\beta \in \mathbb{N},\ \gamma \geq 1,\ c \in \mathbb{R},\ x \in \mathbb{R}^+),$$

where $d_{\gamma,\beta}$ is given by

$$d^{-1}_{\gamma,\beta} := \int_0^\infty \operatorname{sinc}^{2\beta}\left(\frac{\log x}{2\pi\gamma\beta}\right)\frac{dx}{x}.$$

We note that $J^c_{\gamma,1} = F^c_{1/\gamma}$. This kernel is related to a trigonometric kernel due to Jackson and for this reason it has been denoted by $J^c_{\gamma,\beta}$ in the literature.

It is easily seen that $J^c_{\gamma,\beta} \in X^p_c$ for any $p \geq 1$ and $\|J^c_{\gamma,\beta}\|_{X_c} = 1$. Furthermore, it is known that $[J^0_{\gamma,\beta}]^\wedge_{M_0}(iv) = 0$ for $|v| \geq 1/\gamma$.

In the following, we restrict ourselves to $c = 0$, $\gamma = 1$, using the notation $J_\beta := J^0_{1,\beta}$ and $d_\beta := d_{1,\beta}$. By arguing as in the previous example, we find that $J_\beta \in \Phi$.

Now let $\beta \geq 2$ and let $\psi_j(u) := J_\beta(u)(-\log u)^j$. Then, applying the Mellin–Poisson summation formula to the function $t \mapsto \psi_j(tx)$ and proceeding as in the first example, we find for the moments that $m_1(J_\beta, x) = 0$ and that $m_2(J_\beta, x)$ is finite and does not depend on x. Hence condition (iv) of Definition 13.7 is satisfied for $n = 2$. As regards the other two conditions, we choose $n \in \{1, 2\}$, $r > 1$ and observe that

$$\sum_{k=-\infty}^{\infty} \left|J_\beta(e^{-k}u)\right| |k - \log u|^n = S_{1,n}(u, r) + S_{2,n}(u, r),$$

where

$$S_{1,n}(u, r) := d_\beta \sum_{|k-\log u|\leq r} \operatorname{sinc}^{2\beta}\left(\frac{k - \log u}{2\pi\beta}\right) |k - \log u|^n \leq d_\beta\,(2r + 1)r^n$$

and

$$\begin{aligned} S_{2,n}(u, r) :&= d_\beta \sum_{|k-\log u|>r} \operatorname{sinc}^{2\beta}\left(\frac{k - \log u}{2\pi\beta}\right) |k - \log u|^n \\ &\leq d_\beta\,(2\beta)^{2\beta} \sum_{|k-\log u|>r} \frac{1}{|k - \log u|^{2\beta-n}} \\ &\leq d_\beta\, 2(2\beta)^{2\beta} \sum_{j=0}^{\infty} \frac{1}{|r + j|^{2\beta-n}} \leq \frac{d_\beta\, 2(2\beta)^{2\beta}}{(2\beta - n - 1)(r - 1)^{2\beta-n-1}}. \end{aligned}$$

With these estimates, we easily conclude that $M_1(J_\beta) < \infty$, $M_2(J_\beta) < \infty$ and condition (vi) of Definition 13.7 is satisfied for $n \in \{1, 2\}$. Hence $J_\beta \in \Phi_2$ for $\beta \geq 2$. In this case J_β is an admissible kernel function φ in Theorems 13.2–13.4 and 13.8. Generally, if $\beta \geq k \geq 2$, then $J_\beta \in \Phi_{2k-2}$.

13.5 Exponential Sampling with a Multiplier

Theorem 12.4 is of central importance in several ways but it also has some limitations in theoretical as well as in practical use. For example, the assumption that the function f is bandlimited in the sense of Mellin, imposes a severe restriction, as one recognizes by looking at the Mellin version of the Paley–Wiener theorem (see Theorems 8.7, 8.10). It excludes all functions f for which $\limsup_{r\to+\infty} |r^c f(r)| > 0$ or $\limsup_{r\to 0+} |r^c f(r)| > 0$ let alone unbounded functions. Furthermore, the series may converge very slowly. We shall see that these deficiencies can be overcome by incorporating a Mellin bandlimited multiplier in the series.

This technique can be considered within the more general framework of the generalized exponential sampling series of a function f, in which the function $\mathrm{lin}_{c/T}$ is replaced by a general continuous function φ satisfying suitable assumptions (see Definition 13.1). The generalized exponential sampling series takes the form

$$\big(S_w^\varphi f\big)(r) := \sum_{k\in\mathbb{Z}} f\big(e^{k/w}\big)\varphi(e^{-k} r^w) \qquad (r, w \in \mathbb{R}^+). \tag{13.13}$$

The function φ is called the *kernel* of $(S_w^\varphi f)(r)$.

As we have seen, in general, such a series does not reproduce f as the series in Theorem 12.4, but it approximates continuous, bounded functions f in the sense that

$$\lim_{w\to\infty} (S_w^\varphi f)(r) = f(r)$$

(see Theorems 13.2 and 13.3). For finite w, we have $f(r) = (S_w^\varphi f)(r) + (E_w f)(r)$ with an approximation error $(E_w f)(r)$. A further error occurs in computations since we will have to truncate the series $(S_w^\varphi f)(r)$.

In the present section we study of the following questions:

1. How can we find kernels φ for which the corresponding generalized exponential sampling series improves upon the series of Theorem 12.4?
2. What is an appropriate kernel φ for establishing an efficient algorithm by truncating the corresponding generalized exponential sampling series?

We shall answer both questions by adopting a multiplier technique that was effectively developed for the Whittaker–Kotel'nikov–Shannon sampling series.

13.5.1 A Mellin Bandlimited Multiplier

We begin with a lemma that will be useful for studying the convergence of the arising exponential sampling series. For this, we have to recall the definition of the Mellin–Bernstein spaces for polar-analytic functions (see Definition 3.41).

For $f \in \mathscr{B}^p_{c,T}$ and $(r_0, \theta_0) \in \mathbb{H}$, we introduce

$$f_{(r_0,\theta_0)} : (r, \theta) \longmapsto r_0^c f(r_0 r, \theta + \theta_0) \qquad \big((r, \theta) \in \mathbb{H}\big). \tag{13.14}$$

By Propositions 3.43 and 3.45, we have $f_{(r_0,\theta_0)} \in \mathscr{B}^p_{c,T}$ whenever $f \in \mathscr{B}^p_{c,T}$ and also

$$\| f_{(r_0,\theta_0)}(\cdot, 0)\|_{X^p_c} \le e^{T|\theta_0|} \| f(\cdot, 0)\|_{X^p_c}. \tag{13.15}$$

Lemma 13.24 *Let $\Psi \in \mathscr{B}^2_{0,\pi\delta}$, where $\delta \in]0, 1]$. Then, for $w > 0$, we have*

$$\sum_{k\in\mathbb{Z}} \left|\Psi(e^{-k} r^w, w\theta) \operatorname{lin}(e^{-k} r^w e^{iw\theta})\right| \le e^{\pi w(1+\delta)|\theta|} \|\Psi(\cdot, 0)\|_{X^2_0} \quad ((r, \theta) \in \mathbb{H}). \tag{13.16}$$

The series converges uniformly on compact subsets of $\mathbb{H}$.

Proof Using the notation (13.14), we may write

$$\Psi(e^{-k} r^w, w\theta) = \Psi_{(r^w,w\theta)}(e^{-k}, 0).$$

By the translation invariance of the Mellin–Bernstein spaces and the Mellin-Paley–Wiener theorem, we have

$$\Psi_{(r^w,w\theta)}(\cdot, 0) \in B^2_{0,\pi\delta} \subseteq B^2_{0,\pi}.$$

Therefore the Mellin–Parseval formula (see Theorem 12.50) and (13.15) yield

$$\left(\sum_{k\in\mathbb{Z}} \left|\Psi(e^{-k} r^w, w\theta)\right|^2\right)^{1/2} = \left\|\Psi_{(r^w,w\theta)}(\cdot, 0)\right\|_{X^2_0} \le e^{\pi\delta w|\theta|} \|\Psi(\cdot, 0)\|_{X^2_0}. \tag{13.17}$$

Analogously we conclude that

$$\left(\sum_{k\in\mathbb{Z}} \left|\operatorname{lin}(e^{-k} r^w e^{iw\theta})\right|^2\right)^{1/2} \le e^{\pi w|\theta|} \|\operatorname{lin}\|_{X^2_0} = e^{\pi w|\theta|}.$$

Now (13.16) is an immediate consequence of the Cauchy–Schwarz inequality.

It remains to verify the assertion on uniform convergence. For $N \in \mathbb{N}$, we have again by the Cauchy–Schwarz inequality and (13.17) that

$$\sum_{|k|>N} \left|\Psi(e^{-k} r^{w}, w\theta)\mathrm{lin}(e^{-k} r^{w} e^{iw\theta})\right|$$
$$\leq \left(\sum_{|k|>N} \left|\Psi(e^{-k} r^{w}, w\theta)\right|^2\right)^{1/2} \left(\sum_{|k|>N} \left|\mathrm{lin}(e^{-k} r^{w} e^{iw\theta})\right|^2\right)^{1/2}$$
$$\leq e^{\pi\delta w|\theta|} \|\Psi(\cdot, 0)\|_{X_0^2} \left(\sum_{|k|>N} \left|\mathrm{lin}(e^{-k} r^{w} e^{iw\theta})\right|^2\right)^{1/2}.$$

Thus it suffices to prove the assertion on uniform convergence for the last series only.

Now let C be a compact subset of $\mathbb{H}$. There exists an integer $m \in \mathbb{N}$ such that $\sup_{(r,\theta)\in C} w\,|\log r + i\theta| \leq m$. Then for $|k| > m$ we have

$$\left|\mathrm{lin}(e^{-k} r^{w} e^{iw\theta})\right| = \frac{|\sin(\pi w(\log r + i\theta))|}{|\pi(w(\log r + i\theta) - k)|} \leq \frac{e^{\pi m}}{\pi(|k| - m)}$$

and so for $N > m$,

$$\sum_{|k|>N} \left|\mathrm{lin}(e^{-k} r^{w} e^{iw\theta})\right|^2 \leq \frac{2e^{2\pi m}}{\pi^2} \sum_{j=1}^{\infty} \frac{1}{(N + j - m)^2}$$
$$\leq \frac{2e^{2\pi m}}{\pi^2} \int_{N-m}^{\infty} \frac{dt}{t^2} = \frac{2e^{2\pi m}}{\pi^2(N - m)}.$$

Hence, given $\varepsilon > 0$, there exists an $N \in \mathbb{N}$ such that

$$\sum_{|k|>N} \left|\mathrm{lin}(e^{-k} r^{w} e^{iw\theta})\right|^2 < \varepsilon$$

for $(r, \theta) \in C$. This completes the proof. □

Theorem 13.25 *For $\delta \in]0, 1[$, let $\psi \in B^2_{0,\pi\delta}$ such that $\psi(1) = 1$. Suppose that $f \in \mathscr{B}^{\infty}_{c,\pi T}$, where $c \in \mathbb{R}$ and $T \in \mathbb{R}^+$. Then for $w = T/(1 - \delta)$, we have*

$$f(r, 0) = \sum_{k\in\mathbb{Z}} f(e^{k/w}, 0)\psi(e^{-k} r^{w})\mathrm{lin}_{c/w}(e^{-k} r^{w}) \qquad (r \in \mathbb{R}^+). \tag{13.18}$$

The series converges absolutely and uniformly on compact subsets of $\mathbb{R}^+$.

Proof For any $\rho \in \mathbb{R}^+$ serving as a parameter, consider the function $\psi(\rho^w/(\cdot)^w)$. It is easily verified that it belongs to $B^2_{0,\pi\delta w}$. By the Mellin–Paley–Wiener theorem (see Theorem 8.10), there exists a function $\Psi_\rho \in \mathscr{B}^2_{0,\pi\delta w}$ such that $\Psi_\rho(r,0) = \psi(\rho^w/r^w)$ for all $r \in \mathbb{R}^+$. Now we readily see from the definition of Mellin–Bernstein spaces that $f\Psi_\rho \in \mathscr{B}^2_{c,\pi w}$. Again by the Mellin-Paley–Wiener theorem,

$$f(\cdot,0)\Psi_\rho(\cdot,0) = f(\cdot,0)\psi(\rho^w/(\cdot)^w) \in B^2_{c,\pi w}.$$

Therefore Theorem 12.4 with T replaced by w applies to this function and yields

$$f(r,0)\psi\left(\frac{\rho^w}{r^w}\right) = \sum_{k\in\mathbb{Z}} f(e^{k/w},0)\psi(\rho^w e^{-k})\mathrm{lin}_{c/w}(e^{-k}r^w).$$

This holds for any $\rho \in \mathbb{R}^+$. Substituting $\rho = r$, we obtain (13.18).

It remains to verify the assertion on convergence. Since

$$\mathrm{lin}_{c/w}(e^{-k}r^w) = r^{-c}e^{ck/w}\mathrm{lin}(e^{-k}r^w),$$

we may rewrite the right-hand side of (13.18) as

$$r^{-c}\sum_{k\in\mathbb{Z}} e^{ck/w} f(e^{k/w},0)\psi(e^{-k}r^w)\mathrm{lin}(e^{-k}r^w),$$

and so

$$\begin{aligned}
&\sum_{k\in\mathbb{Z}} \left| f(e^{k/w},0)\psi(e^{-k}r^w)\mathrm{lin}_{c/w}(e^{-k}r^w)\right| \\
&\quad\le r^{-c}\sup_{k\in\mathbb{Z}}\left|e^{ck/w}f(e^{k/w},0)\right|\sum_{k\in\mathbb{Z}}\left|\psi(e^{-k}r^w)\mathrm{lin}(e^{-k}r^w)\right| \\
&\quad\le r^{-c}\|f(\cdot,0)\|_{X_c^\infty}\sum_{k\in\mathbb{Z}}\left|\psi(e^{-k}r^w)\mathrm{lin}(e^{-k}r^w)\right|.
\end{aligned}$$

Now the proof is completed by employing Lemma 13.24 with $\theta = 0$. □

Remark 13.26 Note that formula (13.18) of Theorem 13.25 remains true for all $w > T/(1-\delta)$. This is a consequence of the inclusions between Mellin–Bernstein spaces. Indeed, if $f \in \mathscr{B}^\infty_{c,\pi T}$, then $f \in \mathscr{B}^\infty_{c,\pi T_1}$ for all $T_1 > T$ and so Theorem 13.25 allows us to set $w = T_1/(1-\delta)$. A corresponding observation holds for some of the subsequent statements.

In the following we shall show that Theorem 13.25 can be extended in various ways. It will suffice to consider the case $c = 0$ only. For if $f \in \mathscr{B}^\infty_{c,\pi T}$, then

$$g\,:\,(r,\theta)\longmapsto \left(re^{i\theta}\right)^c f(r,\theta)$$

belongs to $\mathscr{B}^{\infty}_{0,\pi T}$. Hence the reconstruction of f can be obtained by applying Theorem 13.25 with $c = 0$ to g and simply multiplying both sides of the corresponding equation (13.18) by r^{-c}.

First we show that Theorem 13.25 can be extended to a reconstruction of f on the whole of $\mathbb{H}$. For this we need a corresponding extension of the multiplier ψ as guaranteed by Theorem 8.10. We denote it by Ψ.

Theorem 13.27 *For $\delta \in]0, 1[$, let $\Psi \in \mathscr{B}^{2}_{0,\pi\delta}$ such that $\Psi(1, 0) = 1$. Suppose that $f \in \mathscr{B}^{\infty}_{0,\pi T}$, where $T \in \mathbb{R}^{+}$. Then for $w = T/(1-\delta)$, we have*

$$f(r,\theta) = \sum_{k\in\mathbb{Z}} f(e^{k/w}, 0)\Psi(e^{-k}r^{w}, w\theta)\mathrm{lin}(e^{-k}r^{w}e^{iw\theta}) \qquad ((r,\theta) \in \mathbb{H}).$$

The series converges absolutely and uniformly on compact subsets of $\mathbb{H}$.

Proof Since

$$\sup_{k\in\mathbb{Z}} \left| f(e^{k/w}, 0)\right| \leq \|f(\cdot, 0)\|_{X^{\infty}_{0}} < \infty,$$

the assertion on the convergence follows from Lemma 13.24. Hence

$$g(r,\theta) := \sum_{k\in\mathbb{Z}} f(e^{k/w}, 0)\Psi(e^{-k}r^{w}, w\theta)\mathrm{lin}(e^{-k}r^{w}e^{iw\theta})$$

exists for all $(r, \theta) \in \mathbb{H}$ and defines a continuous function g. We also note that, as a function of (r, θ), each term of the series defining g is polar-analytic on $\mathbb{H}$.

Now let $\mathcal{R}$ be any rectangle in $\mathbb{H}$ with its edges parallel to the axes of the (r, θ) coordinate system and denote by $\partial\mathcal{R}$ its positively oriented boundary. Then, for any $N \in \mathbb{N}$,

$$\int_{\partial\mathcal{R}} \sum_{|k|\leq N} f(e^{k/w}, 0)\Psi(e^{-k}r^{w}, w\theta)\mathrm{lin}(e^{-k}r^{w}e^{iw\theta})e^{i\theta}(dr + ird\theta) = 0$$

as a consequence of Cauchy's theorem for polar-analytic functions (see Lemma 2.8). On the other hand, given $\varepsilon > 0$, the uniform convergence on compact subsets of $\mathbb{H}$ guarantees the existence of an integer $N \in \mathbb{N}$ such that

$$\left| \int_{\partial\mathcal{R}} \sum_{|k|>N} f(e^{k/w}, 0)\Psi(e^{-k}r^{w}, w\theta)\mathrm{lin}(e^{-k}r^{w}e^{iw\theta})e^{i\theta}(dr + ird\theta) \right| < \varepsilon.$$

This allows us to conclude that

$$\int_{\partial\mathcal{R}} g(r,\theta)e^{i\theta}(dr + ird\theta) = 0.$$

Now Theorem 2.7 implies that g is polar-analytic on $\mathbb{H}$. Then $h := f - g$ is also polar-analytic on $\mathbb{H}$ and, as a consequence of Theorem 13.25, we have $h(r, 0) = 0$ for all $r \in \mathbb{R}^+$. Hence the identity theorem for polar-analytic functions (see Theorem 3.33) implies that h is identically zero on $\mathbb{H}$, as was to be shown. □

Now we want to generalize Theorem 13.25 in another way. Setting $\varphi := \psi\mathrm{lin}$ and denoting $f(\cdot, 0)$ simply by f, we may write the right-hand side of (13.18) for $c = 0$ as

$$\left(S_w^{\varphi} f\right)(r) := \sum_{k\in\mathbb{Z}} f(e^{k/w})\varphi(e^{-k}r^w) \qquad (r \in \mathbb{R}^+). \tag{13.19}$$

This turns out to be a generalized exponential sampling operator as introduced in Sect. 13.1 which has in addition several interesting properties.

Theorem 13.28 *For $\delta \in]0, 1[$, let $\psi \in B^2_{0,\pi\delta}$ such that $\psi(1) = 1$. Set $\varphi := \psi\mathrm{lin}$. Then the following statements hold:*

(a) $\varphi \in \Phi$ *(see Definition 13.1);*
(b) $\varphi \log \in B^2_{0,\pi(1+\delta)}$*;*
(c) *for each bounded function $f : \mathbb{R}^+ \to \mathbb{C}$ the series* (13.19) *converges absolutely and uniformly on compact subsets of $\mathbb{R}^+$ and*

$$\sup_{r\in\mathbb{R}^+} \left|\left(S_w^{\varphi} f\right)(r)\right| \le \|\psi\|_{X_0^2} \sup_{r\in\mathbb{R}^+} |f(r)|\,;$$

(d) *for each bounded function $f : \mathbb{R}^+ \to \mathbb{C}$ the series* (13.19) *is interpolating with respect to the sample points, that is,*

$$\left(S_w^{\varphi} f\right)(e^{\ell/w}) = f(e^{\ell/w}) \quad (\ell \in \mathbb{Z});$$

(e) *for each $f \in \mathscr{B}^{\infty}_{0,\pi w(1-\delta)}$, the restriction to $\mathbb{R}^+$ is a fixpoint of S_w^{φ}, that is,*

$$S_w^{\varphi} f(\cdot, 0) = f(\cdot, 0).$$

Proof Clearly φ is continuous. Next we want to verify the properties (i)–(iii) of Definition 13.1. Since the constant function $g(r, \theta) \equiv 1$ on $\mathbb{H}$ belongs to any of the spaces $\mathscr{B}^{\infty}_{0,T}$, for every $T > 0$, it follows from Theorem 13.25 applied to the function g that

$$1 = \sum_{k\in\mathbb{Z}} \varphi(e^{-k}u)$$

for all $u \in \mathbb{R}^+$. As a consequence of Lemma 13.24 with $w = 1$ and $\theta = 0$, we have

$$\sum_{k\in\mathbb{Z}} \left|\varphi(e^{-k}u)\right| \le \|\psi\|_{X_0^2} \qquad (u \in \mathbb{R}^+).$$

Next, let $\rho \ge 1$ and denote by $\lfloor\rho\rfloor$ the largest integer not exceeding ρ. Then, by the Cauchy–Schwarz inequality and Theorem 12.50,

$$\begin{aligned}
\sum_{|k-\log u|>\rho} \left|\varphi(e^{-k}u)\right| &\le \left(\sum_{k\in\mathbb{Z}} \left|\psi(e^{-k}u)\right|^2\right)^{1/2} \left(\sum_{|k-\log u|>\rho} \left|\mathrm{lin}(e^{-k}u)\right|^2\right)^{1/2} \\
&\le \|\psi\|_{X_0^2} \left(\sum_{|k-\log u|>\rho} \frac{1}{\pi^2(\log u - k)^2}\right)^{1/2} \\
&\le \frac{\sqrt{2}}{\pi} \|\psi\|_{X_0^2} \left(\sum_{n=\lfloor\rho\rfloor+1}^{\infty} \frac{1}{n^2}\right)^{1/2} \\
&\le \frac{\sqrt{2}}{\pi} \|\psi\|_{X_0^2} \left(\int_{\lfloor\rho\rfloor}^{\infty} \frac{dx}{x^2}\right)^{1/2} = \frac{\sqrt{2}}{\pi} \|\psi\|_{X_0^2} \frac{1}{\lfloor\rho\rfloor^{1/2}}.
\end{aligned}$$

This shows that the left-hand side converges uniformly to zero as $\rho \to \infty$. Altogether, we see that $\varphi \in \Phi$, and so (a) is proved.

Next we note that

$$\varphi(r)\log r = \psi(r)\mathrm{lin}\, r \log r = \psi(r)\frac{\sin(\pi \log r)}{\pi} \qquad (r \in \mathbb{R}^+).$$

Since $\psi \in B^2_{0,\pi\delta}$, there exists, by Theorem 8.10, a function $\Psi \in \mathscr{B}^2_{0,\pi\delta}$ such that $\psi(r) = \Psi(r,0)$ and so the function $F(r,\theta) := \Psi(r,\theta)\frac{\sin(\pi(\log r + i\theta))}{\pi}$ belongs to $\mathscr{B}^2_{0,\pi(\delta+1)}$. Again by Theorem 8.10, we see that $\varphi(r)\log r = F(r,0)$ belongs to $B^2_{0,\pi(\delta+1)}$, and so (b) holds.

Assertion (c) is verified with the help of Lemma 13.24.

As regards (d), it suffices to note that if $\ell \in \mathbb{Z}$ and $r = e^{\ell/w}$, then

$$\varphi(e^{-k}r^w) = \varphi(e^{\ell-k}) = \psi(e^{\ell-k})\mathrm{lin}(e^{\ell-k}) = \psi(e^{\ell-k})\mathrm{sinc}(\ell-k) = \delta_{\ell,k}$$

with Kronecker's delta.

Finally assertion (e) is contained in Theorem 13.25. □

It is remarkable that Theorem 13.28 has a converse. For this we need statements (b) and (d) only. In fact, any generalized exponential sampling series (13.19) that is interpolating and has a Mellin bandlimited φ satisfying (b) is a multiplier version of the sampling series of Theorem 12.4.

Theorem 13.29 *A generalized exponential sampling series* (13.19) *has the properties* (b) *and* (d) *of Theorem 13.28 if and only if there exists a* $\psi \in B^2_{0,\pi\delta}$ *such that* $\psi(1) = 1$ *and* $\varphi = \psi \,\mathrm{lin}$.

Proof The sufficiency is already contained in Theorem 13.28.

Necessity. Suppose that statements (b) and (d) of Theorem 13.28 hold. Then, by Theorem 8.10, there exists a function $F \in \mathscr{B}^2_{0,\pi(1+\delta)}$ such that $F(r, 0) = \varphi(r) \log r$ for all $r \in \mathbb{R}^+$. Now consider

$$G(r, \theta) := \frac{\pi F(r, \theta)}{\sin(\pi(\log r + i\theta))}. \tag{13.20}$$

The denominator on the right-hand side has simple zeros at the points $(e^\ell, 0)$ for $\ell \in \mathbb{Z}$ and is different from zero at all other points of $\mathbb{H}$. Hence G is polar-analytic on $\mathbb{H} \setminus \{(e^\ell, 0) : \ell \in \mathbb{Z}\}$. Since lin is bounded on $\mathbb{R}^+$, we conclude from statement (d) with $w = 1$ that

$$\left(S_1^\varphi \mathrm{lin}\right)(e^\ell) = \sum_{k\in\mathbb{Z}} \mathrm{lin}(e^k)\varphi(e^{\ell-k}) = \varphi(e^\ell) = \mathrm{lin}(e^\ell). \tag{13.21}$$

Thus

$$F(e^\ell, 0) = \varphi(e^\ell)\ell = \mathrm{sinc}(\ell)\ell = 0$$

for all $\ell \in \mathbb{Z}$. Hence the points $(e^\ell, 0)$ for $\ell \in \mathbb{Z}$ are all removable isolated singularities of G, and so G has a continuation that is polar-analytic on the whole of $\mathbb{H}$.

It is easily verified that

$$|\sin(\pi(\log r + i\theta))| \geq |\sinh(\pi\theta)| \geq \frac{1 - e^{-2\pi}}{2} e^{\pi|\theta|} \qquad (|\theta| \geq 1). \tag{13.22}$$

As a consequence of statement (b), we have

$$|F(r, \theta)| \leq C_F e^{\pi(1+\delta)|\theta|} \qquad ((r, \theta) \in \mathbb{H}) \tag{13.23}$$

with a constant C_F depending on F only. Hence

$$|G(r, \theta)| \leq \frac{2\pi C_F}{1 - e^{-2\pi}} e^{\pi\delta|\theta|} \qquad (|\theta| \geq 1). \tag{13.24}$$

Next, let $N \in \mathbb{N}$ and let $\mathcal{R}_N$ be a rectangle with vertices at $(e^{\pm(N+1/2)}, \pm 1)$. On its vertical edges, we have $|\sin(\pi(\log r + i\theta))| \geq \cosh(\pi\theta)$, and so (13.23) yields that on these line segments

$$|G(r, \theta)| \leq \frac{2\pi C_F}{1 + e^{-2\pi|\theta|}} e^{\pi\delta|\theta|} \leq \frac{2\pi C_F}{1 + e^{-2\pi}} e^{\pi\delta}. \tag{13.25}$$

On the horizontal edges, it follows from (13.24) that

$$|G(r,\theta)| \le \frac{2\pi C_F}{1-e^{-2\pi}}\, e^{\pi\delta}. \tag{13.26}$$

The bounds (13.25) and (13.26) do not depend on N. By the maximum principle for polar-analytic functions (see Theorem 3.36), it follows that

$$|G(r,\theta)| \le \frac{2\pi C_F}{1-e^{-2\pi}}\, e^{\pi\delta} \qquad (|\theta| \le 1).$$

Combined with (13.24), we see that

$$|G(r,\theta)| \le C_G\, e^{\pi\delta|\theta|} \qquad ((r,\theta) \in \mathbb{H}), \tag{13.27}$$

where

$$C_G = \frac{2\pi C_F}{1-e^{-2\pi}}\, e^{\pi\delta}.$$

We would like to show that $G(\cdot,0) \in X_0^2$ which is a little delicate because of the removable singularities. Therefore, we first show that $G(\cdot,1) \in X_0^2$. By the translation invariance of Mellin–Bernstein spaces, we know that $F(\cdot,1) \in X_0^2$ and by Theorem 3.42 (a),

$$\|F(\cdot,1)\|_{X_0^2} \le e^{\pi(1+\delta)}\|F(\cdot,0)\|_{X_0^2}.$$

From this we conclude with the help of (13.22) that $G(\cdot,1) \in X_0^2$ and

$$\|G(\cdot,1)\|_{X_0^2} \le \frac{2\pi e^{\pi\delta}}{1-e^{-2\pi}}\|F(\cdot,0)\|_{X_0^2}.$$

Now, by employing the Cauchy–Riemann equations in polar form and manipulating (13.27) slightly, we find that $G_1 : (r,\theta) \mapsto G(r,\theta+1)$ belongs to $\mathscr{B}^2_{0,\pi\delta}$. But then, by the translation invariance, G itself belongs to $\mathscr{B}^2_{0,\pi\delta}$.

Finally, setting $\psi := G(\cdot,0)$, we have by the Theorem 8.10, that $\psi \in B^2_{0,\pi\delta}$ and

$$\psi(r)\mathrm{lin}(r) = \frac{F(r,0)}{\log r} = \varphi(r) \qquad (r \in \mathbb{R}^+).$$

Furthermore $\psi(1) = \varphi(1) = \mathrm{lin}(1) = 1$ as a consequence of (13.21) for $\ell = 0$. This completes the proof. □

13.5.2 Approximation of Bounded, Continuous Functions

The approximation power of $S_w^{\psi\,\mathrm{lin}} f$ can be expressed in terms of the best approximation of f by functions from $\mathscr{B}^\infty_{0,\pi T}$, where $T = w(1-\delta)$. For this, we introduce

$$E^\infty_{\pi T}[f] := \inf_{g \in \mathscr{B}^\infty_{0,\pi T}} \|f - g(\cdot, 0)\|_{X_0^\infty}.$$

The announced result can now be stated as follows:

Proposition 13.30 *For $\delta \in]0, 1[$, let $\psi \in B^2_{0,\pi\delta}$ such that $\psi(1) = 1$. Suppose that $f \in C(\mathbb{R}^+) \cap X_0^\infty$. Then for $w \in \mathbb{R}^+$ and $T = w(1-\delta)$, we have*

$$\left\| f - S_w^{\psi\,\mathrm{lin}} f \right\|_{X_0^\infty} \le \left(1 + \|\psi\|_{X_0^2}\right) E^\infty_{\pi T}[f]. \tag{13.28}$$

Proof Given $\varepsilon > 0$, there exists a function $g_\varepsilon \in \mathscr{B}^\infty_{0,\pi T}$ such that

$$\|f - g_\varepsilon(\cdot, 0)\|_{X_0^\infty} \le E^\infty_{\pi T}[f] + \varepsilon.$$

By Theorem 13.25, we have $g_\varepsilon(r, 0) = (S_w^{\psi\,\mathrm{lin}} g_\varepsilon(\cdot, 0))(r)$. Now employing statement (c) of Theorem 13.28, we obtain

$$\begin{aligned}
\left\| f - S_w^{\psi\,\mathrm{lin}} f \right\|_{X_0^\infty} &\le \|f - g_\varepsilon(\cdot, 0)\|_{X_0^\infty} + \left\| S_w^{\psi\,\mathrm{lin}} g_\varepsilon(\cdot, 0) - S_w^{\psi\,\mathrm{lin}} f \right\|_{X_0^\infty} \\
&\le E^\infty_{\pi T}[f] + \varepsilon + \left\| S_w^{\psi\,\mathrm{lin}} (g_\varepsilon(\cdot, 0) - f) \right\|_{X_0^\infty} \\
&\le E^\infty_{\pi T}[f] + \varepsilon + \|\psi\|_{X_0^2} \left(E^\infty_{\pi T}[f] + \varepsilon\right).
\end{aligned}$$

The left-hand side does not depend on ε. Letting $\varepsilon \to 0$, we arrive at (13.28). □

Note that $S_w^{\psi\,\mathrm{lin}} f$ comes close to a best approximation by functions from $\mathscr{B}^\infty_{0,\pi T}$ but misses it in two ways. First $S_w^{\psi\,\mathrm{lin}} f$ is the restriction to $\mathbb{R}^+$ of a function from $\mathscr{B}^\infty_{0,\pi w(1+\delta)}$ while $T = w(1-\delta)$. Secondly, there is a factor $1 + \|\psi\|_{X_0^2}$ on the right-hand side of (13.28), which is at least as big as $1+\delta^{-1/2}$. Indeed, by Theorem 12.50, we have

$$\|\psi\|^2_{X_0^2} = \frac{1}{\delta} \sum_{k\in\mathbb{Z}} \left|\psi\left(e^{k/\delta}\right)\right|^2 \ge \frac{1}{\delta}\, |\psi(1)|^2 = \frac{1}{\delta},$$

where equality is attained for $\psi(r) = \mathrm{lin}(r^\delta)$.

Next we want to estimate the truncation error of the series $(S_w^{\psi \text{lin}} f)(r)$. It will depend on the decay of $|\psi(r)|$ as $r \to 0+$ and $r \to +\infty$. Therefore we will classify multipliers ψ according to their decay.

Definition 13.31 Let $\delta \in]0, 1[$ and let μ be a positive, decreasing function on $[0, +\infty[$ such that $\mu(t) \to 0$ as $t \to +\infty$. Then the class $B^2_{0,\pi\delta}(\mu)$ comprises all functions $\psi \in B^2_{0,\pi\delta}$ such that $\psi(1) = 1$ and

$$|\psi(r) \log r| \le \mu(|\log r|) \qquad (r \in \mathbb{R}^+). \tag{13.29}$$

Example 13.32 For an integer $m \ge 2$, let $\psi(r) := \text{lin}^m(r^{\delta/m})$ and let $\kappa := \pi\delta/m$. Then $\psi \in B^2_{0,\pi\delta}$, $\psi(1) = 1$ and

$$|\psi(r) \log r| = \left| \frac{\sin^m(\log r^\kappa)}{\kappa(\log r^\kappa)^{m-1}} \right| \le \frac{1}{\kappa^m \, |\log r|^{m-1}} \qquad (r \in \mathbb{R}^+ \setminus \{1\}).$$

We also have

$$|\psi(r) \log r| = \left| \text{lin}^{m-1}(r^{\delta/m}) \right| \frac{|\sin(\log r^\kappa)|}{\kappa} \le \frac{1}{\kappa} \qquad (r \in \mathbb{R}^+).$$

The first bound does not extend to $r = 1$, the second is not decreasing, but the harmonic mean of these two bounds leads to

$$|\psi(r) \log r| \le \frac{2}{\kappa(1 + (\kappa \, |\log r|)^{m-1})} \qquad (r \in \mathbb{R}^+).$$

Hence

$$\mu(t) := \frac{2}{\kappa(1 + (\kappa t)^{m-1})} \qquad (t \in [0, +\infty[)$$

defines an admissible function μ that yields (13.29).

When $r \in \mathbb{R}^+$ is given and we want to approximate $(S_w^{\psi \text{lin}} f)(r)$ by $2N + 1$ terms of this series, it seems reasonable to take those terms whose sample points $e^{k/w}$ are closest to r. This means that we associate with (w, r) the integer $N_{w,r} := \lfloor w \log r + 1/2 \rfloor$ and keep only those terms of the series for which

$$k \in \left\{ N_{w,r}, N_{w,r} \pm 1, \ldots, N_{w,r} \pm N \right\}.$$

The truncated series obtained in this way shall be denoted by

$$\left(S_{w,N}^{\psi \text{lin}} f \right)(r) := \sum_{k=N_{w,r}-N}^{N_{w,r}+N} f(e^{k/w}) \psi(e^{-k} r^w) \text{lin}(e^{-k} r^w). \tag{13.30}$$

We also use the notation

$$\| f \|_{w,\infty} := \sup_{k\in\mathbb{Z}} \left| f(e^{k/w}) \right| .$$

The following proposition provides an estimate of the truncation error in terms of μ and N.

Proposition 13.33 *Let $\delta \in]0, 1[$, $\psi \in B^2_{0,\pi\delta}(\mu)$ and $w \in \mathbb{R}^+$. Suppose that f is a function defined on $\mathbb{R}^+$ such that $\| f \|_{w,\infty} < \infty$. Then*

$$\left| \left(S_w^{\psi \text{lin}} f \right)(r) - \left(S_{w,N}^{\psi \text{lin}} f \right)(r) \right| \le |\sin(\pi w \log r)| \, \| f \|_{w,\infty} \frac{2\mu(N+\frac{1}{2})}{\pi N} \tag{13.31}$$

for $N \in \mathbb{N}$ and $r \in \mathbb{R}^+$. This estimate is sharp. Equality is attained for $r = e^{k/w}$ with $k \in \mathbb{Z}$.

Proof Obviously,

$$\begin{aligned} &\left| \left(S_w^{\psi \text{lin}} f \right)(r) - \left(S_{w,N}^{\psi \text{lin}} f \right)(r) \right| \\ &\quad \le \| f \|_{w,\infty} \sum_{|k-N_{w,r}|>N} \left| \psi(e^{-k} r^w) \text{lin}(e^{-k} r^w) \right| \\ &\quad = \| f \|_{w,\infty} \sum_{|j|>N} \left| \psi(\exp(\log r^w - N_{w,r} - j)) \right| \frac{|\sin(\pi w \log r)|}{\pi \left| \log r^w - N_{w,r} - j \right|} . \end{aligned}$$

Since $\log r^w - N_{w,r} \in] - \frac{1}{2}, \frac{1}{2}]$, we find that $\left|\log r^w - N_{w,r} - j\right| \ge |j| - \frac{1}{2}$ for $|j| > N$ and so (13.29) implies

$$\left| \psi(\exp(\log r^w - N_{w,r} - j)) \right| \le \frac{\mu(|j| - \frac{1}{2})}{|j| - \frac{1}{2}} \le \frac{\mu(N + \frac{1}{2})}{|j| - \frac{1}{2}} \qquad (|j| > N).$$

Hence

$$\begin{aligned} &\left| \left(S_w^{\psi \text{lin}} f \right)(r) - \left(S_{w,N}^{\psi \text{lin}} f \right)(r) \right| \\ &\quad \le \frac{2}{\pi} \| f \|_{w,\infty} |\sin(\pi w \log r)| \, \mu \left(N + \frac{1}{2} \right) \sum_{j=N+1}^{\infty} \frac{1}{(j - \frac{1}{2})^2} . \end{aligned}$$

Finally, noting that

$$\frac{1}{(j - \frac{1}{2})^2} < \int_{j-1}^{j} \frac{dx}{x^2} ,$$

we obtain

$$\sum_{j=N+1}^{\infty} \frac{1}{(j-\frac{1}{2})^2} < \int_N^{\infty} \frac{dx}{x^2} = \frac{1}{N}. \tag{13.32}$$

This completes the proof. □

Remark 13.34 The bound on the right-hand side of (13.31) can be simplified by replacing it with

$$\sup_{r\in\mathbb{R}^+} |f(r)| \, \frac{2\mu(N)}{\pi N}.$$

However, the sine in (13.31) shows that the truncated series is still interpolating.

Proposition 13.33 combined with Theorem 13.25 implies the following result.

Corollary 13.35 *Let $\delta \in]0,1[$ and $\psi \in B^2_{0,\pi\delta}(\mu)$. Suppose that $f \in \mathscr{B}^{\infty}_{0,\pi T}$, where $T \in \mathbb{R}^+$. Then, for $w = T/(1-\delta)$, we have*

$$\left| f(r,0) - \left(S^{\psi \mathrm{lin}}_{w,N} f(\cdot,0)\right)(r)\right| \le |\sin(\pi w \log r)| \, \|f\|_{X_0^\infty} \frac{2\mu(N+\frac{1}{2})}{\pi N}$$

for $N \in \mathbb{N}$ and $r \in \mathbb{R}^+$. This estimate is sharp. Equality is attained for $r = e^{k/w}$ with $k \in \mathbb{Z}$.

13.5.3 Extensions to Unbounded Functions

Now we want to show that Proposition 13.33 has a generalization that admits unbounded functions f. Roughly speaking, this is achieved by taking a portion of the decaying multiplier ψ for compensating the growth of f while the remaining portion serves for the decay of the truncation error as $N \to \infty$.

Theorem 13.36 *For $\delta \in]0,1[$, let $\psi \in B^2_{0,\pi\delta}(\mu)$ and let f be a function defined on $\mathbb{R}^+$. Suppose that*

$$|f(r)| \le K(|\log r|) \qquad (r \in \mathbb{R}^+),$$

where K is a non-decreasing, positive function defined on $[0,+\infty[$ such that

$$K(t)\mu(t)^{\lambda} \le K(0)\mu(0)^{\lambda} \qquad (0 \le t < +\infty) \tag{13.33}$$

for some $\lambda \in [0,1]$. Then for $w > 1$ the generalized exponential sampling series $(S^{\psi \mathrm{lin}}_w f)(r)$ converges absolutely for each $r \in \mathbb{R}^+$ and, in the notation (13.30),

$$\left|\left(S_w^{\psi \text{lin}} f\right)(r) - \left(S_{w,N}^{\psi \text{lin}} f\right)(r)\right|$$
$$\leq |\sin(\pi w \log r)|\, K\left(\frac{w\,|\log r| + 1}{w-1}\right) \mu(0)^{\lambda}\, \frac{2\mu(N+\frac{1}{2})^{1-\lambda}}{\pi N} \tag{13.34}$$

for $N \in \mathbb{N}$. This estimate is sharp. Equality is attained for $r = e^{k/w}$ with $k \in \mathbb{Z}$.

Proof The following verification of (13.34) will also show the absolute convergence of the series $(S_w^{\psi \text{lin}} f)(r)$. First we observe that

$$\left|\left(S_w^{\psi \text{lin}} f\right)(r) - \left(S_{w,N}^{\psi \text{lin}} f\right)(r)\right|$$
$$\leq \sum_{|k-N_{w,r}|>N} \left| f(e^{k/w})\psi(e^{-k}r^w)\text{lin}(e^{-k}r^w)\right|$$
$$= \sum_{|j|>N} \left| f(e^{(j+N_{w,r})/w})\psi(e^{w\log r - N_{w,r}-j})\text{lin}(e^{w\log r - N_{w,r}-j})\right|$$
$$\leq |\sin(\pi w \log r)| \sum_{|j|>N} K\left(\frac{\left|j+N_{w,r}\right|}{w}\right) \frac{\mu(\left|w\log r - N_{w,r} - j\right|)}{\pi\left|w\log r - N_{w,r} - j\right|^2}.$$

Next we estimate the term

$$T_j := K\left(\frac{\left|j+N_{w,r}\right|}{w}\right) \frac{\mu(\left|w\log r - N_{w,r} - j\right|)}{\pi\left|w\log r - N_{w,r} - j\right|^2} \qquad (|j| > N).$$

Noting again that $\xi := w\log r - N_{w,r} \in]-\frac{1}{2}, \frac{1}{2}]$, we find that

$$T_j = K\left(\left|\log r + \frac{j-\xi}{w}\right|\right) \frac{\mu(|j-\xi|)}{\pi\,|j-\xi|^2}$$
$$\leq K\left(|\log r| + \frac{|j|+\frac{1}{2}}{w}\right) \frac{\mu(|j|-\frac{1}{2})^{\lambda}}{\pi(|j|-\frac{1}{2})^2}\, \mu(N+\tfrac{1}{2})^{1-\lambda} \qquad (|j| > N).$$

We distinguish two cases: If

$$|\log r| + \frac{|j|+\frac{1}{2}}{w} \leq |j| - \tfrac{1}{2},$$

then

$$K\left(|\log r| + \frac{|j|+\frac{1}{2}}{w}\right) \mu(|j|-\tfrac{1}{2})^{\lambda} \leq K(|j|-\tfrac{1}{2})\mu(|j|-\tfrac{1}{2})^{\lambda} \leq K(0)\mu(0)^{\lambda},$$

and so

$$T_j \le K(0)\mu(0)^\lambda \frac{\mu(N+\frac{1}{2})^{1-\lambda}}{\pi(|j|-\frac{1}{2})^2}. \tag{13.35}$$

If

$$|\log r| + \frac{|j|+\frac{1}{2}}{w} > |j| - \frac{1}{2},$$

then

$$|j| + \frac{1}{2} < \frac{w}{w-1}(|\log r| + 1).$$

Therefore

$$K\left(|\log r| + \frac{|j|+\frac{1}{2}}{w}\right) \le K\left(\frac{w\,|\log r|+1}{w-1}\right),$$

and so

$$T_j \le K\left(\frac{w\,|\log r|+1}{w-1}\right)\frac{\mu(N+\frac{1}{2})}{\pi(|j|-\frac{1}{2})^2}. \tag{13.36}$$

Comparing the estimates (13.35) and (13.36), we see that in any case

$$T_j \le K\left(\frac{w\,|\log r|+1}{w-1}\right)\mu(0)^\lambda \frac{\mu(N+\frac{1}{2})^{1-\lambda}}{\pi(|j|-\frac{1}{2})^2} \qquad (|j| > N).$$

The proof is completed by employing inequality (13.32) for the summation over j. □

Remark 13.37 The choice of λ allows us some flexibility. For $\lambda = 0$, we see from condition (13.33) that only bounded functions f are admissible and then we essentially recover Proposition 13.33 with the rate of convergence to zero of the truncation error being $\mathcal{O}(\mu(N)/N)$ as $N \to \infty$. Note that inequality (13.33) remains true if λ is replaced by any $\lambda' \in]\lambda, 1]$. We may even raise both sides of (13.33) to the power λ'/λ and obtain

$$K(t)^{\lambda'/\lambda}\mu(t)^{\lambda'} \le K(0)^{\lambda'/\lambda}\mu(0)^{\lambda'} \qquad (0 \le t < +\infty).$$

This means that, if $K(0) \ge 1$, we can enlarge the class of admissible functions by requiring that

$$|f(r)| \le K(|\log r|)^{\lambda'/\lambda} \qquad (r \in \mathbb{R}^+),$$

but for the functions added in this way the guaranteed rate of convergence of the truncation error will be $\mathcal{O}(\mu(N)^{1-\lambda'}/N)$ only. When we choose $\lambda' = 1$, we fix a largest class of admissible functions but the speed of convergence in the worst case reduces to $\mathcal{O}(1/N)$.

As a suitable majorant $K(\cdot)$ for the admissible functions we can always take $K(t) := c\mu(t)^{-\lambda}$ with a constant $c > 0$, which satisfies (13.33) trivially.

Theorem 13.36 shows that $(S_w^{\psi \text{lin}} f)(r)$ also exists for certain classes of functions f that are unbounded on $\mathbb{R}^+$, but so far we do not know if this generalized exponential sampling series can reproduce or at least approximate f. We now obtain an extension of Corollary 13.35 to unbounded $f(\cdot, 0)$. For this, we need an extension of Theorem 13.25 to unbounded $f(\cdot, 0)$.

In order to do that, we first introduce a Mellin–Bernstein analogue of the classes $B^2_{0,\pi\delta}(\mu)$.

Definition 13.38 Let $\delta \in]0, 1[$ and let μ be a positive, decreasing function on $[0, +\infty[$ such that $\mu(t) \to 0$ as $t \to +\infty$. Then the class $\mathscr{B}^2_{0,\pi\delta}(\mu)$ comprises all functions $\Psi \in \mathscr{B}^2_{0,\pi\delta}$ such that $\Psi(1, 0) = 1$ and

$$|\Psi(r, \theta)(\log r + i\theta)| \le \mu(|\log r|)e^{\pi\delta|\theta|} \qquad ((r, \theta) \in \mathbb{H}).$$

With the help of the Mellin–Paley–Wiener theorem (Theorem 8.10) it is readily seen that if $\Psi \in \mathscr{B}^2_{0,\pi\delta}(\mu)$, then $\Psi(\cdot, 0) \in B^2_{0,\pi\delta}(\mu)$.

Example 13.39 We want to show that the multiplier ψ in Example 13.32 can be extended to a multiplier $\Psi \in \mathscr{B}^2_{0,\pi\delta}(\mu)$ with the same function μ. Indeed, for an integer $m \ge 2$, let

$$\Psi(r, \theta) := \text{lin}^m\big((re^{i\theta})^{\delta/m}\big) \qquad ((r, \theta) \in \mathbb{H})$$

and let $\kappa := \pi\delta/m$. Then $\Psi \in \mathscr{B}^2_{0,\pi\delta}$ and $\Psi(1, 0) = 1$. Now, using well-known estimates for the sine function and the sinc function in the complex plane, we find that

$$|\Psi(r, \theta)\log(r + i\theta)| = \left|\frac{\sin^m(\kappa(\log r + i\theta))}{\kappa^m(\log r + i\theta)^{m-1}}\right| \le \frac{e^{\pi\delta|\theta|}}{\kappa^m\,|\log r|^{m-1}}$$

valid for $(r, \theta) \in (\mathbb{R}^+ \setminus \{1\}) \times \mathbb{R}$. We also have

$$\begin{aligned}
|\Psi(r, \theta)\log(r + i\theta)| &= \left|\text{lin}^{m-1}\big((re^{i\theta})^{\delta/m}\big)\right| \frac{|\sin(\kappa(\log r + i\theta))|}{\kappa} \\
&= \frac{1}{\kappa}\left|\text{sinc}^{m-1}\left(\frac{\delta}{m}(\log r + i\theta)\right)\right| \big|\sin\big(\kappa(\log r + i\theta)\big)\big| \\
&\le \frac{e^{\pi\delta|\theta|}}{\kappa}
\end{aligned}$$

valid for all $(r, \theta) \in \mathbb{H}$. Taking the harmonic mean of these estimates, we obtain

$$|\Psi(r,\theta)(\log r + i\theta)| \leq \mu(|\log r|)e^{\pi\delta|\theta|} \qquad ((r,\theta) \in \mathbb{H})$$

with the same function μ as in Example 13.32.

Theorem 13.40 *For $\delta \in]0, 1[$, let $\Psi \in \mathscr{B}^2_{0,\pi\delta}(\mu)$ and let f be polar-analytic on $\mathbb{H}$. Suppose that*

$$|f(r,\theta)| \leq K(|\log r|)e^{\pi T|\theta|} \qquad ((r,\theta) \in \mathbb{H}),$$

where $T \in \mathbb{R}^+$ and K is a non-decreasing, positive function defined on $[0, +\infty[$ such that

$$K(t)\mu(t)^{\lambda} \leq K(0)\mu(0)^{\lambda} \qquad (0 \leq t < +\infty)$$

for some $\lambda \in [0, 1]$. Then for $w > 1$ such that $w \geq T/(1-\delta)$, we have

$$f(r,0) = \sum_{k\in\mathbb{Z}} f(e^{k/w}, 0)\Psi(e^{-k}r^w, 0)\text{lin}(e^{-k}r^w) \qquad (r \in \mathbb{R}^+). \tag{13.37}$$

The series converges absolutely for each $r \in \mathbb{R}^+$ and, writing $\psi := \Psi(\cdot, 0)$, there holds the error estimate

$$\left|f(r,0) - \left(S^{\psi\text{lin}}_{w,N} f\right)(r)\right|$$
$$\leq |\sin(\pi w \log r)|\, K\left(\frac{w\,|\log r| + 1}{w-1}\right)\mu(0)^{\lambda}\,\frac{2\mu(N+\frac{1}{2})^{1-\lambda}}{\pi N} \tag{13.38}$$

for $N \in \mathbb{N}$.

Proof For $\rho \in \mathbb{R}^+ \setminus \{1\}$ as a parameter, consider the function

$$F_\rho : (r,\theta) \longmapsto f(r,\theta)\,\Psi\left(\frac{\rho^w}{r^w}, -w\theta\right) \qquad ((r,\theta) \in \mathbb{H}).$$

We want to show that $F_\rho \in \mathscr{B}^2_{0,\pi w}$. For this, we have to verify properties (i)–(iii) of the definition of Mellin–Bernstein spaces (see Definition 3.41).

Clearly, as a product of polar-analytic functions, F_ρ is itself polar-analytic on $\mathbb{H}$, and so (i) holds.

Next, we want to estimate F_ρ on $\mathbb{H}$. Since $\Psi \in \mathscr{B}^2_{0,\pi\delta}(\mu)$, we conclude from property (iii) of the Mellin–Bernstein spaces that

$$\left|\Psi\left(\frac{\rho^w}{r^w}, -w\theta\right)\right| \leq C_\Psi e^{\pi\delta w|\theta|} \tag{13.39}$$

and from the estimate of Definition 13.38 that

$$\left|\Psi\left(\frac{\rho^w}{r^w}, -w\theta\right)\right| \leq \frac{\mu(w\,|\log r - \log\rho|)}{w\,|\log r - \log\rho|}\, e^{\pi\delta w|\theta|} \qquad (r \neq \rho). \tag{13.40}$$

Now we put $a := \frac{w}{w-1}\,|\log\rho|$ and distinguish two cases:

If $r \in [e^{-a}, e^{a}]$, then, using (13.39), we have

$$\left|F_\rho(r,\theta)\right| \leq K(a)C_\Psi e^{\pi(T+\delta w)|\theta|} \leq K\left(\frac{w}{w-1}\,|\log\rho|\right) C_\Psi e^{\pi w|\theta|}. \tag{13.41}$$

If $r \in \mathbb{R}^+ \setminus [e^{-a}, e^{a}]$, then $r \neq \rho$ and $|\log r| < w\,|\log r - \log\rho|$. Therefore we may use (13.40) to conclude that

$$\begin{aligned}
\left|F_\rho(r,\theta)\right| &\leq K(|\log r|)\frac{\mu(w\,|\log r - \log\rho|)}{w\,|\log r - \log\rho|}\, e^{\pi w|\theta|} \\
&\leq K(w\,|\log r - \log\rho|)\frac{\mu(w\,|\log r - \log\rho|)}{w\,|\log r - \log\rho|}\, e^{\pi w|\theta|} \\
&\leq K(0)\mu(0)^\lambda\,\frac{\mu(w\,|\log r - \log\rho|)^{1-\lambda}}{w\,|\log r - \log\rho|}\, e^{\pi w|\theta|} \\
&\leq \frac{K(0)\mu(0)}{w\,|\log r - \log\rho|}\, e^{\pi w|\theta|} \leq \frac{(w-1)K(0)\mu(0)}{w\,|\log\rho|}\, e^{\pi w|\theta|}.
\end{aligned} \tag{13.42}$$

Thus, setting

$$C_{F_\rho} := \max\left\{K\left(\frac{w}{w-1}\,|\log\rho|\right) C_\Psi\,,\ \frac{(w-1)K(0)\mu(0)}{w\,|\log\rho|}\right\},$$

we have

$$\left|F_\rho(r,\theta)\right| \leq C_{F_\rho}\, e^{\pi w|\theta|} \qquad \big((r,\theta) \in \mathbb{H},\ \rho \in \mathbb{R}^+ \setminus \{1\}\big),$$

and so property (iii) of the Mellin–Bernstein spaces is verified for F_ρ.

It remains to verify property (ii), that is, $F_\rho(\cdot, 0) \in X_0^2$. Clearly,

$$\int_{e^{-a}}^{e^{a}} \left|F_\rho(r,0)\right|^2 \frac{dr}{r} < +\infty.$$

For the integration over the complementary subset of $\mathbb{R}^+$, we may use (13.42) to conclude that

$$\int_{\mathbb{R}^+\setminus[e^{-a},e^a]} \left|F_\rho(r,0)\right|^2 \frac{dr}{r} \le \left(\frac{K(0)\mu(0)}{w}\right)^2 \int_{\mathbb{R}^+\setminus[e^{-a},e^a]} \frac{1}{(\log r-\log\rho)^2}\frac{dr}{r}$$
$$= \left(\frac{K(0)\mu(0)}{w}\right)^2 \int_{\mathbb{R}\setminus[-a,a]} \frac{dt}{(t-\log\rho)^2}$$
$$= \left(\frac{K(0)\mu(0)}{w}\right)^2 \left(\frac{1}{a+\log\rho}+\frac{1}{a-\log\rho}\right)$$
$$= \frac{2(w-1)K(0)^2\mu(0)^2}{w(2w-1)\,|\log\rho|} < +\infty.$$

Thus $F_\rho(\cdot,0)\in X_0^2$. Altogether we have shown that $F_\rho\in\mathscr{B}^2_{0,\pi w}$ for each $\rho\in\mathbb{R}^+\setminus\{1\}$. As a consequence of the Mellin–Paley–Wiener theorem (Theorem 8.10), it follows that $F_\rho(\cdot,0)\in B^2_{0,\pi w}$ for these values of ρ. Therefore Theorem 12.4 with $c=0$ and T replaced by w applies to F_ρ for any positive $\rho\ne 1$, and we obtain

$$f(r,0)\Psi\left(\frac{\rho^w}{r^w},0\right)=\sum_{k\in\mathbb{Z}} f(e^{k/w},0)\Psi(\rho^w e^{-k},0)\mathrm{lin}(e^{-k}r^w).$$

For each $r\in\mathbb{R}^+\setminus\{1\}$, we may substitute $\rho=r$ which gives (13.37) except for $r=1$. But for $r=1$ Eq. (13.37) holds trivially since $\Psi(e^{-k},0)\mathrm{lin}(e^{-k})=\delta_{0,k}$ with Kronecker's delta.

Finally, the statement on convergence of the series and the error estimate (13.38) are an immediate consequence of Theorem 13.36. □

Remark 13.41 Theorem 13.36 covers the growth conditions used in Proposition 13.10, namely

$$|f(r)|\le a+b\,|\log r|^m \qquad (m\in\mathbb{N},\ a,b,r\in\mathbb{R}^+),$$

or equivalently,

$$\left|f(e^x)\right|\le a+b\,|x|^m \qquad (m\in\mathbb{N},\ a,b\in\mathbb{R}^+,\ x\in\mathbb{R}).$$

Theorem 13.36 can also be seen as a statement on the domain of $S_w^{\psi\,\mathrm{lin}}$. Theorem 13.40 shows the existence of unbounded functions f which not only lie in the domain of this operator but are reproduced by it, or equivalently, belong to its fixpoints.

13.5.4 A Multiplier Based on a Gaussian Function

In view of Proposition 13.33 and Theorem 13.36 it is desirable to have a rapidly decaying multiplier ψ. This raises the non-trivial question as to how fast ψ can decay. Similarly, prescribing μ as a rapidly decaying function, we may ask if the class $B^2_{0,\pi\delta}(\mu)$ specified in Definition 13.31 will be non-empty?

A complete answer is described in Sect. 13.6 Notes. It will show that if for some $\alpha > 0$ we have $\mu(t) = \mathcal{O}(e^{-\alpha t})$ as $t \to \infty$, then $B^2_{0,\pi\delta}(\mu) = \emptyset$. But there exist $\psi \in B^2_{0,\pi\delta}(\mu)$ under the slightly slower decay

$$\mu(t) = \mathcal{O}\left(e^{-t/(\log t)^\gamma}\right) \qquad (\gamma > 1,\ t \to \infty).$$

Unfortunately such multipliers ψ are not easily available. They can be constructed by an infinite product of lin functions (see [156] for corresponding products of sinc functions) which is quite inconvenient for computations. However, there is a striking observation. When we analyse the proofs of Proposition 13.33 and Theorem 13.36, we find that only the estimate (13.29) is needed but we nowhere use that $\psi \in B^2_{0,\pi\delta}$. For example, we could involve a Gaussian function by taking

$$\psi(r) := e^{-2(\log r)^2}. \tag{13.43}$$

Then inequality (13.29) will hold with $\mu(t) = e^{-t^2}$ and (13.31) will be valid with this function μ. Thus we obtain an exceptionally fast converging truncation error by using a very convenient multiplier. However, since this multiplier ψ is not Mellin bandlimited, Theorem 13.25, Corollary 13.35 and also Theorem 13.40 are not applicable. We produce a so-called aliasing error which may prevent $S_w^{\psi \mathrm{lin}} f$ from being a good approximation to f. We should try to find a balance between the truncation error and the aliasing error.

Setting

$$\psi(r) := \exp\left(-\frac{\alpha}{N}(\log r)^2\right)$$

with a positive α to be fixed later, we consider the truncated generalized exponential sampling series $\left(S_{w,N}^{\psi \mathrm{lin}} f\right)(r)$, defined in (13.30), and call it

$$\left(G_{w,N}^{\alpha} f\right)(r) := \sum_{k=N_{w,r}-N}^{N_{w,r}+N} f(e^{k/w}) \exp\left(-\frac{\alpha}{N}\left(\log(e^{-k} r^w)\right)^2\right) \mathrm{lin}(e^{-k} r^w)$$

$$(N_{w,r} := \lfloor w \log r + 1/2 \rfloor) \tag{13.44}$$

with G referring to Gauss. The following theorem includes an efficient algorithm for approximating the restriction to $\mathbb{R}^+$ of a polar-analytic function from its exponential samples.

Theorem 13.42 *Let f be polar-analytic on $\mathbb{H}$ such that*

$$|f(r,\theta)| \le K(|\log r|)e^{\pi T|\theta|} \qquad ((r,\theta) \in \mathbb{H}),$$

where K is a non-decreasing, nonnegative function defined on $[0, +\infty[$ and $T \ge 0$. Then, for $w > T$, $\alpha = \pi(1 - T/w)/2$, $N \in \mathbb{N}$ and $r \in \mathbb{R}^+$, we have in the notation (13.44),

$$\left|f(r,0) - \left(G^{\alpha}_{w,N} f(\cdot,0)\right)(r)\right| \le |\sin(\pi w \log r)|\, K\left(|\log r| + \frac{N+1}{w}\right) \frac{2e^{-\alpha N}}{\sqrt{\pi \alpha N}}\, \beta_N,$$

where

$$\beta_N = \frac{1}{1 - e^{-2\pi N}} + \frac{2}{\sqrt{\pi \alpha N}} = 1 + \mathcal{O}(N^{-1/2}) \qquad (N \to \infty).$$

Proof First we note that for $r \in \{e^{k/w} : k \in \mathbb{Z}\}$ the assertion is trivially true. For $r \in \mathbb{R}^+ \setminus \{e^{k/w} : k \in \mathbb{Z}\}$, $(\rho, \vartheta) \in \mathbb{H}$ and $\lambda := \alpha/N$, consider

$$F(\rho,\vartheta) := \frac{f(\rho,\vartheta)\exp(-\lambda(\log(\rho^w/r^w) + iw\vartheta)^2)}{(\log(\rho/r) + i\vartheta)\sin(\pi w(\log \rho + i\vartheta))}.$$

This function F is polar-analytic on $\mathbb{H}$ except for isolated singularities at the points $(r, 0)$ and $(e^{k/w}, 0)$ with $k \in \mathbb{Z}$, which are all poles of order at most 1. We want to show by employing the residue theorem for polar-analytic functions (see Theorem 3.39) that the error of the approximation of $f(r,0)$ by $(G^{\alpha}_{w,N} f(\cdot,0))(r)$ can be represented by a contour integral of F. For this, we have to calculate the residues res_0 of the poles of F. It is easily seen that

$$(\mathrm{res}_0 F)(r,0) = \frac{f(r,0)}{\sin(\pi w \log r)}.$$

For calculating the residues at $(e^{k/w}, 0)$, we first factor the sine in the denominator of F as follow:

$$\begin{aligned}
\sin(\pi w(\log \rho + i\vartheta)) &= (-1)^k \sin(\pi w(\log \rho + i\vartheta - k/w)) \\
&= (-1)^k \pi w(\log \rho + i\vartheta - k/w)\mathrm{sinc}(\log \rho^w + iw\vartheta - k) \\
&= (-1)^k \pi w(\log \rho + i\vartheta - k/w)\mathrm{lin}(e^{-k}\rho^w e^{iw\vartheta}).
\end{aligned}$$

With this decomposition, a basic formula for the residues of poles of order at most 1 yields

$$
\begin{aligned}
(\mathrm{res}_0 F)(e^{k/w}, 0) &= \frac{f(e^{k/w}, 0)\exp(-\lambda(\log(e^{-k}r^w))^2)}{(-1)^{k+1}\pi\log(e^{-k}r^w)} \\
&= (-1)^{k+1} f(e^{k/w}, 0)\exp(-\lambda(\log(e^{-k}r^w))^2)\frac{\mathrm{lin}(e^{-k}r^w)}{\sin(\pi\log(e^{-k}r^w))} \\
&= -f(e^{k/w}, 0)\exp(-\lambda(\log(e^{-k}r^w))^2)\frac{\mathrm{lin}(e^{-k}r^w)}{\sin(\pi w\log r)}.
\end{aligned}
$$

Now write $N' := N + \frac{1}{2}$ and denote by $\mathcal{R}_N$ the rectangle with vertices at

$$
\left(e^{(N_{w,r}\pm N')/w}, \frac{N}{w}\right), \quad \left(e^{(N_{w,r}\pm N')/w}, -\frac{N}{w}\right)
$$

and by $\partial\mathcal{R}_N$ its positively oriented boundary. By the residue theorem for polar-analytic functions, we have

$$
\begin{aligned}
&\frac{\sin(\pi w\log r)}{2\pi i}\int_{\partial\mathcal{R}_N} F(\rho, \vartheta)\left(\frac{d\rho}{\rho} + i d\vartheta\right) \\
&\quad = \sin(\pi w\log r)\left[(\mathrm{res}_0 F)(r, 0) + \sum_{|k-N_{w,r}|\le N}(\mathrm{res}_0 F)(e^{k/w}, 0)\right] \\
&\quad = f(r, 0) - \sum_{|k-N_{w,r}|\le N} f(e^{k/w}, 0)\exp(-\lambda(\log(e^{-k}r^w))^2)\mathrm{lin}(e^{-k}r^w) \\
&\quad = f(r, 0) - \big(G^{\alpha}_{w,N} f(\cdot, 0)\big)(r).
\end{aligned}
$$

Next we denote by $I^{\pm}_{\mathrm{hor}}$ the contributions to the integral coming from the two horizontal parts of $\partial\mathcal{R}_N$, where $+$ and $-$ refer to the upper and lower line segment, respectively. Similarly, we denote by $I^{\pm}_{\mathrm{vert}}$ the contributions to the integral coming from the two vertical parts of $\partial\mathcal{R}_N$, where $+$ and $-$ refer to the right and left line segment, respectively. Then

$$
f(r, 0) - (G^{\alpha}_{w,N} f(\cdot, 0))(r) = \frac{\sin(\pi w\log r)}{2\pi i}\big(I^{-}_{\mathrm{hor}} + I^{+}_{\mathrm{vert}} + I^{+}_{\mathrm{hor}} + I^{-}_{\mathrm{vert}}\big). \tag{13.45}
$$

We note that for $(\rho, \vartheta) \in \partial\mathcal{R}_N$, we have

$$
|f(\rho, \vartheta)| \le K\left(|\log r| + \frac{N+1}{w}\right)e^{\pi T|\vartheta|}
$$

and

$$\left|\log\frac{\rho}{r}+i\vartheta\right|\geq\frac{N}{w}.$$

Therefore

$$\begin{aligned}\left|I^{\pm}_{\mathrm{hor}}\right| &\leq K\left(|\log r|+\frac{N+1}{w}\right)\frac{w}{N}e^{\pi TN/w}\\ &\quad\times\left|\int_{e^{(N_{w,r}-N')/w}}^{e^{(N_{w,r}+N')/w}}\frac{\exp(-\lambda(\log(\rho^{w}/r^{w})\pm iN)^{2}))}{\sin(\pi w(\log\rho\pm iN/w))}\frac{d\rho}{\rho}\right|\\ &= K\left(|\log r|+\frac{N+1}{w}\right)\frac{e^{\pi TN/w}}{N}\\ &\quad\times\left|\int_{N_{w,r}-N'}^{N_{w,r}+N'}\frac{\exp(-\lambda(u-\log r^{w}\pm iN)^{2})}{\sin(\pi(u\pm iN))}du\right|.\end{aligned}$$

Since

$$\left|\exp(-\lambda(u-\log r^{w}\pm iN)^{2})\right|=e^{\lambda N^{2}}e^{-\lambda(u-w\log r)^{2}},$$

$$|\sin(\pi(u\pm iN))|\geq\sinh(\pi N)=\tfrac{1}{2}e^{\pi N}(1-e^{-2\pi N}),$$

$$\left|\int_{N_{w,r}-N'}^{N_{w,r}+N'}e^{-\lambda(u-w\log r)^{2}}du\right|\leq\int_{-\infty}^{\infty}e^{-\lambda x^{2}}dx=\sqrt{\frac{\pi}{\lambda}}=\sqrt{\frac{\pi N}{\alpha}}$$

and

$$\pi N\frac{T}{w}-\pi N+\lambda N^{2}=-\alpha N,$$

we conclude that

$$\left|I^{\pm}_{\mathrm{hor}}\right|\leq K\left(|\log r|+\frac{N+1}{w}\right)\frac{2\sqrt{\pi}e^{-\alpha N}}{\sqrt{\alpha N}(1-e^{-2\pi N})}. \tag{13.46}$$

For the contributions coming from the vertical line segments, we have

$$\begin{aligned}\left|I^{\pm}_{\mathrm{vert}}\right| &\leq K\left(|\log r|+\frac{N+1}{w}\right)\frac{w}{N}\\ &\quad\times\left|\int_{-N/w}^{N/w}\frac{\exp(-\lambda(N_{w,r}\pm N'-\log r^{w}+iw\vartheta)^{2})e^{\pi T|\vartheta|}}{\sin(\pi(N_{w,r}\pm N'+iw\vartheta))}d\vartheta\right|.\end{aligned}$$

Here

$$\left|\exp(-\lambda(N_{w,r} \pm N' - \log r^w + iw\vartheta)^2)\right| \le e^{-\lambda N^2} e^{\lambda w^2 \vartheta^2}$$

and

$$\left|\sin(\pi(N_{w,r} \pm N' + iw\vartheta))\right| \ge \cosh(\pi w\vartheta) \ge \tfrac{1}{2} e^{\pi w|\vartheta|}.$$

Substituting $\theta = w\vartheta$ and noting that

$$\lambda\theta^2 + \pi\,|\theta|\,\frac{T}{w} - \pi\,|\theta| = \frac{\alpha}{N}\theta^2 - 2\alpha\,|\theta| \le -\alpha\,|\theta|$$

for $\theta \in [-N, N]$, we arrive at

$$\begin{aligned} \left|I_{\text{vert}}^{\pm}\right| &\le K\left(|\log r| + \frac{N+1}{w}\right)\frac{2e^{-\alpha N}}{N}\int_{-N}^{N} e^{-\alpha|\theta|}d\theta \\ &\le K\left(|\log r| + \frac{N+1}{w}\right)\frac{4e^{-\alpha N}}{\alpha N}. \end{aligned} \tag{13.47}$$

Employing the estimates (13.46) and (13.47) in conjunction with (13.45), we readily complete the proof. □

Note that in Theorem 13.42 the speed of convergence may be lower than $\mathcal{O}(e^{-\alpha N})$ since N also occurs in the argument of the non-decreasing function K. However, when the restriction of f to $\mathbb{R}^+$ is bounded, such a deterioration is not possible. The following statement is an obvious consequence of Theorem 13.42.

Corollary 13.43 *Let $f \in \mathscr{B}^{\infty}_{0,\pi T}$, where $T \in \mathbb{R}^+$. Then, for $w > T$, $\alpha = \frac{\pi}{2}(1 - \frac{T}{w})$, $N \in \mathbb{N}$ and $r \in \mathbb{R}^+$, we have in the notation (13.44),*

$$\left|f(r,0) - \left(G^{\alpha}_{w,N} f(\cdot,0)\right)(r)\right| \le |\sin(\pi w \log r)|\, C_f\, \frac{2e^{-\alpha N}}{\sqrt{\pi\alpha N}}\, \beta_N,$$

where C_f is the constant introduced in Definition 3.41 for $c = 0$ and β_N is as in Theorem 13.42.

When we proved the assertion of Theorem 13.42 for an arbitrarily chosen $N \in \mathbb{N}$, we did not really need that the function f is polar-analytic on the whole of $\mathbb{H}$. It will suffice that it is polar-analytic in a strip

$$S_d := \{(r,\theta) \in \mathbb{H} \,:\, |\theta| < d\}$$

with $d > N/w$. This condition can always be satisfied by choosing $w = (N+\varepsilon)/d$ with $\varepsilon > 0$. This leads us to the following statement in which the hypothesis is designed such that $T = 0$ and the function K is a constant K_f, which allows us to let ε approach zero on both sides of the error estimate.

Corollary 13.44 *Let f be polar-analytic in a strip S_d with $d > 0$. Suppose that*

$$K_f := \sup_{(r,\theta)\in S_d} |f(r,\theta)| < \infty.$$

Then for $N \in \mathbb{N}$ and $r \in \mathbb{R}^+$, we have in the notation (13.44),

$$\left|f(r,0) - \left(G^{\pi/2}_{N/d,N} f(\cdot,0)\right)(r)\right| \le \left|\sin\left(\frac{\pi N}{d}\log r\right)\right| K_f \frac{2\sqrt{2}\,e^{-\pi N/2}}{\pi\sqrt{N}}\beta_N,$$

where

$$\beta_N = \frac{1}{1-e^{-2\pi N}} + \frac{2\sqrt{2}}{\pi\sqrt{N}} = 1 + \mathcal{O}(N^{-1/2}) \qquad (N \to \infty).$$

13.5.5 Numerical Experiments

As a first example, we consider the function

$$f\ :\ (r,\theta) \longmapsto \cos\bigl(\pi(\log r + i\theta)\bigr) \tag{13.48}$$

which belongs to $\mathscr{B}^{\infty}_{0,\pi}$. It qualifies for Corollaries 13.35 and 13.43. In the case of Corollary 13.35, we use the multiplier ψ of Example 13.32 with $\delta = 1/2$ and $m = 8$. Then the constructed function μ becomes

$$\mu(t) = \frac{32}{\pi(1+(\pi t/16)^7)}.$$

In view of Remark 13.26, we may choose $w = 16$ in both corollaries. This enables us to compare the two statements by computing approximations of $f(r_j,0)$ at $r_j = e^{(j+1/2)/w}$ in both cases. Results for the true absolute errors are shown in the columns of Table 13.1. For the chosen points r_j the error bounds of the two corollaries do not depend on j. They are presented in the last line. We see that the approximation by the operator $G^{\alpha}_{w,N}$ of Corollary 13.43 is much better than the approximation by $S^{\psi\,\mathrm{lin}}_{w,N}$ of Corollary 13.35. Moreover, the error bound of Corollary 13.43 is closer to the true errors.

Our second example is the function

$$f\ :\ (r,\theta) \longmapsto \frac{re^{i\theta} + r^{-1}e^{-i\theta}}{2}, \tag{13.49}$$

which is polar-analytic on $\mathbb{H}$ but unbounded on $\mathbb{R}^+$. It satisfies the hypotheses of Theorem 13.42 with $K(t) = \cosh(t)$, $T = 0$ and $\alpha = \pi/2$. We choose $w = 16/\pi$ and compute approximations of $f(r_j,0)$ for $r_j = e^{(j+1/2)/w}$. The true absolute errors and their bounds are shown in Tables 13.2 and 13.3 for four different values

Table 13.1 Computations for the function (13.48) according to Corollaries 13.35 and 13.43, respectively

$N = 10$	Corollary 13.35	Corollary 13.43	$N = 20$	Corollary 13.35	Corollary 13.43	$N = 30$	Corollary 13.35	Corollary 13.43
j	Abs. error	Abs. error	j	Abs. error	Abs. error	j	Abs. error	Abs. error
0	7.15e−06	5.69e−08	0	4.04e−09	1.61e−14	0	3.48e−13	5.57e−21
2	5.19e−06	5.13e−08	2	4.96e−09	1.41e−14	2	2.73e−13	4.92e−21
4	1.67e−05	3.80e−08	4	1.32e−08	9.95e−15	4	1.56e−13	3.52e−21
6	2.57e−05	1.88e−08	6	1.94e−08	4.29e−15	6	1.57e−14	1.58e−21
8	3.08e−05	3.21e−09	8	2.27e−08	2.02e−15	8	1.27e−13	5.93e−22
Bound	4.07e−03	1.53e−07	Bound	1.89e−05	4.06e−14	Bound	7.82e−07	1.29e−20

Table 13.2 Computations for the function (13.49) according to Theorem 13.42

N=5	Theorem 13.42			
j	Abs. error	Bound	Rel. error	Rel. bound
0	7.14e−05	4.23e−04	7.11e−05	4.21e−04
8	2.08e−04	1.90e−03	7.58e−05	6.90e−04
16	9.73e−04	9.09e−03	7.61e−05	7.11e−04
24	4.68e−03	4.37e−02	7.62e−05	7.12e−04
32	2.25e−02	2.10e−01	7.62e−05	7.12e−04
40	1.08e−01	1.01e+00	7.62e−05	7.12e−04
48	5.21e−01	4.87e+00	7.62e−05	7.12e−04

N=10	Theorem 13.42			
j	Abs. error	Bound	Rel. error	Rel. bound
0	1.63e−08	2.66e−07	1.62e−08	2.65e−07
8	5.27e−08	1.27e−06	1.92e−08	4.62e−07
16	2.48e−07	6.10e−06	1.94e−08	4.77e−07
24	1.19e−06	2.93e−05	1.94e−08	4.78e−07
32	5.74e−06	1.41e−04	1.94e−08	4.78e−07
40	2.76e−05	6.79e−04	1.94e−08	4.78e−07
48	1.33e−04	3.27e−03	1.94e−08	4.78e−07

Table 13.3 Computations for the function (13.49) according to Theorem 13.42, continued

N=20	Theorem 13.42			
j	Abs. error	Bound	Rel. error	Rel. bound
0	1.73e−15	1.87e−13	1.72e−15	1.86e−13
8	9.33e−15	9.00e−13	3.40e−15	3.28e−13
16	4.51e−14	4.33e−12	3.53e−15	3.39e−13
24	2.17e−13	2.08e−11	3.53e−15	3.39e−13
32	1.04e−12	1.00e−10	3.53e−15	3.39e−13
40	5.02e−12	4.82e−10	3.53e−15	3.39e−13
48	2.42e−11	2.32e−09	3.53e−15	3.39e−13

N=30	Theorem 13.42			
j	Abs. error	Bound	Rel. error	Rel. bound
0	1.83e−21	1.59e−19	1.82e−21	1.58e−19
8	1.66e−21	7.65e−19	6.03e−22	2.78e−19
16	6.49e−21	3.68e−18	5.08e−22	2.88e−19
24	3.09e−20	1.77e−17	5.03e−22	2.88e−19
32	1.49e−19	8.51e−17	5.03e−22	2.88e−19
40	7.15e−19	4.10e−16	5.03e−22	2.88e−19
48	3.44e−18	1.97e−15	5.03e−22	2.88e−19

Table 13.4 Computations for the function (13.50) according to Corollary 13.44

$N=5$	Corollary 13.44	$N=10$	Corollary 13.44	$N=20$	Corollary 13.44	$N=30$	Corollary 13.44
j	Abs. error	j	Abs. error	j	Abs. error	j	Abs. error
0	6.44e−05	0	1.72e−08	0	1.56e−15	0	1.74e−22
2	1.39e−04	4	3.94e−08	8	4.13e−15	12	5.05e−22
4	2.05e−04	8	5.82e−08	16	6.16e−15	24	7.57e−22
6	2.38e−04	12	6.84e−08	24	7.32e−15	36	9.01e−22
8	2.36e−04	16	6.92e−08	32	7.46e−15	48	9.21e−22
10	2.00e−04	20	6.02e−08	40	6.55e−15	60	8.11e−22
Bound	3.64e−04	Bound	9.14e−08	Bound	9.11e−15	Bound	1.09e−21

of N. The absolute errors grow with increasing $|\log r_j|$ as suggested by the bound of Theorem 13.42 which contains $|\log r|$ in the argument of the increasing function K. However, the relative errors

$$\left|\frac{f(r_j,0)-\big(G_w^\alpha f(\cdot,0)\big)(r_j)}{f(r_j,0)}\right|$$

and their bounds are nearly constant. In dependence of N, the bound for the absolute error is asymptotically equal to $\mathcal{O}\big(\frac{e^{-7\pi N/16}}{\sqrt{N}}\big)$ as $N\to\infty$.

As a third example, we consider the function

$$f\,:\,(r,\theta)\longmapsto\sqrt{1+(\log r+i\theta)^2}\,\mathrm{lin}\big((re^{i\theta})^{1/\pi}\big). \tag{13.50}$$

It is polar-analytic and bounded on the strip S_1 with

$$\sup_{(r,\theta)\in\mathcal{S}_1}|f(r,\theta)|\le 2^{1/2}3^{-1/4}\cosh(1)=:K_f.$$

Therefore it may serve for illustrating Corollary 13.44. The columns of Table 13.4 show the index j and the true absolute errors at the points $r_j = e^{(j+1/2)/N}$ for four choices of N. For these points, the error bounds provided by Corollary 13.44 do not depend on j. They are presented in the last line. It turns out that they are very realistic. The factor of overestimation of the largest absolute error in a column decreases with increasing N. For $N=5$ it is 1.53; for $N=30$ it has decreased to 1.18.

13.6 Notes

Modifications of the classical sampling series of Whittaker–Kotel'nikov–Shannon (WKS series for short) have been considered since a long time. P.L. Butzer and his collaborators proposed and studied generalizations in which the sinc function of the WKS series is replaced by a more general function φ, called a *kernel function*; see Ries and Stens [260], Butzer and Stens [97, 98]. Depending on the choice of φ, such a sampling operator may no longer reconstruct bandlimited signals but may gain advantages for time-limited signals, which cannot be bandlimited as a consequence of the Paley–Wiener theorem. This concept stimulated a wide field of research on the convergence properties of these sampling series and prediction theory; see, e.g., Butzer and Stens [97, 98], Butzer et al. [50], Bardaro and Mantellini [38, 39], Angeloni and Vinti [13], Angeloni et al. [14] and the references therein. In particular, in [50] the sampling approximation of a discontinuous signal is studied in L^p spaces using suitable moduli of smoothness.

Further generalizations, inspired by those for Bernstein polynomials in which the sampled values are replaced by local integrals, use some kind of convolution operator instead of the values of f at the nodes. This leads to the theory of Kantorovich and Durrmeyer type sampling operators, studied by several authors in view of their applications in various fields, such as signal and image processing, engineering, and many others; see, e.g., Acar et al. [2], Angeloni et al. [15], Bardaro et al. [48], Bardaro and Mantellini [41, 42], Cantarini et al. [123], Costarelli et al. [134], Costarelli and Vinti [131–133].

General (abstract) approaches, which unify convergence properties of integral and discrete operators, including the generalized sampling series, can be found in Bardaro and Vinti [44, 45], Bardaro and Mantellini [29], Vinti and Zampogni [296] and the references therein.

Section 13.2 is based on Bardaro et al. [57, Sect. 3] who defined and studied a generalized exponential sampling series. Employing a technique analogous to that used for Mellin–Fejér convolution operators (see Sect. 6.4.3), Balsamo and Mantellini [28] obtained pointwise and uniform convergence results for certain linear combinations of generalized exponential sampling operators. In this way the order of approximation can be increased.

The results of Sect. 13.3 originate from Bardaro et al. [63]. When $f \in X_c^p$, then a generalized exponential sampling series need not converge to f in the norm of X_c^p. Therefore a special challenge is to identify a large subspace for which convergence can be guaranteed. For this a weaker form of bounded variation, called *bounded coarse variation* turns out to be very helpful. This notion was first introduced by Haber and Shisha [162] for characterizing the approximability of improper integrals by Riemann sums. Further contributions are due to Osgood [231] and Osgood and Shisha [232].

The examples of Sect. 13.4 were designed in Bardaro et al. [57, Sect. 4] as Mellin analogues of some familiar kernels of approximation theory and Fourier analysis; see, e.g., de Boor [136, Chap. 9], Butzer et al. [101] for B-splines and Butzer and Nessel [95, Chap. 3, Subsect. 3.1.2] for Fejér kernels. Applications to the study of seismic waves with a bivariate version of the generalized exponential sampling series were given in Bardaro et al. [61].

Other generalized versions of the exponential sampling series were studied in semi-discrete forms by Angamuthu and Bajpeyi [7], Acar [3], Aral et al. [19], Bajpeyi and Kumar [23], Bardaro and Mantellini [43]. In particular, in [43] a notion of Durrmeyer type exponential sampling series was introduced and developed in Bajpeyi et al. [24].

The results of Sect. 13.5 originate from Bardaro et al. [68], except for Theorem 13.40, which has not been published elsewhere. For the WKS sampling series the multiplier technique has been used since a long time. For some developments in chronological order, see Helms and Thomas [169], Jagerman [179], Campbell [122], Lee [197], Rahman and Schmeisser [252], Qian [249], Qian and Creamer [250], Schmeisser and Stenger [273], Schmeisser [272].

As regards exponential sampling, Kivinukk and Tamberg [193] favored a generalization which amounts to using a multiplier with poles at the nodes. However, other than for a multiplier that attains finite values on the set of nodes and 1 at the node 1, their sampling series cannot interpolate.

In view of Definition 13.31 and the subsequent results in Sects. 13.5.2 and 13.5.3, the non-trivial question arises as to how fast $\psi(r)$ can decay as $r \to 0+$ and $r \to +\infty$ when ψ is a Mellin bandlimited function satisfying $\psi(1) = 1$. For a corresponding question about entire functions of exponential type, an answer can be found in the book of Koosis [195, p. 101], which reads as follows: *Let $S(r) \geq 1$ be increasing. A necessary and sufficient condition for there to exist entire functions $\varphi \not\equiv 0$ of exponential type with*

$$|\varphi(x)| \leq \frac{1}{S(|x|)} \qquad \text{on } \mathbb{R}$$

is that

$$\int_0^\infty \frac{\log S(r)}{1+r^2}\,dr < \infty.$$

If that condition is met, there are entire functions $\varphi \not\equiv 0$ of arbitrarily small exponential type satisfying the inequality in question.

We recall that there is a correspondence between Mellin bandlimited functions and entire functions of exponential type. If $\psi \in B^2_{0,\pi\delta}$ and $f(x) := \psi(e^x)$, then f is the restriction to $\mathbb{R}$ of an entire function of exponential type $\pi\delta$ that belongs to the Lebesgue class L^2 on the real line. The converse is also true. Given f of the latter kind, then ψ defined by $\psi(r) := f(\log r)$ belongs to $B^2_{0,\pi\delta}$. Therefore the cited result implies the following criterion for the class $B^2_{0,\pi\delta}(\mu)$ in Definition 13.31.

Proposition 13.45 *The class $B^2_{0,\pi\delta}(\mu)$ is non-empty if and only if*

$$-\int_0^\infty \frac{\log \mu(t)}{1+t^2}\,dt < \infty.$$

This shows that in Proposition 13.33 and Theorem 13.36 we can have $\mathcal{O}(e^{-N/(\log N)^\gamma})$ with $\gamma > 1$ for the convergence of the truncion error as $N \to \infty$ but it is not possible to have such a rate of convergence with $\gamma \leq 1$. In particular, we cannot have $\mathcal{O}(e^{-\alpha N})$ for some positive α.

In case of the WKS sampling series, some scientists experimented with Gaussian multipliers, that is, with functions of the form $t \mapsto c_1 e^{-c_2 t^2}$, where c_1 and c_2 are positive constants, despite the fact that these functions are not bandlimited. It was reported that the Chinese chemist G.W. Wei and his collaborators tackled diverse dynamical problems arising in physics and chemistry by using such multipliers. Inspired by their excellent numerical results, Qian [249] (also see Creamer and Qian [250]) provided a mathematical justification for this approach with rigorous error estimates. Using methods of Fourier analysis, these authors showed that a bandlimited function can be approximated by $2N+1$ terms of its classical sampling series equipped with a cleverly scaled Gaussian multiplier with an error that decays like

$$\mathcal{O}\left(\frac{e^{-\alpha N}}{\sqrt{N}}\right) \qquad (N \to \infty), \tag{13.51}$$

where α is a positive constant. This is an amazing result since it shows that, by allowing in addition an aliasing error, one can achieve a better approximation by $2N + 1$ samples of f than with the truncated series of an exact formula. This result was extended by Schmeisser and Stenger [273] using methods of complex analysis. Multidimensional versions were presented by Asharabi and Prestin [22] and Asharabi and Al-Haddad [21]. Micchelli et al. [226] considered the sequence of samples of a bandlimited function f occurring in the classical sampling formula and asked for an optimal algorithm that approximates $f(x)$ by using $2N + 1$ of these samples. They found that such an algorithm converges for the worst case of the function f as fast as (13.51) with some $\alpha > 0$ but not faster. Hence sampling with a Gaussian multiplier as initiated by Wei and Qian has the rate of convergence of an optimal algorithm and is therefore the method of choice for computations. A very early use of Gaussian multipliers in a somewhat different context can be found in E.T. Whittaker's construction of "cotabular" functions; see Whittaker [299].

The distinguished role of Gaussian multipliers for the WKS sampling series can be made plausible by the following reflection: For the truncation error to converge fast one should have a rapidly decaying multiplier ψ. For the aliasing error to converge fast one should have a multiplier ψ whose Fourier transform decays rapidly. This leads us naturally to Gaussian functions since they decay rapidly and their Fourier transforms are again Gaussian functions.

Chapter 14
Quadrature over the Positive Real Axis

The Mellin–Poisson summation formula is an exact quadrature formula for Mellin bandlimited functions. Indeed, Corollary 6.100 implies the following statement which is substantial for the purposes of the present chapter.

Corollary 14.1 *Let $c \in \mathbb{R}$ and $\sigma \in \mathbb{R}^+$. Suppose that $f \in B^1_{c,2\pi\sigma}$. Then*

$$\sigma [f]^{\wedge}_{M_c}(c) = \sum_{k\in\mathbb{Z}} f(e^{k/\sigma}) e^{kc/\sigma},$$

or equivalently,

$$\int_0^\infty f(x)x^{c-1}dx = \frac{1}{\sigma}\sum_{k=-\infty}^{\infty} f(e^{k/\sigma})e^{kc/\sigma}. \tag{14.1}$$

Formula (14.1) can be interpreted as an exact quadrature formula.

If $f \notin B^1_{c,2\pi\sigma}$, this quadrature formula is no longer exact but it holds with a remainder which can be expressed in terms of the Mellin transform of f. Indeed, we can rewrite the formula of Corollary 6.100 as

$$\int_0^\infty f(x)x^{c-1}dx - \frac{1}{\sigma}\sum_{k=-\infty}^{\infty} f(e^{k/\sigma})e^{kc/\sigma} = -\sum_{k\in\mathbb{Z}\setminus\{0\}} [f]^{\wedge}_{M_c}(c+2\pi i\sigma k).$$

Thus, defining

$$R_{c,\sigma}[f] := -\sum_{k\in\mathbb{Z}\setminus\{0\}} [f]^{\wedge}_{M_c}(c+2\pi i\sigma k), \tag{14.2}$$

C. Bardaro et al., *Mellin Analysis, Transform Theory, and Applications*,
https://doi.org/10.1007/978-3-031-96672-9_14

we obtain the approximate quadrature formula

$$\int_0^{\infty} f(x)x^{c-1}dx = \frac{1}{\sigma} \sum_{k=-\infty}^{\infty} f(e^{k/\sigma})e^{kc/\sigma} + R_{c,\sigma}[f]. \tag{14.3}$$

In the next sections we shall study estimates of the remainder $R_{c,\sigma}[f]$ when the function f belongs to suitable function spaces.

14.1 Estimates of the Remainders

In this section our main aim is to estimate the series (14.2) which defines the remainder $R_{c,\sigma}$. First we introduce a class of functions for which the Mellin–Poisson summation formula is meaningful.

Definition 14.2 For $c \in \mathbb{R}$, the class $\mathcal{K}_c$ comprises all functions $f \in W_c^{1,1}(\mathbb{R}^+) \cap C(\mathbb{R}^+)$ such that

$$\sum_{k=-\infty}^{\infty} \left|[f]^{\wedge}_{M_c}(c + i2\pi k\sigma)\right| < \infty \tag{14.4}$$

for all $\sigma \in \mathbb{R}^+$.

Remark 14.3 In view of Corollary 6.100, the assumption $f \in \mathcal{K}_c$ implies that (14.3) holds for all $\sigma \in \mathbb{R}^+$. Moreover, the series in (14.2) and (14.3) both converge absolutely for all $\sigma \in \mathbb{R}^+$.

For guaranteeing (14.4) for all $\sigma \in \mathbb{R}^+$, it suffices to verify it for all small positive values of σ. Indeed, suppose that (14.4) holds for all $\sigma \in]0, \sigma_0]$. Take any $\sigma_1 > \sigma_0$. Then there exists an integer $m \in \mathbb{N}$ such that $\sigma^* := \sigma_1/m \in]0, \sigma_0]$. By the hypothesis,

$$\sum_{k=-\infty}^{\infty} \left|[f]^{\wedge}_{M_c}(c + i2\pi k\sigma^*)\right| < \infty.$$

Then also

$$\sum_{k=-\infty}^{\infty} \left|[f]^{\wedge}_{M_c}(c + i2\pi km\sigma^*)\right| < \infty$$

since this is a partial series of the previous series with the summation extending only over those integers which are multiples of m. But $m\sigma^* = \sigma_1$. Thus we have verified the absolute convergence for σ_1.

We shall establish estimates of the remainder of the quadrature formula (14.3) under the basic assumption that the involved function f belongs to the class $\mathcal{K}_c$.

We divide this section in two parts. In the first one we give estimates of the remainder in Mellin–Sobolev type spaces of integer order and in the second one we discuss a fractional case.

14.1.1 Estimates in Mellin–Sobolev Spaces of Integer Order

Let us consider the Mellin–Lipschitz classes (see Sect. 3.3)

$$\mathrm{Lip}_r(\alpha, X_c) := \left\{ f \in X_c \, : \, \omega_r(f, \delta, X_c) = \mathcal{O}(\delta^\alpha), \; \delta \to 0+ \right\},$$

where $\alpha \in]0, r]$. In this setting the remainder $R_{c,\sigma}[f]$ of the quadrature formula (14.3) can be estimated asymptotically as follows.

Theorem 14.4 *Let $f \in \mathcal{K}_c$ with $c \in \mathbb{R}$. If in addition $f \in \mathrm{Lip}_r(\alpha, X_c)$, where $r \geq 2$ is an integer and $\alpha \in]1, r]$, then*

$$\left| R_{c,\sigma}[f] \right| = \mathcal{O}(\sigma^{-\alpha}) \qquad (\sigma \to \infty). \tag{14.5}$$

Proof The hypotheses imply that inequality (10.4) holds, and so

$$\left| [f]^\wedge_{M_c}(c + iv) \right| \leq \frac{1}{2^r} \, \omega_r \left(f, \frac{\pi}{|v|}, X_c \right).$$

Therefore, by (14.2),

$$\left| R_{c,\sigma}[f] \right| \leq \sum_{k \in \mathbb{Z} \setminus \{0\}} \left| [f]^\wedge_{M_c}(c + 2\pi i \sigma k) \right| \leq \sum_{k \in \mathbb{Z} \setminus \{0\}} \frac{1}{2^r} \, \omega_r \left(f, \frac{1}{2\sigma |k|}, X_c \right).$$

Since $f \in \mathrm{Lip}_r(\alpha, X_c)$, there exists a constant $C > 0$ such that for sufficiently large $\sigma > 0$, we have

$$\begin{aligned} \left| R_{c,\sigma}[f] \right| &\leq \frac{1}{2^{r-1}} \sum_{k=1}^{\infty} \omega_r \left(f, \frac{1}{2\sigma k}, X_c \right) \leq \frac{C}{2^{r-1}} \sum_{k=1}^{\infty} \frac{1}{(2\sigma k)^\alpha} \\ &= \frac{C}{2^{r+\alpha-1} \sigma^\alpha} \, \zeta(\alpha) = \mathcal{O}(\sigma^{-\alpha}) \qquad (\sigma \to \infty), \end{aligned}$$

where $\zeta(\cdot)$ denotes the Riemann zeta function (see Sect. 1.5.2). □

Under the existence of derivatives of higher order, the previous asymptotic estimate can be improved.

Theorem 14.5 *Let $f \in \mathcal{K}_c$ with $c \in \mathbb{R}$. If in addition $f \in W_c^{j,1}(\mathbb{R}^+)$ and $\Theta_c^j f \in$* $\mathrm{Lip}_r(\alpha, X_c)$*, where $j, r \in \mathbb{N}$ and $\alpha \in]0, r[$, then*

$$|R_{c,\sigma}[f]| = \mathcal{O}(\sigma^{-\alpha-j}) \qquad (\sigma \to \infty). \tag{14.6}$$

Proof For $f \in W_c^{j,1}(\mathbb{R}^+)$ Theorem 6.14 shows that

$$[\Theta_c^j f]^\wedge_{M_c}(c + iv) = (-iv)^j [f]^\wedge_{M_c}(c + iv) \qquad (v \in \mathbb{R})$$

or

$$[f]^\wedge_{M_c}(c + iv) = (-iv)^{-j} [\Theta_c^j f]^\wedge_{M_c}(c + iv) \qquad (v \in \mathbb{R} \setminus \{0\}).$$

Using inequality (10.4) with f replaced by $\Theta_c^j f$, we obtain

$$|[f]^\wedge_{M_c}(c + iv)| = |v|^{-j} |[\Theta_c^j f]^\wedge_{M_c}(c + iv)| \leq \frac{1}{2^r} |v|^{-j} \omega_r \left(\Theta_c^j f, \frac{\pi}{|v|}, X_c \right)$$

for $v \neq 0$. Therefore

$$\begin{aligned}
|R_{c,\sigma}[f]| &\leq \sum_{k \in \mathbb{Z} \setminus \{0\}} |[f]^\wedge_{M_c}(c + 2\pi i \sigma k)| \\
&\leq \frac{1}{2^r} \sum_{k \in \mathbb{Z} \setminus \{0\}} \frac{1}{(2\pi \sigma k)^j} \omega_r \left(\Theta_c^j f, \frac{\pi}{2\pi \sigma |k|} X_c \right) \\
&= \frac{1}{2^{r+j-1} (\pi \sigma)^j} \sum_{k=1}^{\infty} \frac{1}{k^j} \omega_r \left(\Theta_c^j f, \frac{1}{2\sigma k}, X_c \right).
\end{aligned}$$

Since $\Theta_c^j f \in \mathrm{Lip}_r(\alpha, X_c)$, there exists a constant $C > 0$ such that for sufficiently large σ, we have

$$\begin{aligned}
|R_{c,\sigma}[f]| &\leq \frac{1}{2^{r+j-1} (\pi \sigma)^j} \sum_{k=1}^{\infty} \frac{C}{k^j (2\sigma k)^\alpha} \\
&= \frac{C}{2^{r+j+\alpha-1} \pi^j \sigma^{j+\alpha}} \zeta(j + \alpha) = \mathcal{O}(\sigma^{-\alpha-j}) \qquad (\sigma \to \infty),
\end{aligned}$$

which is (14.6). □

14.1.2 Estimates in Mellin–Sobolev Spaces of Fractional Order

Now we discuss the fractional case. Let $\alpha > 0$ be a fixed real number. We recall that the Mellin (fractional) difference of order α is defined by

$$(\Delta_h^{\alpha,c} f)(x) := \sum_{j=0}^{\infty}(-1)^j \binom{\alpha}{j} h^{-cj} f(h^{-j}x) \qquad (x,\, h \in \mathbb{R}^+).$$

By Proposition 9.29, we know that for $f \in X_c$ and any $h > 0$ the fractional difference exists almost everywhere and

$$\|\Delta_h^{\alpha,c} f\|_{X_c} \le \|f\|_{X_c} \sum_{j=0}^{\infty} \left|\binom{\alpha}{j}\right|$$

with a convergent series on the right-hand side.

In the next theorem, we shall use the pointwise fractional derivative $\Theta_c^{\alpha} f$ as specified in Definition 9.31. By Theorem 9.33, we know that for $f \in \mathscr{W}_c^{\alpha,1}$ this derivative exists as a function in X_c and its Mellin transform is given by

$$[\Theta_c^{\alpha} f]^{\wedge}_{M_c}(c+it) = (-it)^{\alpha}[f]^{\wedge}_{M_c}(c+it) \qquad (t \in \mathbb{R}), \tag{14.7}$$

where the power of exponent α is defined by its principal value, that is, $(-it)^{\alpha} = |t|^{\alpha}\, e^{-i\,\mathrm{sgn}(t)\pi\alpha/2}$. For the remainder $R_{c,\sigma}[f]$, we can now deduce the following asymptotic estimate.

Theorem 14.6 *Let $f \in \mathcal{K}_c$ with $c \in \mathbb{R}$. If in addition $f \in \mathscr{W}_c^{\alpha,1}$ and $\Theta_c^{\alpha} f \in \mathrm{Lip}_r(\beta, X_c)$, where $\alpha \in \mathbb{R}^+$, $r \in \mathbb{N}$ and $\beta \in]0, r]$ such that $\alpha + \beta > 1$, then*

$$|R_{c,\sigma}[f]| = \mathcal{O}(\sigma^{-\alpha-\beta}) \qquad (\sigma \to \infty).$$

Proof By using (14.7), the proof can be performed analogously to that of Theorem 14.5. □

Since the remainder $R_{c,\sigma}[f]$ vanishes as soon as f belongs to the Mellin–Paley–Wiener space $B^1_{c,2\pi\sigma}$, it is of interest to estimate $|R_{c,\sigma}[f]|$ in terms of the distance of f (or its Mellin derivatives) from $B^1_{c,2\pi\sigma}$. Recalling Theorems 10.2 and 10.5, Corollary 10.6 and Lemma 10.8, we can state for $p = 1$ and $q = \infty$ the following relations between distances and Mellin transforms.

Corollary 14.7 *Let $f \in \mathcal{M}^1_c$, where $c \in \mathbb{R}$. Then*

$$\mathrm{dist}_{\infty}(f, B^1_{c,\sigma}) = \sup_{|v|\ge\sigma} \left|[f]^{\wedge}_{M_c}(c+iv)\right| \qquad (\sigma \in \mathbb{R}^+).$$

If in addition $(\cdot)^\alpha [f]^\wedge_{M_c}(c+i\cdot) \in L^1(\mathbb{R}) \cap C(\mathbb{R})$ *for some* $\alpha \in \mathbb{R}^+$, *then* $\Theta_c^\alpha f$ *exists and*

$$\operatorname{dist}_\infty(\Theta_c^\alpha f, B^1_{c,\sigma}) = \sup_{|v|\ge\sigma} \left|v^\alpha [f]^\wedge_{M_c}(c+iv)\right| \qquad (\sigma \in \mathbb{R}^+).$$

Concerning the remainder in (14.3), we can now deduce the following result generalized to translated functions $\tau_h^c f$.

Theorem 14.8 *Let* $f \in \mathcal{K}_c \cap \mathcal{M}_c^1$ *with* $c \in \mathbb{R}$. *If in addition* $(\cdot)^\alpha [f]^\wedge_{M_c}(c+i\cdot) \in L^1(\mathbb{R}) \cap C(\mathbb{R})$ *for some* $\alpha > 1$, *then* $\Theta_c^\alpha f$ *exists and*

$$|R_{c,\sigma}[\tau_h^c f]| \le \frac{2\zeta(\alpha)}{(2\pi\sigma)^\alpha} \operatorname{dist}_\infty(\Theta_c^\alpha f, B^1_{c,2\pi\sigma}) \qquad (h,\sigma \in \mathbb{R}^+). \tag{14.8}$$

Proof By Theorem 6.10 we have $[\tau_h^c f]^\wedge_{M_c}(c+it) = h^{-it}[f]^\wedge_{M_c}(c+it)$. Therefore it is not difficult to show that the hypothesis $f \in \mathcal{K}_c$ implies $\tau_h^c f \in \mathcal{K}_c$. Using (14.2) for $\tau_h^c f$, we obtain

$$\begin{aligned} |R_{c,\sigma}[\tau_h^c f]| &\le \sum_{k\in\mathbb{Z}\setminus\{0\}} \frac{1}{|2\pi k\sigma|^\alpha} \left|(2\pi k\sigma)^\alpha [f]^\wedge_{M_c}(c+i2\pi k\sigma)\right| \\ &\le \frac{2\zeta(\alpha)}{(2\pi\sigma)^\alpha} \sup_{|v|\ge 2\pi\sigma} \left|v^\alpha [f]^\wedge_{M_c}(c+iv)\right| = \frac{2\zeta(\alpha)}{(2\pi\sigma)^\alpha} \operatorname{dist}_\infty(\Theta_c^\alpha f, B^1_{c,2\pi\sigma}), \end{aligned}$$

where only in the last step we have used Corollary 14.7. Thus (14.8) is verified. □

Note that for $h = 1$ we obtain an estimate for $R_{c,\sigma}[f]$ since $\tau_1^c f = f$.

For the spaces $W_c^{\alpha,*}(\mathbb{R}^+)$ (see Definition 10.19) an estimate of $|R_{c,\sigma}[f]|$ which does not involve the distance functional seems to be more appropriate. Slightly more generally, we again estimate the remainder for the Mellin translated function $\tau_h^c f$ with $h > 0$.

Theorem 14.9 *Let* $f \in \mathcal{K}_c \cap W_c^{\alpha,*}(\mathbb{R}^+)$, *where* $c \in \mathbb{R}$ *and* $\alpha \in]1,\infty[$. *Then*

$$\begin{aligned} |R_{c,\sigma}[\tau_h^c f]| &\le \frac{2\zeta(\alpha)}{(2\pi\sigma)^\alpha} \sup_{|v|\ge 2\pi\sigma} \left|v^\alpha [f]^\wedge_{M_c}(c+iv)\right| \\ &\le \frac{2\zeta(\alpha)}{(2\pi\sigma)^\alpha} \|f\|_{W_c^{\alpha,*}} \qquad (h,\sigma \in \mathbb{R}^+). \end{aligned}$$

The first inequality is best possible in the sense that it is no longer true if its right-hand side is multiplied by a constant less than 1. *In the second estimate equality can be attained.*

Proof Using again that $\tau_h^c f \in \mathcal{K}_c$ and $[\tau_h^c f]^\wedge_{M_c}(c+it) = h^{-it}[f]^\wedge_{M_c}(c+it)$, we conclude that

$$\begin{aligned}|R_{c,\sigma}[\tau_h^c f]| &\le \sum_{k\in\mathbb{Z}\setminus\{0\}} \frac{1}{|2\pi k\sigma|^\alpha}\left|(2\pi k\sigma)^\alpha[f]^\wedge_{M_c}(c+2\pi ik\sigma)\right| \\ &\le \frac{2\zeta(\alpha)}{(2\pi\sigma)^\alpha}\sup_{|v|\ge 2\pi\sigma}\left|v^\alpha[f]^\wedge_{M_c}(c+iv)\right| \le \frac{2\zeta(\alpha)}{(2\pi\sigma)^\alpha}\|f\|_{W_c^{\alpha,*}}.\end{aligned}$$

The statement on the precision of these estimates in the case $\alpha = 4$ can be verified by Example 4 in Sect. 14.3. □

Remark 14.10 Note that, under the hypotheses of Theorem 14.9, the fractional Mellin derivative $\Theta_c^\alpha f$ need not exist. Therefore, other than in the proof of Theorem 14.8, the supremum cannot be expressed as $\mathrm{dist}_\infty(\Theta_c^\alpha f, B^1_{c,2\pi\sigma})$.

14.2 Equivalence Theorems

The results of the previous section show that high regularity of the function f yields rapid convergence to zero of the remainders of the quadrature formula (14.3) as $\sigma \to \infty$. One may ask if the converse is also true. Does rapid convergence to zero of the remainders imply high regularity of the function? The answer is *no* for a simple reason. Due to certain symmetries of the function f, the remainders may be zero even if the graph of f is very erratic as far as regularity is concerned. In fact, we shall see subsequently that f can be split into two parts, called the Mellin-even and the Mellin-odd part, such that the remainders of the Mellin-odd part are always zero. In order to eliminate this phenomenon, there are two possibilities: One may either study the remainders in terms of the Mellin-even part only or, alternatively, one may require that a certain speed of convergence not only holds for f itself but also under the action of the Mellin translation operator τ_h^c on f with values of h near to 1, which will destroy possible symmetries. Both approaches will be considered in this section.

First we characterize the speed of convergence of the remainders in terms of distances from Mellin–Bernstein spaces. Then, using the results of Sect. 10.2, which characterize distances from Mellin–Bernstein spaces by functions spaces guaranteeing a certain regularity, we can deduce a characterization of the speed of convergence by specific function spaces.

14.2.1 Characterization of Speed of Convergence by Distances

Case of the Mellin-Even Part

For $f : \mathbb{R}^+ \to \mathbb{C}$, we define

$$f_{c+}(x) := \frac{1}{2}(x^c f(x) + x^{-c} f(1/x)),$$

$$f_{c-}(x) := \frac{1}{2}(x^c f(x) - x^{-c} f(1/x)) \quad (x \in \mathbb{R}^+),$$

so that $f(x) = x^{-c}(f_{c+}(x) + f_{c-}(x))$ for every $x \in \mathbb{R}^+$. We call f_{c+} and f_{c-} the c-Mellin-even and the c-Mellin-odd part of f, respectively. Moreover, for any function $f \in X_c$, we have $f_{c+}, f_{c-} \in X_0$ and, for $t \in \mathbb{R}$,

$$[f_{c+}]^{\wedge}_{M_0}(\pm it) = \frac{1}{2}\big([f]^{\wedge}_{M_c}(c + it) + [f]^{\wedge}_{M_c}(c - it)\big)$$

and

$$[f_{c-}]^{\wedge}_{M_0}(it) = \frac{1}{2}\big([f]^{\wedge}_{M_c}(c + it) - [f]^{\wedge}_{M_c}(c - it)\big).$$

Using these formulae, we can see that the remainders in (14.3) satisfy the following relations.

Proposition 14.11 *Let $f \in \mathcal{K}_c$, where $c \in \mathbb{R}$. Then*

$$R_{c,\sigma}[f] = R_{0,\sigma}[f_{c+}]$$

and $R_{0,\sigma}[f_{c-}] = 0$ for all $\sigma \in \mathbb{R}^+$.

This leads us to the representation

$$R_{c,\sigma}[f] = -2 \sum_{k=1}^{\infty} [f_{c+}]^{\wedge}_{M_0}(i2\pi k\sigma). \tag{14.9}$$

With the help of the Möbius function $\mu(\cdot)$, this formula can be inverted. In fact, the following statement holds.

Lemma 14.12 *Let $f \in \mathcal{K}_c$, where $c \in \mathbb{R}$, and let $\sigma \in \mathbb{R}^+$. If*

$$\sum_{n=1}^{\infty} |R_{c,n\sigma}[f]| < \infty, \tag{14.10}$$

then

$$[f_{c+}]^{\wedge}_{M_0}(i2\pi n\sigma) = -\frac{1}{2}\sum_{k=1}^{\infty}\mu(k)R_{c,nk\sigma}[f] \tag{14.11}$$

for all $n \in \mathbb{N}$. In particular,

$$|[f_{c+}]^{\wedge}_{M_0}(i2\pi\sigma)| \leq \frac{1}{2}\sum_{k=1}^{\infty}|R_{c,k\sigma}[f]|. \tag{14.12}$$

Proof For all $n \in \mathbb{N}$, define

$$a_n := -2[f_{c+}]^{\wedge}_{M_0}(i2\pi n\sigma) \quad \text{and} \quad b_n := R_{c,n\sigma}[f].$$

Then the hypotheses of the lemma together with (14.9) imply that the sequences $(a_n)_{n\in\mathbb{N}}$ and $(b_n)_{n\in\mathbb{N}}$ satisfy the assumptions of Proposition 1.50. Therefore

$$a_n = \sum_{k=1}^{\infty}\mu(k)\,b_{nk} \qquad (n \in \mathbb{N}),$$

which shows that (14.11) holds. Finally, choosing $n = 1$ and noting that $|\mu(k)| \leq 1$ for all $k \in \mathbb{N}$, we easily obtain (14.12). □

Next we describe an important situation where condition (14.10) is satisfied.

Lemma 14.13 *Under the assumptions of Theorem 14.8 we have*

$$\sum_{n=1}^{\infty}|R_{c,n\sigma}[f]| < \infty \qquad (\sigma \in \mathbb{R}^+).$$

Proof Noting that $B^1_{c,2\pi\sigma} \subset B^1_{c,2\pi n\sigma}$ for $n \in \mathbb{N}$, we deduce from the estimate in Theorem 14.8 that

$$|R_{c,n\sigma}[f]| \leq \frac{2\zeta(\alpha)}{(2\pi n\sigma)^{\alpha}}\operatorname{dist}_{\infty}(\Theta^{\alpha}_c f, B^1_{c,2\pi\sigma}).$$

From this the assertion follows immediately. □

We are now ready for the main result of this subsection.

Theorem 14.14 *Let λ be a nonnegative, non-increasing function on an interval $[t_0, \infty[$ with $t_0 > 0$. Then, under the hypotheses of Theorem 14.8, the following statements are equivalent:*

(i) $\operatorname{dist}_{\infty}(\Theta^{\alpha} f_{c+}, B^1_{0,2\pi\sigma}) = \mathcal{O}(\lambda(\sigma)) \qquad (\sigma \to \infty)$;
(ii) $R_{c,\sigma}[f] = \mathcal{O}(\sigma^{-\alpha}\lambda(\sigma)) \qquad (\sigma \to \infty)$.

Proof Suppose that (i) holds. Since f_{c+} satisfies the same assumptions as f, but with $c = 0$, we see that (ii) follows from Theorem 14.8 in conjunction with Proposition 14.11.

Conversely, suppose that (ii) holds. Then there exist a constant $C > 0$ and $\sigma_0 > t_0$ such that in view of Proposition 14.11 we have

$$\left|R_{0,\sigma}[f_{c+}]\right| = \left|R_{c,\sigma}[f]\right| \leq C\,\frac{\lambda(\sigma)}{\sigma^\alpha}$$

for all $\sigma \geq \sigma_0$. Since Lemma 14.12 is applicable, we obtain with the help of (14.12) that

$$\left|[f_{c+}]^\wedge_{M_0}(i2\pi\sigma)\right| \leq \frac{1}{2}\sum_{k=1}^{\infty}\left|R_{c,k\sigma}[f]\right| \leq \frac{C}{2}\sum_{k=1}^{\infty}\frac{\lambda(k\sigma)}{(k\sigma)^\alpha}.$$

Therefore, by the hypotheses on the function λ,

$$\left|(2\pi\sigma)^\alpha[f_{c+}]^\wedge_{M_0}(i2\pi\sigma)\right| \leq \frac{C}{2}(2\pi)^\alpha\zeta(\alpha)\lambda(\sigma),$$

for every $\sigma \geq \sigma_0$. This implies that

$$\sup_{|v|\geq 2\pi\sigma}\left\{\left|v^\alpha[f_{c+}]^\wedge_{M_0}(iv)\right|\right\} \leq \frac{C}{2}(2\pi)^\alpha\zeta(\alpha)\lambda(\sigma) \qquad (\sigma \geq \sigma_0),$$

and (i) follows by Corollary 14.7. □

Case of the Translated Function $\tau_h^c f$

Now we establish a characterization of the speed of convergence of the remainder for the translated function $\tau_h^c f$ in terms of the distance $\mathrm{dist}_\infty(f, B^1_{c,2\pi\sigma})$. As a counterpart of Lemma 14.12 for remainders of translated functions, we state the following auxiliary result.

Lemma 14.15 *Let $f \in \mathcal{K}_c$, where $c \in \mathbb{R}$. Then*

$$\left|[f]^\wedge_{M_c}(c \pm i2\pi\sigma)\right| \leq \sup\left\{\left|R_{c,\sigma}\left[\tau_h^c f\right]\right| \,:\, e^{-1/(2\sigma)} \leq h \leq e^{1/(2\sigma)}\right\} \quad (\sigma \in \mathbb{R}^+).$$

Proof For fixed c, let us write the Mellin translation as $f_h := \tau_h^c f$ for short. We recall that by Theorem 6.10

$$[f_h]^\wedge_{M_c}(c+it) = h^{-it}[f]^\wedge_{M_c}(c+it),$$

which implies by (14.2) that

$$R_{c,\sigma}[f_h] = -\sum_{k\in\mathbb{Z}\setminus\{0\}} h^{-i2\pi k\sigma}[f]^{\wedge}_{M_c}(c+i2\pi k\sigma).$$

Setting $x := h^{2\pi\sigma}$, we may rewrite this equation as

$$-\frac{x^{-c}}{2\pi}R_{c,\sigma}[f_{x^{1/(2\pi\sigma)}}] = \frac{x^{-c}}{2\pi}\sum_{k\in\mathbb{Z}\setminus\{0\}} [f]^{\wedge}_{M_c}(c+i2\pi k\sigma)x^{-ik}.$$

The right-hand side is a Mellin–Fourier series in x (see Sect. 5.2). When h traverses the interval $[e^{-1/(2\sigma)}, e^{1/(2\sigma)}]$, then x traverses the interval $[e^{-\pi}, e^{\pi}]$. Therefore, by Theorem 5.23, we have

$$\sum_{k\in\mathbb{Z}\setminus\{0\}} \left|[f]^{\wedge}_{M_c}(c+i2\pi k\sigma)\right|^2 = \frac{1}{2\pi}\int_{e^{-\pi}}^{e^{\pi}} \left|R_{c,\sigma}[f_{x^{1/(2\pi\sigma)}}]\right|^2 \frac{dx}{x}. \tag{14.13}$$

From this we deduce that

$$\sum_{k\in\mathbb{Z}\setminus\{0\}} \left|[f]^{\wedge}_{M_c}(c+i2\pi k\sigma)\right|^2 \le \sup\left\{\left|R_{c,\sigma}\,[f_h]\right|^2 \,:\, e^{-1/(2\sigma)} \le h \le e^{1/(2\sigma)}\right\}.$$

Now, considering on the left-hand side the terms for $k = \pm 1$ only, we obtain the assertion. □

As an analogue of Theorem 14.14, we present the following result.

Theorem 14.16 *Let λ be a nonnegative, non-increasing function on $[t_0, \infty[$ with $t_0 > 0$. Then, under the hypotheses of Theorem 14.8, the following statements are equivalent:*

(i) $\operatorname{dist}_\infty(\Theta_c^\alpha f, B^1_{c,2\pi\sigma}) = \mathcal{O}(\lambda(\sigma))$ *as* $\sigma \to \infty$;
(ii) $R_{c,\sigma}[\tau_h^c f] = \mathcal{O}(\sigma^{-\alpha}\lambda(\sigma))$ *as* $\sigma \to \infty$ *uniformly for* $h \in [e^{-1/(2\sigma)}, e^{1/(2\sigma)}]$.

Proof If (i) holds, then Theorem 14.8 implies immediately that (ii) is true.

Conversely, if (ii) holds, then, by Lemma 14.15,

$$\left|[f]^{\wedge}_{M_c}(c \pm i2\pi\sigma)\right| = \mathcal{O}\left(\sigma^{-\alpha}\lambda(\sigma)\right)$$

as $\sigma \to \infty$, which implies that

$$\operatorname{dist}_\infty(\Theta_c^\alpha f, B^1_{c,2\pi\sigma}) = \sup_{|v|\ge 2\pi\sigma} \left|v^\alpha[f]^{\wedge}_{M_c}(c+iv)\right| = \mathcal{O}(\lambda(\sigma)) \qquad (\sigma \to \infty),$$

and so (i) is true. □

14.2.2 *Characterization of the Speed of Convergence by Function Spaces*

Combining the results of Sect. 14.2.1 with those of Sect. 10.3, we obtain a characterization of the speed of convergence by function spaces.

Case of the Mellin-Even Part

The following statement is obtained by combining Proposition 14.11, Lemma 14.12 and Theorem 10.17.

Proposition 14.17 *Let $\varphi \in \mathcal{K}_c \cap \mathcal{M}_c^1$, where $c \in \mathbb{R}$. Then*

$$\varphi_{c+} \in B_{0,2\pi\sigma}^1 \iff R_{c,\rho}[\varphi] = 0 \text{ for all } \rho > \sigma \in \mathbb{R}^+.$$

Next we characterize exponential speed of convergence.

Proposition 14.18 *Let $\varphi \in \mathcal{K}_c \cap \mathcal{M}_c^1$, where $c \in \mathbb{R}$. Then for $a > 0$, we have*

$$\left.\begin{array}{l} f \in H_0^*(\mathbb{H}_a) \\ f(\cdot, 0) \equiv \varphi_{c+} \end{array}\right\} \iff R_{c,\sigma}[\varphi] = \mathcal{O}(e^{-2\pi a\sigma}) \qquad (\sigma \to \infty).$$

Proof Let $f \in H_0^*(\mathbb{H}_a)$ such that $f(\cdot, 0) \equiv \varphi_{c+}$. Then, by Theorem 10.18, we have $\operatorname{dist}_\infty(\varphi_{c+}, B_{0,2\pi\sigma}^1) = \mathcal{O}(e^{-2\pi a\sigma})$ as $\sigma \to \infty$. Hence there exist $\sigma_0 > 0$ and $C > 0$ such that

$$\sup_{|v| \geq 2\pi\sigma} \left|[\varphi_{c+}]^\wedge_{M_0}(iv)\right| \leq Ce^{-2\pi a\sigma} \qquad (\sigma \geq \sigma_0).$$

This implies that

$$\left|[\varphi_{c+}]^\wedge_{M_0}(i2\pi k\sigma)\right| \leq Ce^{-2\pi ak\sigma}$$

for all $k \in \mathbb{N}$ and $\sigma \geq \sigma_0$. Now, using (14.9), we readily conclude that $R_{c,\sigma}[\varphi] = \mathcal{O}(e^{-2\pi a\sigma})$ as $\sigma \to \infty$.

Conversely, if $R_{c,\sigma}[\varphi] = \mathcal{O}(e^{-2\pi a\sigma})$ as $\sigma \to \infty$, then there exist $\sigma_0 > 0$ and $C > 0$ such that

$$\left|R_{c,k\sigma}[\varphi]\right| \leq Ce^{-2\pi ak\sigma} \qquad (k \in \mathbb{N},\ \sigma \geq \sigma_0).$$

Therefore Lemma 14.12 applies and (14.12) yields

$$[\varphi_{c+}]^\wedge_{M_0}(i2\pi\sigma) = \mathcal{O}(e^{-2\pi a\sigma}) \qquad (\sigma \to \infty).$$

Hence $\operatorname{dist}_\infty(\varphi_{c+}, B_{0,\sigma}^1) = \mathcal{O}(e^{-a\sigma})$ as $\sigma \to \infty$. Now Theorem 10.18 yields the desired assertion. □

Polynomial speed of convergence can be characterized as follows.

Proposition 14.19 *Let $\varphi \in \mathcal{K}_c$, where $c \in \mathbb{R}$. Then for $\alpha > 1$, we have*

$$\varphi_{c+} \in W_0^{\alpha,*}(\mathbb{R}^+) \iff R_{c,\sigma}[\varphi] = \mathcal{O}(\sigma^{-\alpha}) \quad (\sigma \to \infty).$$

Proof As an immediate consequence of Theorem 14.9 in conjunction with Proposition 14.11, we obtain that the statement on the left-hand side implies the one on the right-hand side.

Conversely, assume that $R_{c,\sigma}[\varphi] = \mathcal{O}(\sigma^{-\alpha})$ as $\sigma \to \infty$. Then there exist $\sigma_0 > 0$ and $C > 0$ such that

$$|R_{c,\sigma}[\varphi]| \leq \frac{C}{\sigma^\alpha} \qquad (\sigma \geq \sigma_0).$$

We note that condition (14.10) of Lemma 14.12 is satisfied for all $\sigma \geq \sigma_0$ and so we obtain with the help of (14.12) that

$$\left|[\varphi_{c+}]^\wedge_{M_0}(i2\pi\sigma)\right| \leq \frac{C}{2}\sum_{k=1}^\infty \frac{1}{(k\sigma)^\alpha} = \frac{C\zeta(\alpha)}{2\sigma^\alpha} \qquad (\sigma \geq \sigma_0).$$

This yields

$$\sup_{|v|\geq 2\pi\sigma_0} \left|v^\alpha [\varphi_{c+}]^\wedge_{M_0}(iv)\right| \leq \frac{C}{2}(2\pi)^\alpha \zeta(\alpha). \tag{14.14}$$

The hypothesis on φ implies that $(\cdot)^\alpha [\varphi_{c+}]^\wedge_{M_0}(i\cdot) \in C(\mathbb{R})$. Hence (14.14) allow us to conclude that

$$\sup_{v\in\mathbb{R}} \left|v^\alpha [\varphi_{c+}]^\wedge_{M_0}(iv)\right| < \infty,$$

and so $\varphi_{c+} \in W_0^{\alpha,*}(\mathbb{R}^+)$. □

Case of the Translated Function $\tau_h^c f$

The propositions of Sect. 10.3 in conjunction with the results in Sect. 14.2.1 lead us to the following characterizations.

Proposition 14.20 *Let $\varphi \in \mathcal{K}_c \cap \mathcal{M}_c^1$, where $c \in \mathbb{R}$. Then*

$$\varphi \in B^1_{c,2\pi\sigma} \iff \begin{cases} R_{c,\rho}\left[\tau_h^c \varphi\right] = 0 \\ \textit{for all } \rho > \sigma,\ h \in [e^{-1/(2\rho)}, e^{1/(2\rho)}]. \end{cases}$$

Proposition 14.21 *Let $\varphi \in \mathcal{K}_c \cap \mathcal{M}_c^1$, where $c \in \mathbb{R}$. Then for $a > 0$, we have*

$$\left.\begin{array}{l} f \in H_c^*(\mathbb{H}_a) \\ f(\cdot, 0) \equiv \varphi \end{array}\right\} \iff \left\{\begin{array}{l} R_{c,\sigma}\left[\tau_h^c \varphi\right] = \mathcal{O}(e^{-2\pi a \sigma}) \quad (\sigma \to \infty) \\ \textit{uniformly for } h \in [e^{-1/(2\sigma)}, e^{1/(2\sigma)}]. \end{array}\right.$$

Finally we present a characterization of polynomial speed of convergence. Its proof is analogous to that of Proposition 14.19 but with the role of Lemma 14.12 being taken by Lemma 14.15.

Proposition 14.22 *Let $\varphi \in \mathcal{K}_c$, where $c \in \mathbb{R}$. Then for $\alpha > 1$, we have*

$$\varphi \in W_c^{\alpha,*}(\mathbb{R}^+) \iff \left\{\begin{array}{l} R_{c,\sigma}\left[\tau_h^c \varphi\right] = \mathcal{O}(\sigma^{-\alpha}) \quad (\sigma \to \infty) \\ \textit{uniformly for } h \in [e^{-1/(2\sigma)}, e^{1/(2\sigma)}]. \end{array}\right.$$

14.3 Numerical Experiments

In this section we give some examples illustrating the theory developed in previous sections. The computations were performed with the help of Maple 16.

Example 1: A Mellin Bandlimited Integrand
For $m \in \mathbb{N}$, we consider

$$f_{2m}(r) := \begin{cases} \left(\dfrac{\sin(\pi \log r)}{\pi \log r}\right)^{2m} & \text{if } r \neq 1, \\ 1 & \text{if } r = 1. \end{cases}$$

We want to compute the integral

$$I_{2m} := \int_0^\infty f_{2m}(r) \frac{dr}{r}.$$

Its exact value is given by (see [160, p. 494, § 3.836/2])

$$I_{2m} = \frac{m}{2^{2m-3}} \sum_{j=0}^{m-1} (-1)^j \frac{(2m-2j)^{2m-2}}{j!(2m-1-j)!}.$$

In particular,

$$I_2 = 1, \quad I_4 = \frac{2}{3}, \quad I_6 = \frac{11}{20}, \quad I_8 = \frac{151}{315}.$$

Note that f_{2m} is a 0-Mellin-even function which is Mellin bandlimited to $[-2\pi m, 2\pi m]$. The quadrature formula (14.3) now yields

$$I_{2m} = \frac{1}{\sigma}\left(1 + 2\sum_{k=1}^{\infty} f_{2m}\left(e^{k/\sigma}\right)\right) + R_{0,\sigma}[f_{2m}] \tag{14.15}$$

for any positive σ. Note that we have $R_{0,\sigma}[f_{2m}] = 0$ as soon as $\sigma \geq m$. However, in computations we have to truncate the series in (14.15), and so there is always a truncation error of the form

$$T_{m,\sigma,K} := \frac{2}{\sigma} \sum_{k=K+1}^{\infty} f_{2m}\left(e^{k/\sigma}\right),$$

if we take only the first K terms. By standard estimates and the integral comparison method, we find that

$$T_{m,\sigma,K} \leq \frac{2}{\sigma} \sum_{k=K+1}^{\infty} \left(\frac{\sigma}{\pi k}\right)^{2m} \leq \frac{2\sigma^{2m-1}}{\pi^{2m}} \int_K^{\infty} x^{-2m}\, dx = \frac{2}{(2m-1)\pi}\left(\frac{\sigma}{K\pi}\right)^{2m-1}.$$

Thus, in order to guarantee that the truncation error does not exceed $10^{-\ell}$, say, we should choose

$$K \geq \frac{\sigma}{\pi}\left(\frac{2 \cdot 10^{\ell}}{(2m-1)\pi}\right)^{1/(2m-1)}.$$

The total error is given by

$$E_{m.\sigma,K} := I_{2m} - \frac{1}{\sigma}\left(1 + 2\sum_{k=1}^{K} f_{2m}\left(e^{k/\sigma}\right)\right).$$

For $m = 4$ and $\ell = 12$ computations provided the results shown in Table 14.1. It is clearly seen that the total error reduces to the truncation error as soon as $\sigma \geq 4$.

Example 2: Integrand with a Branch Point

We modify the function f_8 of the previous example by multiplying it with a square root, which creates a branch point. More precisely, for $a > 0$ we put

$$g_a(r) := f_8(r)\sqrt{a^2 + (\log r)^2}.$$

We want to compute the integral

$$J_a := \int_0^{\infty} g_a(r)\, \frac{dr}{r}.$$

Table 14.1 Example 1 for $m = 4$

σ	K	$E_{4,\sigma,K}$	σ	K	$E_{4,\sigma,K}$
0.50	6	−1.520635e+00	4.50	53	2.267646e−13
1.00	12	−5.206349e−01	5.00	59	2.266369e−13
1.50	18	−1.884456e−01	5.50	65	2.266218e−13
2.00	24	−4.761905e−02	6.00	71	2.266187e−13
2.50	30	−6.755332e−03	6.50	77	2.266178e−13
3.00	36	−3.968254e−04	7.00	82	2.267143e−13
3.50	41	−3.100198e−06	7.50	88	2.266709e−13
4.00	47	2.329011e−13	8.00	94	2.266480e−13

First we look for a polar-analytic extension of g_a. It is easy to see that

$$\widetilde{f_8}(r,\theta) := \begin{cases} \left(\dfrac{\sin(\pi(\log r + i\theta))}{\pi(\log r + i\theta)}\right)^8 & \text{if } (r,\theta) \in \mathbb{H} \setminus \{(1,0)\}, \\ 1 & \text{if } (r,\theta) = (1,0) \end{cases}$$

is a polar-analytic extension of f_8 to $\mathbb{H}$. Therefore $\widetilde{g}_a$, defined by

$$\widetilde{g}_a(r,\theta) := \widetilde{f_8}(r,\theta)\sqrt{a^2 + (\log r + i\theta)^2},$$

is the right candidate for a polar-analytic extension of g_a. The expression under the root vanishes if and only if $(r,\theta) = (1, \pm a)$. Hence $\mathbb{H}_a$ is the largest strip in $\mathbb{H}$ on which $\widetilde{g}_a$ is polar-analytic. Now it is easily verified that $\widetilde{g}_a$ belongs to the Mellin–Hardy space $H_0^1(\mathbb{H}_a)$, which is a subspace of $H_0^*(\mathbb{H}_a)$, but it does not belong to $H_0^*(\mathbb{H}_b)$ for $b > a$.

Since $g_a \in X_0$, we have $M_0^*[g_a](it) \equiv M_0[g_a](it)$. This allows us to conclude with the help of Theorem 8.20 that $g_a \in \mathcal{K}_0 \cap \mathcal{M}_0^1$. Hence the theory established in the previous sections is applicable to g_a.

We note that g_a is again a 0-Mellin-even function. By the quadrature formula (14.3) we have

$$J_a = \frac{1}{\sigma}\left[a + 2\sum_{k=1}^{\infty} f_8(e^{k/\sigma})\sqrt{a^2 + (k/\sigma)^2}\right] + R_{0,\sigma}[g_a].$$

The exact value of J_a is not known. For numerically given a, we used Maple for gaining J_a up to 40 decimal places. The total error in our computation by formula (14.3) will be

$$E_{a,\sigma,K} := J_a - \frac{1}{\sigma}\left[a + 2\sum_{k=1}^{K} f_8(e^{k/\sigma})\sqrt{a^2 + (k/\sigma)^2}\right] = R_{0,\sigma}[g_a] + T_{a,\sigma,K},$$

where

$$T_{a,\sigma,K} := \frac{2}{\sigma} \sum_{k=K+1}^{\infty} f_8(e^{k/\sigma}) \sqrt{a^2 + (k/\sigma)^2}$$

is the truncation error. Since

$$f_8(e^{k/\sigma}) \le \frac{1}{(\pi k/\sigma)^8} \quad \text{and} \quad \sqrt{a^2 + (k/\sigma)^2} \le \frac{k}{\sigma} + \frac{a^2\sigma}{2k},$$

we find by using the integral comparison method that

$$\begin{aligned} T_{a,\sigma,K} &\le \frac{2}{\pi^8} \int_K^{\infty} \left[\left(\frac{\sigma}{x}\right)^7 + \frac{a^2}{2} \left(\frac{\sigma}{x}\right)^9 \right] \frac{dx}{\sigma} \\ &= \frac{1}{\pi^8} \left[\frac{1}{3} \left(\frac{\sigma}{K}\right)^6 + \frac{a^2}{8} \left(\frac{\sigma}{K}\right)^8 \right]. \end{aligned}$$

Proposition 14.18 tells us that $R_{0,\sigma}[g_a] = \mathcal{O}(e^{-2\pi a\sigma})$ as $\sigma \to \infty$. To ensure that the truncation error does not exceed the remainder asymptotically, we choose K such that

$$\frac{1}{3} \left(\frac{\sigma}{K}\right)^6 + \frac{a^2}{8} \left(\frac{\sigma}{K}\right)^8 \le \frac{\pi^8}{10} e^{-2\pi a\sigma}.$$

The asymptotic result for the remainder and the choice of K for controlling the truncation error suggest that

$$E_{a,\sigma,K} \approx C\, e^{-2\pi a\sigma}$$

for large σ. In our numerical experiments we check this behavior in two ways by computing

$$C := E_{a,\sigma,K}\, e^{2\pi a\sigma} \quad \text{and} \quad \text{rate} := -\frac{\log |E_{a,\sigma,K}|}{\sigma}. \tag{14.16}$$

While C should remain bounded, the rate should approach $2\pi a$ as $\sigma \to \infty$. As an immediate consequence of (14.16), we have

$$\text{rate} = 2\pi a - \frac{\log |C|}{\sigma}.$$

Hence for bounded σ, say $2 \le \sigma \le 15$, the numbers rate can be close to $2\pi a$ only if $|C|$ is close to 1. This explains the behavior of the numbers in the last two columns of Tables 14.2, 14.3, and 14.4. In view of Theorem 8.20, we expect that C depends on $\|g_a\|_{H_0^*(\mathbb{H}_a)}$. It seems that this expression is growing considerably when a moves from 1/2 to 1.

Table 14.2 Example 2 for $a = \frac{1}{2}$, rate $\to \pi = 3.141592\ldots$

$J_a = 0.25523736847216208683891588\ldots$

σ	K	$E_{a,\sigma,K}$	C	rate
2	8	−1.385e−02	−7.415e+00	2.139842
3	4	4.441e−04	5.503e+00	2.573168
4	9	6.830e−06	1.959e+00	2.973533
5	19	1.453e−07	9.641e−01	3.148914
6	37	4.117e−09	6.321e−01	3.218038
7	73	1.274e−10	4.526e−01	3.254852
8	141	4.216e−12	3.467e−01	3.274013
9	267	1.478e−13	2.812e−01	3.282555
10	500	5.372e−15	2.365e−01	3.285766
11	927	1.995e−16	2.033e−01	3.286429
12	1707	7.525e−18	1.774e−01	3.285687
13	3122	2.883e−19	1.573e−01	3.283862
14	5675	1.120e−20	1.414e−01	3.281298
15	10,264	4.400e−22	1.285e−01	3.278358

Table 14.3 Example 2 for $a = \frac{5}{8}$, rate $\to \frac{5\pi}{4} = 3.926990\ldots$

$J_a = 0.31237704377490102358516251\ldots$

σ	K	$E_{a,\sigma,K}$	C	rate
2	3	−2.172e−02	−5.596e+01	1.914666
3	6	2.043e−04	2.671e+01	2.831941
4	15	1.962e−06	1.302e+01	3.285439
5	36	1.730e−08	5.827e+00	3.574493
6	81	2.082e−10	3.558e+00	3.715444
7	182	2.856e−12	2.478e+00	3.797346
8	400	4.219e−14	1.858e+00	3.849564
9	865	6.546e−16	1.463e+00	3.884719
10	1849	1.051e−17	1.192e+00	3.909403
11	3912	1.733e−19	9.976e−01	3.927212
12	8212	2.916e−21	8.520e−01	3.940341

Table 14.4 Example 2 for $a = 1$, rate $\to 2\pi = 6.283185\ldots$

$J_a = 0.48761056549919471345809158\ldots$

σ	K	$E_{a,\sigma,K}$	C	rate
2	5	−4.256e−02	−1.220e+04	1.578402
3	19	−1.580e−04	−2.426e+04	2.917691
4	71	1.248e−07	1.026e+04	3.974063
5	250	7.809e−11	3.438e+03	4.654642
6	854	8.050e−14	1.898e+03	5.025082
7	2838	9.886e−17	1.248e+03	5.264689
8	9241	1.333e−19	9.014e+02	5.432688
9	29,625	1.907e−22	6.905e+02	5.556805
10	93,800	2.841e−25	5.509e+02	5.652030

Example 3: An Integral Representing the Gamma Function
For complex arguments z with $\Re z > 0$ let us consider the gamma function (see Sect. 1.5)

$$\Gamma(z) = \int_0^\infty e^{-r} r^{z-1} dr. \tag{14.17}$$

Writing $f(r) := e^{-r}$ and $z = c + it$ with $c > 0$, $t \in \mathbb{R}$, we see that $\Gamma(z) = [f]^\wedge_{M_c}(c+it)$. We note that $\widetilde{f}$, defined by

$$\widetilde{f}(r,\theta) := \exp\left(-re^{i\theta}\right) = e^{-r\cos\theta} e^{-ir\sin\theta} \qquad \left((r,\theta) \in \mathbb{H}\right),$$

is a polar-analytic extension of f. It can be easily verified that for any $a \in]0, \pi/2[$, we have $\widetilde{f} \in H^1_c(\mathbb{H}_a)$ but $\widetilde{f} \notin H^*_c(\mathbb{H}_{\pi/2})$ since condition (d) of Definition 8.11 fails.

As in the previous example, we can conclude with the help of Theorem 8.20 that $f \in \mathcal{K}_c \cap \mathcal{M}^1_c$. Now Proposition 14.21 yields that

$$R_{c,\sigma}[f] = \mathcal{O}(e^{-2\pi a\sigma}) \qquad (\sigma \to \infty)$$

for each $a \in]0, \pi/2[$. For $c = 1/2$ the quadrature formula (14.3) reads as

$$\int_0^\infty f(r) r^{1/2} \frac{dr}{r} = \frac{1}{\sigma} \sum_{k=-\infty}^{\infty} \exp\left(-e^{k/\sigma}\right) e^{k/(2\sigma)} + R_{1/2,\sigma}[f]. \tag{14.18}$$

By (14.17), the left-hand side is equal to $\Gamma(1/2) = \sqrt{\pi}$. This time the integrand is not Mellin-even. Therefore we prefer an appropriate asymmetric truncation of the series in (14.18), which gives a total error

$$E_{\sigma,N,K} := \sqrt{\pi} - \frac{1}{\sigma} \sum_{k=-N}^{K} \exp\left(-e^{k/\sigma}\right) e^{k/(2\sigma)}.$$

By standard estimates including the integral comparison method, we find that the choice

$$N := \left\lceil 2\pi^2\sigma^2 + 2\sigma \log \frac{15}{4} \right\rceil, \quad K := \left\lceil \sigma \log \frac{N}{2\sigma} \right\rceil$$

will produce a truncation error that does not exceed the remainder asymptotically. We recall that $\lceil\cdot\rceil$ denotes the ceiling function mapping x to the smallest integer that is greater than or equal to x. Analogously to the previous example, we also compute

$$C := E_{\sigma,N,K}\, e^{\pi^2\sigma} \quad \text{and} \quad \text{rate} := -\frac{\log|E_{\sigma,N,K}|}{\sigma}.$$

By our theory, the numbers rate should become larger than $2\pi a$ for any $a < \pi/2$ as $\sigma \to \infty$. On the other hand, it can be shown that the 1/2-Mellin-even part $f_{1/2+}$ does not have an extension belonging to $H_0^*(\mathbb{H}_{\pi/2})$. Therefore Propositions 14.18 and 14.21 imply that the numbers rate must converge to π^2 as $\sigma \to \infty$.

Table 14.5 shows that for small values of σ and, consequently, with relatively short sums, we obtain already satisfactory approximations to the integral. However, the numbers rate are still considerably smaller than π^2. Therefore we continued with larger values of σ, setting Maple's environment variable *Digits* := 80 in order to suppress round-off errors. Table 14.6 shows that the approximations to the integral become excellent and the numbers rate get much closer to π^2. Note that the last line of Table 14.5 and the first line of Table 14.6 do not contain the same numbers in columns 4–6. The true reason is that in the latter table the round-off errors have been reduced by increasing the number of digits in the computation by Maple considerably.

Table 14.5 Example 3 for small σ, rate $< \pi^2 = 9.869604\ldots$

σ	N	K	$E_{\sigma,N,K}$	C	rate
0.25	2	1	−3.038e−01	−3.583	4.765302
0.50	7	1	−3.125e−02	−4.346	6.931166
0.75	14	2	2.746e−03	4.502	7.863614
1.00	23	3	−1.219e−04	−2.356	9.012572
1.25	35	4	1.160e−05	2.644	9.091811
1.50	49	5	−1.103e−06	−2.966	9.144985
1.75	66	6	1.600e−07	5.090	8.941767
2.00	85	7	−1.100e−08	−4.200	9.162685

Table 14.6 Example 3, rate $\to \pi^2 = 9.869604\ldots$

σ	N	K	$E_{\sigma,N,K}$	C	rate
2	85	7	−1.135e−08	−4.242	9.147070
3	186	11	−2.546e−13	−1.840	9.666308
4	327	15	−2.862e−17	−3.999	9.523119
5	507	20	−1.539e−22	−0.416	10.045137
6	727	25	−1.680e−26	−0.877	9.891425
7	986	30	4.948e−30	4.996	9.639809
8	1285	36	1.358e−34	2.651	9.747734
9	1623	41	5.354e−39	2.021	9.791439
10	2001	47	5.676e−43	4.142	9.727498
11	2418	52	2.957e−47	4.171	9.739769
12	2875	58	−1.102e−51	−3.007	9.777868
13	3371	64	−7.861e−57	−0.415	9.937342
14	3906	70	2.173e−60	2.215	9.812793
15	4481	76	−9.514e−65	−1.875	9.827685

Example 4: Integrand Belonging to a Mellin–Sobolev Space
Consider the function $g : \mathbb{R}^+ \to \mathbb{R}$ defined by

$$g(r) = \begin{cases} r \log^2 r, & 0 < r < 1, \\ r^{-1} \log^2 r, & r \geq 1. \end{cases}$$

It is seen to be 0-Mellin-even. By a straightforward calculation, we find that

$$[g]^{\wedge}_{M_0}(iv) = 4 \frac{1 - 3v^2}{(1 + v^2)^3} \qquad (v \in \mathbb{R}),$$

and so

$$\int_0^{\infty} g(r) \frac{dr}{r} = [g]^{\wedge}_{M_0}(0) = 4.$$

Since we know g and its Mellin transform explicitly, we can easily verify that $g \in \mathcal{K}_0 \cap W_0^{4,*}$.

The remainder of the quadrature formula (14.3) can now be written as

$$\begin{aligned} R_{0,\sigma}[g] &= \int_0^{\infty} g(r) \frac{dr}{r} - \frac{1}{\sigma} \sum_{k=-\infty}^{\infty} g(e^{k/\sigma}) \\ &= 4 - \frac{2}{\sigma} \sum_{k=1}^{\infty} \left(\frac{k}{\sigma}\right)^2 e^{-k/\sigma} \\ &= 4 - \frac{2}{\sigma^3} \cdot \frac{e^{-2/\sigma} + e^{-1/\sigma}}{(1 - e^{-1/\sigma})^3} . \end{aligned}$$

An expansion of the right-hand side yields

$$R_{0,\sigma}[g] = \frac{1}{60\,\sigma^4} - \frac{1}{756\,\sigma^6} + \mathcal{O}\left(\sigma^{-8}\right) \qquad (\sigma \to \infty). \tag{14.19}$$

Next we want to consider the estimate of Theorem 14.9 for $c = 0$, $f = g$ and $h = 1$. We have to determine the supremum of

$$\left|v^4 [g]^{\wedge}_{M_0}(iv)\right| = \left|4v^4 \frac{1 - 3v^2}{(1 + v^2)^3}\right|$$

for $v \in \mathbb{R}$. By standard calculus we find that for any $\sigma > 0$,

$$\sup_{|v| \geq 2\pi\sigma} \left|4v^4 \frac{1 - 3v^2}{(1 + v^2)^3}\right| = \lim_{v \to \infty} \left|4v^4 \frac{1 - 3v^2}{(1 + v^2)^3}\right| = 12 = \|g\|_{W_0^{4,*}}. \tag{14.20}$$

Table 14.7 Example 4: $C \to \frac{1}{60} = 0.01666\ldots$

σ	$R_{0,\sigma}[g]$	upper bound (14.21)	overestimation	C
2	1.021267e−03	1.041667e−03	1.019974824393	0.016340272591
4	6.478229e−05	6.510417e−05	1.004968650783	0.016584265244
8	4.063969e−06	4.069010e−06	1.001240599981	0.016646015620
16	2.542343e−07	2.543132e−07	1.000310052376	0.016661500729
32	1.589334e−08	1.589457e−08	1.000077506994	0.016665374984
64	9.933915e−10	9.934107e−10	1.000019376367	0.016666343733
128	6.208787e−11	6.208817e−11	1.000004844068	0.016666585933
256	3.880506e−12	3.880511e−12	1.000001211016	0.016666646483
512	2.425318e−13	2.425319e−13	1.000000302754	0.016666661621
1024	1.515824e−14	1.515825e−14	1.000000075688	0.016666665405
2048	9.473903e−16	9.473903e−16	1.000000018922	0.016666666351
4096	5.921189e−17	5.921189e−17	1.000000004731	0.016666666588
8192	3.700743e−18	3.700743e−18	1.000000001183	0.016666666647

Thus

$$\left|R_{0,\sigma}[g]\right| \le \frac{2\zeta(4)}{(2\pi\sigma)^4}\,\|g\|_{W_0^{4,*}} = \frac{1}{60\,\sigma^4}\,. \tag{14.21}$$

Now comparison with (14.19) shows that the first inequality in the conclusion of Theorem 14.9 is best possible in the described sense and (14.20) shows that equality is attained in the second estimate.

The precision of this estimate is illustrated in Table 14.7. The third column shows that the upper bound in (14.21) becomes very close to the true value of the remainder as σ grows. Consequently, the factor of overestimation, defined by

$$\text{overestimation} := \frac{\text{upper bound}}{\text{remainder}} = \frac{\frac{1}{60\sigma^4}}{R_{0,\sigma}[g]}\,,$$

becomes very close to 1. The fifth column shows $C := R_{0,\sigma}[g]\,\sigma^4$, which converges rapidly to $1/60$ as $\sigma \to \infty$.

14.4 Features of a Gaussian Quadrature Formula

In this last section we will show some connections with the classical quadrature formulae of Gaussian type (see, e.g., Szegő [282, p. 47, Sect. 3.4]).

In this respect the following phenomenon occurs. Let $n \ge 2$ be a positive integer and denote by $x_{n,1}, \ldots, x_{n,n}$ the zeros of the Legendre polynomial of degree n. Then by Lagrange interpolation, we can reconstruct each polynomial f of degree at most $n-1$ from its values

$$f(x_{n,1}), \ldots, f(x_{n,n}). \tag{14.22}$$

But it is not possible to reconstruct a polynomial f of degree n or higher from that data. On the other hand, the Gaussian quadrature formula shows that the same data (14.22) allows us to obtain the value of the integral

$$\int_{-1}^{1} f(x)dx$$

for all polynomials f up to degree $2n-1$. An explanation of this phenomenon can be given by Hermite interpolation. In addition to (14.22), consider the values of the derivative

$$f'(x_{n,1}), \ldots, f'(x_{n,n}). \tag{14.23}$$

Hermite interpolation with the data (14.22) and (14.23) yields a reconstruction formula

$$f(x) = \sum_{k=1}^{n} \left[f(x_{n,k}) A_{n,k}(x) + f'(x_{n,k}) B_{n,k}(x) \right], \tag{14.24}$$

where $A_{n,k}(x)$ and $B_{n,k}(x)$ are certain polynomials of degree $2n-1$. This formula is valid for all polynomials f of degree at most $2n-1$. When we integrate both sides of (14.24) over the interval $[-1, 1]$, it happens that

$$\int_{-1}^{1} B_{n,k}(x)dx = 0 \qquad (k = 1, \ldots, n)$$

as a consequence of our particular choice of the so-called nodes $x_{n,1}, \ldots, x_{n,n}$. Thus we obtain the Gaussian quadrature formula in the form

$$\int_{-1}^{1} f(x)dx = \sum_{k=1}^{n} f(x_{n,k}) \int_{-1}^{1} A_{n,k}(x)dx.$$

Similar observations hold for the quadrature formula of this chapter. First we note that the exponential sampling formula of Theorem 12.4 yields that

$$f(x) = \sum_{k=-\infty}^{\infty} f(e^{k/\sigma}) \mathrm{linc}_{c/\sigma}(e^{-k} x^{\sigma}) \tag{14.25}$$

holding for all $f \in B^1_{c,\pi\sigma}$. On the other hand, Corollary 14.1 shows that

$$\int_0^\infty f(x)x^{c-1}dx = \frac{1}{\sigma}\sum_{k=-\infty}^{\infty} f(e^{k/\sigma})e^{ck/\sigma} \tag{14.26}$$

holding for all $f \in B^1_{c,2\pi\sigma}$. We see that both formulae (14.25) and (14.26) involve the same samples of f on the right-hand side, but (14.26) is valid for functions up to twice the bandwidth of those admissible in (14.25). An explanation of this phenomenon can be given by derivative sampling (see Sect. 12.3). It follows from Corollary 12.19 that

$$\begin{aligned} f(x) = \sum_{k=-\infty}^{\infty} \Big[& f(e^{k/\sigma})\mathrm{lin}^2_{c/(2\sigma)}(e^{-k}x^\sigma) \\ & + \frac{(-1)^k}{\pi\sigma}(\Theta_c f)(e^{k/\sigma})\sin(\log x^{\pi\sigma})\mathrm{lin}_{c/\sigma}(e^{-k}x^\sigma)\Big] \end{aligned}$$

holding for all $f \in B^1_{c,2\pi\sigma}$. Now we multiply both sides by x^{c-1} and integrate over $\mathbb{R}^+$. By straightforward calculations, we find that

$$\int_0^\infty \mathrm{lin}^2_{c/(2\sigma)}(e^{-k}x^\sigma)x^{c-1}dx = \frac{e^{ck/\sigma}}{\sigma} \qquad (k \in \mathbb{Z})$$

and

$$\int_0^\infty \sin(\log x^{\pi\sigma})\mathrm{lin}_{c/(\sigma)}(e^{-k}x^\sigma)x^{c-1}dx = 0 \qquad (k \in \mathbb{Z}).$$

Thus, if we justify that integration and summation may be interchanged, we have obtained an alternative proof of (14.26) and see that this formula has features of a Gaussian formula in which the role of polynomials is taken by Mellin bandlimited functions and that of the degree by the bandwidth. For the desired justification, one can imitate approaches as given in Higgins et al. [176, pp. 349–351] or Butzer et al. [113, proof of Theorem 4.3] in the corresponding case of Fourier analysis.

14.5 Notes

The contents of Sects. 14.1–14.3 is borrowed from Bardaro et al. [60]. Partly this paper in turn presents or completes some earlier results. Furthermore it was inspired by analogous studies of Butzer et al. [120] for quadrature on $\mathbb{R}$, where the corresponding formulae can be identified as the compound trapezoidal rule with equidistantly spaced nodes. For the estimate corresponding to (14.8) it was shown in [120, Sect. 9] by a sophisticated construction that for each $\sigma > 0$ and each $\alpha > 1$ there exists a function $\phi_{\alpha,\sigma}$ for which equality is attained. This stronger form of sharpness can also be verified for (14.8) by modifying the extremal function $\phi_{\alpha,\sigma}$ suitably. Möbius inversion (see Proposition 1.50), which is crucial in the proof of Lemma 14.12, was first used in numerical analysis by Lyness [203], Brass [77], and Loxton and Sanders [199, 200]. A preliminary form of the results in Sect. 14.2.2 for the Mellin-even part was given in Schmeisser [270, Sects. VI–VII]. However, the corresponding Proposition 14.18 (see [270, Theorem 6.5]) needed the restriction $a \in]0, 2\pi^2]$ in order to avoid the use of Riemann surfaces. Due to the concept of polar-analytic functions this condition can be dropped. Equivalence theorems based on translated functions were first obtained by Dryanov et al. [142] for the compound trapezoidal rule on $\mathbb{R}$ and further studied by Butzer et al. [120].

That the quadrature formulae of this chapter have features of a Gaussian formula with the role of the polynomials being taken by Mellin bandlimited functions was already demonstrated by Schmeisser [270, Sects. II–III]. There is another characterization of Gaussian quadrature formulae which has not yet been studied in the Mellin setting. Consider the classical Gaussian quadrature formula for the integral $\int_{-1}^{1} f(x)dx$ with nodes $x_1, \ldots, x_n$ and call the monic polynomial $P(x) = \prod_{j=1}^{n}(x - x_j)$ the associated *nodal polynomial*. Then, among all monic polynomials of degree n, the nodal polynomial is the only one of smallest $L^2[-1, 1]$ norm. With the compound trapezoidal rule on $\mathbb{R}$ one has analogously associated an appropriate nodal function and shown that, within a class of entire functions of exponential type, it minimizes a weighted $L^2(\mathbb{R})$ norm; see Rahman and Schmeisser [253], Schmeisser [269]. It would be desirable to study an analogous extremal problem for the quadrature formulae of this chapter with nodal functions belonging to a class of polar-analytic functions whose restrictions to $\mathbb{R}^+$ are Mellin bandlimited.

In a widest sense, a starting point for the considerations of this chapter can be seen in the desire to extend the compound trapezoidal rule from compact intervals to the whole of $\mathbb{R}$. Three different approaches deserve to be mentioned: (i) Transforming the classical Euler–Maclaurin formula to an interval $[-a, a]$, one performs the limit $a \to \infty$. (ii) Using an appropriately designed cotangent kernel, one proceeds by contour integration. (iii) One simply observes that the general classical Poisson summation formula can be interpreted as a compound trapezoidal rule on $\mathbb{R}$. The method of contour integration allows a lot of flexibility. For example, it gives access to generalizations of the compound trapezoidal rule which involve also derivatives; see Rahman and Schmeisser [255]. However, it requires functions f that are analytic on a strip of the complex plane, symmetrical and parallel to the real line. The Poisson summation formula needs no analyticity and provides a convenient representation of the remainder which enables estimates of high precision and equivalence theorems.

Now one may ask what is the most appropriate analogue of the compound trapezoidal rule for quadrature over $\mathbb{R}^+$. A genuine compound trapezoidal rule on $\mathbb{R}^+$ with equidistantly spaced nodes is contained in the Abel–Plana summation formula. Unfortunately the remainder appears in a representation which is quite unformatable for numerical purposes. By transforming the Euler–Maclaurin formula to an interval $[0, a]$ and performing the limit $a \to \infty$, one can obtain the same formula with a more convenient representation of the remainder by a series involving Bernoulli numbers and derivatives of f of odd order at the origin; see Rahman and Schmeisser [256]. In this context, we mention that the summation formulae of Euler–Maclaurin and Abel–Plana as well as the Poisson summation formula can be deduced from each other as was shown by Rahman and Schmeisser [257] and further studied by Butzer et al. [114]. In general, the compound trapezoidal rule with equidistantly spaced nodes on $\mathbb{R}^+$ yields poor approximations of the integral unless one adds some initial terms of the series obtained as remainder. This leads to considerable improvements and amounts to using the origin as a multiple node. It seems that for quadrature on $\mathbb{R}^+$ one should have nodes which become more dense as they approach the origin and more sparse as they approach infinity.

Stenger [278] proposed the formula

$$\int_0^\infty f(x)dx = \frac{1}{\sigma}\sum_{k=-\infty}^{\infty} f(e^{k/\sigma})e^{k/\sigma} + R_\sigma[f] \qquad (\sigma > 0); \tag{14.27}$$

also see Stenger's survey article [279, Sect. 4.2] and his monograph [280, Sect. 4.2]. It was deduced from the compound trapezoidal rule on $\mathbb{R}$ by a conformal transformation that maps a strip symmetrical to the real line of the complex plane onto a sector symmetrical to the positive real axis. Furthermore, estimates for the remainder $R_\sigma[f]$ were obtained in the case when f extends to an analytic function on that sector. Note that (14.27) is no longer a *real* compound trapezoidal rule but may be considered as an analogue. In fact, if we approximate the integral $\int_{e^{k/\sigma}}^{e^{(k+1)/\sigma}} f(x)dx$ by the area T_k of a trapezoid and sum over $k \in \mathbb{Z}$, we obtain $\sinh(1/\sigma)$ instead of $1/\sigma$ in front of the summation in (14.27). Obviously this formula is the case $c = 1$ of (14.3). The Mellin version of the Poisson summation formula for arbitrary $c \in \mathbb{R}$ was first described by Butzer and Jansche [91, Theorem 7.3] and used as a quadrature formula by Schmeisser [270].

For further information on sampling and quadrature, we refer to Butzer et al. [104, Sect. 2.11]. For some interesting contributions to quadrature on $\mathbb{R}^+$ involving the classical Gaussian quadrature formulae, namely the ones with respect to polynomials, see Gautschi [155], Mastroianni and Monegato [216], and Mastroianni et al. [217].

Bibliography

1. M. Abramowitz, I.A. Stegun (eds.), *Handbook of Mathematical Functions*, 10th edn. National Bureau of Standards. Applied Math. Series 55 (1972)
2. T. Acar, O. Alagoz, A. Aral, D. Costarelli, M. Turgay, G. Vinti, Convergence of generalized sampling series in weighted spaces. Demonstr. Math. **55**(1), 153–162 (2022)
3. T. Acar, S. Kursun, M. Turgay, Multidimensional Kantorovich modifications of exponential sampling series. Quaestiones Math. Published online (January 2022)
4. L.V. Ahlfors, *Complex Analysis*, 3rd edn. McGraw-Hill Int. Eds (1979)
5. N.B. Andersen, On real Paley–Wiener theorems for certain integral transforms. J. Math. Anal. Appl. **288**, 124–135 (2003)
6. N.B. Andersen, M. de Jeu, Real Paley–Wiener theorems and local spectral radius formula. Trans. Am. Math. Soc. **362**, 3613–3640 (2010)
7. S.K. Angamuthu, S. Bajpeyi, Direct and inverse results for Kantorovich type exponential sampling series. Results Math. **75**(3) (2020), Paper 119
8. L. Angeloni, G. Vinti, Approximation in variation by homothetic operators in multidimensional setting. Differ. Integr. Equ. **26**(5–6), 655–674 (2013)
9. L. Angeloni, G. Vinti, Variation and approximation in multidimensional setting for Mellin integral operators, in *New Perspectives on Approximation and Sampling Theory*, ed. by G. Schmeisser, A. Zayed. Festschrift in Honor of Prof. Butzer's 85th birthday, Applied and Numerical Harmonic Analysis (Birkhäuser, Basel, 2014)
10. L. Angeloni, G. Vinti, Convergence and rate of approximation in $BV^{\phi}(\mathbb{R}^N_+)$ for a class of Mellin integral operators. Atti Accad. Naz. Lincei Rend. Lincei Mat. Appl. **25**(3), 217–232 (2014)
11. L. Angeloni, G. Vinti, A characterization of absolute continuity by means of Mellin integral operators. Z. Anal. Anwend. **34**(3), 343–356 (2015)
12. L. Angeloni, G. Vinti, A concept of absolute continuity and its characterization in terms of convergence in variation. Math. Nachr. **289**(16), 1986–1994 (2016)
13. L. Angeloni, G. Vinti, Discrete operators of sampling type and approximation in φ-variation. Math. Nachr. **291**(4), 546–555 (2018)
14. L. Angeloni, D. Costarelli, G. Vinti, A characterization of the convergence in variation for the generalized sampling series. Ann. Acad. Sci. Fenn. Math. **43**(2), 755–767 (2018)
15. L. Angeloni, D. Costarelli, G. Vinti, Quantitative estimates for sampling type operators with respect to Jordan variation. Atti Accad. NAz. Lincei, Rend. Lincei Mat. Appl. **31**(2), 269–284 (2020)
16. M.H. Annaby, P.L. Butzer, Mellin type differential equations and associated sampling expansions. Numer. Funct. Anal. Optim. **21**, 1–24 (2000)

C. Bardaro et al., *Mellin Analysis, Transform Theory, and Applications*,
https://doi.org/10.1007/978-3-031-96672-9

17. T.M. Apostol, *Introduction to Analytic Number Theory* (Springer, New York, 1976)
18. T.M. Apostol, *Modular Functions and Dirichlet Series in Number Theory* (Springer, New York, 1992)
19. A. Aral, T. Acar, S. Kursun, Generalized Kantorovich forms of exponential sampling series. Anal. Math. Phys. **12**(2) (2022). Paper 50
20. E. Artin, *The Gamma Function* (Holt, Rinehart and Winston, New York, 1964)
21. R.M. Asharabi, F.H. Al-Haddad, On multidimensional sinc-Gauss sampling formulas for analytic functions. Electron. Trans. Numer. Anal. **55**, 242–262 (2022)
22. R.M. Asharabi, J. Prestin, On two-dimensional classical and Hermite sampling. IMA J. Numer. Anal. **36**, 851–871 (2016)
23. S. Bajpeyi, A.S. Kumar, On approximation by Kantorovich exponential sampling operators. Numer. Funct. Anal. Optim. **42**(9), 1096–1113 (2021)
24. S. Bajpeyi, A.S. Kumar, I. Mantellini, Approximation by Durrmeyer type exponential sampling operators. Numer. Funct. Anal. Optim. **43**(1), 16–34 (2022)
25. J. Bak, D.J. Newman, *Complex Analysis*, 3rd edn. (Springer-Verlag, New York, 2017)
26. A. Bakan, S. Kaijser, Hardy spaces for the strip. J. Math. Anal. Appl. **333**(1), 347–364 (2007)
27. D. Baleanu, K. Diethelm, E. Scalas, J.J. Trujillo, *Fractional Calculus: Models and Numerical Methods*, vol. 3. Series on Complexity, Nonlinearity and Chaos (World Scientific, Singapore, 2012)
28. S. Balsamo, I. Mantellini, On linear combinations of generalized exponential sampling series. Results Math. **74**, Paper 180 (2019)
29. C. Bardaro, I. Mantellini, Approximation properties in abstract modular spaces for a class of general sampling-type operators. Appl. Anal. **85**(4), 383–413 (2006)
30. C. Bardaro, I. Mantellini, Pointwise convergence theorems for nonlinear Mellin convolution operators. Int. J. Pure Appl. Math. **27**(4), 431–447 (2006)
31. C. Bardaro, I. Mantellini, Voronovskaya-type estimates for Mellin convolution operators. Result Math. **50**, 1–16 (2007)
32. C. Bardaro, I. Mantellini, A quantitative Voronovskaja formula for Mellin convolution operators. Mediterr. J. Math. **7**(4), 483–501 (2010)
33. C. Bardaro, I. Mantellini, Asymptotic behaviour of Mellin–Fejer convolution operators. East J. Approx. **17**(2), 127–147 (2011)
34. C. Bardaro, I. Mantellini, A note on the Voronovskaja theorem for Mellin–Fejer convolution operators. Appl. Math. Lett. **24**, 2064–2067 (2011)
35. C. Bardaro, I. Mantellini, Approximation properties for linear combinations of moment type operators. Comput. Math. Appl. **62**, 213–229 (2011)
36. C. Bardaro, I. Mantellini, On the iterates of Mellin–Fejer convolution operators. Acta Appl. Math., **121**, 213–229 (2012)
37. C. Bardaro, I. Mantellini, On Voronovskaja formula for linear combinations of Mellin–Gauss–Weierstrass operators. Appl. Math. Comput. **218**, 10171–10179 (2012)
38. C. Bardaro, I. Mantellini, Asymptotic formulae for linear combinations of generalized sampling type operators. Z. Anal. Anwend. **32**(3), 279–298 (2013)
39. C. Bardaro, I. Mantellini, On linear combinations of multivariate generalized sampling type series. Mediterr. J. Math. **10**, 1833–1852 (2013)
40. C. Bardaro, I. Mantellini, On Mellin convolution operators: a direct approach to the asymptotic formulae. Integral Transform Spec. Funct. **25**(3), 182–195 (2014)
41. C. Bardaro, I. Mantellini, Asymptotic expansion of generalized Durrmeyer sampling type series. Jaen J. Approx. **6**(2), 143–165 (2014)
42. C. Bardaro, I. Mantellini, Quantitative Voroniovskaja formulae for generalized Durrmeyer sampling type series. Math. Nachr. **289**(14–15), 1702–1720 (2016)
43. C. Bardaro, I. Mantellini, On a Durrmeyer type modification of the exponential sampling series. Rend. Circ. Mat. Palermo Ser. 2 **70**, 1289–1304 (2021)
44. C. Bardaro, G. Vinti, An abstract approach to sampling-type operators inspired by the work of P. L. Butzer. I. Linear operators. Sampl. Theory Signal Image Process. **2**(3), 271–295 (2003)

45. C. Bardaro, G. Vinti, An abstract approach to sampling-type operators inspired by the work of P. L. Butzer. II. Nonlinear operators. Sampl. Theory Signal Image Process. **3**(1), 29–44 (2004)
46. C. Bardaro, J. Musielak, G. Vinti, *Nonlinear Integral Operators and Applications*. De Gruyter Series in Nonlinear Analysis and Applications 9 (Walter De Gruyter, Berlin, 2003)
47. C. Bardaro, P.L. Butzer, R.L. Stens, G. Vinti, Approximation error of the Whittaker cardinal series in terms of an averaged modulus of smoothness covering discontinuous signals. J. Math. Anal. Appl. **316**, 269–306 (2006)
48. C. Bardaro, P.L. Butzer, R.L. Stens, G. Vinti, Kantorovich-type generalized sampling series in the setting of Orlicz spaces. Sampl. Theory Signal Image Process. **6**(1), 29–52 (2007)
49. C. Bardaro, G. Vinti, H. Karsli, On pointwise convergence of linear integral operators with homogeneous kernels. Integral Transform Spec. Funct. **19**(6), 429–439 (2008)
50. C. Bardaro, P.L. Butzer, R.L. Stens, G. Vinti, Prediction by samples from the past with error estimates covering discontinuous signals. IEEE Trans. Inf. Theory **56**(1), 614–633 (2010)
51. C. Bardaro, A. Boccuto, X. Dimitriou, I. Mantellini, Modular filter convergence theorems for abstract sampling type operators. Appl. Anal. **92**(11), 2404–2423 (2013)
52. C. Bardaro, P.L. Butzer, I. Mantellini, The exponential sampling theorem of signal analysis and the reproducing kernel formula in the Mellin Transform setting. Sampl. Theory Signal Image Process. **13**(1), 35–66 (2014)
53. C. Bardaro, P.L. Butzer, I. Mantellini, The foundations of fractional calculus in the Mellin transform setting with applications. J. Fourier Anal. Appl. **21**, 961–1017 (2015)
54. C. Bardaro, P.L. Butzer, I. Mantellini, The Mellin–Parseval formula and its interconnections with the exponential sampling theorem of optical physics. Integral Transform Spec. Funct. **27**(1), 17–29 (2016)
55. C. Bardaro, P.L. Butzer, I. Mantellini, G. Schmeisser, On the Paley–Wiener theorem in the Mellin transform setting. J. Approx. Theory **207**, 60–75 (2016)
56. C. Bardaro, P.L. Butzer, I. Mantellini, G. Schmeisser, Mellin analysis and its basic associated metric. Applications to sampling theory. Anal. Math. **42**(4), 297–321 (2016)
57. C. Bardaro, L. Faina, I. Mantellini, A generalization of the exponential sampling series and its approximation properties. Math. Slovaca **67**(6), 1481–1496 (2017)
58. C. Bardaro, P.L. Butzer, I. Mantellini, G. Schmeisser, A fresh approach to the Paley–Wiener theorem for Mellin transforms and the Mellin–Hardy spaces. Math. Nachr. **290**, 2759–2774 (2017)
59. C. Bardaro, P.L. Butzer, I. Mantellini, G. Schmeisser, A generalization of the Paley–Wiener theorem for Mellin transforms and metric characterization of function spaces. Frac. Calc. Appl. Anal. **20**(5), 1216–1238 (2017)
60. C. Bardaro, P.L. Butzer, I. Mantellini, G. Schmeisser, Quadrature formulae for the positive real axis in the setting of Mellin analysis: sharp error estimates in terms of the Mellin distance. Calcolo **55**(3), Paper 26 (2018)
61. C. Bardaro, G. Bevignani, I. Mantellini, M. Seracini, Bivariate generalized exponential sampling series and applications to seismic waves. Constr. Math. Anal. **2**(4), 153–167 (2019)
62. C. Bardaro, P.L. Butzer, I. Mantellini, G. Schmeisser, Development of a new concept of polar-analytic functions useful in Mellin analysis. Complex Var. Elliptic Eqs. **64**(12), 2040–2062 (2019)
63. C. Bardaro, I. Mantellini, G. Schmeisser, Exponential sampling series: convergence in Mellin–Lebesgue spaces. Results Math. **74**, Paper 119 (2019)
64. C. Bardaro, P.L. Butzer, I. Mantellini, G. Schmeisser, Integration of polar-analytic functions and applications to Boas' differentiation formula and Bernstein's inequality in Mellin frame. Boll. Unione Mat. Ital. **13**(4), 503–514 (2020)
65. C. Bardaro, P.L. Butzer, I. Mantellini, G. Schmeisser, Valiron's interpolation formula and a derivative sampling formula in the Mellin setting acquired via polar-analytic functions. Comput. Methods Funct. Theory **20**(3–4), 629–652 (2020)
66. C. Bardaro, P.L. Butzer, I. Mantellini, G. Schmeisser, Polar-analytic functions: old and new results, applications. Results Math. **77**, Paper 64 (2022)

67. C. Bardaro, P.L. Butzer, I. Mantellini, G. Schmeisser, R.L. Stens, Classical and approximate exponential sampling formula: their interconnections in uniform and Mellin–Lebesgue norms, in *Sampling, Approximation and Signal Analysis—Harmonic Analysis in the Spirit of J. Rowland Higgins*. Appl. Numer. Harmonic Anal. (Birkhauser/Springer, Cham, 2023), pp. 3–22
68. C. Bardaro, I. Mantellini, G. Schmeisser, Exponential sampling with a multiplier. Sampl. Theory Signal Process. Data Anal. **21**, Paper 8 (2023)
69. M.G. Beaty, M.M. Dodson, Abstract harmonic analysis and the sampling theorem, in *Sampling Theory in Fourier and Signal Analysis: Advances Topics*, ed. by J.R. Higgins, R.L. Stens. Oxford Science Publ. (Oxford University Press, Oxford, 1999)
70. M. Bertero, E.R. Pike, Exponential sampling method for Laplace and other dilationally invariant transforms I. Singular-system analysis. II. Examples in photon correction spectroscopy and Frauenhofer diffraction. Inverse Problems **7**, 1–20, 21–41 (1991)
71. J. Bertrand, P. Bertrand, J.P. Ovarlez, The Mellin transform, in *The Transforms and Applications. Handbook*, 2nd edn., ed. by A.D. Poularkas (CRC Press, Boca Raton, 2000)
72. R.P. Boas, The derivative of a trigonometric integral. J. Lond. Math. Soc. **12**, 164–165 (1937)
73. R.P. Boas, *Entire Functions* (Academic Press, New York, 1954)
74. A. Boccuto, D. Candeloro, A. Sambucini, Vitali-type theorems for filter convergence related to Riesz space-valued modulars and applications to stochastic processes. J. Math. Anal. Appl. **419**(2), 818–838 (2014)
75. L. Boyadjiev, Yu. Luchko, Mellin integral transform approach to analyze the multidimentional diffusion-wave equations. Chaos Solitons Fractals **102**, 127–134 (2017)
76. R.N. Bracewell, *The Fourier Transform and Its Applications* (McGraw-Hill, New York, 1986)
77. H. Brass, Der Wertebereich des Trapezverfahrens, in *Numerische Integration*, ed. by G. Hämmerlin. Proc. Conf. Math. Res. Inst., Oberwolfach, October 1–7, 1978 (Birkhäuser, Basel, 1979), pp. 98–108
78. A. Browder, On Bernstein's inequality and the norm of Hermitian operators. Am. Math. Mon. **78**(8), 871–873 (1971)
79. Yu.A. Brychkov, *The Mellin Transforms. I* (Russian). Reports in Applied Mathematics, Rossiiskaya Akademiya Nauk, Vychislitel'nyi Tsentrim. A. A. Dorodnitsyna, Moscow (1997)
80. Yu.A. Brychkov, H.-J. Glaeske, O.I. Marichev, Factorization of integral transforms of convolution type. Itogi Nauki i Tekhniki, Ser. Mat. Anal. **21**, VINITI, 3–41 (1983) (Russian); English transl.: J. Soviet Math. **30**(3), 2071–2094 (1985)
81. Yu.A. Brychkov, H.-J. Glaeske, A.P. Prudnikov, K.T. Vu, *Multidimensional Integral Transformations* (Gordon and Breach, Philadelphia, 1992)
82. Yu.A. Brychkov, O.I. Marichev, N.V. Savischenko, *Handbook of Mellin Transforms* (CRC Press, Boca Raton, 2019)
83. R.B. Burckel, *An Introduction to Classical Complex Analysis*, vol. 1 (Birkhäuser, Basel, 1979)
84. P.L. Butzer, Linear combinations of Bernstein polynomials. Can. J. Math. **5**, 559–567 (1953)
85. P.L. Butzer, The sampling theorem of signal processing; applications to combinatorial analysis and number theory, an overview, in *Trends in Industrial and Applied Mathematics*, ed. by A.H. Siddiqi, M. Kocvara (Kluwer Academic Publishers, Dordrecht, 2002), pp. 23–61
86. P.L. Butzer, H. Berens, *Semi-groups of Operators and Approximation*. Grundlehren Math. Wiss., vol. 145 (Springer, Berlin, 1967)
87. P.L. Butzer, M. Hauss, Stirling functions of first and second kind; some new applications, in *Approximation, Interpolation and Summability*. Proc. Conf. in honor of Prof. Jakimovski, Research Inst. of Math. Science, Bar Ilan University, Tel Aviv. Weizmann Press, Israel, 4 (1991), pp. 89–108
88. P.L. Butzer, S. Jansche, Mellin transform theory and the role of its differential and integral operators, in *Proc. Second Int. Workshop "Transform methods and special functions"*, Varna (1996), pp. 63–83
89. P.L. Butzer, S. Jansche, A direct approach to the Mellin transform. J. Fourier Anal. Appl. **3**, 325–376 (1997)

90. P.L. Butzer, S. Jansche, The exponential sampling theorem of signal analysis. Atti Sem. Mat. Fis. Univ. Modena, Suppl. **46**, 99–122 (1998). Special issue dedicated to Professor Calogero Vinti
91. P.L. Butzer, S. Jansche, The finite Mellin Transform, Mellin–Fourier series and the Mellin–Poisson summation formula. Rend. Circ. Mat. Palermo, Ser. II, Suppl. **52**, 55–81 (1998)
92. P.L. Butzer, S. Jansche, A self-contained approach to Mellin transform analysis for square integrable functions, applications. Integral Transform Spec. Funct. **8**, 175–198 (1999)
93. P.L. Butzer, S. Jansche, Mellin–Fourier series and the classical Mellin transform. Approximation in Mathematics (Memphis, 1997). Comput. Math. Appl. **40**(1), 49–62 (2000)
94. P.L. Butzer, G. Nasri-Roudsari, Kramer's sampling theorem in signal analysis and its role in mathematics, in *Image Processing: Mathematical Methods and Applications*, ed. by J.M. Blackedge. Proc. Conf. on Image Processing, Cranfield Univ. (1994) (Oxford University Press, Oxford, 1997), pp. 49–95
95. P.L. Butzer, R.J. Nessel, *Fourier Analysis and Approximation. Vol. I* (Academic Press, New York, 1971)
96. P.L. Butzer, R.L. Stens, The Poisson summation formula, Whittaker's cardinal series and approximate integration, in *Approximation Theory*. Proc. Conf. Edmonton, 1982, Canadian Math. Soc. Conf. Proceedings, vol. 3, Providence (1983), pp. 19–36
97. P.L. Butzer, R.L. Stens, Prediction of non-bandlimited signals in terms of splines of low degree. Math. Nachr. **132**, 115–130 (1987)
98. P.L. Butzer, R.L. Stens, Linear prediction by samples from the past, in *Advances Topics in Shannon Sampling and Interpolation Theory*, ed. by R.J. Marks II (Springer-Verlag, Berlin, 1993), pp. 157–183
99. P.L. Butzer, U. Westphal, An access to fractional differentiation via fractional difference quotients, in *Fractional Calculus and Its Applications*. Proc. Conf. New Haven. Lecture Notes in Math, vol. 457 (Springer, Heidelberg, 1975), pp. 116–145
100. P.L. Butzer, U. Westphal, An introduction to fractional calculus, in *Applications of Fractional Calculus in Physics*, ed. by H. Hifler (World Scientific Publ., Singapore, 2000), pp. 1–85
101. P.L. Butzer, M. Schmidt, E.L. Stark, Observations on the history of central B-splines. Arch. Hist. Exact Sci. **39**, 137–156 (1988)
102. P.L. Butzer, W. Splettstosser, R.L. Stens, The sampling theorem and linear prediction in signal analysis. Jahresber. Deutsch. Math.-Verein. **90**, 1–70 (1988)
103. P.L. Butzer, M. Hauss, R.L. Stens, The sampling theorem and its unique role in various branches of mathematics. Mitt. Math. Ges. Hamburg **12**, 523–547 (1991)
104. P.L. Butzer, G. Schmeisser, R.L. Stens, An introduction to sampling analysis, in *Nonuniform Sampling: Theory and Practice*, ed. by F. Marvasti (Kluwer, New York, 2001), pp. 17–121
105. P.L. Butzer, A.A. Kilbas, J.J. Trujillo, Fractional calculus in the Mellin setting and Hadamard-type fractional integral. J. Math. Anal. Appl. **269**, 1–27 (2002)
106. P.L. Butzer, A.A. Kilbas, J.J. Trujillo, Compositions of Hadamard-type fractional integration operators and the semigroup property. J. Math. Anal. Appl. **269**, 387–400 (2002)
107. P.L. Butzer, A.A. Kilbas, J.J. Trujillo, Mellin transform analysis and integration by parts for Hadamard-type fractional integrals. J. Math. Anal. Appl. **270**, 1–15 (2002)
108. P.L. Butzer, A.A. Kilbas, J.J. Trujillo, Stirling functions of the second kind in the setting of difference and fractional calculus. Numer. Funct. Anal. Optim. **4**(7–8), 673–711 (2003)
109. P.L. Butzer, A.A. Kilbas, J.J. Trujillo, Generalized Stirling functions of second type and representation of fractional order difference via derivatives. J. Differ. Equ. Appl. **9**, 503–533 (2003)
110. P.L. Butzer, J.R. Higgins, R.L. Stens, Classical and approximate sampling theorems: studies in the $L^p(\mathbb{R})$ and the uniform norm. J. Approx. Theory **137**, 250–263 (2005)
111. P.L. Butzer, A.A. Kilbas L. Rodrigues-Germá, J.J. Trujillo, Stirling functions of first kind in the setting of fractional calculus and generalized differences. J. Differ. Equ. Appl. **13**(8–9), 683–721 (2007)

112. P.L. Butzer, P.J.S.G. Ferreira, J.R. Higgins, S. Saitoh, G. Schmeisser, R.L. Stens, Interpolation and sampling: E.T. Whittaker, K. Ogura and their followers. J. Fourier Anal. Appl. **17**, 320–354 (2011)
113. P.L. Butzer, P.J.S.G. Ferreira, J.R. Higgins, G. Schmeisser, R.L. Stens, The sampling theorem, Poisson's summation formula, general Parseval formula, reproducing kernel formula and the Paley–Wiener theorem for bandlimited signals: their interconnections. Appl. Anal. **90**(3–4), 431–461 (2011)
114. P.L. Butzer, P.J.S.G. Ferreira, G. Schmeisser, R.L. Stens, The summation formulae of Euler–Maclaurin, Abel–Plana, Poisson, and their interconnections with the approximate sampling formula of signal analysis. Results Math. **59**, 359–400 (2011)
115. P.L. Butzer, G. Schmeisser, R.L. Stens, Shannon's sampling theorem for bandlimited signls and their Hilbert transform, Boas-type formulae for higer order derivatives: the aliasing error involved by their extensions from bandlimited to non-bandlimited signals. Entropy **14**(11), 2192–2226 (2012)
116. P.L. Butzer, G. Schmeisser, R.L. Stens, Basic relations valid for the Bernstein space B_σ^p and their extensions to functions from larger spaces with error estimates in terms of their distances from B_σ^p. J. Fourier Anal. Appl. **19**, 333–375 (2013)
117. P.L. Butzer, M.M. Dodson, P.J.S.G. Ferreira, J.R. Higgins, G. Schmeisser, R.L. Stens, Seven pivotal theorems of Fourier analysis, signal analysis, numerical analysis and number theory: their interconnections. Bull. Math. Sci. (3), **4**, 481–525 (2014)
118. P.L. Butzer, G. Schmeisser, R.L. Stens, Basic relations valid for the Bernstein space B_σ^p and their extensions to larger functions spaces via a unified distance concept, in *Function Spaces X*. Banach Centre Publications, vol. 102 (Warszawa, 2014), pp. 41–55
119. P.L. Butzer, G. Schmeisser, R.L. Stens, The classical and approximate sampling theorems and their equivalence for entire functions of exponential type. J. Approx. Theory **179**, 94–111 (2014)
120. P.L. Butzer, G. Schmeisser, R.L. Stens, The distance between the general Poisson summation formula and that for bandlimited functions; applications to quadrature fomulae. Appl. Comput. Harmon. Anal. **45**(3), 597–615 (2018)
121. E. Cahen, Sur la fonction $\zeta(s)$ de Riemann et sur des fonctions analogues. Ann. Sci. Ecole Norm. Sup. (3) **11**, 75–164 (1894)
122. L.L. Campbell, Sampling theorem for the Fourier transform of a distribution with bounded support. SIAM J. Appl. Math. **16**, 626–636 (1968)
123. M. Cantarini, D. Costarelli, G. Vinti, Approximation of differentiable and not differentiable signals by the first derivative of sampling Kantorovich operators. J. Math. Anal. Appl. **509**(1) Paper 125913 (2022)
124. M. Caputo, Mean fractional order derivatives: differential equations and filters. Ann. Univ. Ferrara, Sez VII **41**, 73–84 (1995)
125. D. Casasent (ed.), *Optical Data Processing* (Springer, Berlin, 1978), pp. 241–282
126. S.-Y. Choi, S. Veng, J.-H. Kim, J.-H. Yoon, A Mellin transform approach to the pricing of options with default risk. Comput. Econ. **59**(3), 1113–1134 (2022)
127. R.V. Churchill, J.W. Brown, R.F. Verhey, *Complex Variables and Applications*, 3rd edn. (McGraw-Hill, New York, 1974)
128. S. Colombo, *Les transformations de Mellin et de Hankel* (Centre National de la Recherche Scientifique, Paris, 1959)
129. S. Colombo, J. Lavoine, *Transformations de Laplace et de Mellin, Formulaires, Mode d'utilisation*. Memorial Sci Math., vol. 169 (Centre National de la Recherche Scientifique, Paris, 1959)
130. L. Comtet, *Advanced Combinatorics* (Reidel Publishing, Dordrecht, 1974)
131. D. Costarelli, G. Vinti, Saturation by the Fourier transform method for the sampling Kantorovich series based on banblimited kernels. Anal. Math. Phys. **9**(4), 2263–2280 (2019)
132. D. Costarelli, G. Vinti, Convergence of sampling Kantorovich operators in modular spaces with applications. Rend. Circ. Mat. Palermo (2) **70**(2), 1115–1136 (2021)

133. D. Costarelli, G. Vinti, Approximation properties of the sampling Kantorovich operators: regularization, saturation, inverse results and Favard classes in L^p-spaces. J. Fourier Anal. Appl. **29**(3), Paper 49 (2022)
134. D. Costarelli, M. Piconi, G. Vinti, On the convergence properties of Durrmeyer-Sampling type operators in Orlicz spaces. Math. Nachr. **296**, 588–609 (2023)
135. B. Davies, *Integral Transforms and Their Applications* (Springer, New York, 1984)
136. C. de Boor, *A Practical Guide to Splines*. Applied Mathematical Sciences, vol. 27 (Springer, New York, 1978)
137. G. Doetsch, *Theorie und Anwendungen der Laplace-Transformation* (Springer, Berlin, 1937) (Reprint: Dover, New York, 1943)
138. G. Doetsch, *Handbuch der Laplace-Transformation*, vol. I (Birkhäuser, Basel, 1950)
139. G. Doetsch, *Einführung in Theorie und Anwendung der Laplace-Transformation*. Lehrbücher und Monographien aus dem Gebiete der exakten Wissenschaften. Mathematische Reihe 24 (Birkhäuser, Basel, 1958)
140. P. Domański, M. Langenbruch, Euler type partial differential operators on real analytic functions. J. Math. Anal. Appl. **443**(2), 652–674 (2016)
141. P. Domański, M. Langenbruch, Interpolation of holomorphic functions and surjectivity of Taylor coefficient multipliers. Adv. Math. **293**, 782–855 (2016)
142. D.P. Dryanov, Q.I. Rahman, G. Schmeisser, Converse theorems in the theory of approximate integration. Constr. Approx. **6**, 321–334 (1990)
143. N. Dunford, J.T. Schwartz, *Linear Operators Vol. I: General Theory*. Interscience Publ. (1958)
144. H.M. Edwards, *Riemann's Zeta Function*. Pure Appl. Math. (Academic Press, New York, 1974)
145. G. Elfving, *The History of Mathematics in Finland 1828–1918* (Frenckell, Helsinki, 1981)
146. J. Elschner, I.G. Graham, Numerical methods for integral equations of Mellin type. J. Comput. Appl. Math. **125**, 423–437 (2000)
147. W. Erdélyi, A. Magnus, F. Oberhettinger et al., *Higher Trascendental Functions*, vol. 1 (McGraw-Hill, New York, 1953)
148. T. Estermann, *Complex Numbers and Functions* (Athlone Press, London, 1962)
149. L. Euler, Remarques sur un beau rapport entre les series des puissances taut directes que reciproques. Mere. Acad. Sci. Berlin **17**, 83–106 (1761) (= Opera (1) **15**, 70–90)
150. G.B. Folland, *Real Analysis. Modern Techniques and Their Applications*. Pure Appl. Math. Wiley Interscience Publ. (John Wiley & Sons, New York, 1984)
151. G.B. Folland, *Fourier Analysis and Its Applications*. Pure and Applied Undergraduated Text (AMS, Providence, 2009)
152. G.B. Folland, E.M. Stein, *Hardy Spaces on Homogeneous Groups*. Mathematical Notes, vol. 28 (Princeton University Press, Princeton, 1982)
153. R. Frontczak, Pricing options in jump diffusion models using Mellin transforms. J. Math. Finance **3**, 366–373 (2013)
154. S. Garcia, Group quantization of quadratic Hamiltonians in finance. arXiv, 2102.05338 v2 (17 Feb 2021)
155. W. Gautschi, Quadrature formulae on half-infinite intervals. BIT **31**, 438–446 (1991)
156. R. Gervais, Q.I. Rahman, G. Schmeisser, A bandlimited function simulating a duration-limited one, in *Anniversery Volume on Approximation Theory and Functional Analysis*, ed. by P.L. Butzer, R.L. Stens, B. Sz.-Nagy, vol. 65. ISNM (Birkhäuser Verlag, Basel, 1984), pp. 355–362
157. H.-J. Glaeske, A.P. Prudnikov, K.A. Skornik, *Operational Calculus and Related Topics* (Chapman and Hall, CRC, Boca Raton, 2006)
158. A.V. Glushak, T.A. Manaenkova, Direct and inverse problems for an abstract differential equation containing Hadamard fractional derivatives. Diff. Equ. **47**(9), 1307–1317 (2011)
159. F. Gori, Sampling in optics, in *Advances Topics in Shannon Sampling and Interpolation Theory*, ed. by R.J. Marks II (Springer, New York, 1993), pp. 37–83

160. I.S. Gradshteyn, I.M. Ryzhik, *Table of Integrals, Series and Products*, 5th edn. (Academic Press, Cambridge, 1994)
161. C. Guardasoni, M.R. Rodrigo, S. Sanfelici, A Mellin transform approach to barrier option pricing. IMA J. Manag. Math. **31**(1), 49–67 (2020)
162. S. Haber, O. Shisha, Improper integrals, simple integrals and numerical quadrature. J. Approx. Theory **11**, 1–15 (1974)
163. J. Hadamard, Essai sur l'etude des fonctions donnees par leur developpement de Taylor. J. Math. Pures Appl. Ser. 4, **8**, 101–186 (1892)
164. H. Hamburger, Über einige Beziehungen, die mit der Funktionalgleichung der Riemannschen ζ-Funktion equivalent sind. Math. Ann. **85**, 129–140 (1922)
165. G.H. Hardy, J.E. Littlewood, New proofs of the prime-number theorem and similar theorems. Quart. J. Math. Oxford **46**, 215–219 (1915)
166. G.H. Hardy, J.E. Littlewood, Contributions to the theory of the Riemann zeta-function and the theory of the distribution of primes. Acta Math. **41**, 119–196 (1918)
167. G.H. Hardy, J.E. Littlewood, G. Pólya, *Inequalities* (Cambridge University Press, Cambridge, 1934)
168. M. Hauss, *Verallgemeinerte Stirling, Bernoulli und Euler Zahlen, deren Anwendungen und schnell konvergente Reihen für Zeta funktionen*. Dissertation RWTH Aachen (1995)
169. D. Helms, J.B. Thomas, Truncation error of sampling-theorem expansions. Proc. IRE **50**, 179–184 (1962)
170. E. Hewitt, K.A. Ross, *Abstract Harmonic Analysis*. Vol. II: Structure and Analysis for Compact Groups. Analysis on Locally Compact Abelian Groups (Springer, New York, 1970)
171. E. Hewitt, K. Stromberg, *Real and Abstract Analysis*. Graduate Texts in Math., vol. 25 (Springer, New York, 1965)
172. J.R. Higgins, Five short stories about the cardinal series. Bull. Am. Math. Soc. (N.S.) **12**(1), 45–89 (1985)
173. J.R. Higgins, *Sampling Theory in Fourier and Signal Analysis, Foundations* (Oxford University Press, Oxford, 1996)
174. J.R. Higgins, Historical origins of interpolation and sampling, up to 1841. Sampl. Theory Signal Image Process. **2**(2), 117–128 (2003)
175. J.R. Higgins, R.L. Stens (eds.), *Sampling Theory in Fourier and Signal Analysis: Advanced Topics*. Oxford Science Publications (Oxford University Press, Oxford, 1999)
176. J.R. Higgins, G. Schmeisser, J.J. Voss, The sampling theorem and several equivalent results in analysis. J. Comput. Anal. Appl. **2**(4), 333–371 (2000)
177. R. Hilfer, Fractional time evolution, in *Applications of Fractional Calculus in Physics* (World scientific Publ., River Edge, 2000), pp. 187–130
178. A. Ivić, *The Riemann Zeta-Function* (Wiley-Interscience, New York, 1985)
179. D. Jagerman, Bounds for truncation error of the sampling expansion. SIAM J. Appl. Math. **14**, 714–723 (1966)
180. J. Jeon, G. Kim, Pricing European continuous-installement currency options with mean-reversion. North. Am. J. Econ. Finance **59**, Paper 101605 (2022)
181. C. Jordan, *Calculus of Finite Differences* (Chelsea Publishing, New York, 1965)
182. T. Kaiser, The Dirichlet problem in the plane with semianalytic raw data, quasi analyticity, and o-minimal structure. Duke Math J. **147**(2), 285–314 (2009)
183. T. Kaiser, The Riemann mapping theorem for semianalytic domains and o-minimality. Proc. Lond. Math. Soc. **98**(3), 427–444 (2009)
184. T. Kaiser, J.P. Rolin, P. Speissegger, Transition maps at non-resonat hyperbolic singularities are o-minimal. J. Reine Angew Math. **636**, 1–45 (2009)
185. F. Kang, Generalized Mellin transforms I. Sc. Sin. **7**, 582–605 (1958)
186. M. Katsurada, On Mellin–Barnes type integrals and sums associated with Riemann-Zeta function. Publ. Inst. Math. (N.S.) Tome **62**(76), 13–25 (1997)
187. U.N. Katugampola, New approach to a general fractional integral. Appl. Math. Comput. **218**, 860–865 (2011)

188. U.N. Katugampola, A new approach to generalized fractional derivatives. Bull. Math. Anal. Appl. **6**(4), 1–15 (2014)
189. U.N. Katugampola, Mellin transform of generalized fractional integrals and derivatives. Appl. Math. Comput. **257**, 566–580 (2015)
190. Y. Katznelson, *An Introduction to Harmonic Analysis* (John Wiley, New York, 1968)
191. A.A. Kilbas, Hadamard-type fractional calculus. J. Korean Math. Soc. **38**(6), 1191–1204 (2001)
192. A.A. Kilbas, H.M. Srivastava, J.J. Trujillo, *Theory and Applications of Fractional Differential Equations* (Elsevier, Amsterdam, 2006)
193. A. Kivinukk, G. Tamberg, A note on the cosine operator function for Mellin sampling series. Sampl. Theory Signal Process. Data Anal. **22**, Paper 7 (2024)
194. W. Kolbe, R.J. Nessel, Saturation theory in connections with the Mellin transform methods. SIAM J. Math. Anal. **3**, 246–262 (1972)
195. P. Koosis, *The Logarithmic Integral*, vol. I (Cambridge University Press, Cambridge, 1988)
196. C. Kou, J. Liu, Y. Ye, Existence and uniqueness of solutions for the Cauchy-type problems of fractional differential equations. Discrete Dyn. Nat. Soc. **2010**, Paper 142175 (2010)
197. J. Lee, Approximate interpolation and sampling theory. SIAM J. Appl. Math. **32**, 731–744 (1977)
198. E. Lindelöf, Robert Hjalmar Mellin. Acta Math. **61**, I–VI (1933)
199. J.H. Loxton, J.W. Sanders, The kernel of a rule of approximate integration. J. Austral. Math. Soc. Ser. B **21**(3), 257–267 (1980)
200. J.H. Loxton, J.W. Sanders, On an inversion theorem of Möbius. J. Austral. Math. Soc. (Series A) **30**, 15–32 (1980)
201. Yu. Luchko, Integral transforms of the Mellin convolution type and their generating operators. Integral Transform Spec. Funct. **19**(11), 809–851 (2008)
202. Yu. Luchko, V. Kiryakova, The Mellin integral transform in fractional calculus. Fract. Calc. Appl. Anal. **16**(2), 405–430 (2013)
203. J.N. Lyness, The calculation of Fourier coefficients by the Möbius inversion of the Poisson summation formula. I. Functions whose early derivatives are continuous. Math. Comput. **24**, 101–135 (1970)
204. G. Lysik, On the structure of Mellin distributions. Ann. Polon. Math. **51**(1), 219–228 (1990)
205. G. Lysik, Z. Szmydt (1923–2010) (in Polish). Wiad. Mat. **51**(2), 283–297 (2015)
206. F. Mainardi, G. Pagnini, Mellin–Barnes integrals for stable distributions and their convolutions. Frac. Calc. Appl. Anal. **11**(4), 443–452 (2008)
207. F. Mainardi, G. Pagnini, R. Gorenflo, Mellin convolution for subordinated stable processes. J. Math. Sci. (N.Y.) **132**(5), 637–642 (2006)
208. R.G. Mamedov, *The Mellin Transform and Approximation Theory* (in Russian) ("Elm", Baku, 1991)
209. R.G. Mamedov, G.N. Orudhzev, The approximation of functions by singular integrals of the Mellin type (Russian) (Azerbaidzhan. Inst. Nefti i Khimii, Baku, 1979), pp. 1–76
210. R.G. Mamedov, G.N. Orudhzev, Some characteristics of classes of functions that have fractional derivatives (Russian), in *Investigations on Some Questions of the Constructive Theory of Functions and Differential Equations* (Azerbaidzhan. Inst. Nefti i Khimii, Baku, 1981), pp. 3–11
211. R.G. Mamedov, G.N. Orudhzev, Some classes of functions, their interconnection and characteristics (Russian), in *Investigations on Some Questions of the Constructive Theory of Functions and Differential Equations* (Azerbaidzhan. Inst. Nefti i Khimii, Baku, 1981), pp. 12–15
212. I. Mantellini, On the asymptotic behaviour of linear combinations of Mellin–Picard type operators. Math. Nachr. **286**(17–18), 1820–1832 (2013)
213. O.I. Marichev, *Handbook of Integral Transforms of Higher Transcendental Functions: Theory and Algorithmic Tables* (Ellis Horwood, Chichester, 1983)
214. M.J. Marsden, F.B. Richards, S.D. Riemenschneider, Cardinal spline interpolation operators on ℓ^p data. Indiana Univ. Math. J. **24**(7), 677–689 (1975)

215. C. Martinez, M. Sanz, D. Martinez, About fractional integrals in the space of locally integrable functions. J. Math. Anal. Appl. **167**, 111–122 (1992)
216. G. Mastroianni, G. Monegato, Truncated quadrature rules over $(0, \infty)$ and Nyström-type methods. SIAM J. Numer. Anal. **41**(5), 1870–1892 (2003)
217. G. Mastroianni, I. Notarangelo, G.V. Milovanović, Gaussian quadrature rules with an exponential weight on the real semiaxis. IMA J. Numer. Anal. **34**, 1654–1685 (2014)
218. H. Mellin, On the fundamental importance of Cauchy's theorem for the theory of gamma and hypergeometric functions. Acta Soc. Sci. Fennicae **21**(1), 115 pp. (1896; Finnish)
219. H. Mellin, Ueber eine Verallgemeinerung der Riemann'schen Function $\zeta(s)$. Acta Soc. Sci. Fennicae **24**(10), 50 pp. (1899)
220. H. Mellin, Über den Zusammenhang zwischen den linearen Differential- and Differenzengleichungen. Acta Math. **25**, 139–164 (1902)
221. H. Mellin, Eine Formel für den Logarithmus transcendenter Funktionen von endlichem Geschlecht. Acta Math. **25**, 165–184 (1902)
222. H. Mellin, Die Dirichletschen Reihen, die zahlentheoretischen Funktionen und die unendlichen Produkte von endlichem Geschlecht. Acta Soc. Sci. Fennicae **31**, 48 pp. (1903)
223. H. Mellin, Die Dirichletschen Reihen, die zahlentheoretischen Funktionen und die unendlichen Produkte von endlichem Geschlecht. Acta Math. **28**, 37–64 (1904)
224. H. Mellin, Abriß einer einheitlichen Theorie der Gamma- und der hypergeometrischen Funktionen. Math. Ann. **68**, 305–337 (1910)
225. H. Mellin, Remarks following the proof of a theorem by Hardy about the zeta function. Ann. Acad. Sci. Fennicae (A) **11**(3), 17 pp. (1917; Finnish)
226. C.A. Micchelli, Y. Xu, H. Zhang, Optimal learning of bandlimited functions from localized sampling. J. Complexity **25**, 85–114 (2009)
227. O.P. Misra, J.L. Lavoine, *Transform Analysis of Generalized Functions*. North-Holland Mathematics Studies, vol. 119 (Elsevier, Amsterdam, 1986)
228. T.H. Nguyen, S.B. Yakubovich, *The Double Mellin–Barnes Type Integrals and Their Applications to Convolution Theory*. Series on Soviet and East European Mathematics, vol. 6 (World Scientific Publishing Co., Inc., River Edge, 1992)
229. S.M. Nikol'skii, *Approximation of Functions of Several Variables and Imbedding Theorems*. Die Grundlehren der Mathematischen Wissenschaften, vol. 205 (Springer-Verlag, Berlin, 1975)
230. F. Oberhettinger, *Tables of Mellin Transform* (Springer-Verlag, Berlin-Heidelberg-New York, 1974)
231. C.F. Osgood, Obtaining a function of bounded coarse variation by a change of variable. J. Approx. Theory **44**(1), 14–20 (1985)
232. C.F. Osgood, O. Shisha, On simple integrability and bounded coarse variation, in *Approximation Theory II* (Proc. Intern. Sympos. Univ Texas, Austin, 1976) (Academic Press, New York, 1976), pp. 491–501
233. N. Ostrowsky, D. Sornette, P. Parker, E.R. Pike, Exponential sampling method for light scattering polydispersity analysis. Opt. Acta **28**, 1059–1070 (1981)
234. R. Paley, N. Wiener, *Fourier Transform in the Complex Domain*. Amer. Math. Soc. Colloq. Publ., vol. 19 (American Mathematical Society, New York, 1934)
235. R. Panini, R.P. Srivastav, Option pricing with Mellin transnforms. Math. Comput. Model. **40**, 43–56 (2004)
236. R.B. Paris, D. Kaminski, *Asymptotic and Mellin–Barnes Integrals*. Encyclopedia of Mathematics and Its Applications, vol. 85 (Cambridge University Press, Cambridge, 2001)
237. S.J. Patterson, *An Introduction to the Theory of the Riemann-Zeta-Function*. Cambridge Studies Advanced Math., vol. 14 (Cambridge University Press, Cambridge, 1988)
238. O. Perron, Zur Theorie der Dirichletschen Reihen. J. Reine Angew. Math. **134**, 95–143 (1908)
239. I.Z. Pesenson, Boas-type formulas and sampling in Banach spaces with applications to analysis on manifolds, in *New Perspectives on Approximation and Sampling Theory*, ed. by A.I. Zayed, G. Schmeisser (Springer, Cham, 2014), pp. 39–61

240. I.Z. Pesenson, Sampling formulas for groups of operators in Banach spaces. Sampl. Theory Signal Image Process. **14**(1), 1–16 (2015)
241. I.Z. Pesenson, To multidimensional Mellin analysis: Besov spaces, K-functor, approximations, frames. Sampl. Theory Signal Process. Data Anal. **21**, Paper 7 (2023)
242. I.Z. Pesenson, Bernstein spaces, sampling, and Riesz–Boas interpolation formulas in Mellin Analysis, in *Sampling, Approximation and Signal Analysis—Harmonic Analysis in the Spirit of J. Rowland Higgins*. Appl. Numer. Harmonic Anal. (Birkhauser/Springer, Cham, 2023), pp. 69–88
243. S. Pincherle, Sulle funzioni ipergeometriche generalizzate. Rend. Accad. Naz. Lincei **4**, 792–799 (1888)
244. M.L. Platonov, On the numbers of a combinatorial structure. Sibirsk. Mat. Zh. **5**, 1317–1325 (1964; Russian)
245. D. Popa, The complete asymptotic evaluation for Mellin convolution operators. Constr. Approx. **58**, 253–270 (2023)
246. S. Prössdorf, B. Silbermann, *Numerical Analysis for Integral and Related Operator Equations* (Akademie Verlag, Berlin, 1991)
247. A.P. Prudnikov, Yu.A. Bryckhov, O.I. Marichev, Calculation of integrals and the Mellin transform (Russian). Translated in J. Soviet. Math. **54**(6), 1239–1341 (1991)
248. M.D. Qassim, K.M. Furati, N.-E. Tatar, On a differential equation involving Hilfer–Hadamard fractional derivative. Abstr. Appl. Anal. **2012**, Paper 391062 (2012)
249. L.W. Qian, On the regularized Whittaker–Kotel'nikov–Shannon sampling formula. Proc. Am. Math. Soc. **131**, 1169–1176 (2003)
250. L.W. Qian, D.B. Creamer, A modification of the sampling series with a Gaussian multiplier. Sampl. Theory Signal Image Process. **5**, 307–325 (2006)
251. H. Rademacher, *Topics in Analytic Number Theory*. Grundlehren Math. Wiss., vol. 169 (Springer, Berlin, 1973)
252. Q.I. Rahman, G. Schmeisser, Reconstruction and approximation of functions from samples, in *Delay Equations, Approximation and Application*, ed. by G. Meinardus, G. Nürnberger, vol. 74. ISNM (Birkhäuser Verlag, Basel, 1985), pp. 213–233
253. Q.I. Rahman, G. Schmeisser, On a Gaussian quadrature formula for entire functions of exponential type, in *Numerical Methods in Approximation Theory, VIII*, ed. by L. Collatz, G. Meinardus, G. Nürnberger. September–October 1986. ISNM, vol. 81 (Birkhäuser, Basel, 1987), pp. 169–183
254. Q.I. Rahman, G. Schmeisser, Lp inequalities for entire functions of exponential type. Trans. Am. Math. Soc. **320**(1), 91–103 (1990)
255. Q.I. Rahman, G. Schmeisser, Quadrature formulae and functions of exponential type. Math. Comput. **54**, 245–270 (1990)
256. Q.I. Rahman, G. Schmeisser, A quadrature formula for entire functions of exponential type. Math. Comput. **63**, 215–227 (1994)
257. Q.I. Rahman, G. Schmeisser, The summation formulae of Poisson, Plana, Euler–Maclaurin and their relationships (Part I). J. Math. Sci. **28**, 151–171 (1994)
258. E.D. Rainville, *Special Functions* (McMillan Co., New York, 1960)
259. B. Riemann, Ueber die Anzahl der Primzahlen unter einer gegebenen Grösse, Gesammelte Mathematische Werke, Wissenschaftlicher Nachlaß und Nachträge, *Collected Papers* (nach der Ausgabe von H. Weber und R. Dedekind, neu herausgegeben von R. Narasimhan, Ed.) (Springer, Berlin, Teubner, Leipzig, 1990), pp. 177–185
260. S. Ries, R.L. Stens, Approximation by generalized sampling series, in *Constructive Theory of Functions*, ed. by Bl. Sendov, P. Petrushev, R. Maalev, S. Tashev (Publ. House Bulgarian Acad. Sci., Sofia, 1984), pp. 746–756
261. M. Riesz, Formule d'interpolation pour la dérivée d'un polynôme. C. R. Acad. Sci. Paris **158**, 1152–1154 (1914)
262. M. Riesz, Sur les fonctions conjugées. Math. Z. **27**, 218–244 (1927)
263. M. Rosenblum, J. Rovnyak, *Hardy Classes and Operator Theory*. Oxford Mathematical Monographs (The Clarendon Press, Oxford, 1985)

264. H.L. Royden, *Real Analysis*, 3rd edn. (Macmillan Publ. Co., New York, 1988)
265. W. Rudin, *Real and Complex Analysis*, 3rd edn. (McGraw-Hill, New York, 1986)
266. W. Rudin, *Functional Analysis*, 2nd edn. (McGraw-Hill, New York, 1991)
267. S.G. Samko, A.A. Kilbas, O.I. Marichev, *Fractional Integrals and Derivatives. Theory and Applications* (Gordon and Breach, Yverdon, Amsterdam, 1993)
268. R.J. Sasiela, *Electromagnetic Wave Propagation in Turbulence: Evaluation and Application of Mellin Transforms* (Springer, Berlin, 1994)
269. G. Schmeisser, Sampling, Gaussian quadrature, and Poisson summation formula, in *Proceedings of the 1997 Int. Workshop on Sampling Theory and Applications* (Univ. Aveiro, Aveiro, 1997), pp. 327–332
270. G. Schmeisser, Quadrature over a semi-infinite interval and Mellin transform, in *Proceedings of the 1999 International Workshop on Sampling Theory and Applications*, ed. by Y. Lyubarskii. ISBN 82-7151-0991 (Norwegian University of Science and Technology, Trondheim, 1999), pp. 203–208
271. G. Schmeisser, Numerical differentiation inspired by a formula of R.P. Boas. J. Approx. Theory **160**, 202–222 (2009)
272. G. Schmeisser, Interconnections between multiplier methods and window methods in generalized sampling. Sampl. Theory Signal Image Process. **9**(1–3), 1–24 (2010)
273. G. Schmeisser, F. Stenger, Sinc approximation with a Gaussian multiplier. Sampl. Theory Signal Image Process. **6**(2), 199–221 (2007)
274. W.R. Schneider, W. Wyss, Fractional diffusion and wave equations. J. Math. Phys. **30**(1), 134–145 (1988)
275. I.N. Sneddon, Functional analysis, in *Encyclopedia of Physics, Mathematical Methods*, ed. by S. Fluegge, vol. II (Springer, Berlin, 1955), pp. 198–348
276. I.N. Sneddon, *The Use of Integral Transforms* (McGraw-Hill, New York, 1972)
277. E.M. Stein, G. Weiss, *Introduction to Fourier Analysis on Euclidean Spaces* (Princeton University Press, Princeton, 1971)
278. F. Stenger, Interpolation formulae based on the trapezoidal formula. J. Inst. Maths. Appl. **12**, 103–114 (1973)
279. F. Stenger, Numerical methods based on Whittaker cardinal, or sinc functions. SIAM Rev. **23**, 165–224 (1981)
280. F. Stenger, *Numerical Methods Based on Sinc and Analytic Functions* (Springer, New York, 1993)
281. A.T. Suzuki, E.S. Santos, A.G.M. Schmidt, One loop n-point equivalence among negative dimensional, Mellin–Barnes and Feynman parametrization aproaches to Feynman integrals. J. Phys. A **36**(47), 1185–1187 (2003)
282. G. Szegő, *Orthogonal Polynomials*, 4th edn., vol. 23 (Amer. Math. Soc. Colloquium Publications, Providence, 1975)
283. Z. Szmydt, Paley–Wiener theorems for the Mellin transformation. Ann. Polon. Math. **51**(1), 313–324 (1990)
284. Z. Szmydt, B. Ziemian, *The Mellin Transformation and Fuchsian Type Partial Differential Equations* (Kluwer, Dordrecht, 1992)
285. W. Szpankowski, *Average Case Analysis of Algorithms on Sequences* (John Wiley & Sons, New York, 2001)
286. E.C. Titchmarsh, *Introduction to the Theory of Fourier Integrals* (Oxford University Press, Oxford, 1948)
287. E.C. Titchmarsh, *The Theory of the Riemann Zeta-Function* (Clarendon Press, Oxford, 1986)
288. H. Triebel, *Higher Analysis* (Huthig Publ. Ltd, Basel, 1997)
289. L. Tschakaloff, Zweite Lösung der Aufgabe 105. Jahresber. Dtsch. Math-Ver. **43**, 11–12 (1933)
290. V.K. Tuan, New type Paley–Wiener theorems for the modified multidimensional Mellin transform. J. Fourier Anal. Appl. **4**, 317–328 (1998)

291. V.K. Tuan, O.I. Marichev, S.B. Yakubovich, Composition structure of integral transforms. Dokl. AN SSSR **286**, 786–790 (1986; Russian). English transl. in J. Soviet Math. **33**, 166–169 (1986)
292. O. Tyr, D. Radouan, Mellin integral transforms and generalized Lipschitz and Zygmund spaces. J. Pseudo-Differ. Oper. Appl. **14**(2), Paper 18 (2023)
293. O. Tyr, D. Radouan, Jackson's inequality in Mellin Analysis. Ann. Univ. Ferrara, Sez. VII, Sci. Mat. **70**(1), 141–160 (2024)
294. Y.A. Uflyand, *Integral Transforms in Problems of the Theory of Elasticity*. Izdat, Akad. Nauk SSSR (1963; Russian)
295. G. Valiron, Sur la formule d'interpolation de Lagrange. Bull. Sci. Math. **49**, 181–192, 203–224 (1925)
296. G. Vinti, L. Zampogni, A general approximation approach for the simultaneous treatment of integral and discrete operators. Adv. Nonlinear Stud. **18**(4), 705–724 (2018)
297. A. Vretblad, *Fourier Analysis and Its Applications* (Springer-Verlag, Berlin, 2003)
298. A. Weil (ed.), *Essais historiques sur la theorie des nombres*. Monographie No. 22 de l'Enseignement Mathematique, Genève (1975)
299. E.T. Whittaker, On the functions which are represented by the expansion of the interpolation-theory. Proc. R. Soc. Edin. Sect. A **35**, 181–194 (1915)
300. J.M. Whittaker, On the cardinal function of interpolation theory. Proc. Edin. Math. Soc. (2) **1**, 41–46 (1927)
301. J.M. Whittaker, *Interpolatory Function Theory* (Cambridge University Press, Cambridge, 1935) (Reprint: Stechert–Hafner, Inc., New York, 1964)
302. D.V. Widder, *The Laplace Transform* (University of Princeton Press, Princeton, 1946)
303. D.V. Widder (ed.), *An Introduction to Transform Theory*. Pure Appl. Math., vol. 42 (Academic Press, New York, 1971)
304. W. Wyss, The fractional diffusion equation. J. Math. Phys. **27**(11), 2782–2786 (1986)
305. S.B. Yakubovich, A class of index transforms generated by the Mellin and Laplace operators. J. Math. Anal. Appl. **403**(2), 33–343 (2013)
306. S.B. Yakubovich, Y.F. Luchko, *The Hypergeometric Approach to Integral Transforms and Convolutions*. Mathematics and Its Applications, vol. 287 (Kluwer Academic Publishers, Springer, Dordrecht, 1994)
307. K. Yosida, *Functional Analysis*, 6th edn. (Springer-Verlag, Berlin, 1980)
308. A.I. Zayed, *Advances in Shannon's Sampling Theory* (CRC Press, Boca Raton, 1993)
309. A.I. Zayed, *Handbook of Function and Generalized Function Transformations*. Mathematical Sciences Reference Series (CRC Press, Boca Raton, 1996)
310. A.H. Zemanian, *Generalized Integral Transformations* (Interscience, New York, 1968)
311. B. Ziemian, Generalized analytic functions with applications to singular ordinary partial differential equations. Dissertationes Math. **354**, 1–100 (1996)
312. A. Zygmund, *Trigonometric Series*, 3rd edn. (Cambridge University Press, Cambridge, 2002)

Lists of Symbols, Rules, and Transforms

List of Symbols for Spaces

Symbol	Space	Page
$\mathrm{AC}_{\mathrm{loc}}(\mathbb{R}^+)$	Locally absolutely continuous on $\mathbb{R}^+$	64
$\mathrm{AC}^{r-1}_{\mathrm{loc}}(\mathbb{R})$	Locally r-times absolutely continuous functions	9
$\mathrm{AC}^{m}_{c,\mathrm{loc}}$	Subspaces of AC_{loc}	254
$B^p_{c,T}$	Mellin–Paley–Wiener spaces	217
$B^2_{0,\pi\delta}(\mu)$	Subspaces of Mellin–Paley–Wiener spaces	409
$\widetilde{B}^2_{c,T}$	Mellin–Bernstein spaces	67
$\mathscr{B}^p_{c,T}$	Polar Mellin–Bernstein spaces	80
B^p_σ	Bernstein spaces	20
$\widehat{B}^p_\sigma$	Paley–Wiener spaces	31
$\mathrm{BCV}(\mathbb{R})$	Functions with bounded coarse variation	8
$\mathrm{BCV}^p_c(\mathbb{R}^+)$	Mellin version of BCV spaces	392
$\mathrm{BV}(J)$	Function with bounded variation on J	8
$C(A)$	Continuous functions on A	3
$C_{\mathrm{u}}(A)$	Uniformly continuous functions on A	3
$C_{\mathrm{ub}}(A)$	Uniformly continuous and bounded functions	3
$C^k(A)$	k-Times diff. functions, $f^{(k)} \in C(A)$	3
$C_{\mathrm{comp}}(A)$	Functions in $C(A)$ with compact support	3
$C^k_{\mathrm{comp}}(A)$	Functions in $C^k(A)$ with compact support	3
$\mathcal{C}(\mathbb{R}^+)$	Log-uniformly continuous functions	59
G^q_c	Spaces of inverse Mellin transforms	279
$H^p(\mathcal{S}_d)$	Hardy spaces on strips $\mathcal{S}_d$	21
$H^p_c(\mathbb{H}_a)$	Mellin–Hardy spaces	87
$H^*_c(\mathbb{H}_a)$	Intermediate Mellin–Hardy spaces	227

(continued)

C. Bardaro et al., *Mellin Analysis, Transform Theory, and Applications*,
https://doi.org/10.1007/978-3-031-96672-9

Symbol	Space	Page
$\mathbb{H}$	Polar plane	36
$\mathbb{H}_{\alpha,\beta}$	Strip of the polar plane	41
$\mathbb{H}_a$	Strip $\mathbb{H}_{-a,a}$ of the polar plane	87
$\mathcal{K}_c$	Subspaces of $W_c^{1,1}(\mathbb{R}^+)$	434
$L^p(A)$	Lebesgue spaces	2
$L^p(c+i\mathbb{R})$	Lebesgue spaces on a line in $\mathbb{C}$	170
$L^p(\mathrm{St}]a,b[)$	Lebesgue spaces on strip in $\mathbb{C}$	170
L^1_{loc}	Locally integrable functions	3
$\mathrm{Lip}_r(\alpha, X_c^p)$	Lipschitz spaces	61
$\mathcal{L}_c^p(Q)$	Subspaces of X_c^p	393
$\ell^p(\mathbb{A})$	Lebesgue sequence spaces	3
ℓ_c^p	Mellin–Lebesgue sequence spaces	359
$\ell_c^p(\Sigma)$	Function space associated with a partition Σ	360
$\mathcal{M}_c^p$	Mellin inversion classes	285
$\mathbb{M}_c$	Mellin test functions space	215
$\mathbb{M}_c^*$	Dual space of $\mathbb{M}_c$	215
$\mathbb{N}$	Positive integers	1
$\mathbb{N}_0$	Nonnegative integers	1
R_{comp}	Riemann integrable on compact intervals of $\mathbb{R}^+$	361
$\mathbb{R}$	Real line	1
$\mathbb{R}^+$	Positive real line	1
$\mathbb{R}_0^+$	Nonnegative real numbers	1
$\mathrm{St}]a,b[$	Open vertical strip in $\mathbb{C}$	126
$\mathcal{S}$	Schwartz space	212
$\mathcal{S}^*$	Dual space of $\mathcal{S}$	213
$\mathcal{T}_c$	Mellin test functions space	214
$W^{r,p}(\mathbb{R})$	Sobolev spaces	10
$W_c^{r,p}(\mathbb{R}^+) \equiv W_c^{r,p}$	Mellin–Sobolev spaces	65
$W^{m,1}_{c,\mathrm{loc}}$	Local Mellin–Sobolev spaces	255
$W_c^{\gamma,*}(\mathbb{R}^+)$	Fractional Mellin–Sobolev space	297
$W(X_c, D^\alpha_{0+,c})$	Fractional Mellin–Hadamard–Sobolev spaces	255
$W(X_J, D^\alpha_{0+})$	Fractional Mellin–Hadamard–Sobolev spaces	256
$W(X_{c,\mathrm{loc}}, D^\alpha_{0+,c})$	Local fractional Mellin–Hadamard–Sobolev spaces	256
$W(X_{J,\mathrm{loc}}, D^\alpha_{0+})$	Local fractional Mellin–Hadamard–Sobolev spaces	256
$W(X_c, \Theta_c^\alpha)$	Fractional Mellin–Sobolev spaces	265
$W(X_J, \Theta^\alpha)$	Fractional Mellin–Sobolev spaces	266
$W(X_c, \text{s-}\Theta_c^\alpha)$	Strong fractional Mellin–Sobolev spaces	265
$W(X_J, \text{s-}\Theta^\alpha)$	Strong fractional Mellin–Sobolev spaces	266
$\mathscr{W}_c^{\alpha,1}$	Subspace of Fractional Mellin–Sobolev spaces	266
$\mathscr{W}_J^{\alpha,1}$	Subspace of Fractional Mellin–Sobolev spaces	266
X_c	Weighted Lebesgue space	51
X_c^p	Weighted Lebesgue spaces	51

(continued)

Symbol	Space	Page
X_I	Global weighted Lebesgue spaces	52
X_I^p	Global weighted Lebesgue spaces	52
$X_{c,\mathrm{loc}}$,	Weighted Lebesgue spaces (local version)	52
$\widetilde{X}_{c,\mathrm{loc}}$	Local weighted spaces	245
Y_c	Space of recurrent functions	54
Y_c^p	Space of recurrent functions	56
$\mathbb{Z}$	Integers	1
Λ_c^p	An intermediate space between $W_c^{r,p}$ and X_c^p	360

List of Symbols for Operators

Symbol	Operator	Page
$\mathcal{C}\mathcal{D}_{0+}^{\alpha}$	Caputo fractional derivative	278
$\mathcal{D}_{0+}^{\alpha}$	Riemann–Liouville fractional derivative	278
$D_{0+,c}^{\alpha}f$	Left-sided Mellin–Hadamard fractional derivative	232
$D_{-,c}^{\alpha}f$	Right-sided Mellin–Hadamard fractional derivative	276
$\{D_n^c\}_{n\in\mathbb{N}_0}$	Finite Mellin–Dirichlet kernel	104
$D_{\mathrm{pol}}f$	Polar derivative	37
$\mathrm{dist}_q(f, g)$	Mellin distance in G_c^q	279
$\mathrm{d}_p(f, g)$	Distance in $L^p(\mathbb{R})$	298
$\mathrm{dist}_p(f, g)$	Distance (via Fourier transform)	299
$\mathrm{dist}_q(f, B_\sigma^p)$	Distance in B_σ^p	299
$\{F_n^c\}_{n\in\mathbb{N}_0}$	Finite Mellin–Fejer kernel	109
$f * g$	Mellin convolutions	91
$f\tilde{*}g$	Mellin convolutions (alternative)	94
f_{c+}, f_{c-}	Mellin-even and Mellin-odd part of f	440
$\widehat{f}$	Fourier transform of f	28
$(\widehat{f_k})_{k\in\mathbb{Z}}$	Finite Fourier transform of f	27
Hz	Discrete Hilbert transform	366
$\mathcal{H}z$	Related discrete Hilbert transform	366
$H_M y$	Mixed Mellin–Hilbert transform	367
I_d	Identity operator	62
I_{0+}^{α}	Riemann–Liouville fractional integral	278
$J_c^r f$	Mellin integrals	190
$J_{0+,\mu}^{\alpha}$	Left-sided Mellin fractional integral	239
$J_{-,c}^{\alpha}$	Right-sided Mellin fractional integral	276
$\{\kappa_\rho\}_{\rho>0}$	Local approximate identity	134
$\mathcal{L}$	Laplace transform	196
$\mathcal{L}_B$	Bilateral Laplace transform	196
MB	Mellin–Barnes integral	199
$M_\alpha(\varphi, x)$ $\alpha > 0$	Absolute discrete moment of order α	383
$M_c[f] \equiv [f]^{\wedge}_{M_c}$	Mellin transform in X_c	125

(continued)

Symbol	Operator	Page
$M_j(K)$	Absolute logarithmic moment of order j	148
$M_c^{-1}[g]$	Inverse Mellin transform	170
$M_c^2[f] \equiv [f]^{\wedge}_{M_c^2}$	Mellin transform in X_c^2	203
$M_c^{2,-1}[g]$	Inverse Mellin transform in $L^2(c + i\mathbb{R})$	205
$M_c^p[f] \equiv [f]^{\wedge}_{M_c^p}$	Mellin transform in X_c^p, $1 < p < 2$	210
$M_\alpha(\varphi)$	Uniform absolute moment	383
$\mathscr{M}^c[f] = [f]^{\wedge}_{\mathscr{M}^c}$	Finite Mellin transform	101
$m_j(K)$	Logarithmic moment	148
$m_j(\varphi, x)$	Discrete moment of order j	383
$\operatorname{sgn} x$	Sign function	2
$T\rho f$	Mellin–Fejér operators	142
$T_{\rho,\nu} f$	Linear combinations of T_ρ	157
$U_\rho f$	Mellin convolution operators	134
$V(f, I)$	Total variation on I	7
$V(f, \mathbb{R})$	Total variation on $\mathbb{R}$	8
$V_\epsilon(f)$	Coarse variation	8
$\Delta_h^{r,c} f$	Differences of order r	59
$\varDelta_h^{\alpha,c} f$	Mellin fractional differences	261
$\Theta_c f$, $\Theta_c^r f$	Mellin derivatives	62
$\varTheta_{-,c}^{\alpha}$	Right-sided Mellin fractional derivative	277
$\widetilde{\Theta}_c f$, $\widetilde{\Theta}_c^r f$	Polar Mellin derivatives	69
$\varTheta_c^\alpha f$	Mellin fractional derivative	265
s-$\varTheta_c^\alpha f$	Strong Mellin fractional derivative	265
$\tau_h^c f$	Mellin translation	57
$\omega(f, \eta) \equiv \omega(\eta)$	Logarithmic modulus of continuity	155
$\omega_r(f, \delta, X_c^p)$	Modulus of smoothness	60
$\\|f\\|_p$	L^p norm of f	2
$\\|f\\|_\infty^*$	Norm of bounded functions f	3
$\\|(a_n)\\|_p$	Norm of sequences $(a_n) \in \ell^p$	3
$\\|f\\|_{W^{r,p}(\mathbb{R})}$	Norm of Sobolev spaces	10
$\\|T\\|_{[D,Y]}$	Norm of the operator T	10
$\\|f\\|_{H^p(\mathcal{S}_d)}$	Norm of Hardy spaces	21
$\\|f\\|_{X_c^p}$	Norm in X_c^p	51
$\\|f\\|_{X_c^\infty}^*$	Norm of bounded functions $(\cdot)^c f(\cdot)$	51
$\\|f\\|_{Y_c^p}$	Norm in Y_c^p	56
$\\|f\\|_{H_c^p(\mathbb{H}_a)}$	Norm of Mellin–Hardy spaces	87
$\\|f\\|_{H_c^*(\mathbb{H}_a)}$	Norm of intermediate Mellin–Hardy spaces	230
$[f]_q$	Norm of spaces G_c^q	279
$\\|\cdot\\|_{W_c^{\gamma,*}}$	Norm of Sobolev spaces $W_c^{\gamma,*}$	297
$\\|x\\|_{\ell_c^p}$	Norm of Mellin–Lebesgue sequence spaces ℓ_c^p	359
$\lvert\cdot\rvert_{\ell_c^p(\Sigma)}$	Functional on the space $\ell_c^p(\Sigma)$	360
$\lvert f \rvert_{w,\infty}$	Discrete seminorm of f	410

Basic Finite Mellin Transform Rules

Rule	$f(x)$	$[f]^{\wedge}_{\mathscr{M}^c}(k) \equiv \mathscr{M}^c[f](c+ik)$	Ref
Linearity	$(\alpha f + \beta g)(x)$	$\alpha[f]^{\wedge}_{\mathscr{M}^c}(k) + \beta[g]^{\wedge}_{\mathscr{M}^c}(k)$ $(k \in \mathbb{Z})$	Proposition 5.2 (a)
Representation	$f(x)$	$[f]^{\wedge}_{\mathscr{M}^c}(k) = [(\cdot)^c f(\cdot)]^{\wedge}_{\mathscr{M}^0}(k)$ $(k \in \mathbb{Z})$	Proposition 5.2 (b)
Scaling	$f(\alpha x)$ $(\alpha > 0)$	$\alpha^{-(c+ik)}[f]^{\wedge}_{\mathscr{M}^c}(k)$ $(k \in \mathbb{Z})$	Proposition 5.2 (c)
Multiplication	$x^{ij} f(x)$ $(j \in \mathbb{Z})$	$[f]^{\wedge}_{\mathscr{M}^c}(k+j)$ $(k \in \mathbb{Z})$	Proposition 5.2 (d)
Convolution	$(f * g)(x)$	$[f]^{\wedge}_{\mathscr{M}^c}(k)\,[g]^{\wedge}_{\mathscr{M}^c}(k)$ $(k \in \mathbb{Z})$	Proposition 5.3
Orthogonality	$\varphi_{c,k}(x) = x^{-(c+ik)}$	$[\varphi_{c,j}]^{\wedge}_{\mathscr{M}^c}(k) = 2\pi\delta_{j,k}$ $(j, k \in \mathbb{Z})$	Proposition 5.5
Inner product preservation	$f, g \in Y_c^2$	$\langle f, g\rangle_{Y_c^2} = \frac{1}{2\pi}\sum_{k\in\mathbb{Z}}[f]^{\wedge}_{\mathscr{M}^c}(k)\,[g]^{\wedge}_{\mathscr{M}^c}(k)$	Theorem 5.23

Basic Mellin Transform Rules

Rule	$f(x)$	$[f]^{\wedge}_{M_c}(s) = [f]^{\wedge}_{M_c}(c+it)$	Ref
Linearity	$(\alpha f + \beta g)(x)$	$\alpha[f]^{\wedge}_{M_c}(s) + \beta[g]^{\wedge}_{M_c}(s)$	Proposition 6.8 (a)
Scaling	$f(\alpha x)$ $(\alpha > 0)$	$\alpha^{-s}[f]^{\wedge}_{M_c}(s)$	Proposition 6.8 (b)
Multiplication	$x^{\alpha} f(x)$	$[f]^{\wedge}_{M_c}(s+\alpha)$	Proposition 6.8 (c)
Translation	$\tau_h^c(x)$ $(c \in \mathbb{R}, h > 0)$	$h^{-it}[f]^{\wedge}_{M_c}(s)$	Theorem 6.10
Convolution	$(f * g)(x)$	$[f]^{\wedge}_{M_c}(s)\,[g]^{\wedge}_{M_c}(s)$	Theorem 6.12
Derivative	$\Theta_c^r f(x)$ $(c \in \mathbb{R},\ r \in \mathbb{N})$	$(-it)^r[f]^{\wedge}_{M_c}(s)$	Theorem 6.14
Integral	$J_c^r f(x)$ $(c \in \mathbb{R},\ r \in \mathbb{N})$	$(-it)^{-r}[f]^{\wedge}_{M_c}(s)$	Theorem 6.95
Fractional integral	$J^{\alpha}_{0+,c} f(x)$ $(\alpha > 0,\ c \in \mathbb{R})$	$(-it)^{-\alpha}[f]^{\wedge}_{M_c}(s)$ $(t \neq 0)$	Proposition 9.14
M–H fractional derivative	$D^{\alpha}_{0+,c} f(x)$, $(c \in \mathbb{R},\ \alpha > 0)$	$(-it)^{\alpha}[f]^{\wedge}_{M_c}(s)$	Proposition 9.23
Fractional differences	$\Delta_h^{\alpha,c} f(x)$ $(c \in \mathbb{R}, \alpha, h > 0)$	$(1-h^{it})^{\alpha}[f]^{\wedge}_{M_c}(s)$	Proposition 9.29
M-fractional derivative	$\Theta_c^{\alpha} f(x)$ $(c \in \mathbb{R}, \alpha > 0)$	$(-it)^{\alpha}[f]^{\wedge}_{M_c}(s)$	Theorem 9.33

Mellin Transforms of Particular Functions

$f(x)$	$[f]^{\wedge}_{M_c}(s)$	Domain	Ref
$\chi_{[a,b]}(x)$	$\frac{b^s - a^s}{s}$	$(s \neq 0)$	
$x^{1-c}\chi_{]0,1]}(x)\ (c \in \mathbb{R})$	$\frac{1}{1+it}$	$s = c + it$	Sect. 6.7
$(1-x)^{\nu}\chi_{]0,1]}(x)\ (\nu > -1)$	$B(\nu+1, s)$	$s \in \mathrm{St}]0, \infty[$	[230, p. 16, (2.20)]
$\frac{1}{(1+x^a)^b}\ (ab > 0)$	$\frac{\Gamma(s/a)\Gamma(b - s/a)}{a\Gamma(b)}$	$s \in \mathrm{St}]0, ab[$	[160, p. 1195]
$\frac{1}{(\beta+\alpha x)^{\nu}}\ (\alpha, \beta > 0, \nu > 0)$	$\left(\frac{\beta}{\alpha}\right)^s \beta^{-\nu} B(s, \nu - s)$	$s \in \mathrm{St}]0, \nu[$	[230, p. 15, (2.19)]
e^{-x}	$\Gamma(s)$	$s \in \mathrm{St}]0, \infty[$	Remark 6.7
e^{-x^2}	$\frac{1}{2}\Gamma\left(\frac{s}{2}\right)$	$s \in \mathrm{St}]0, \infty[$	[160, p. 1195]
$\sin x$	$\Gamma(s)\sin\left(\frac{\pi s}{2}\right)$	$s \in \mathrm{St}]-1, 0[$	[160, p. 1195]
$\arctan x$	$-\frac{\pi}{2s}\sec\left(\frac{\pi s}{2}\right)$	$s \in \mathrm{St}]-1, 0[$	[160, p. 1196]
$x^{\alpha}e^{-x}\ (\alpha > 0)$	$\Gamma(\alpha + s)$	$s \in \mathrm{St}]-\alpha, \infty[$	
$\log\lvert 1-x\rvert$	$\pi s^{-1}\cot(\pi s)$	$s \in \mathrm{St}]-1, 0[$	[230, p. 35, (4.9)]
$\mathrm{lin}_c(x)\ (c \in \mathbb{R})$	$\chi_{[-\pi,\pi]}(t)$	$s = c + it$	Lemma 12.2 (c)
$x\log^2 x\,\chi_{]0,1[}(x) +$ $x^{-1}\log^2 x\,\chi_{[1,\infty[}(x)$	$4\frac{1-3t^2}{(1+t^2)^3}$	$s = it$	Sect. 14.3, Example 4
$P_{\rho}(x)\ (\rho > 0)$ (Mellin–Abel–Poisson kernel)	$\left\{1 - \left(\frac{s}{2}\right)^2\right\}^{-1}$	$s \in \mathrm{St}]-\rho, \rho[$	Example 6.29
$G_{\rho}(x)\ (\rho > 0)$ (Mellin–Gauss–Weierstrass kernel)	$\exp\left(\left(\frac{s}{\rho}\right)^2\right)$	$s \in \mathbb{C}$	Example 6.25
$F^c_{\rho}(x)\ (c \in \mathbb{R}, \rho > 0)$ (Mellin–Fejér kernel)	$\left(1 - \frac{\lvert t\rvert}{\rho}\right)\chi_{]-\rho,\rho[}(t)$	$s = c + it$	Definition 6.83
$B_n(x)\ (n \in \mathbb{N})$ (Mellin B-splines)	$\mathrm{sinc}^n\left(\frac{t}{2\pi}\right)$	$s = it$	Sect. 13.4

Index

C. Bardaro et al., *Mellin Analysis, Transform Theory, and Applications*,
https://doi.org/10.1007/978-3-031-96672-9

G

H

I

GPSR Compliance

The European Union's (EU) General Product Safety Regulation (GPSR) is a set of rules that requires consumer products to be safe and our obligations to ensure this.

If you have any concerns about our products, you can contact us on ProductSafety@springernature.com

In case Publisher is established outside the EU, the EU authorized representative is:

Springer Nature Customer Service Center GmbH
Europaplatz 3
69115 Heidelberg, Germany

Batch number: 10755312

Printed by Printforce, the Netherlands